SCHAUM'S OUTLINE OF

THEORY AND PROBLEMS

of

BASIC CIRCUIT ANALYSIS

•

by

JOHN O'MALLEY, Ph.D.

Professor of Electrical Engineering
University of Florida

SCHAUM'S OUTLINE SERIES

McGRAW-HILL BOOK COMPANY

New York   St. Louis   San Francisco   Auckland   Bogotá   Guatemala   Hamburg   Johannesburg
Lisbon   London   Madrid   Mexico   Montreal   New Delhi   Panama   Paris
San Juan   São Paulo   Singapore   Sydney   Tokyo   Toronto

JOHN R. O'MALLEY is a Professor of Electrical Engineering at the University of Florida in Gainesville, Florida, where he has been a faculty member since 1959. He received his B.S. (1951) and M.S. (1952) degrees from Purdue University, his LL.B. (1956) degree from Georgetown University, and his Ph.D. (1964) degree from the University of Florida. He is the author of two books on circuit analysis and one on the digital computer.

Schaum's Outline of Theory and Problems of
BASIC CIRCUIT ANALYSIS

1 2 3 4 5 6 7 8 9 10 11 12 13 14 15 16 17 18 19 20 SH SH 8 6 5 4 3 2 1

Sponsoring Editor, John Aliano
Editing Supervisor, John Fitzpatrick
Production Manager, Nick Monti

Library of Congress Cataloging in Publication Data

O'Malley, John.
    Schaum's outline of basic circuit analysis.

    (Schaum's outline series)
    Includes index.
1. Electric circuits. 2. Electric circuit analysis.
I. Title.
TK454.046   621.319'2   80-26925
ISBN 0-07-047820-1

Preface

Studying from this book will help both electrical technology and electrical engineering students learn circuit analysis. Since this book begins with the analysis of dc resistive circuits and continues to that of ac circuits, as do the popular circuit analysis textbooks, a student can, from the start, use this book as a supplement to a circuit analysis textbook.

The reader does not need a knowledge of differential or integral calculus, even though this book has derivatives in the chapters on capacitors, inductors, and transformers, as is required for the voltage-current relations. The few problems with derivatives have clear physical explanations of them, and there is not a single integral anywhere in the book. Despite its lack of higher mathematics, this book can be very useful to an electrical engineering reader since most material in introductory electrical engineering circuit analysis textbooks—the dc and ac circuit analyses—requires only a knowledge of algebra.

One of the special features of this book is the use of the International System of Units (SI) throughout. Also, the problems are practically oriented, and the solutions are so short that a reader can master concepts quickly without becoming embroiled in mathematics. Further, where there are different definitions in the electrical technology and engineering fields, as for capacitive reactances and phasors, the reader is cautioned and the various definitions are explained. There is even some transient analysis—that of the dc-excited *RC* circuits (and also *RL* circuits) that are so important for timers. Finally, the book has many problems involving transistor models.

I want to thank Dr. B. E. Cherrington, Chairman of the Department of Electrical Engineering at the University of Florida, for his nurturing of a professional environment conducive to the writing of books. Thanks are also due to my wife, Lois Anne, and to my children who were at home during the period of writing—Tim, Margaret, Cecilia, and Mathew—for their constant support and encouragement, without which I could not have written this book.

JOHN R. O'MALLEY

Contents

Chapter 1   POWERS-OF-10 NOTATION AND THE INTERNATIONAL SYSTEM OF UNITS 1

Digit Grouping 1

Powers-of-10 Notation 1

Arithmetic Operations with Numbers in Powers-of-10 Notation 2

International System of Units 3

Conversion 3

Chapter 2   ELECTRIC CHARGE, CURRENT, VOLTAGE, AND POWER 9

Electric Charge 9

Electric Current 9

Voltage 10

Power 11

Electric Energy 12

Chapter 3   RESISTANCE 21

Ohm's Law 21

Resistivity 21

Temperature Effects 22

Resistors 23

Resistor Power Absorption 23

Nominal Values and Tolerances 23

Color Code 24

Open and Short Circuits 24

Internal Resistance 25

Chapter 4   SERIES AND PARALLEL DC CIRCUITS 35

Branches, Nodes, Loops, and Meshes 35

Kirchhoff's Voltage Law and Series DC Circuits 35

Kirchhoff's Current Law and Parallel DC Circuits 36

Current Division 38

Chapter 5   DC CIRCUIT ANALYSIS 53

Cramer's Rule 53

Source Conversions 54

Mesh Analysis 55

Loop Analysis 56

Nodal Analysis 56

Dependent Sources and Circuit Analysis 57

Chapter 6   DC EQUIVALENT CIRCUITS, NETWORK THEOREMS, AND BRIDGE CIRCUITS 79

Introduction 79

Thevenin's and Norton's Theorems 79

Maximum Power Transfer Theorem 80

Superposition Theorem 81

Millman's Theorem 81

Y-Δ and Δ-Y Conversions 82

Bridge Circuits 83

Input and Output Resistances 84

CONTENTS

Chapter 7 CAPACITIES AND CAPACITANCE **104**

Introduction 104

Capacitance 104

Capacitor Construction 104

Total Capacitance 105

Energy Storage 106

Time-Varying Voltages and Currents 106

Capacitor Current 107

Single-Capacitor DC-Excited Circuits 107

RC Timers and Oscillators 108

Chapter 8 INDUCTORS AND INDUCTANCE **124**

Introduction 124

Magnetic Flux 124

Inductance and Inductor Construction 125

Inductor Voltage and Current Relation 125

Total Inductance 126

Energy Storage 127

Single-Inductor DC-Excited Circuits 127

Chapter 9 SINUSOIDAL ALTERNATING VOLTAGE AND CURRENT **140**

Introduction 140

Sine and Cosine Waves 141

Phase Relations 143

Average Value 144

Resistor Sinusoidal Response 144

Effective or RMS Values 144

Inductor Sinusoidal Response 145

Capacitor Sinusoidal gresponse 146

Chapter 10 COMPLEX ALGEBRA AND PHASORS **162**

Introduction 162

Imaginary Numbers 162

Complex Numbers and the Rectangular Form 163

Polar Form 164

Phasors 165

Chapter 11 BASIC AC CIRCUIT ANALYSIS, IMPEDANCE, AND ADMITTANCE **178**

Introduction 178

Frequency-Domain Circuit Elements 178

Ac Series Circuit Analysis 180

Impedance 180

Voltage Division 182

AC Parallel Circuit Analysis 183

Admittance 183

Current Division 184

Chapter 12 MESH, LOOP, AND NODAL ANALYSES OF AC CIRCUITS **205**

Introduction 205

Source Conversions 205

Mesh and Loop Analyses 205

Nodal Analysis 207

CONTENTS

Chapter 13   AC EQUIVALENT CIRCUITS, NETWORK THEOREMS, AND BRIDGE CIRCUITS **228**

Introduction 228

Thevenin's and Norton's Theorems 228

Maximum Power Transfer Theorem 229

Superposition Theorem... 229

AC Y-Δ and Δ-Y Conversions 230

AC Bridge Circuits 230

Chapter 14   POWER IN AC CIRCUITS **257**

Introduction 257

Circuit Power Absorption 257

Wattmeters 258

Reactive Power 259

Complex Power and Apparent Power 259

Power Factor Correction 260

Chapter 15   TRANSFORMERS **281**

Introduction 281

Right-Hand Rule 281

Dot Convention 282

The Ideal Transformer 282

The Air-Core Transformer 284

The Autotransformer 286

Chapter 16   THREE-PHASE CIRCUITS **307**

Introduction 307

Subscript Notation... 307

Three-Phase Voltage Generation 307

Generator Winding Connections 308

Phase Sequence 309

Balanced Y Circuit 310

Balanced Δ Load 312

Parallel Loads 313

Three-Phase Power Measurements 314

Unbalanced Circuits 315

INDEX **335**

Chapter 1

Powers-of-10 Notation and the International System of Units

DIGIT GROUPING

To make numbers easier to read, some international scientific committees have recommended the practice of separating digits into groups of three to the right and to the left of decimal points, as in 64 325.473 53. No separation is necessary, however, for just four digits, and they are preferably not separated. For example, either 4138 or 4 138 is acceptable, as is 0.1278 or 0.127 8, with 4138 and 0.1278 preferred. The international committees did not approve of the use of the comma to separate digits because in some countries the comma is used in place of the decimal point.

POWERS-OF-10 NOTATION

Some common electrical quantities may have very small values, and some others may have very large ones. For instance, a capacitor value may be 0.000 000 002, or much smaller. And a frequency may have a value of 70 000 000, or much greater. Because writing all the leading and trailing zeros in these numbers is inconvenient, *powers-of-10 notation* is often used instead. This notation refers to the number 10 with an integer exponent. Some examples are shown in Table 1-1.

Notice that a power of 10 with a value of 1 or greater has an exponent equal to the number of zeros: 100 has two zeros and is equal to 10^2, 10 000 has four zeros and is equal to 10^4, and so on. A power of 10 with a value less than 1 has an exponent equal to the negative of the number of zeros, including the zero to the left of the decimal point: 0.001 has three zeros and is equal to 10^{-3}, 0.000 01 has five zeros and is equal to 10^{-5}, and so on.

Placing a number in powers-of-10 notation often requires shifting the decimal point. Keeping the value the same in this shifting requires adding a positive 1 to the exponent of 10 for each shift to the left and adding a negative 1 for each shift to the right. As an illustration, suppose that 0.000 056 2 is to be put into powers-of-10 notation. The result depends on the final decimal point location, which is somewhat arbitrary. Placing it between the 5 and 6, for example, requires a decimal point shift of five places to the right. So, the exponent of 10 must be -5: $0.000 056 2 = 5.62 \times 10^{-5}$.

Table 1-1

$1 = 10^0$	$\frac{1}{10} = 0.1 = 10^{-1}$
$10 = 10^1$	$\frac{1}{100} = 0.01 = 10^{-2}$
$100 = 10^2$	$\frac{1}{1000} = 0.001 = 10^{-3}$
$1000 = 10^3$	$\frac{1}{10\,000} = 0.0001 = 10^{-4}$
$10\,000 = 10^4$	$\frac{1}{100\,000} = 0.000\,01 = 10^{-5}$
$100\,000 = 10^5$	

The powers-of-10 notation with one nonzero digit to the left of the decimal point, as in 5.62×10^{-5}, has the special name of *scientific notation*. Using this notation is sometimes convenient, but seldom essential.

Sometimes it is preferable to eliminate the power of 10 as in, for example, 0.045×10^4. Of course the rules for eliminating the powers of 10 are just the opposite for forming them. So, if the exponent of 10 is positive, the decimal point shift is to the right a number of places equal to the value of the exponent: $0.045 \times 10^4 = 450$. If the exponent is negative, as in 3254.62×10^{-2}, the decimal point shift is to the left instead: $3254.62 \times 10^{-2} = 32.5462$. If the power of 10 is in the denominator, the decimal point shift is to the left for a positive exponent and to the right for a negative exponent. As an illustration,

$$\frac{625.47}{10^2} = 6.2547 \quad \text{and} \quad \frac{0.0489}{10^{-3}} = 48.9$$

These rules follow from the fact that $1/10^n = 10^{-n}$ and $1/10^{-n} = 10^n$.

The procedure is similar if the multiplier or divisor is a power of 10 but not in powers-of-10 notation. If a power-of-10 multiplier has a value greater than 1, the decimal point shift is to the right a number of places equal to the number of zeros. If, instead, this power of 10 is a divisor, the decimal point shift is to the left. For example,

$$46.2 \times 1000 = 46\,200 \quad \text{and} \quad \frac{632.8}{10\,000} = 0.063\,28$$

If the power-of-10 multiplier or divisor is less than 1, the decimal point shift is in the opposite direction, as in

$$152.3 \times 0.001 = 0.1523 \quad \text{and} \quad \frac{0.0267}{0.000\,01} = 2670$$

ARITHMETIC OPERATIONS WITH NUMBERS IN POWERS-OF-10 NOTATION

To be added or subtracted, numbers in powers-of-10 notation must have the same power of 10—that is, the 10s must have the same exponent. Then the sum or difference has this same power of 10, and the adding or subtracting applies to the other parts of the numbers. As an illustration,

$$3.2 \times 10^5 + 0.64 \times 10^6 = 3.2 \times 10^5 + 6.4 \times 10^5 = (3.2 + 6.4) \times 10^5 = 9.6 \times 10^5$$

Alternatively,

$$3.2 \times 10^5 + 0.64 \times 10^6 = 0.32 \times 10^6 + 0.64 \times 10^6 = 0.96 \times 10^6$$

In multiplication the exponents add, and in division they subtract. The other parts of the numbers multiply or divide in the usual fashion. For example,

$$(4.1 \times 10^4)(2.5 \times 10^5) = (4.1 \times 2.5) \times 10^{4+5} = 10.25 \times 10^9$$

and

$$\frac{24.48 \times 10^7}{7.2 \times 10^3} = \frac{24.48}{7.2} \times 10^{7-3} = 3.4 \times 10^4$$

In general,

$$(A \times 10^m)(B \times 10^n) = AB \times 10^{m+n} \quad \text{and} \quad \frac{A \times 10^m}{B \times 10^n} = \frac{A}{B} \times 10^{m-n}$$

The rules for finding powers and roots are

$$(A \times 10^m)^n = A^n \times 10^{m \times n} \quad \text{and} \quad (A \times 10^m)^{1/n} = A^{1/n} \times 10^{m/n}$$

For example,

$$(3 \times 10^4)^2 = 3^2 \times 10^{4 \times 2} = 9 \times 10^8$$

and
$$(225 \times 10^8)^{1/2} = 225^{1/2} \times 10^{8/2} = 15 \times 10^4$$

because $225^{1/2}$ is the square root of 225, also designated by $\sqrt{225}$, and is equal to 15 because $15 \times 15 = 225$.

INTERNATIONAL SYSTEM OF UNITS

The *International System of Units* (*SI*), being adopted throughout the world, is the international measurement language.   SI has seven base units: meter for length, kilogram for mass, kelvin for temperature, second for time, ampere for current, mole for amount of substance, and candela for luminous intensity.   Units of all other physical quantities are derived from these.

A meter is approximately 39.37 in (inches)—a little more than a yard.   A kilogram is the mass of anything that weighs 2.205 lb (pounds).   A temperature in kelvin is 273.15 more than the same temperature in the more common degrees Celsius, an alternate SI temperature unit.   And the relation between a temperature in degrees Celsius (T_C) and the same temperature in degrees Fahrenheit (T_F), not an SI unit, is $T_C = (T_F - 32)/1.8$.   The second is the common second.   The ampere is one of the topics of the next chapter.   And the mole and candela are unimportant in circuit analysis.

SI has a unique and well-defined set of symbols.   For the base units the symbols are m for meter, kg for kilogram, K for kelvin, s for second, A for ampere, mol for mole, and cd for candela.   (The symbol for degrees Celsius is °C.)

There is a *decimal* relation, indicated by prefixes, among multiples and submultiples of each base unit.   An SI prefix is a term attached to the beginning of an SI unit name to form either a decimal multiple or submultiple.   For example, since "kilo" is the prefix for one thousand, a kilometer equals 1000 m.   And because "micro" is the SI prefix for one-millionth, one microsecond equals 0.000 001 s.

The SI prefixes have symbols as shown in Table 1-2, which also shows the corresponding powers of 10.

For most circuit analyses, only mega, kilo, milli, micro, nano, and pico are important.

The proper location for a prefix symbol is in front of a unit symbol, as in km for kilometer and cm for centimeter.

CONVERSION

For additions and subtractions of quantities expressed with different prefix symbols, one approach is to convert all quantities to basic unit sizes, using powers of 10.   For example, to add 0.325 ms and 24.2 μs, both can be converted to seconds: 0.325×10^{-3} s and 24.2×10^{-6} s.   Then,

Table 1-2

Multiplier	Prefix	Symbol	Multiplier	Prefix	Symbol
10^{18}	exa	E	10^{-1}	deci	d
10^{15}	peta	P	10^{-2}	centi	c
10^{12}	tera	T	10^{-3}	milli	m
10^{9}	giga	G	10^{-6}	micro	μ
10^{6}	mega	M	10^{-9}	nano	n
10^{3}	kilo	k	10^{-12}	pico	p
10^{2}	hecto	h	10^{-15}	femto	f
10^{1}	deka	da	10^{-18}	atto	a

the power of 10 of one number must be made the same as the other by shifting the decimal point: $0.325 \times 10^{-3} = 325 \times 10^{-6}$. With both powers of 10 the same, the addition can be performed:

$$325 \times 10^{-6} + 24.2 \times 10^{-6} = 349.2 \times 10^{-6} \text{ s} = 349.2 \ \mu\text{s}$$

Usually, though, it is easier to convert one number such that the prefix symbols are the same. For example, $0.325 \text{ ms} = 325 \ \mu\text{s}$, and so

$$0.325 \text{ ms} + 24.2 \ \mu\text{s} = 325 \ \mu\text{s} + 24.2 \ \mu\text{s} = 349.2 \ \mu\text{s}$$

In such conversions, deciding whether to multiply or to divide by the correct power of 10 may be a problem. The orderly conversion procedure of *dimensional analysis* eliminates this problem. In dimensional analysis, unit symbols are used with factors in multiplications and are considered to be algebraic symbols that are divided out where convenient. Specifically, in the conversion of one quantity to another having a more desired unit, the quantity is multiplied by 1 expressed as the ratio of equal quantities with the undesired unit in the denominator and the desired unit in the numerator. The unit symbols are included and the undesired ones are divided out. To illustrate, in the conversion of 624 μs to milliseconds, a multiplier is formed from 1000 μs = 1 ms expressed as a ratio equal to 1: $1 = 1 \text{ ms}/1000 \ \mu\text{s}$. The milliseconds quantity is in the numerator because it is the desired final unit. Then,

$$624 \ \cancel{\mu\text{s}} \times \frac{1 \text{ ms}}{1000 \ \cancel{\mu\text{s}}} = 0.624 \text{ ms}$$

Note the dividing out of the μs symbols as if they were algebraic symbols.

Dimensional analysis is also useful in the multiplication of numbers corresponding to different physical quantities—as, for example, in finding the total number of revolutions of a synchronous motor that rotates at 3600 r/min (revolutions per minute) for 2.4 h (hours). Multiplying the rate of revolution by the specified time does not give the correct answer because the time units do not match. With dimensional analysis it is easy to see that there must be an additional factor of 60 min/1 h to get the minutes in the rate to divide out:

$$\text{Number of revolutions} = 3600 \ \frac{\text{r}}{\cancel{\text{min}}} \times \frac{60 \ \cancel{\text{min}}}{1 \ \cancel{\text{h}}} \times 2.4 \ \cancel{\text{h}} = 518\,400$$

Solved Problems

1.1 Show examples of conversions to powers of 10.

> $0.000\,000\,01 = 10^{-8}$
> $10\,000\,000 = 10^7$
> $0.000\,000\,000\,000\,000\,01 = 10^{-17}$
> $1\,000\,000\,000\,000\,000\,000 = 10^{18}$

1.2 Show examples of conversions to scientific notation.

> $562\,000\,000 = 5.62 \times 10^8$
> $0.001\,205 = 1.205 \times 10^{-3}$
> $6542.3 = 6.5423 \times 10^3$
> $0.000\,000\,043\,21 = 4.321 \times 10^{-8}$

1.3 Using powers of 10, show examples of addition and subtraction.

> $5.48 \times 10^2 + 32 \times 10^1 = 5.48 \times 10^2 + 3.2 \times 10^2 = (5.48 + 3.2) \times 10^2 = 8.68 \times 10^2$

$$0.032 + 2.6 \times 10^{-2} = 3.2 \times 10^{-2} + 2.6 \times 10^{-2} = (3.2 + 2.6) \times 10^{-2} = 5.8 \times 10^{-2}$$
$$42\,000 - 6.43 \times 10^3 = 42 \times 10^3 - 6.43 \times 10^3 = (42 - 6.43) \times 10^3 = 35.57 \times 10^3$$
$$0.000\,062\,3 - 0.000\,91 = 6.23 \times 10^{-5} - 91 \times 10^{-5} = (6.23 - 91) \times 10^{-5} = -84.77 \times 10^{-5}$$

1.4   Using powers of 10, show examples of multiplication.

$$100(10\,000) = 10^2 \times 10^4 = 10^{2+4} = 10^6$$
$$10^5(10\,000)(0.001) = 10^5 \times 10^4 \times 10^{-3} = 10^{5+4-3} = 10^6$$
$$10^6(10^{-3})(0.000\,01) = 10^6 \times 10^{-3} \times 10^{-5} = 10^{6-3-5} = 10^{-2}$$
$$(10^{-4})^2(10\,000)(10^3)^4 = 10^{-4\times2} \times 10^4 \times 10^{3\times4} = 10^{-8} \times 10^4 \times 10^{12} = 10^{-8+4+12} = 10^8$$
$$600(4.3 \times 10^{-3})(0.021) = (6 \times 10^2)(4.3 \times 10^{-3})(2.1 \times 10^{-2}) = (6 \times 4.3 \times 2.1) \times 10^{2-3-2}$$
$$= 54.18 \times 10^{-3}$$
$$2.2^2(64 \times 10^{-4})^{1/2}(4 \times 10^4)^4 = 2.2^2(64^{1/2} \times 10^{-4/2})(4^4 \times 10^{4\times4}) = (4.84 \times 8 \times 256) \times 10^{-2+16}$$
$$= 9912.32 \times 10^{14}$$

1.5   Using powers of 10, show examples of division.

$$\frac{100}{10\,000} = \frac{10^2}{10^4} = 10^2 \times 10^{-4} = 10^{2-4} = 10^{-2}$$

$$\frac{0.001}{100} = \frac{10^{-3}}{10^2} = 10^{-3} \times 10^{-2} = 10^{-3-2} = 10^{-5}$$

$$\frac{1000}{0.000\,01} = \frac{10^3}{10^{-5}} = 10^3 \times 10^{-(-5)} = 10^{3+5} = 10^8$$

$$\frac{0.000\,000\,01}{0.000\,01} = \frac{10^{-8}}{10^{-5}} = 10^{-8-(-5)} = 10^{-8+5} = 10^{-3}$$

$$\frac{10^3 \times 0.000\,01}{100 \times 10^7} = \frac{10^3 \times 10^{-5}}{10^2 \times 10^7} = \frac{10^{3-5}}{10^{2+7}} = \frac{10^{-2}}{10^9} = 10^{-2-9} = 10^{-11}$$

$$\frac{36.3 \times 0.0045}{2.1 \times 0.032} = \frac{(3.63 \times 10^1)(4.5 \times 10^{-3})}{2.1 \times (3.2 \times 10^{-2})} = \frac{3.63 \times 4.5}{2.1 \times 3.2} \times \frac{10^1 \times 10^{-3}}{10^{-2}}$$
$$= 2.43 \times \frac{10^{1-3}}{10^{-2}} = 2.43 \times \frac{10^{-2}}{10^{-2}} = 2.43$$

$$\frac{620 \times 0.000\,02 \times 4800}{9200 \times 430 \times 0.000\,000\,008\,5} = \frac{(6.2 \times 10^2)(2 \times 10^{-5})(4.8 \times 10^3)}{(9.2 \times 10^3)(4.3 \times 10^2)(8.5 \times 10^{-9})} = \frac{6.2 \times 2 \times 4.8}{9.2 \times 4.3 \times 8.5} \times \frac{10^{2-5+3}}{10^{3+2-9}}$$
$$= \frac{59.52}{336.26} \times \frac{10^0}{10^{-4}} = 0.177 \times 10^{0-(-4)} = 0.177 \times 10^4 = 1770$$

1.6   Show examples of powers with powers of 10.

$$1000^4 = (10^3)^4 = 10^{3\times4} = 10^{12}$$
$$0.000\,01^3 = (10^{-5})^3 = 10^{-5\times3} = 10^{-15}$$
$$100^{-3} = (10^2)^{-3} = 10^{2\times(-3)} = 10^{-6}$$
$$4000^2 = (4 \times 10^3)^2 = 4^2 \times 10^{3\times2} = 16 \times 10^6$$
$$2000^{-2} = (2 \times 10^3)^{-2} = 2^{-2} \times 10^{3\times(-2)} = \frac{1}{2^2} \times 10^{-6} = \tfrac{1}{4} \times 10^{-6} = 0.25 \times 10^{-6} = 2.5 \times 10^{-7}$$
$$\frac{5000^2(700)}{42 \times 0.06^3} = \frac{(5 \times 10^3)^2(7 \times 10^2)}{(4.2 \times 10^1)(6 \times 10^{-2})^3} = \frac{(5^2 \times 10^{3\times2})(7 \times 10^2)}{(4.2 \times 10^1)(6^3 \times 10^{-2\times3})} = \frac{5^2 \times 7}{4.2 \times 6^3} \times \frac{10^6 \times 10^2}{10^1 \times 10^{-6}} = \frac{175}{907.2} \times \frac{10^{6+2}}{10^{1-6}}$$
$$= 0.1929 \times \frac{10^8}{10^{-5}} = 0.1929 \times 10^{8-(-5)} = 0.1929 \times 10^{13} = 1.929 \times 10^{12}$$

1.7   Show examples of roots with powers of 10.

$$\sqrt{10^4} = (10^4)^{1/2} = 10^{4\times0.5} = 10^2$$
$$(10^8)^{1/4} = 10^{8\times0.25} = 10^2$$
$$\sqrt{0.000\,001} = (10^{-6})^{1/2} = 10^{-6\times0.5} = 10^{-3}$$
$$\sqrt{900} = 900^{1/2} = (3^2 \times 10^2)^{1/2} = 3^{2\times0.5} \times 10^{2\times0.5} = 3^1 \times 10^1 = 30$$
$$8000^{1/3} = (2^3 \times 10^3)^{1/3} = 2^{3\times1/3} \times 10^{3\times1/3} = 2^1 \times 10^1 = 20$$

1.8   Express (a) 0.06 s in milliseconds, (b) 862.3 mm in meters, (c) 0.0465 m in milli-meters, (d) 0.0235 mA in microamperes, (e) 0.000 056 3 km in millimeters, (f) 0.062 m² (square meters) in square centimeters,   and   (g) 5.74×10^8 cm³ (cubic centimeters) in cubic meters.

(a) $0.06 \text{ s} \times \dfrac{1000 \text{ ms}}{1 \text{ s}} = 0.06 \times 1000 \text{ ms} = 60 \text{ ms}$

(b) $862.3 \text{ mm} \times \dfrac{1 \text{ m}}{1000 \text{ mm}} = 0.8623 \text{ m}$

(c) $0.0465 \text{ m} \times \dfrac{1000 \text{ mm}}{1 \text{ m}} = 46.5 \text{ mm}$

(d) $0.0235 \text{ mA} \times \dfrac{1000 \text{ } \mu\text{A}}{1 \text{ mA}} = 23.5 \text{ } \mu\text{A}$

(e) $0.000 056 3 \text{ km} \times \dfrac{10^6 \text{ mm}}{1 \text{ km}} = 56.3 \text{ mm}$

(f) $0.062 \text{ m}^2 \times \left(\dfrac{100 \text{ cm}}{1 \text{ m}}\right)^2 = 0.062 \times (10^2)^2 \text{ cm}^2 = 0.062 \times 10^{2 \times 2} \text{ cm}^2$

$= 0.062 \times 10^4 \text{ cm}^2 = 620 \text{ cm}^2$

(g) $5.74 \times 10^8 \text{ cm}^3 \times \left(\dfrac{1 \text{ m}}{100 \text{ cm}}\right)^3 = \dfrac{5.74 \times 10^8}{(10^2)^3} \text{ m}^3 = \dfrac{5.74 \times 10^8}{10^6} \text{ m}^3 = 574 \text{ m}^3$

1.9   Find (a) 0.03 s + 6.4 ms − 720 μs in milliseconds, and (b) 2050 m + 4.3 km + 6.8×10^4 cm in kilometers.

(a) $0.03 \text{ s} \times \dfrac{1000 \text{ ms}}{1 \text{ s}} + 6.4 \text{ ms} - 720 \text{ } \mu\text{s} \times \dfrac{1 \text{ ms}}{1000 \text{ } \mu\text{s}} = 30 + 6.4 - 0.72 \text{ ms} = 35.68 \text{ ms}$

(b) $2050 \text{ m} \times \dfrac{1 \text{ km}}{1000 \text{ m}} + 4.3 \text{ km} + 6.8 \times 10^4 \text{ cm} \times \dfrac{1 \text{ m}}{100 \text{ cm}} \times \dfrac{1 \text{ km}}{1000 \text{ m}} = 2.05 + 4.3 + \dfrac{6.8 \times 10^4}{10^5} \text{ km}$

$= 2.05 + 4.3 + 0.68 \text{ km} = 7.03 \text{ km}$

1.10   Find the maximum legal speed limit for automobiles expressed in kilometers per second (km/s) given that   1 mi = 1.609 344 km,   exactly.

By dimensional analysis,

$$55 \dfrac{\text{mi}}{\text{h}} \times \dfrac{1.609 \text{ km}}{1 \text{ mi}} \times \dfrac{1 \text{ h}}{3600 \text{ s}} = 0.0246 \text{ km/s}$$

1.11   Find the length of a football field in meters given that   1 in = 2.54 cm,   exactly.

By dimensional analysis,

$$100 \text{ yd} \times \dfrac{36 \text{ in}}{1 \text{ yd}} \times \dfrac{2.54 \text{ cm}}{1 \text{ in}} \times \dfrac{1 \text{ m}}{100 \text{ cm}} = 91.44 \text{ m}$$

1.12   Find the volume in cubic centimeters of an electrical conductor that has a diameter of 2 mm and a length of 50 m.

Since the cross-sectional area equals $\pi d^2/4$ and the volume of a conductor is its cross-sectional area times its length, the volume is

$$\dfrac{\pi (2 \text{ mm})^2}{4} \times \left(\dfrac{1 \text{ cm}}{10 \text{ mm}}\right)^2 \times 50 \text{ m} \times \dfrac{100 \text{ cm}}{1 \text{ m}} = 157 \text{ cm}^3$$

1.13   Given that a cubic meter of copper has 8.5×10^{28} free electrons, find the number of free electrons in 1000 ft of No. 12 AWG (American Wire Gauge) copper wire.   This size wire has a diameter of 80.81 mils.   (A mil is one one-thousandth of an inch.)

The number of free electrons equals the volume of the wire in cubic meters times the specified electron density.   Of course this volume is the cross-sectional area times the length.   Also,   1 in = 2.54 cm = 0.0254 m.   So, the number of free electrons is

$$\frac{\pi(80.81 \times 10^{-3} \text{ in})^2}{4} \times \left(\frac{0.0254 \text{ m}}{1 \text{ in}}\right)^2 \times 1000 \text{ ft} \times \frac{12 \text{ in}}{1 \text{ ft}} \times \frac{0.0254 \text{ m}}{1 \text{ in}} \times 8.5 \times 10^{28} \frac{\text{electrons}}{\text{m}^3}$$
$$= 8.6 \times 10^{25} \text{ electrons}$$

1.14   The circular mil is a circular unit of area.   One circular mil (1 cmil) is the cross-sectional area of a wire that has a diameter of 1 mil.   In general, the cross-sectional area of a wire in circular mils is the square of the number of mils in the diameter.   Find the area in circular mils of a wire having a diameter of 0.025 in.

The diameter in mils is

$$0.025 \text{ in} \times \frac{1000 \text{ mils}}{1 \text{ in}} = 25 \text{ mils}$$

And so the area is $25^2 = 625$ cmils.

1.15   Given that the No. 14 AWG wire used in houses has a cross-sectional area of 4110 cmils, find the cross-sectional area of this wire in square inches.

The diameter of the wire is $\sqrt{4110}$ mils.   So, the cross-sectional area is, from $\pi d^2/4$,

$$\frac{\pi(\sqrt{4110} \text{ mils})^2}{4} \times \left(\frac{1 \text{ in}}{1000 \text{ mils}}\right)^2 = 0.003\,23 \text{ in}^2$$

Supplementary Problems

1.16   Express the following numbers in powers-of-10 notation:

(a)   0.000 000 000 001   (b)   100 000 000 000   (c)   0.000 000 000 000 01   (d)   1 000 000 000 000
Ans.   (a) 10^{-12},   (b) 10^{11},   (c) 10^{-14},   (d) 10^{12}

1.17   Put the following numbers into scientific notation:

(a)   0.004 78   (b)   8 297 000   (c)   0.000 000 000 000 643 5   (d)   4 278 000 000 000 000
Ans.   (a) 4.78×10^{-3},   (b) 8.297×10^6,   (c) 6.435×10^{-13},   (d) 4.278×10^{15}

1.18   Add and subtract as indicated:

(a) $4.6 \times 10^3 + 38.2 \times 10^2$
(b) $726 - 5.8 \times 10^2 + 9.2 \times 10^3$
(c) $56\,728 + 42.3 \times 10^2 - 363 \times 10^2 + 0.068 \times 10^5$
(d) $0.006\,32 - 4.2 \times 10^{-3} + 0.047 \times 10^{-2} + 832 \times 10^{-5}$
Ans.   (a) 8.42×10^3,   (b) 9.346×10^3,   (c) 3.1458×10^4,   (d) 1.091×10^{-2}

1.19   Multiply and divide as indicated:

(a) $10^3(100\,000)(0.000\,01)(10^7)(10^{-5})$
(b) $892(0.000\,58)(7.2 \times 10^3)(2.9 \times 10^{-2})$
(c) $4.2^3(3.7 \times 10^3)(16 \times 10^3)^3(0.04 \times 10^{-3})^2(0.09 \times 10^{-6})^{-2}(4 \times 10^8)^{1/2}$
(d) $\dfrac{400(5000)^2}{0.08}$

(e) $\dfrac{0.000\,064^{1/2}}{40\,000}$

(f) $\dfrac{0.004^2(0.000\,06)^3(7200)^2}{[600(0.0008)(4.2 \times 10^2)]^{1/3}}$ (calculator required)

Ans. (a) 10^5, (b) 108,   (c) 4.44×10^{27},   (d) 1.25×10^{11},   (e) 2×10^{-7},   (f) 3.06×10^{-11}

1.20 Express

(a) 4372 m in kilometers

(b) 0.000 49 s in microseconds

(c) 0.000 043 2 km in millimeters

(d) 92.8 μA in milliamperes

(e) 0.000 049 2 mA in nanoamperes

(f) 0.0632 m^2 in square centimeters

(g) 2.92×10^7 cm^3 in cubic meters

(h) 72.3×10^{-5} m^3 in cubic centimeters

Ans. (a) 4.372 km, (b) 490 μs, (c) 43.2 mm, (d) 0.0928 mA, (e) 49.2 nA, (f) 632 cm^2, (g) 29.2 m^3, (h) 723 cm^3

1.21 Express

(a) 30 mi/h in meters per second

(b) 75 ft/s in kilometers per hour

(c) 0.047 km/s in miles per hour

(d) 212 000 cmils in square inches

Ans. (a) 13.41 m/s, (b) 82.3 km/h, (c) 105.1 mi/h, (d) 0.1665 in^2

1.22 Find the number of free electrons in 1 mi of No. 10 AWG copper wire, given that a cubic meter of copper has 8.5×10^{28} free electrons and that No. 10 AWG wire has a cross-sectional area of 10 400 cmils.   *Ans.* 7.21×10^{26} electrons

Chapter 2

Electric Charge, Current, Voltage, and Power

ELECTRIC CHARGE

Scientists have discovered two kinds of electric charge: *positive* and *negative charge*. Positive charge is carried by subatomic particles called *protons*, and negative charge by subatomic particles called *electrons*. All amounts of charge are integer multiples of these elemental charges. Scientists have also found that charges produce forces on each other: Charges of the same sign repel each other, but charges of opposite sign attract each other. Moreover, in an electric circuit there is *conservation of charge*, which means that the net electric charge remains constant—charge is neither created nor destroyed. (Electric components interconnected to form at least one closed path comprise an *electric circuit* or *network*.)

The charge of an electron or proton is much too small to be the basic charge unit. Instead, the SI unit of charge is the *coulomb* with unit symbol C. The quantity symbol is Q for a constant charge and q for a charge that varies with time. The charge of an electron is -1.602×10^{-19} C and that of a proton is 1.602×10^{-19} C. Put another way, the combined charge of 6.241×10^{18} electrons equals -1 C, and that of 6.241×10^{18} protons equals 1 C.

Each atom of matter has a positively charged nucleus consisting of protons and uncharged particles called *neutrons*. Electrons orbit around the nucleus under the attraction of the protons. For an undisturbed atom the number of electrons equals the number of protons, making the atom electrically neutral. But if an outer electron receives energy from, say, heat, it can gain enough energy to overcome the force of attraction of the protons and become a *free electron*. The atom then has more positive than negative charge and is a *positive ion*. Some atoms can also "capture" free electrons to gain a surplus of negative charge and become *negative ions*.

ELECTRIC CURRENT

Electric current results from the movement of electric charge. The SI unit of current is the *ampere* with unit symbol A. The quantity symbol is I for a constant current and i for a time-varying current. If a steady flow of 1 C of charge passes a given point in a conductor in 1 s, the resulting current is 1 A. In general,

$$I(\text{amperes}) = \frac{Q(\text{coulombs})}{t(\text{seconds})}$$

in which t is the quantity symbol for time.

Current has an associated direction. By convention the direction of current flow is in the direction of positive charge movement and against the direction of negative charge movement. In solids only free electrons move to produce current flow—the ions cannot move. But in gases and liquids, ions *can* move to produce current flow. Since electric circuits consist almost entirely of solids, only electrons produce current flow in almost all circuits. But this fact is seldom important in circuit analyses because the analyses are almost always at the current level and not the charge level.

In a circuit diagram each I (or i) usually has an associated arrow to indicate the *current reference direction*, as shown in Fig. 2-1. This arrow specifies the direction of positive current flow, but not necessarily the direction of actual flow. If, after calculations, I is found to be positive, then actual current flow is in the direction of the arrow. But if I is negative, current flow is in the opposite direction.

A current that flows in only one direction all the time is a *direct current* (dc), while a current that alternates in direction of flow is an *alternating current* (ac). Usually, though, direct current

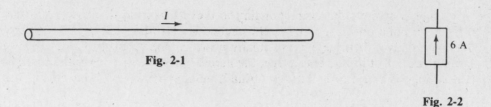

Fig. 2-1

Fig. 2-2

refers only to a constant current, and alternating current refers only to a current that varies sinusoidally with time.

A *current source* is a circuit element that causes a specified current to flow through the source. Figure 2-2 shows the circuit diagram symbol for a current source. This source causes 6 A to flow in the direction of the arrow.

VOLTAGE

The concept of voltage involves work, which in turn involves force and distance. The SI unit of work is the *joule* with unit symbol J, the SI unit of force is the *newton* with unit symbol N, and of course the SI unit for distance is the meter with unit symbol m.

Work is required for moving an object against a force that opposes the motion. For example, lifting something against the force of gravity requires work. In general the work required in joules is the product of the force in newtons and the distance moved in meters:

$$W(\text{joules}) = F(\text{newtons}) \times s(\text{meters})$$

where W, F, and s are the quantity symbols for work, force, and distance, respectively.

Energy is the capacity to do work. One of its forms is *potential energy*, which is the energy a body has because of its position.

The *voltage difference* (also called the *potential difference*) between two points is the work in joules required to move 1 C of charge from one point to the other. The SI unit of voltage is the *volt* with unit symbol V. The quantity symbol is also V or v, although E and e are also popular. In general,

$$V(\text{volts}) = \frac{W(\text{joules})}{Q(\text{coulombs})}$$

The voltage quantity symbol V sometimes has subscripts to designate the two points to which the voltage corresponds. If the letter a designates one point and b the other, and if W joules of work are required to move Q coulombs from point b to a, then $V_{ab} = W/Q$. Note that the first subscript is the point to which the charge is moved. The work quantity symbol sometimes also has subscripts as in $V_{ab} = W_{ab}/Q$.

If moving a positive charge from b to a (or a negative charge from a to b) actually requires work, then point a is positive with respect to point b. This is the *voltage polarity*. In a circuit diagram this voltage polarity is indicated by a positive sign (+) at point a and a negative sign (−) at point b, as shown in Fig. 2-3a for 6 V. Terms used to designate this voltage are a 6-V *voltage* or *potential rise* from b to a or, equivalently, a 6-V *voltage* or *potential drop* from a to b.

If the voltage is designated by a quantity symbol as in Fig. 2-3b, the positive and negative signs are reference polarities and not necessarily actual polarities. Also, if subscripts are used, the

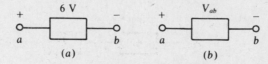

Fig. 2-3

positive polarity sign is at the point corresponding to the first subscript (*a* here) and the negative polarity sign is at the point corresponding to the second subscript (*b* here).   If after calculations, V_{ab} is found to be positive, then point *a* is actually positive with respect to point *b*, in agreement with the polarity signs.   But if V_{ab} is negative, the actual polarities are opposite those shown.

A constant voltage is called a *dc voltage*.   And a voltage that varies sinusoidally with time is called an *ac voltage*.

A *voltage source*, such as a battery or generator, produces a voltage that, ideally, does not depend on the current flow through the source.   Figure 2-4*a* shows the circuit symbol for a battery.   This source produces a dc voltage of 12 V.   This symbol is also often used for a dc voltage source that may not be a battery.   Usually, the + and − signs are not shown because, by convention, the long end-line designates the positive terminal and the short end-line the negative terminal.   Another circuit symbol for a dc voltage source is shown in Fig. 2-4*b*.   A battery uses chemical energy to move negative charges from the attracting positive terminal, where there is a surplus of protons, to the repulsing negative terminal, where there is a surplus of electrons. A voltage generator supplies this energy from mechanical energy that rotates a magnet past coils of wire.

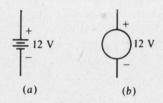

Fig. 2-4

POWER

The *rate* at which something either absorbs or produces energy is the *power* absorbed or produced.   A source of energy produces or delivers power and a load absorbs it.   The SI unit of power is the *watt* with unit symbol W.   The quantity symbol is *P* for constant power and *p* for time-varying power.   If 1 J of work is either absorbed or delivered at a constant rate in 1 s, the corresponding power is 1 W.   In general,

$$P(\text{watts}) = \frac{W(\text{joules})}{t(\text{seconds})}$$

The power *absorbed* by an electric component is the product of voltage and current if the current reference arrow is into the positively referenced terminal, as shown in Fig. 2-5:

$$P(\text{watts}) = V(\text{volts}) \times I(\text{amperes})$$

Such references are called *associated references*.   If the references are not associated (the current arrow is into the negatively referenced terminal), the power absorbed is $P = -VI$.

If the calculated *P* is positive with either formula, the component actually *absorbs* power.   But if *P* is negative, the component *produces* power—it is a *source* of electric energy.

The power output rating of motors is usually expressed in a power unit called the *horsepower* (hp) even though this is not an SI unit.   The relation between horsepower and watts is   1 hp = 745.7 W.

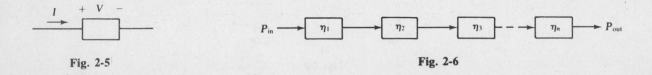

Fig. 2-5   Fig. 2-6

Electric motors and other systems have an *efficiency* (η) of operation defined by

$$\text{Efficiency} = \frac{\text{power output}}{\text{power input}} \times 100\% \quad \text{or} \quad \eta = \frac{P_{\text{out}}}{P_{\text{in}}} \times 100\%$$

Efficiency can also be based on work output divided by work input. In calculations, efficiency is usually expressed as a decimal fraction that is the percentage divided by 100.

The overall efficiency of a cascaded system as shown in Fig. 2-6 is the product of the individual efficiencies:

$$\frac{P_{\text{out}}}{P_{\text{in}}} = \eta_1 \eta_2 \eta_3 \cdots \eta_n$$

ELECTRIC ENERGY

Electric energy used or produced is the product of the electric power input or output and the time over which this input or output occurs:

$$W(\text{joules}) = P(\text{watts}) \times t(\text{seconds})$$

Electric energy is what customers purchase from electric utility companies. These companies do not use the joule as an energy unit, but instead use the much larger and more convenient *kilowatthour* (kWh) even though it is not an SI unit. The number of kilowatthours consumed equals the product of the power absorbed in kilowatts and the time in hours over which it is absorbed:

$$W(\text{kilowatthours}) = P(\text{kilowatts}) \times t(\text{hours})$$

Solved Problems

2.1 Find the charge in coulombs of (*a*) 5.31×10^{20} electrons, and (*b*) 2.9×10^{22} protons.

(*a*) Since the charge of an electron is -1.602×10^{-19} C, the total charge is

$$5.31 \times 10^{20} \text{ electrons} \times \frac{-1.602 \times 10^{-19} \text{ C}}{1 \text{ electron}} = -85.1 \text{ C}$$

(*b*) Similarly, the total charge is

$$2.9 \times 10^{22} \text{ protons} \times \frac{1.602 \times 10^{-19} \text{ C}}{1 \text{ proton}} = 4.65 \text{ kC}$$

2.2 How many protons have a combined charge of 6.8 pC?

Because the combined charge of 6.241×10^{18} protons is 1 C, the number of protons is

$$6.8 \times 10^{-12} \text{ C} \times \frac{6.241 \times 10^{18} \text{ protons}}{1 \text{ C}} = 4.24 \times 10^7 \text{ protons}$$

2.3 Find the current flow through a light bulb from a steady movement of (*a*) 60 C in 4 s, (*b*) 15 C in 2 min, and (*c*) 10^{22} electrons in 1 h.

Current is the rate of charge movement in coulombs per second. So,

(*a*) $I = \dfrac{Q}{t} = \dfrac{60 \text{ C}}{4 \text{ s}} = 15 \text{ C/s} = 15 \text{ A}$

(b) $I = \dfrac{15 \text{ C}}{2 \text{ min}} \times \dfrac{1 \text{ min}}{60 \text{ s}} = 0.125 \text{ C/s} = 0.125 \text{ A}$

(c) $I = \dfrac{10^{22} \text{ electrons}}{1 \text{ h}} \times \dfrac{1 \text{ h}}{3600 \text{ s}} \times \dfrac{-1.602 \times 10^{-19} \text{ C}}{1 \text{ electron}} = -0.455 \text{ C/s} = -0.445 \text{ A}$

The negative sign in the answer indicates that the current flows in a direction opposite that of electron movement. But this sign is unimportant here and can be omitted because the problem statement does not give the direction of electron movement.

2.4   Electrons pass to the right through a wire cross section at the rate of 6.4×10^{21} electrons per minute. What is the current in the wire?

Because current is the rate of charge movement in coulombs per second,

$$I = \dfrac{6.4 \times 10^{21} \text{ electrons}}{1 \text{ min}} \times \dfrac{-1 \text{ C}}{6.241 \times 10^{18} \text{ electrons}} \times \dfrac{1 \text{ min}}{60 \text{ s}} = -17.1 \text{ C/s} = -17.1 \text{ A}$$

The negative sign in the answer indicates that the current is to the left, opposite the direction of electron movement.

2.5   In a liquid, negative ions, each with a single surplus electron, move to the left at a steady rate of 2.1×10^{20} ions per minute and positive ions, each with two surplus protons, move to the right at a steady rate of 4.8×10^{19} ions per minute. Find the current to the right.

The negative ions moving to the left and the positive ions moving to the right both produce a current to the right because current flow is in a direction opposite that of negative charge movement and the same as that of positive charge movement. For a current to the right, the movement of electrons to the left is a negative movement. Also, each positive ion, being doubly ionized, has double the charge of a proton. So,

$$I = -\dfrac{2.1 \times 10^{20} \text{ electrons}}{1 \text{ min}} \times \dfrac{-1.602 \times 10^{-19} \text{ C}}{1 \text{ electron}} \times \dfrac{1 \text{ min}}{60 \text{ s}} + \dfrac{2 \times 4.8 \times 10^{19} \text{ protons}}{1 \text{ min}} \times \dfrac{1.602 \times 10^{-19} \text{ C}}{1 \text{ proton}}$$
$$\times \dfrac{1 \text{ min}}{60 \text{ s}} = 0.817 \text{ A}$$

2.6   Will a 10-A fuse blow for a steady rate of charge flow through it of 45 000 C/h?

The current is

$$\dfrac{45\,000 \text{ C}}{1 \text{ h}} \times \dfrac{1 \text{ h}}{3600 \text{ s}} = 12.5 \text{ A}$$

which is more than the 10-A rating. So the fuse will blow.

2.7   Assuming a steady current flow through a switch, find the time required for   (a) 20 C to flow if the current is 15 mA,   (b) 12 μC to flow if the current is 30 pA,   and   (c) 2.58×10^{15} electrons to flow if the current is -64.2 nA.

Since $I = Q/t$ solved for t is $t = Q/I$,

(a) $t = \dfrac{20}{15 \times 10^{-3}} = 1.33 \times 10^3 \text{ s} = 22.2 \text{ min}$

(b) $t = \dfrac{12 \times 10^{-6}}{30 \times 10^{-12}} = 4 \times 10^5 \text{ s} = 111 \text{ h}$

(c) $t = \dfrac{2.58 \times 10^{15} \text{ electrons}}{-64.2 \times 10^{-9} \text{ A}} \times \dfrac{-1 \text{ C}}{6.241 \times 10^{18} \text{ electrons}} = 6.44 \times 10^3 \text{ s} = 1.79 \text{ h}$

2.8   The total charge that a battery can deliver is usually specified in ampere-hours (Ah).   An ampere-hour is the quantity of charge corresponding to a current flow of 1 A for 1 h. Find the number of coulombs corresponding to 1 Ah.

Since from $Q = It$, 1 C is equal to one ampere second (As),

$$Q = 1 \text{ Ah} \times \frac{3600 \text{ s}}{1 \text{ h}} = 3600 \text{ As} = 3600 \text{ C}$$

2.9 A certain car battery is rated at 70 Ah at 3.5 A, which means that the battery can deliver 3.5 A for approximately 70/3.5 = 20 h. However, the larger the current, the less the charge that can be drawn. How long can this battery deliver 2 A?

The time that the current can flow is approximately equal to the ampere-hour rating divided by the current:

$$t = \frac{70 \text{ Ah}}{2 \text{ A}} = 35 \text{ h}$$

Actually, the battery can deliver 2 A for longer than 35 h because the ampere-hour rating for this smaller current is greater than that for 3.5 A.

2.10 Find the average drift velocity of electrons in a No. 14 AWG copper wire carrying a 10-A current, given that copper has 1.38×10^{24} free electrons per cubic inch and that the cross-sectional area of No. 14 AWG wire is 3.23×10^{-3} in^2.

The average drift velocity (v) equals the current divided by the product of the cross-sectional area and the electron density:

$$v = \frac{10 \text{ C}}{1 \text{ s}} \times \frac{1}{3.23 \times 10^{-3} \text{ in}^2} \times \frac{1 \text{ in}^3}{1.38 \times 10^{24} \text{ electrons}} \times \frac{0.0254 \text{ m}}{1 \text{ in}} \times \frac{1 \text{ electron}}{-1.602 \times 10^{-19} \text{ C}}$$
$$= -3.56 \times 10^{-4} \text{ m/s}$$

The negative sign in the answer indicates that the electrons move in a direction opposite that of current flow. Notice the low velocity. An electron travels only 1.28 m in 1 h, on the average, even though the electric impulses produced by the electron movement travel at near the speed of light (2.998×10^8 m/s).

2.11 Find the work required to lift a 4500-kg elevator a vertical distance of 50 m.

The work required is the product of the distance moved and the force needed to overcome the weight of the elevator. Since this weight in newtons is 9.8 times the mass in kilograms,

$$W = Fs = (9.8 \times 4500)(50) \text{ J} = 2.2 \text{ MJ}$$

2.12 Find the potential energy in joules gained by a 180-lb man in climbing a 6-ft ladder.

The potential energy gained by the man equals the work he had to do to climb the ladder. The force involved is his weight, and the distance is the height of the ladder. The conversion factor from weight in pounds to a force in newtons is 1 N = 0.225 lb. Thus,

$$W = 180 \text{ lb} \times 6 \text{ ft} \times \frac{1 \text{ N}}{0.225 \text{ lb}} \times \frac{12 \text{ in}}{1 \text{ ft}} \times \frac{0.0254 \text{ m}}{1 \text{ in}} = 1.46 \times 10^3 \text{ N} \cdot \text{m} = 1.46 \text{ kJ}$$

2.13 How much chemical energy must a 12-V car battery expend in moving 8.93×10^{20} electrons from its positive terminal to its negative terminal?

The appropriate formula is $W = QV$. Although the signs of Q and V are important, obviously here the product of these quantities must be positive because energy is required to move the electrons. So, the easiest approach is to ignore the signs of Q and V. Or, if signs are used, V is negative because the charge moves to a more negative terminal, and of course Q is negative because electrons have a negative charge.   Thus,

$$W = QV = 8.93 \times 10^{20} \text{ electrons} \times -12 \text{ V} \times \frac{-1 \text{ C}}{6.241 \times 10^{18} \text{ electrons}} = 1.72 \times 10^{3} \text{ VC} = 1.72 \text{ kJ}$$

2.14   If moving 16 C of positive charge from point b to point a requires 0.8 J, find V_{ab}, the voltage drop from point a to point b.

$$V_{ab} = \frac{W_{ab}}{Q} = \frac{0.8}{16} = 0.05 \text{ V}$$

2.15   In moving from point a to point b, 2×10^{19} electrons *do* 4 J of work.   Find V_{ab}, the voltage drop from point a to point b.

Work done *by* the electrons is equivalent to *negative* work done *on* the electrons, and voltage depends on work done *on* charge.   So, $W_{ba} = -4$ J,   but $W_{ab} = -W_{ba} = 4$ J.   Thus,

$$V_{ab} = \frac{W_{ab}}{Q} = \frac{4 \text{ J}}{2 \times 10^{19} \text{ electrons}} \times \frac{6.241 \times 10^{18} \text{ electrons}}{-1 \text{ C}} = -1.25 \text{ J/C} = -1.25 \text{ V}$$

The negative sign indicates that there is a voltage rise from a to b instead of a voltage drop.

2.16   Find V_{ab}, the voltage drop from point a to point b, if 24 J are required to move charges of   (a) 3 C,   (b) −4 C,   and   (c) 20×10^{19} electrons from point a to point b.

If 24 J are required to move the charges from point a to point b, then −24 J are required to move them from point b to point a.   In other words, $W_{ab} = -24$ J.   So,

(a) $V_{ab} = \dfrac{W_{ab}}{Q} = \dfrac{-24}{3} = -8$ V

The negative sign in the answer indicates that point a is more negative than point b—there is a voltage rise from a to b.

(b) $V_{ab} = \dfrac{W_{ab}}{Q} = \dfrac{-24}{-4} = 6$ V

(c) $V_{ab} = \dfrac{W_{ab}}{Q} = \dfrac{-24 \text{ J}}{20 \times 10^{19} \text{ electrons}} \times \dfrac{6.241 \times 10^{18} \text{ electrons}}{-1 \text{ C}} = 0.749$ V

2.17   Find the energy stored in a 12-V car battery rated at 70 Ah.

From $W = QV$ and that fact that   1 As = 1 C,

$$W = 70 \text{ Ah} \times \frac{3600 \text{ s}}{1 \text{ h}} \times 12 \text{ V} = 252\,000 \text{ As} \times 12 \text{ V} = 3.02 \text{ MJ}$$

2.18   Find the voltage drop across a light bulb if a 0.5-A current flowing through it for 4 s causes the light bulb to give off 240 J of light and heat energy.

Since the charge that flows is $Q = It = 0.5 \times 4 = 2$ C,

$$V = \frac{W}{Q} = \frac{240}{2} = 120 \text{ V}$$

2.19   Find the average input power to a radio that consumes 3600 J in 2 min.

$$P = \frac{W}{t} = \frac{3600 \text{ J}}{2 \text{ min}} \times \frac{1 \text{ min}}{60 \text{ s}} = 30 \text{ J/s} = 30 \text{ W}$$

2.20   How many joules does a 60-W light bulb consume in 1 h?

From rearranging $P = W/t$ and from the fact that 1 Ws $= 1$ J,

$$W = Pt = 60 \text{ W} \times 1 \text{ hr} \times \frac{3600 \text{ s}}{1 \text{ hr}} = 216\,000 \text{ Ws} = 216 \text{ kJ}$$

2.21 How long does a 100-W light bulb take to consume 13 kJ?

From rearranging $P = W/t$,

$$t = \frac{W}{P} = \frac{13\,000}{100} = 130 \text{ s}$$

2.22 How much power does a stove element absorb if it draws 10 A when connected to a 115-V line?

$$P = VI = 115 \times 10 \text{ W} = 1.15 \text{ kW}$$

2.23 What current does a 1200-W toaster draw from a 120-V line?

From rearranging $P = VI$,

$$I = \frac{P}{V} = \frac{1200}{120} = 10 \text{ A}$$

2.24 Figure 2-7 shows a circuit diagram of a voltage source of V volts connected to a current source of I amperes.   Find the power absorbed by the voltage source for

(a) $V = 2$ V, $I = 4$ A

(b) $V = 3$ V, $I = -2$ A

(c) $V = -6$ V, $I = -8$ A.

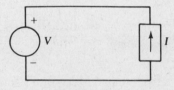

Fig. 2-7

Because the reference arrow for I is into the positively referenced terminal for V, the current and voltage references for the voltage source are associated. This means that there is a positive sign (or the absence of a negative sign) in the relation between power absorbed and the product of voltage and current: $P = VI$. With the given values inserted,

(a) $P = VI = 2 \times 4 = 8$ W

(b) $P = VI = 3 \times (-2) = -6$ W

The negative sign for the power indicates that the voltage source delivers rather than absorbs power.

(c) $P = VI = -6 \times (-8) = 48$ W

2.25 How long can a 12-V car battery supply 40 A to a starter motor if the battery has 4×10^6 J of chemical energy that can be converted to electric energy?

The best approach is to use $t = W/P$. Here,

$$P = VI = 12 \times 40 = 480 \text{ W}$$

And so

$$t = \frac{W}{P} = \frac{4 \times 10^6}{480} = 8.33 \times 10^3 \text{ s} = 2.31 \text{ h}$$

2.26 Find the current drawn from a 115-V line by a dc electric motor that delivers 1 hp.   Assume 100 percent efficiency of operation.

From rearranging $P = VI$ and from the fact that   1 W/V = 1 A,

$$I = \frac{P}{V} = \frac{1 \text{ hp}}{115 \text{ V}} \times \frac{745.7 \text{ W}}{1 \text{ hp}} = 6.48 \text{ W/V} = 6.48 \text{ A}$$

2.27 Find the efficiency of operation of an electric motor that delivers 1 hp while absorbing an input of 900 W.

$$\eta = \frac{P_{\text{out}}}{P_{\text{in}}} \times 100\% = \frac{1 \text{ hp}}{900 \text{ W}} \times \frac{745.7 \text{ W}}{1 \text{ hp}} \times 100\% = 82.9\%$$

2.28 What is the operating efficiency of a fully loaded 2-hp dc electric motor that draws 19 A at 100 V?   (The power rating of a motor specifies the output power and not the input power.)

Since the input power is

$$P_{\text{in}} = VI = 100 \times 19 = 1900 \text{ W}$$

the efficiency is

$$\eta = \frac{P_{\text{out}}}{P_{\text{in}}} \times 100\% = \frac{2 \text{ hp}}{1900 \text{ W}} \times \frac{745.7 \text{ W}}{1 \text{ hp}} \times 100\% = 78.5\%$$

2.29 Find the input power to a fully loaded 5-hp motor that operates at 80 percent efficiency.

For almost all calculations, the efficiency is better expressed as a decimal fraction that is the percentage divided by 100—0.8 here.   Then from $\eta = P_{\text{out}}/P_{\text{in}}$,

$$P_{\text{in}} = \frac{P_{\text{out}}}{\eta} = \frac{5 \text{ hp}}{0.8} \times \frac{745.7 \text{ W}}{1 \text{ hp}} = 4.66 \text{ kW}$$

2.30 Find the current drawn by a dc electric motor that delivers 2 hp while operating at 85 percent efficiency from a 110-V line.

From $P_{\text{in}} = VI = P_{\text{out}}/\eta$,

$$I = \frac{P_{\text{out}}}{\eta V} = \frac{2 \text{ hp}}{0.85 \times 110 \text{ V}} \times \frac{745.7 \text{ W}}{1 \text{ hp}} = 15.95 \text{ A}$$

2.31 Maximum received solar power is about 1 kW/m². If solar panels, which convert solar energy to electric energy, are 13 percent efficient, how many square meters of solar cell panels are needed to supply the power to a 1600-W toaster?

The power from each square meter of solar panels is

$$P_{\text{out}} = \eta P_{\text{in}} = 0.13 \times 1000 = 130 \text{ W}$$

So, the total solar panel area needed is

$$\text{Area} = 1600 \text{ W} \times \frac{1 \text{ m}^2}{130 \text{ W}} = 12.3 \text{ m}^2$$

2.32 What horsepower must an electric motor develop to pump water up 40 ft at the rate of 2000 gallons per hour (gal/h) if the pumping system operates at 80 percent efficiency?

One way to solve for the power is to use the work done by the pump in 1 h, which is the weight of the water lifted in 1 h times the height through which it is lifted. This work divided by the time taken is the power output of the pumping system. And this power divided by the efficiency is the input power to the pumping system, which is the required output power of the electric motor. Some needed data are that 1 gal of water weighs 8.33 lb, and that   1 hp = 550 (ft · lb)/s.   Thus,

$$P = \frac{2000 \text{ gal}}{1 \text{ h}} \times 40 \text{ ft} \times \frac{1}{0.8} \times \frac{1 \text{ h}}{3600 \text{ s}} \times \frac{8.33 \text{ lb}}{1 \text{ gal}} \times \frac{1 \text{ hp}}{550 \text{ (ft · lb)/s}} = 0.42 \text{ hp}$$

2.33 Two systems are in cascade. One operates with an efficiency of 75 percent and the other with an efficiency of 85 percent. If the input power is 5 kW, what is the output power?

$$P_{\text{out}} = \eta_1 \eta_2 P_{\text{in}} = 0.75(0.85)(5000) \text{ W} = 3.19 \text{ kW}$$

2.34 Find the conversion relation between kilowatthours and joules.

The approach here is to convert from kilowatthours to watt-seconds, and then use the fact that   1 J = 1 Ws:

$$1 \text{ kWh} = 1000 \text{ W} \times 3600 \text{ s} = 3.6 \times 10^6 \text{ Ws} = 3.6 \text{ MJ}$$

2.35 For an electric rate of 7¢/kilowatthour, what does it cost to leave a 60-W light bulb on for 8 h?

The cost equals the total energy used times the cost per energy unit:

$$\text{Cost} = 60 \text{ W} \times 8 \text{ h} \times \frac{1 \text{ kWh}}{1000 \text{ Wh}} \times \frac{7¢}{1 \text{ kWh}} = 3.36¢$$

2.36 An electric motor delivers 5 hp while operating with an efficiency of 85 percent. Find the cost for operating it continuously for one day (d) if the electric rate is 6¢/kilowatthour.

The total energy used is the output power times the time of operation, all divided by the efficiency. The product of this energy and the electric rate is the total cost:

$$\text{Cost} = 5 \text{ hp} \times 1 \text{ d} \times \frac{1}{0.85} \times \frac{6¢}{1 \text{ kWh}} \times \frac{0.7457 \text{ kW}}{1 \text{ hp}} \times \frac{24 \text{ h}}{1 \text{ d}} = 632¢ = \$6.32$$

Supplementary Problems

2.37 Find the charge in coulombs of   (a) 6.28×10^{21} electrons   and   (b) 8.76×10^{20} protons.
Ans.   (a) −1006 C,   (b) 140 C

2.38 How many electrons have a total charge of −4 nC?   Ans. 2.5×10^{10} electrons

2.39 Find the current flow through a switch from a steady movement of   (a) 90 C in 6 s,   (b) 900 C in 20 min,   and   (c) 4×10^{23} electrons in 5 h.   Ans.   (a) 15 A, (b) 0.75 A, (c) 3.56 A

2.40   A capacitor is an electric circuit component that stores electric charge. If a capacitor charges at a steady rate to 10 mC in 0.02 ms, and if it discharges in 1 μs at a steady rate, what are the magnitudes of the charging and discharging currents?   *Ans.* 500 A, 10 000 A

2.41   In a gas, if doubly ionized negative ions move to the right at a steady rate of 3.62×10^{20} ions per minute and if singly ionized positive ions move to the left at a steady rate of 5.83×10^{20} ions per minute, find the current to the right.   *Ans.* −3.49 A

2.42   Find the shortest time that 120 C can flow through a 20-A circuit breaker without tripping it.   *Ans.* 6 s

2.43   If a steady current flows to a capacitor, find the time required for the capacitor to   (*a*) charge to 2.5 mC if the current is 35 mA,   (*b*) charge to 36 pC if the current is 18 μA,   and   (*c*) store 9.36×10^{17} electrons if the current is 85.6 nA.   *Ans.*   (*a*) 71.4 ms,   (*b*) 2 μs,   (*c*) 20.3 d

2.44   How long can a 4.5-Ah, 1.5-V flashlight battery deliver 100 mA?   *Ans.* 45 h

2.45   Find the potential energy in joules lost by a 1.2-lb book in falling off a desk that is 31 in high.   *Ans.* 4.2 J

2.46   How much chemical energy must a 1.25-V flashlight battery expend in producing a current flow of 130 mA for 5 min?   *Ans.* 48.8 J

2.47   Find the work done by a 9-V battery in moving 5×10^{20} electrons from its positive terminal to its negative terminal.   *Ans.* 721 J

2.48   Find the total energy available from a rechargeable 1.25-V flashlight battery with a 1.2-Ah rating.   *Ans.* 5.4 kJ

2.49   If all the energy in a 9-V transistor radio battery rated at 0.392 Ah is used to lift a 150-lb man, how high in feet will he be lifted?   *Ans.* 62.5 ft

2.50   If a charge of −4 C in moving from point *a* to point *b* gives up 20 J of energy, what is V_{ab}?   *Ans.* −5 V

2.51   Moving 6.93×10^{19} electrons from point *b* to point *a* requires 98 J of work. Find V_{ab}.   *Ans.* −8.83 V

2.52   How much power does an electric clock require if it draws 27.3 mA from a 110-V line?   *Ans.* 3 W

2.53   Find the current drawn by a 1000-W steam iron from a 120-V line.   *Ans.* 8.33 A

2.54   Find the average input power to a radio that consumes 4500 J in 3 min.   *Ans.* 25 W

2.55   Find the voltage drop across a toaster that gives off 7500 J of heat when a 13.64-A current flows through it for 5 s.   *Ans.* 110 V

2.56   How many joules does a 40-W light bulb consume in 1 d?   *Ans.* 3.46 MJ

2.57   How long can a 12-V car battery supply 30 A to a starter motor if the battery has 3.46 MJ of chemical energy that can be converted to electric energy?   *Ans.* 2.67 h

2.58   How long does it take a 420-W color TV set to consume   (*a*) 2 kWh   and   (*b*) 15 kJ?
Ans.   (*a*) 4.76 h,   (*b*) 35.7 s

2.59   Find the current drawn by a 110-V dc electric motor that delivers 2 hp.   Assume 100 percent efficiency of operation.   *Ans.* 13.6 A

2.60   Find the efficiency of operation of an electric motor that delivers 5 hp while absorbing an input of 4190 W.   *Ans.* 89 percent

2.61   What is the operating efficiency of a dc electric motor that delivers 1 hp while drawing 7.45 A from a 115-V line?   *Ans.* 87 percent

2.62   Find the current drawn by a 100-V dc electric motor that operates at 85 percent efficiency while delivering 0.5 hp.   *Ans.* 4.39 A

2.63   What is the horsepower produced by an automobile starter motor that draws 40 A from a 12-V battery while operating at an efficiency of 90 percent?   *Ans.* 0.579 hp

2.64   What horsepower must an electric motor develop to operate a pump that pumps water at a rate of 24 000 liters per hour (L/h) up a vertical distance of 50 m if the efficiency of the pump is 90 percent?   The gravitational force on 1 L of water is 9.78 N.   *Ans.* 4.86 hp

2.65   An ac electric motor drives a dc electric voltage generator.   If the motor operates at an efficiency of 90 percent and the generator at an efficiency of 80 percent, and if the input power to the motor is 5 kW, find the output power from the generator.   *Ans.* 3.6 kW

2.66   Find the cost for one year (365 d) to operate a 20-W transistor FM-AM radio 5 h a day if electrical energy costs 8¢/kilowatthour.   *Ans.* $2.92

2.67   For a cost of $5, how long can a fully loaded 5-hp electric motor be run if the motor operates at an efficiency of 85 percent and if the electric rate is 6¢/kilowatthour?   *Ans.* 19 h

2.68   If electric energy costs 6¢/kilowatthour, calculate the utility bill for one month for operating eight 100-W light bulbs for 50 h each, ten 60-W light bulbs for 70 h each, one 2-kW air conditioner for 80 h, one 3-kW range for 45 h, one 420-W color TV set for 180 h, and one 300-W refrigerator for 75 h.   *Ans.* $28.51

Chapter 3

Resistance

OHM'S LAW

In flowing through a conductor, free electrons collide with conductor atoms and lose some kinetic energy that is converted into heat. An applied voltage will cause them to regain energy and speed, but subsequent collisions will slow them down again. This speeding up and slowing down occur continually as free electrons move among conductor atoms.

Resistance is this property of materials that opposes or resists the movement of electrons and makes it necessary to apply a voltage to cause current to flow. The SI unit of resistance is the *ohm* with symbol Ω, the Greek uppercase letter omega. The quantity symbol is R.

In metallic and some other types of conductors, the current is proportional to the applied voltage: Doubling the voltage doubles the current, tripling the voltage triples the current, and so on. If the applied voltage V and resulting current I have associated references, the relation between V and I is

$$I(\text{amperes}) = \frac{V(\text{volts})}{R(\text{ohms})}$$

in which R is the constant of proportionality. This relation is known as *Ohm's law*. For time-varying voltages and currents, $i = v/R$. And for nonassociated references, $I = -V/R$ or $i = -v/R$.

From Ohm's law it is evident that, the greater the resistance, the less the current for any applied voltage. Also, the electric resistance of a conductor is $1\ \Omega$ if an applied voltage of 1 V causes a current of 1 A to flow.

The inverse of resistance is often useful. It is called *conductance* and its quantity symbol is G. The SI unit of conductance is the *siemens* with symbol S, which is replacing the popular non-SI unit *mho* with symbol $\mho$ (inverted omega). Since conductance is the inverse of resistance, $G = 1/R$. In terms of conductance, Ohm's law is

$$I(\text{amperes}) = G(\text{siemens}) \times V(\text{volts})$$

which shows that the greater the conductance of a conductor, the greater the current for any applied voltage.

RESISTIVITY

The resistance of a conductor of uniform cross section is directly proportional to the length of the conductor and inversely proportional to the cross-sectional area. Resistance is also a function of the temperature of the conductor, as is explained in the next section. At a fixed temperature the resistance of a conductor is

$$\Omega = (\Omega \cdot m)\left(\frac{m}{m^2}\right) \Rightarrow \qquad R = \rho \frac{l}{A} \Rightarrow \text{Resistance} = (\text{resistivity})\left(\frac{\text{length}}{\text{x-sect. Area}}\right)$$
$$\rho = \frac{1}{\sigma}$$

where l is the conductor length in meters and A is the cross-sectional area in square meters. The constant of proportionality ρ, the Greek lowercase letter rho, is the quantity symbol for *resistivity*, the factor that depends on the type of material.

The SI unit of resistivity is the *ohm meter* with unit symbol $\Omega \cdot m$. Table 3-1 shows the resistivities of some materials at 20°C.

A good conductor has a resistivity close to $10^{-8}\ \Omega \cdot m$. Silver, the best conductor, is too expensive for most uses. Copper is a common conductor, as is aluminum. Materials with

Table 3-1

Material	Resistivity ($\Omega \cdot$ m at 20°C)	Material	Resistivity ($\Omega \cdot$ m at 20°C)
Silver	1.64×10^{-8}	Nichrome	100×10^{-8}
Copper, annealed	1.72×10^{-8}	Silicon	2500
Aluminum	2.83×10^{-8}	Paper	10^{10}
Iron	12.3×10^{-8}	Mica	5×10^{11}
Constantan	49×10^{-8}	Quartz	10^{17}

resistivities greater than 10^{10} $\Omega \cdot$ m are *insulators*. They can provide physical support without significant current leakage. Also, insulating coatings on wires prevent current leaks between wires that touch. Materials with resistivities in the range of 10^{-4} to 10^{-7} $\Omega \cdot$ m are *semiconductors*, from which transistors are made.

The relationship among conductance, length, and cross-sectional area is

$$\mho = \left(\frac{\mho}{m}\right)\left(\frac{m^2}{m}\right) \Rightarrow \qquad (\mho)\, G = \sigma \frac{A}{l} \Rightarrow Conductance = (conductivity)\left(\frac{x\text{-}sect.\ Area}{length}\right)$$

where the constant of proportionality σ, the Greek lowercase sigma, is the quantity symbol for *conductivity*. The SI unit of conductivity is the *siemens per meter* with symbol $S \cdot m^{-1}$.

TEMPERATURE EFFECTS

The resistances of most good conducting materials increase almost linearly with temperature over the range of normal operating temperatures, as shown by the solid line in Fig. 3-1. However, some materials, and common semiconductors in particular, have resistances that decrease with temperature increases.

If the straight-line portion in Fig. 3-1 is extended to the left, it crosses the temperature axis at a temperature T_0 at which the resistance appears to be zero. This temperature T_0 is the *inferred zero resistance temperature*. (The actual zero resistance temperature is -273°C.) If T_0 is known and if the resistance R_1 at another temperature T_1 is known, then the resistance R_2 at another temperature T_2 is, from straight-line geometry;

$$R_2 = \frac{T_2 - T_0}{T_1 - T_0} R_1$$

Table 3-2 has inferred zero resistance temperatures for some common conducting materials.

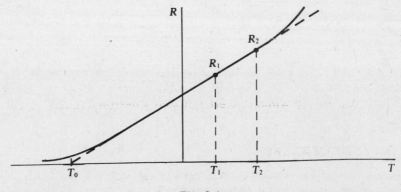

Fig. 3-1

A different but equivalent way of finding the resistance R_2 is from

$$R_2 = R_1[1 + \alpha_1(T_2 - T_1)]$$

where α_1, with the Greek lowercase alpha, is the *temperature coefficient of resistance* at the temperature T_1. Often T_1 is 20°C. Table 3-3 has temperature coefficients of resistance at 20°C for some common conducting materials. Note that the unit of α is *per degree Celsius* with symbol °C^{-1}.

<table>
<tr><td colspan="2" align="center">Table 3-2</td></tr>
<tr><th>Material</th><th>Inferred zero resistance temperature (°C)</th></tr>
<tr><td>Tungsten</td><td>−202</td></tr>
<tr><td>Copper</td><td>−234.5</td></tr>
<tr><td>Aluminum</td><td>−236</td></tr>
<tr><td>Silver</td><td>−243</td></tr>
<tr><td>Constantan</td><td>−125 000</td></tr>
</table>

<table>
<tr><td colspan="2" align="center">Table 3-3</td></tr>
<tr><th>Material</th><th>Temperature coefficient (°C^{-1} at 20°C)</th></tr>
<tr><td>Tungsten</td><td>0.0045</td></tr>
<tr><td>Copper</td><td>0.003 93</td></tr>
<tr><td>Aluminum</td><td>0.003 91</td></tr>
<tr><td>Silver</td><td>0.0038</td></tr>
<tr><td>Constantan</td><td>0.000 008</td></tr>
<tr><td>Carbon</td><td>−0.0005</td></tr>
</table>

RESISTORS

In a practical sense a *resistor* is a circuit component that is used because of its resistance. Mathematically, a resistor is a circuit component for which there is a relation between its instantaneous voltage and instantaneous current such as $v = iR$, the voltage-current relation for a resistor that obeys Ohm's law—a *linear resistor*. Any other type of voltage-current relation ($v = 4i^2 + 6$, for example) is for a *nonlinear resistor*. The term "resistor" usually designates a linear resistor. Nonlinear resistors are specified as such. Figure 3-2a shows the circuit symbol for a linear resistor, and Fig. 3-2b that for a nonlinear resistor.

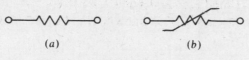

(a)   (b)

Fig. 3-2

RESISTOR POWER ABSORPTION

Substitution from $V = IR$ into $P = VI$ gives the power absorbed by a linear resistor in terms of resistance:

$$P = \frac{V^2}{R} = I^2R$$

Although only electric *energy* is actually absorbed or dissipated, these terms are commonly applied to power. Every resistor has a *power rating*, also called *wattage rating*, that is the maximum power that the resistor can absorb without overheating to a destructive temperature.

NOMINAL VALUES AND TOLERANCES

Manufacturers print resistance values on resistor casings either in numerical form or in a color code. These values, though, are only *nominal values*: They are only approximately equal to the actual resistances. The possible percentage variation of resistance about the nominal value is called the *tolerance*. The popular carbon-composition resistor has tolerances of 20, 10, and 5 percent, which means that the actual resistances can vary from the nominal values by as much as ± 20, ± 10, and ± 5 percent of the nominal values.

COLOR CODE

The most popular resistance color code has nominal resistance values and tolerances indicated by the colors of either three or four bands around the resistor casing, as shown in Fig. 3-3.

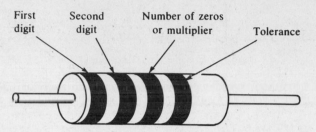

Fig. 3-3

Each color has a corresponding numerical value as specified in Table 3-4. The colors of the first and second bands correspond, respectively, to the first two digits of the nominal resistance value. The color of the third band, except for silver and gold, corresponds to the number of zeros that follow the first two digits. A third band of silver corresponds to a multiplier of 10^{-2}, and a third band of gold to a multiplier of 10^{-1}. The fourth band indicates the tolerance and is either gold- or silver-colored, or is missing. Gold corresponds to a tolerance of 5 percent, silver to 10 percent, and a missing band to 20 percent.

Table 3-4

Color	Number	Color	Number
Black	0	Blue	6
Brown	1	Violet	7
Red	2	Gray	8
Orange	3	White	9
Yellow	4	Gold	0.1
Green	5	Silver	0.01

OPEN AND SHORT CIRCUITS

An *open circuit* has an infinite resistance, which means that it has zero current flow through it for any finite voltage across it. On a circuit diagram it is indicated by two terminals not connected to anything—no path is shown for current to flow through. An open circuit is sometimes called an *open*.

A *short circuit* is the opposite of an open circuit. It has zero voltage across it for any finite current flow through it. On a circuit diagram a short circuit is designated by an ideal conducting wire—a wire with zero resistance. A short circuit is often called a *short*.

Not all open and short circuits are desirable. Frequently, one or the other is a circuit defect that occurs as a result of a component failure from an accident or the misuse of a circuit.

INTERNAL RESISTANCE

Every practical voltage or current source has an *internal resistance* that adversely affects the operation of the source. Except for an open-circuit load, a voltage source has a loss of voltage across its internal resistance. And except for a short-circuit load, a current source has a loss of current through its internal resistance.

In a practical voltage source the internal resistance has almost the same effect as a resistor in series with an ideal voltage source, as shown in Fig. 3-4a. (Components in series carry the same current.) In a practical current source the internal resistance has almost the same effect as a resistor in parallel with an ideal current source, as shown in Fig. 3-4b. (Components in parallel have the same voltage across them.)

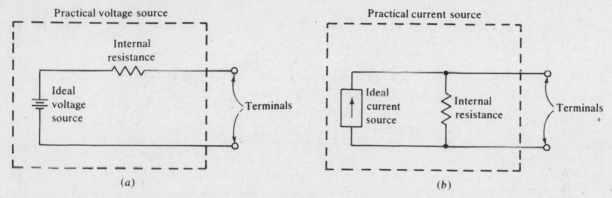

Fig. 3-4

Solved Problems

3.1   If an oven has a 240-V heating element with a resistance of 24 Ω, what is the minimum rating of a fuse that can be used in the lines to the heating element?

The fuse must be able to carry the current of the heating element:

$$I = \frac{V}{R} = \frac{240}{24} = 10 \text{ A}$$

3.2   What is the resistance of a soldering iron that draws 0.8333 A at 120 V?

$$R = \frac{V}{I} = \frac{120}{0.8333} = 144 \text{ } \Omega$$

3.3   A toaster with 8.27 Ω of resistance draws 13.9 A. Find the applied voltage.

$$V = IR = 13.9 \times 8.27 = 115 \text{ V}$$

3.4   What is the conductance of a 560-kΩ resistor?

$$G = \frac{1}{R} = \frac{1}{560 \times 10^3} \text{ S} = 1.79 \text{ } \mu\text{S}$$

3.5   What is the conductance of an ammeter that indicates 20 A when 0.01 V is across it?

$$G = \frac{I}{V} = \frac{20}{0.01} = 2000 \text{ S}$$

3.6   Find the resistance at 20°C of an annealed copper bus bar 3 m in length and 0.5 cm by 3 cm in rectangular cross section.

The cross-sectional area of the bar is $(0.5 \times 10^{-2})(3 \times 10^{-2}) = 1.5 \times 10^{-4} \text{ m}^2$. Table 3-1 has the resistivity of annealed copper: $1.72 \times 10^{-8} \ \Omega \cdot \text{m}$ at 20°C. So,

$$R = \rho \frac{l}{A} = \frac{(1.72 \times 10^{-8})(3)}{1.5 \times 10^{-4}} \Omega = 344 \ \mu\Omega$$

3.7   Find the resistance of an aluminum wire that has a length of 1000 m and a diameter of 1.626 mm. The wire is at 20°C.

The cross-sectional area of the wire is πr^2, in which $r = d/2 = 1.626 \times 10^{-3}/2 = 0.813 \times 10^{-3}$ m. From Table 3-1 the resistivity of aluminum is $2.83 \times 10^{-8} \ \Omega \cdot \text{m}$. So,

$$R = \rho \frac{l}{A} = \frac{(2.83 \times 10^{-8})(1000)}{\pi(0.813 \times 10^{-3})^2} = 13.6 \ \Omega$$

3.8   The resistance of a certain wire is 15 Ω. Another wire of the same material and at the same temperature has a diameter one-third as great and a length twice as great. Find the resistance of the second wire.

The resistance of a wire is proportional to the length and inversely proportional to the area. Also, the area is proportional to the square of the diameter. So, the resistance of the second wire is

$$R = \frac{15 \times 2}{(1/3)^2} = 270 \ \Omega$$

3.9   What is the resistivity of platinum if a cube of it 1 cm along each edge has a resistance of 10 $\mu\Omega$ across opposite faces?

From $R = \rho l/A$ and the fact that $A = 10^{-2} \times 10^{-2} = 10^{-4} \text{ m}^2$ and $l = 10^{-2}$ m,

$$\rho = \frac{RA}{l} = \frac{(10 \times 10^{-6})(10^{-4})}{10^{-2}} = 10 \times 10^{-8} \ \Omega \cdot \text{m}$$

3.10   A 15-ft length of wire with a cross-sectional area of 127 cmils has a resistance of 8.74 Ω at 20°C. What material is the wire made from?

The material can be found from calculating the resistivity and comparing it with the resistivities given in Table 3-1. For this calculation it is convenient to use the fact that, by the definition of a circular mil, the corresponding area in square inches is the number of circular mils times $\pi/4 \times 10^{-6}$. From rearranging $R = \rho l/A$,

$$\rho = \frac{AR}{l} = \frac{[127(\pi/4 \times 10^{-6}) \text{ in}^2](8.74 \ \Omega)}{15 \text{ ft}} \times \frac{1 \text{ ft}}{12 \text{ in}} \times \frac{0.0254 \text{ m}}{1 \text{ in}} = 12.3 \times 10^{-8} \ \Omega \cdot \text{m}$$

Since iron has this resistivity in Table 3-1, the material must be iron.

3.11   What is the length of No. 28 AWG (0.000 126 in^2 in cross-sectional area) Nichrome wire required for a 24-Ω resistor at 20°C?

From rearranging $R = \rho l / A$ and using the resistivity of Nichrome given in Table 3-1,

$$l = \frac{AR}{\rho} = \frac{(0.000\,126\text{ in}^2)(24\text{ }\Omega)}{100 \times 10^{-8}\text{ }\Omega \cdot \text{m}} \times \frac{0.0254\text{ m}}{1\text{ in}} \times \frac{0.0254\text{ m}}{1\text{ in}} = 1.95\text{ m}$$

3.12 A certain aluminum wire has a resistance of 5 Ω at 20°C. What is the resistance of an annealed copper wire of the same size and at the same temperature?

For the copper and aluminum wires, respectively,

$$R = \rho_c \frac{l}{A} \quad \text{and} \quad 5 = \rho_a \frac{l}{A}$$

Taking the ratio of the two equations causes the length and area quantities to divide out, with the result that the ratio of the resistances is equal to the ratio of the resistivities:

$$\frac{R}{5} = \frac{\rho_c}{\rho_a} \quad \text{or} \quad R = \frac{\rho_c}{\rho_a} \times 5$$

Then with the insertion of resistivities from Table 3-1,

$$R = \frac{1.72 \times 10^{-8}}{2.83 \times 10^{-8}} \times 5 = 3.04\text{ }\Omega$$

3.13 A wire 50 m in length and 2 mm² in cross section has a resistance of 0.56 Ω. A 100-m length of wire of the same material has a resistance of 2 Ω at the same temperature. Find the diameter of this wire.

From the data given for the first wire, the resistivity of the conducting material is

$$\rho = \frac{RA}{l} = \frac{0.56(2 \times 10^{-6})}{50} = 2.24 \times 10^{-8}\text{ }\Omega \cdot \text{m}$$

Therefore the cross-sectional area of the second wire is

$$A = \frac{\rho l}{R} = \frac{(2.24 \times 10^{-8})(100)}{2} = 1.12 \times 10^{-6}\text{ m}^2$$

and, from $A = \pi(d/2)^2$, the diameter is

$$d = 2\sqrt{\frac{A}{\pi}} = 2\sqrt{\frac{1.12 \times 10^{-6}}{\pi}} = 1.19 \times 10^{-3}\text{ m} = 1.19\text{ mm}$$

3.14 A wirewound resistor is to be made from 0.2-mm diameter constantan wire wound around a cylinder that is 1 cm in diameter. How many turns of wire are required for a resistance of 50 Ω at 20°C?

The number of turns equals the wire length divided by the circumference of the cylinder. From $R = \rho l / A$ and the resistivity of constantan given in Table 3-1, the length of the wire that has a resistance of 50 Ω is

$$l = \frac{RA}{\rho} = \frac{R\pi r^2}{\rho} = \frac{50\pi(0.1 \times 10^{-3})^2}{49 \times 10^{-8}} = 3.21\text{ m}$$

The circumference of the cylinder is $2\pi r$, in which $r = 10^{-2}/2 = 0.005$ m, the radius of the cylinder. So, the number of turns is

$$\frac{l}{2\pi r} = \frac{3.21}{2\pi(0.005)} = 102\text{ turns}$$

3.15 A No. 14 AWG standard annealed copper wire is 0.003 23 in² in cross section and has a resistance of 2.58 mΩ/ft at 25°C. What is the resistance of 500 ft of No. 6 AWG wire of the same material at 25°C? The cross-sectional area of this wire is 0.0206 in².

Perhaps the best approach is to calculate the resistance of a 500-ft length of the No. 14 AWG wire,

$$(2.58 \times 10^{-3})(500) = 1.29 \ \Omega$$

and then take the ratio of the two $R = \rho l/A$ equations. Since the resistivities and lengths are the same, they divide out, with the result that

$$\frac{R}{1.29} = \frac{0.003\ 23}{0.0206} \quad \text{or} \quad R = \frac{0.003\ 23}{0.0206} \times 1.29 = 0.202 \ \Omega$$

3.16 The conductance of a certain wire is 0.5 S. Another wire of the same material and at the same temperature has a diameter twice as great and a length three times as great. What is the conductance of the second wire?

The conductance of a wire is proportional to the area and inversely proportional to the length. Also, the area is proportional to the square of the diameter. Therefore the conductance of the second wire is

$$G = \frac{0.5 \times 2^2}{3} = 0.667 \ \text{S}$$

3.17 Find the conductance of 100 ft of No. 14 AWG iron wire, which has a diameter of 64 mils. The temperature is 20°C.

The conductance formula is $G = \sigma A/l$, in which $\sigma = 1/\rho$ and $A = \pi(d/2)^2$. Of course, the resistivity of iron can be gotten from Table 3-1. Thus,

$$G = \sigma \frac{A}{l} = \frac{1}{12.3 \times 10^{-8}} \frac{\text{S}}{\text{m}} \times \frac{\pi(64 \times 10^{-3} \ \text{in}/2)^2}{100 \ \text{ft}} \times \frac{1 \ \text{ft}}{12 \ \text{in}} \times \frac{0.0254 \ \text{m}}{1 \ \text{in}} = 0.554 \ \text{S}$$

3.18 The resistance of a certain copper power line is 100 Ω at 20°C. What is its resistance when the sun heats the line up to 38°C?

From Table 3-2 the inferred absolute zero resistance temperature of copper is -234.5°C, which is T_0 in the formula $R_2 = R_1(T_2 - T_0)/(T_1 - T_0)$. Also, from the given data, $T_2 = 38$°C, $R_1 = 100 \ \Omega$, and $T_1 = 20$°C. So, the wire resistance at 38°C is

$$R_2 = \frac{T_2 - T_0}{T_1 - T_0} R_1 = \frac{38 - (-234.5)}{20 - (-234.5)} \times 100 = 107 \ \Omega$$

3.19 When 120 V is applied across a certain light bulb, a 0.5-A current flows, causing the temperature of the tungsten filament to increase to 2600°C. What is the resistance of the light bulb at the normal room temperature of 20°C?

The resistance of the energized light bulb is $120/0.5 = 240 \ \Omega$. And since from Table 3-2 the inferred zero resistance temperature for tungsten is -202°C, the resistance at 20°C is

$$R_2 = \frac{T_2 - T_0}{T_1 - T_0} R_1 = \frac{20 - (-202)}{2600 - (-202)} \times 240 = 19 \ \Omega$$

3.20 A certain unenergized copper transformer winding has a resistance of 30 Ω at 20°C. Under rated operation, however, the resistance increases to 35 Ω. Find the temperature of the energized winding.

The formula $R_2 = R_1(T_2 - T_0)/(T_1 - T_0)$ solved for T_2 becomes

$$T_2 = \frac{R_2(T_1 - T_0)}{R_1} + T_0$$

From the specified data, $R_2 = 35 \ \Omega$, $T_1 = 20$°C, and $R_1 = 30 \ \Omega$. Also, from Table 3-2, $T_0 =$

$-234.5°C$. So,

$$T_2 = \frac{35[20 - (-234.5)]}{30} - 234.5 = 62.4°C$$

3.21 The resistance of a certain aluminum power line is 150 Ω at 20°C.   Find the line resistance when the sun heats the line up to 42°C.   First use the inferred zero resistance temperature formula and then the temperature coefficient of resistance formula to show that the two formulas are equivalent.

From Table 3-2 the zero resistance temperature of aluminum is −236°C.   Thus,

$$R_2 = \frac{T_2 - T_0}{T_1 - T_0} R_1 = \frac{42 - (-236)}{20 - (-236)} \times 150 = 163 \ \Omega$$

From Table 3-3 the temperature coefficient of resistance of aluminum is 0.003 91 °C^{-1} at 20°C.   So,

$$R_2 = R_1[1 + \alpha_1(T_2 - T_1)] = 150[1 + 0.003\,91(42 - 20)] = 163 \ \Omega$$

3.22 Find the resistance at 35°C of an aluminum wire that has a length of 200 m and a diameter of 1 mm.

The wire resistance at 20°C can be found and used in the temperature coefficient of resistance formula.   (Alternatively, the inferred zero resistance temperature formula can be used.)   Since the cross-sectional area of the wire is $\pi(d/2)^2$, where $d = 10^{-3}$ m, and since from Table 3-1 the resistivity of aluminum is 2.83×10^{-8} Ω · m, the wire resistance at 20°C is

$$R = \rho \frac{l}{A} = (2.83 \times 10^{-8}) \times \frac{200}{\pi(10^{-3}/2)^2} = 7.21 \ \Omega$$

The only other quantity needed to calculate the wire resistance at 35°C is the temperature coefficient of resistance of aluminum at 20°C.   From Table 3-3 it is 0.003 91 °C^{-1}.   So,

$$R_2 = R_1[1 + \alpha(T_2 - T_1)] = 7.21[1 + 0.003\,91(35 - 20)] = 7.63 \ \Omega$$

3.23 Derive a formula for calculating the temperature coefficient of resistance from the temperature T_1 of a material and T_0, its inferred zero resistance temperature.

In $R_2 = R_1[1 + \alpha_1(T_2 - T_1)]$ select $T_2 = T_0$.   Then $R_2 = 0 \ \Omega$, by definition.   The result is $0 = R_1[1 + \alpha_1(T_0 - T_1)]$, from which

$$\alpha_1 = \frac{1}{T_1 - T_0}$$

3.24 Calculate the temperature coefficient of resistance of aluminum at 30°C and use it to find the resistance of an aluminum wire at 70°C if the wire has a resistance of 40 Ω at 30°C.

From Table 3-2, aluminum has an inferred zero resistance temperature of −236°C.   With this value inserted, the formula derived in the solution to Prob. 3.23 gives

$$\alpha_{30°C} = \frac{1}{T_1 - T_0} = \frac{1}{30 - (-236)} - 0.003\,759 \ °C^{-1}$$

So

$$R_2 = R_1[1 + \alpha_1(T_2 - T_1)] = 40[1 + 0.003\,759(70 - 30)] = 46 \ \Omega$$

3.25 Find the resistance of an electric heater that absorbs 2400 W when connected to a 120-V line.

From $P = V^2/R$,

$$R = \frac{V^2}{P} = \frac{120^2}{2400} = 6 \ \Omega$$

3.26 Find the internal resistance of a 2-kW water heater that draws 8.33 A.

From $P = I^2 R$,

$$R = \frac{P}{I^2} = \frac{2000}{8.33^2} = 28.8 \ \Omega$$

3.27 What is the greatest voltage that can be applied across a $\frac{1}{8}$-W, 2.7-MΩ resistor without causing it to overheat?

From $P = V^2/R$,

$$V = \sqrt{RP} = \sqrt{(2.7 \times 10^6)(1/8)} = 581 \ \text{V}$$

3.28 If a nonlinear resistor has a voltage-current relation of $V = 3I^2 + 4$, what current does it draw when energized by 61 V? Also, what power does it absorb?

Inserting the applied voltage into the nonlinear equation results in $61 - 3I^2 + 4$, from which

$$I = \sqrt{\frac{61 - 4}{3}} = 4.36 \ \text{A}$$

Then from $P = VI$,

$$P = 61 \times 4.36 = 266 \ \text{W}$$

3.29 At 20°C a pn junction silicon diode has a current-voltage relation of $I = 10^{-14}(e^{40V} - 1)$. What is the diode voltage when the current is 50 mA?

From the given formula,

$$50 \times 10^{-3} = 10^{-14}(e^{40V} - 1)$$

Multiplying both sides by 10^{14} and adding 1 to both sides results in

$$50 \times 10^{11} + 1 = e^{40V}$$

Then from the natural logarithm of both sides,

$$V = \frac{1}{40} \ln(50 \times 10^{11} + 1) = 0.73 \ \text{V}$$

3.30 What is the resistance range for (a) a 10 percent, 470-Ω resistor, and (b) a 20 percent, 2.7-MΩ resistor? (*Hint*: 10 percent corresponds to 0.1 and 20 percent to 0.2.)

(a) The resistance can be as much as $0.1 \times 470 = 47 \ \Omega$ from the 470-Ω nominal value. So, the resistance can be as small as $470 - 47 = 423 \ \Omega$ or as great as $470 + 47 = 517 \ \Omega$.

(b) Since the maximum resistance variation from the nominal value is $0.2(2.7 \times 10^6) = 0.54 \ \text{M}\Omega$, the resistance can be as small as $2.7 - 0.54 = 2.16 \ \text{M}\Omega$ or as great as $2.7 + 0.54 = 3.24 \ \text{M}\Omega$.

3.31 A voltage of 110 V is across a 5 percent, 20-kΩ resistor. What range must the current be in? (*Hint*: 5 percent corresponds to 0.05.)

The resistance can be as much as $0.05(20 \times 10^3) = 10^3 \ \Omega$ from the nominal value, which means that the resistance can be as small as $20 - 1 = 19 \ \text{k}\Omega$ or as great as $20 + 1 = 21 \ \text{k}\Omega$. Therefore, the current can be as small as

$$\frac{110}{21 \times 10^3} = 5.24 \ \text{mA}$$

or as great as

Green Blue Gold Silver

$$\frac{110}{19 \times 10^3} = 5.79 \text{ mA}$$

3.32 What are the colors of the bands on a 10 percent, 5.6-Ω resistor?

Since $5.6 = 56 \times 0.1$, the resistance has a first digit of 5, a second digit of 6, and a multiplier of 0.1. From Table 3-4, green corresponds to 5, blue to 6, and gold to 0.1. Also, silver corresponds to the 10 percent tolerance. So, the color bands and arrangement are green-blue-gold-silver from an end to the middle of the resistor casing.

3.33 Determine the colors of the bands on a 20 percent, 2.7-MΩ resistor.

The numerical value of the resistance is 2 700 000, which is a 2 and a 7 followed by five zeros. From Table 3-4 the corresponding color code is red for the 2, violet for the 7, and green for the five zeros. Also, there is a missing color band for the 20 percent tolerance. So, the color bands from an end of the resistor casing to the middle are red-violet-green-missing.

3.34 What are the nominal resistance and tolerance of a resistor with color bands in the order of green-blue-yellow-silver from an end of the resistor casing toward the middle?

From Table 3-4, green corresponds to 5, blue to 6, and yellow to 4. The 5 is the first digit and 6 the second digit of the resistance value, and 4 is the number of trailing zeros. Consequently, the resistance is 560 000 Ω or 560 kΩ. The silver band designates a 10 percent tolerance.

2 4 0 5%

3.35 Find the resistance corresponding to color bands in the order of red-yellow-black-gold.

From Table 3-4, red corresponds to 2, yellow to 4, and black to 0 (no trailing zeros). The fourth band of gold corresponds to a 5 percent tolerance. So, the resistance is 24 Ω with a 5 percent tolerance.

3.36 If a 12-V car battery has a 0.04-Ω internal resistance, what is the battery terminal voltage when the battery delivers 40 A?

The battery terminal voltage is the generated voltage minus the voltage drop across the internal resistance:

$$V = 12 - IR = 12 - 40(0.04) = 10.4 \text{ V}$$

3.37 If a 12-V car battery has a 0.1-Ω internal resistance, what terminal voltage causes a 4-A current to flow *into* the positive terminal?

The applied voltage must equal the battery generated voltage plus the voltage drop across the internal resistance:

$$V = 12 + IR = 12 + 4(0.1) = 12.4 \text{ V}$$

3.38 If a 10-A current source has a 100-Ω internal resistance, what is the current flow from the source when the terminal voltage is 200 V?

The current flow from the source is the 10 A minus the current flow through the internal resistance:

$$I = 10 - \frac{V}{R} = 10 - \frac{200}{100} = 8 \text{ A}$$

Supplementary Problems

3.39 What is the resistance of a 240-V electric clothes dryer that draws 23.3 A?   *Ans.* 10.3 Ω

3.40 If a voltmeter has 500 kΩ of internal resistance, find the current flow through it when it indicates 86 V.   *Ans.* 172 μA

3.41 If an ammeter has 2 mΩ of internal resistance, find the voltage across it when it indicates 10 A.
Ans. 20 mA

3.42 What is the conductance of a 39-Ω resistor?   *Ans.* 25.6 mS

3.43 What is the conductance of a voltmeter that indicates 150 V when 0.3 mA flows through it?
Ans. 2 μS

3.44 Find the resistance at 20°C of an annealed copper bus bar 2 m long and 1 cm by 4 cm in rectangular cross section.   *Ans.* 86 μΩ

3.45 What is the resistance of an annealed copper wire that has a length of 500 m and a diameter of 0.404 mm?   *Ans.* 67.1 Ω

3.46 The resistance of a wire is 25 Ω.   Another wire of the same material and at the same temperature has a diameter twice as great and a length six times as great.   Find the resistance of the second wire.
Ans. 37.5 Ω

3.47 What is the resistivity of tin if a cube of it 10 cm along each edge has a resistance of 1.15 μΩ across opposite faces?   *Ans.* 11.5×10^{-8} Ω · m

3.48 A 40-m length of wire with a diameter of 0.574 mm has a resistance of 75.7 Ω at 20°C.   What material is the wire made from?   *Ans.* Constantan

3.49 What is the length of No. 30 AWG (10.0-mil diameter) constantan wire at 20°C required for a 200-Ω resistor?   *Ans.* 20.7 m

3.50 If No. 29 AWG annealed copper wire at 20°C has a resistance of 83.4 Ω per 1000 ft, what is the resistance per 100 ft of Nichrome wire of the same size and at the same temperature?
Ans. 485 Ω per 100 ft

3.51 A wire with a resistance of 5.16 Ω has a diameter of 45 mils and a length of 1000 ft.   Another wire of the same material has a resistance of 16.5 Ω and a diameter of 17.9 mils.   What is the length of this second wire if both wires are at the same temperature?   *Ans.* 506 ft

3.52 A wirewound resistor is to be made from No. 30 AWG (10.0-mil diameter) constantan wire wound around a cylinder that is 0.5 cm in diameter.   How many turns are required for a resistance of 25 Ω at 20°C?   *Ans.* 165 turns

3.53 The conductance of a wire is 2.5 S.   Another wire of the same material and at the same temperature has a diameter one-fourth as great and a length twice as great.   Find the conductance of the second wire.   *Ans.* 78.1 mS

3.54 Find the conductance of 5 m of Nichrome wire that has a diameter of 1 mm.   *Ans.* 157 mS

3.55 If an aluminum power line has a resistance of 80 Ω at 30°C, what is its resistance when cold air lowers its temperature to −10°C?   *Ans.* 68 Ω

3.56 If the resistance of a constantan wire is 2 MΩ at −150°C, what is its resistance at 200°C?
Ans. 2.006 MΩ

3.57 The resistance of an aluminum wire is 2.4 Ω at −5°C.   At what temperature will it be 2.8 Ω?
Ans. 33.5°C

3.58 What is the resistance at 90°C of a carbon rod that has a resistance of 25 Ω at 20°C?   *Ans.* 24.1 Ω

3.59 Find the temperature coefficient of resistance of iron at 20°C if iron has an inferred zero resistance temperature of −162°C.   *Ans.* 0.0055 °C^{-1}

3.60 What is the maximum current that a 1-W, 56-kΩ resistor can safely conduct?   *Ans.* 4.23 mA

3.61 What is the maximum voltage that can be safely applied across a $\frac{1}{2}$-W, 91-Ω resistor?   *Ans.* 6.75 V

3.62 What is the resistance of a 240-V, 5600-W electric heater?   *Ans.* 10.3 Ω

3.63 A nonlinear resistor has a voltage-current relation of $V = 2I^2 + 3I + 10$.   Find the current drawn by this resistor when 37 V is applied across it.   *Ans.* 3 A

3.64 If a diode has a current-voltage relation of $I = 10^{-14}(e^{40V} - 1)$,   what is the diode voltage when the current is 150 mA?   *Ans.* 0.758 V

3.65 What is the resistance range for a 5 percent, 75-kΩ resistor?   *Ans.* 71.25 to 78.75 kΩ

3.66 A 12.1-mA current flows through a 10 percent, 2.7-kΩ resistor.   What range must the resistor voltage be in?   *Ans.* 29.4 to 35.9 V

3.67 What are the resistor color codes for tolerances and nominal resistances of   (a) 10 percent, 0.18 Ω;   (b) 5 percent, 39 kΩ;   and   (c) 20 percent, 20 MΩ?
Ans.   (a) Brown-gray-silver-silver,   (b) orange-white-orange-gold,   (c) red-black-blue-missing

3.68 Find the tolerances and nominal resistances corresponding to color codes of   (a) brown-brown-silver-gold,   (b) green-brown-brown-missing,   and   (c) blue-gray-green-silver.
Ans.   (a) 5 percent, 0.11 Ω;   (b) 20 percent, 510 Ω;   (c) 10 percent, 6.8 MΩ

3.69 A battery that produces 6 V on open circuit produces 5.4 V when delivering 6 A.   What is the internal resistance of the battery?   *Ans.* 0.1 Ω

3.70 A $\frac{1}{2}$-hp automobile electric starter motor operates at 85 percent efficiency from a 12-V battery.   What is the battery internal resistance if the battery terminal voltage drops to 10.5 V when energizing the starter motor?   *Ans.* 0.036 Ω

3.71 A short circuit across a current source draws 20 A. When the current source has an open circuit across it, the terminal voltage is 600 V. Find the internal resistance of the source.   *Ans.* 30 Ω

3.72 A short circuit across a current source draws 15 A. If a 10-Ω resistor across the source draws 13 A, what is the internal resistance of the source?   *Ans.* 65 Ω

Chapter 4

Series and Parallel DC Circuits

BRANCHES, NODES, LOOPS, AND MESHES

Strictly speaking, a *branch* of a circuit is a single component such as a resistor or a source. Sometimes, though, this term is used for a group of components that carry the same current—components in *series*, especially when they are of the same type.

A *node* is a connection point between two or more branches. On a circuit diagram a node is sometimes indicated by a dot that may be a solder point in the actual circuit. The node also includes all wires connected to the point. In other words, it includes all points at the same potential. If a short circuit connects two nodes, these two nodes are equivalent to and in fact are just a single node, even if two dots are shown.

A *loop* is any closed path in a circuit. A *mesh* is a loop that does not have a closed path in its interior. No components are inside a mesh.

KIRCHHOFF'S VOLTAGE LAW AND SERIES DC CIRCUITS

Kirchhoff's voltage law, abbreviated KVL, has three equivalent versions: At any instant around a loop, in either a clockwise or counterclockwise direction,

1. The algebraic sum of the voltage drops is zero.
2. The algebraic sum of the voltage rises is zero.
3. The algebraic sum of the voltage drops equals the algebraic sum of the voltage rises.

In all these versions, the word "algebraic" means that the signs of the voltage drops and rises are included in the additions. Remember that a voltage rise is a negative voltage drop, and that a voltage drop is a negative voltage rise. For loops with no current sources, the most convenient KVL version is often the third one, restricted such that the voltage drops are only across resistors and the voltage rises only across voltage sources.

In the application of KVL, a loop current is usually referenced clockwise, as shown in the series circuit of Fig. 4-1, and KVL is applied in the direction of the current. (This is a series circuit because the same current I flows through all components.) The sum of the voltage drops across the resistors, $V_1 + V_2 + V_3$, is set equal to the voltage rise V_S across the voltage source: $V_1 + V_2 + V_3 = V_S$. Then the IR Ohm's law relations are substituted for the resistor voltages:

$$V_S = V_1 + V_2 + V_3 = IR_1 + IR_2 + IR_3 = I(R_1 + R_2 + R_3) = IR_T$$

from which $I = V_S/R_T$ and $R_T = R_1 + R_2 + R_3$. This R_T is the *total resistance* of the series-connected resistors. Another term used is *equivalent resistance*, with symbol R_{eq}.

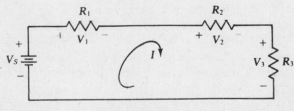

Fig. 4-1

From this it should be evident that, in general, the total resistance of series-connected resistors (series resistors) equals the sum of the individual resistances:

$$R_T = R_1 + R_2 + R_3 \cdots$$

Further, if the resistances are the same (R), and if there are N of them, then $R_T = NR$. Finding the current in a series circuit is easier using total resistance than applying KVL directly.

If a series circuit has more than one voltage source, then

$$I(R_1 + R_2 + R_3 + \cdots) = V_{S_1} + V_{S_2} + V_{S_3} + \cdots$$

in which each V_S term is positive for a voltage rise and is negative for a voltage drop in the direction of I.

KVL is seldom applied to a loop containing a current source because the voltage across the current source is not known and there is no formula for it.

VOLTAGE DIVISION

The *voltage division* or *voltage divider rule* applies to resistors in series. It gives the voltage across any resistor in terms of the resistances and the total voltage across the series combination— the step of finding the resistor current is eliminated. The voltage division formula is easy to find from the circuit shown in Fig. 4-1. Consider finding the voltage V_2. By Ohm's law, $V_2 = IR_2$. Also, $I = V_S/(R_1 + R_2 + R_3)$. Eliminating I results in

$$V_2 = \frac{R_2}{R_1 + R_2 + R_3} V_S$$

In general, for any number of series resistors with a total resistance of R_T and with a voltage of V_S across the series combination, the voltage V_X across one of the resistors R_X is

$$V_X = \frac{R_X}{R_T} V_S$$

This is the formula for the voltage division or divider rule. For this formula, V_S and V_X must have opposing polarities; that is, around a closed path one must be a voltage drop and the other a voltage rise. If both are rises or both are drops, the formula requires a negative sign. The voltage V_S need not be that of a source. It is just the total voltage across the series resistors.

KIRCHHOFF'S CURRENT LAW AND PARALLEL DC CIRCUITS

Kirchhoff's current law, abbreviated KCL, has three equivalent versions:
At any instant in a circuit,

1. The algebraic sum of the currents leaving a closed surface is zero.

2. The algebraic sum of the currents entering a closed surface is zero.

3. The algebraic sum of the currents entering a closed surface equals the algebraic sum of those leaving.

The word "algebraic" means that the signs of the currents are included in the additions. Remember that a current entering is a negative current leaving, and that a current leaving is a negative current entering.

In almost all circuit applications, the closed surfaces of interest are those enclosing nodes. So, there is little loss of generality in using the word "node" in place of "closed surface" in each KCL version. Also, for a node to which no voltage sources are connected the most convenient KCL version is often the third one, restricted such that the currents entering are from current sources and the currents leaving are through resistors.

In the application of KCL, one node is selected as the *ground* or *reference* or *datum node*, which is often indicated by the ground symbol ($\perp$). Usually, the node at the bottom of the circuit is the ground node, as shown in the parallel circuit of Fig. 4-2. (This is a parallel circuit because the same voltage V is across all circuit components.) The voltages on other nodes are almost always referenced positive with respect to the ground node. At the nongrounded node in the circuit shown in Fig. 4-2, the sum of the currents leaving through resistors, $I_1 + I_2 + I_3$, equals the current I_S entering this node from the current source: $I_1 + I_2 + I_3 = I_S$. The substitution of the $I = GV$ Ohm's law relations for the resistor currents results in

$$I_S = I_1 + I_2 + I_3 = G_1V + G_2V + G_3V = (G_1 + G_2 + G_3)V = G_TV$$

from which $V = I_S/G_T$ and $G_T = G_1 + G_2 + G_3 = 1/R_1 + 1/R_2 + 1/R_3$. This G_T is the *total conductance* of the circuit. Another term used is *equivalent conductance*, with symbol G_{eq}.

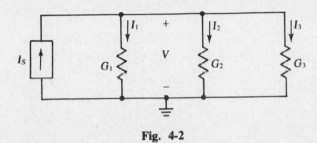

Fig. 4-2

From this it should be evident that the total conductance of parallel-connected resistors (parallel resistors) equals the sum of the individual conductances:

$$G_T = G_1 + G_2 + G_3 + \cdots$$

If the conductances are the same (G), and if there are N of them, then $G_T = NG$ and $R_T = 1/G_T = 1/NG = R/N$. Finding the voltage in a parallel circuit is easier using total conductance than applying KCL directly.

Sometimes working with resistances is preferable to conductances. Then from $R_T = 1/G_T = 1/(G_1 + G_2 + G_3 + \cdots)$,

$$R_T = \frac{1}{1/R_1 + 1/R_2 + 1/R_3 + \cdots}$$

An important check on calculations with this formula is that R_T must always be less than the least resistance of the parallel resistors.

For the special case of just two parallel resistors,

$$R_T = \frac{1}{1/R_1 + 1/R_2} = \frac{R_1R_2}{R_1 + R_2}$$

So, the total or equivalent resistance of two parallel resistors is the product of the resistances divided by the sum.

The symbol $\|$ as in $R_1\|R_2$ indicates the resistance of two parallel resistors: $R_1\|R_2 = R_1R_2/(R_1 + R_2)$. It is also sometimes used to indicate that two resistors are in parallel.

If a parallel circuit has more than one current source,

$$(G_1 + G_2 + G_3 + \cdots)V = I_{S_1} + I_{S_2} + I_{S_3} + \cdots$$

in which each I_S term is positive for a current entering the nongrounded node and is negative for a current leaving this node.

KCL is seldom applied to a node to which a voltage source is connected. The reason is that the current through a voltage source is not known and there is no formula for it.

CURRENT DIVISION

The *current division* or *current divider rule* applies to resistors in parallel. It gives the current through any resistor in terms of the conductances and the current into the parallel combination— the step of finding the resistor voltage is eliminated. The current division formula is easy to derive from the circuit shown in Fig. 4-2. Consider finding the current I_2. By Ohm's law, $I_2 = G_2 V$. Also, $V = I_S/(G_1 + G_2 + G_3)$. Eliminating V results in

$$I_2 = \frac{G_2}{G_1 + G_2 + G_3} I_S$$

In general, for any number of parallel resistors with a total conductance G_T and with a current I_S entering the parallel combination, the current I_X through one of the resistors with conductance G_X is

$$I_X = \frac{G_X}{G_T} I_S$$

This is the formula for the current division or divider rule. For this formula, I_S and I_X must be referenced in the same direction, with I_X referenced away from the node of the parallel resistors that I_S is referenced into. If both currents enter this node, then the formula requires a negative sign. The current I_S need not be that of a source. It is just the total current entering the parallel resistors.

For the special case of two parallel resistors, the current division formula is usually expressed in resistances instead of conductances. If the two resistances are R_1 and R_2, the current I_1 in the resistor with resistance R_1 is

$$I_1 = \frac{G_1}{G_1 + G_2} I_S = \frac{1/R_1}{1/R_1 + 1/R_2} I_S = \frac{R_2}{R_1 + R_2} I_S$$

In general, as this formula indicates, the current flowing in one of two parallel resistors equals the resistance of the other resistor divided by the sum of the resistances, all times the current flowing into the parallel combination.

Solved Problems

4.1   Determine the number of nodes and branches in the circuit shown in Fig. 4-3.

Dots 1 and 2 are one node, as are dots 3 and 4 and also dots 5 and 6, all with connecting wires. Dot 7 and the two wires on both sides are another node, as are dot 8 and the two wires on both sides of it. So, there are five nodes. Each of the shown components A through H is a branch—eight branches in all.

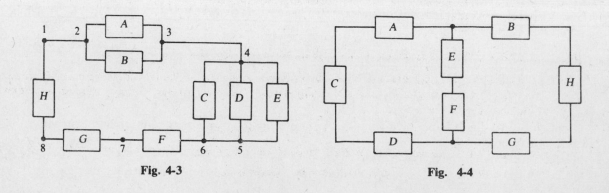

Fig. 4-3   Fig. 4-4

4.2   Which components in Fig. 4-3 are in series and which are in parallel?

Components F, G, and H are in series because they carry the same current. Components A and B, being connected together at both ends, have the same voltage and so are in parallel. The same is true for components C, D, and E—they are in parallel. Further, the parallel group of A and B is in series with the parallel group of C, D, and E, and both groups are in series with components F, G, and H.

4.3   Identify all the loops and all the meshes for the circuit shown in Fig. 4-4.   Also, specify which components are in series and which are in parallel.

There are three loops: one of components A, E, F, D, and C; a second of components B, H, G, F, and E; and a third of A, B, H, G, D, and C. The first two loops are also meshes, but the third is not because components E and F are inside it.   Components A, C, and D are in series because they carry the same current.   For the same reason, components E and F are in series, as also are components B, H, and G.   No components are in parallel.

4.4   Repeat Prob. 4.3 for the circuit shown in Fig. 4-5.

The three loops of components A, B, and C; C, D, and E; and F, D, and B are also meshes—the only meshes.   All other loops are not meshes because components are inside them.   Components A, B, D, and E form one of these other loops; components A, F, and E another one; components A, F, D, and C a third; and components F, E, C, and B a fourth.   The circuit has three meshes and seven loops.   No components are in series or in parallel.

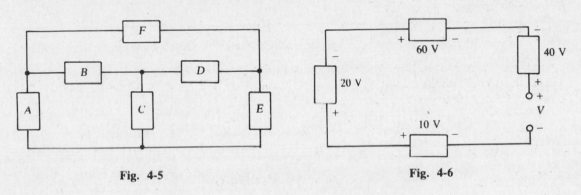

Fig. 4-5   Fig. 4-6

4.5   What is V across the open circuit in the circuit shown in Fig. 4-6?

The sum of the voltage drops in a clockwise direction is, starting from the upper left corner,

$$60 - 40 + V - 10 + 20 = 0 \quad \text{from which} \quad V = -30\text{ V}$$

In the summation, the 40 and 10 V are negative because they are voltage rises in a clockwise direction. The negative sign in the answer indicates that the actual open-circuit voltage has a polarity opposite the shown reference polarity.

4.6   Find the unknown voltages in the circuit shown in Fig. 4-7.   Find V_1 first.

The basic KVL approach is to use loops having only one unknown voltage apiece.   Such a loop for V_1 includes the 10-, 8-, and 9-V components.   The sum of the voltage drops in a clockwise direction around this loop is

$$10 - 8 + 9 - V_1 = 0 \quad \text{from which} \quad V_1 = 11\text{ V}$$

Similarly, for V_2 the sum of the voltage drops clockwise around the top mesh is

$$V_2 + 8 - 10 = 0 \quad \text{from which} \quad V_2 = 2\text{ V}$$

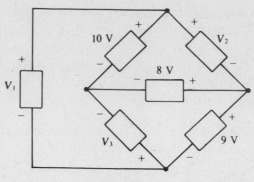

Fig. 4-7

Clockwise around the bottom mesh, the sum of the voltage drops is

$$-8 + 9 + V_3 = 0 \qquad \text{from which} \qquad V_3 = -1 \text{ V}$$

The negative sign for V_3 indicates that the polarity of the actual voltage is opposite the reference polarity.

4.7 What is the total resistance of 2-, 5-, 8-, 10-, and 17-Ω resistors connected in series?

The total resistance of series resistors is the sum of the individual resistances: $R_T = 2 + 5 + 8 + 10 + 17 = 42 \ \Omega$.

4.8 What is the total resistance of thirty 6-Ω resistors connected in series?

The total resistance is the number of resistors times the common resistance of 6 Ω: $R_T = 30 \times 6 = 180 \ \Omega$.

4.9 What is the total conductance of 4-, 10-, 16-, 20-, and 24-S resistors connected in series?

The best approach is to convert the conductances to resistances, add the resistances to get the total resistance, and then invert the total resistance to get the total conductance:

$$R_T = \tfrac{1}{4} + \tfrac{1}{10} + \tfrac{1}{16} + \tfrac{1}{20} + \tfrac{1}{24} = 0.504 \ \Omega$$

and

$$G_T = \frac{1}{R_T} = \frac{1}{0.504} = 1.98 \text{ S}$$

4.10 A string of Christmas tree lights consists of eight 6-W, 15-V bulbs connected in series. What current flows when the string is plugged into a 120-V outlet, and what is the hot resistance of each bulb?

The total power is $P_T = 8 \times 6 = 48$ W. From $P_T = VI$, the current is $I = P_T/V = 48/120 = 0.4$ A. And from $P = I^2R$, the hot resistance of each bulb is $R = P/I^2 = 6/0.4^2 = 37.5 \ \Omega$.

4.11 A 3-V, 300-mA flashlight bulb is to be used as the dial light in a 120-V radio. What is the resistance of the resistor that should be connected in series with the flashlight bulb to limit the current?

Since 3 V is to be across the flashlight bulb, there will be $120 - 3 = 117$ V across the series resistor. The current is the rated 300 mA. Consequently, the resistance is $117/0.3 = 390 \ \Omega$.

4.12 A person wants to move a 20-W FM-AM transistor radio from a junked car with a 6-V battery to a new car with a 12-V battery.   What is the resistance of the resistor that should be connected in series with the radio to limit the current, and what is its minimum power rating?

From $P = VI$,   the radio requires $20/6 = 3.33$ A.   The resistor, being in series, has the same current. Also, it has the same voltage because $12 - 6 = 6$ V. As a result, $R = 6/3.33 = 1.8\ \Omega$. With the same voltage and current, the resistor must dissipate the same power as the radio, and so has a 20-W minimum power rating.

4.13 A series circuit consists of a 240-V source and 12-, 20-, and 16-Ω resistors.   Find the current out of the positive terminal of the voltage source.   Also find the resistor voltages.   Assume associated references, as should always be done when there is no specification of references.

The current is the applied voltage divided by the equivalent resistance:

$$I = \frac{240}{12 + 20 + 16} = 5\ \text{A}$$

Each resistor voltage is this current times the corresponding resistance: $V_{12} = 5 \times 12 = 60$ V, $V_{20} = 5 \times 20 = 100$ V, and $V_{16} = 5 \times 16 = 80$ V. As a check, the sum of the resistor voltages is $60 + 100 + 80 = 240$ V, the same as the applied voltage.

4.14 A resistor in series with an 8-Ω resistor absorbs 100 W when the two are connected across a 60-V line.   Find the unknown resistance R.

The total resistance is $8 + R$, and thus the current is $60/(8 + R)$.   From $I^2R = P$,

$$\left(\frac{60}{8 + R}\right)^2 R = 100 \qquad \text{or} \qquad 3600R = 100(8 + R)^2$$

which simplifies to $R^2 - 20R + 64 = 0$.   The quadratic formula can be used to find R. Recall that for the equation $ax^2 + bx + c = 0$, this formula is

$$x = \frac{-b \pm \sqrt{b^2 - 4ac}}{2a}$$

So
$$R = \frac{-(-20) \pm \sqrt{(-20)^2 - 4(1)(64)}}{2(1)} = \frac{20 \pm 12}{2} = 16\ \Omega \text{ or } 4\ \Omega$$

A resistor with a resistance of either 16 or 4 Ω will dissipate 100 W when connected in series with an 8-Ω resistor across a 60-V line.

This particular quadratic equation can be factored without using the quadratic formula. By inspection, $R^2 - 20R + 64 = (R - 16)(R - 4) = 0$, from which $R = 16\ \Omega$ or $R = 4\ \Omega$, the same as before.

4.15 Resistors R_1, R_2, and R_3 are in series with a 100-V source.   The total voltage drop across R_1 and R_2 is 50 V, and that across R_2 and R_3 is 80 V.   Find the three resistances if the total resistance is 50 Ω.

The current is the applied voltage divided by the total resistance: $I = 100/50 = 2$ A.   Since the voltage across resistors R_1 and R_2 is 50 V, there must be $100 - 50 = 50$ V   across R_3. By Ohm's law, $R_3 = 50/2 = 25\ \Omega$. Resistors R_2 and R_3 have 80 V across them, leaving $100 - 80 = 20$ V   across R_1. Thus, $R_1 = 20/2 = 10\ \Omega$. The resistance of R_2 is the total resistance minus the resistances of R_1 and R_3: $R_2 = 50 - 10 - 25 = 15\ \Omega$.

4.16 What is the maximum voltage that can be applied across the series combination of a 150-Ω, 2-W resistor and a 100-Ω, 1-W resistor without exceeding the power rating of either resistor?

From $P = I^2R$, the maximum safe current for the 150-Ω resistor is $I = \sqrt{P/R} = \sqrt{2/150} = 0.115$ A. That for the 100-Ω resistor is $\sqrt{1/100} = 0.1$ A. The maximum current cannot exceed the lesser of these two currents and so is 0.1 A. For this current, $V = I(R_1 + R_2) = 0.1(150 + 100) = 25$ V.

4.17 In a series circuit, a current flows from the positive terminal of a 180-V source through two resistors, one of which has 30 Ω of resistance and the other of which has 45 V across it. Find the current and the unknown resistance.

The 30-Ω resistor has $180 - 45 = 135$ V across it and thus a $135/30 = 4.5$-A current through it. The other resistance is $45/4.5 = 10$ Ω.

4.18 Find the current and unknown voltages in the circuit shown in Fig. 4-8.

The total resistance is the sum of the resistances: $10 + 15 + 6 + 8 + 11 = 50$ Ω. The total voltage rise from the voltage sources in the direction of I is $12 - 5 + 8 = 15$ V. The current I is this voltage divided by the total resistance: $I = 15/50 = 0.3$ A. By Ohm's law, $V_1 = 0.3 \times 10 = 3$ V, $V_2 = 0.3 \times 15 = 4.5$ V, $V_3 = -0.3 \times 6 = -1.8$ V, $V_4 = 0.3 \times 8 = 2.4$ V, and $V_5 = -0.3 \times 11 = -3.3$ V. The equations for V_3 and V_5 have negative signs because the references for these voltages are not associated with the reference for the current I.

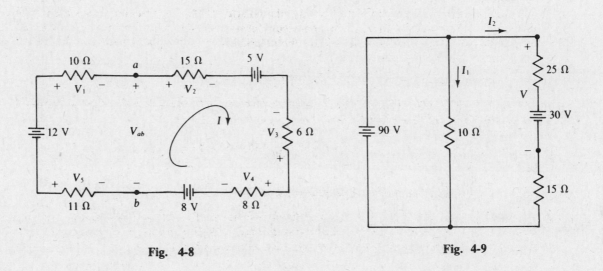

Fig. 4-8   Fig. 4-9

4.19 Find the voltage V_{ab} in the circuit shown in Fig. 4-8.

V_{ab} is the voltage drop from node a to node b, which is the sum of the voltage drops across the components connected between nodes a and b either to the right or to the left of node a. It is convenient to choose the path to the right because this is the direction of the $I = 0.3$-A current found in the solution of Prob. 4.18. Thus we have

$$V_{ab} = (0.3 \times 15) + 5 + (0.3 \times 6) + (0.3 \times 8) - 8 = 5.7 \text{ V}$$

Note that *an IR drop is always positive in the direction of I.* A voltage reference, and that of V_3 in particular here, has no effect on this.

4.20 Find I_1, I_2, and V in the circuit shown in Fig. 4-9.

Since the 90-V source is across the 10-Ω resistor, $I_1 = 90/10 = 9$ A. Around the outside loop in a clockwise direction, the voltage drop across the two resistors is $(25 + 15)I_2 = 40I_2$. This is equal to the sum of the voltage rises across the voltage sources in this outside loop:

$$40I_2 = -30 + 90 \qquad \text{from which} \qquad I_2 = 60/40 = 1.5 \text{ A}$$

The voltage V is equal to the sum of the drops across the 25-Ω resistor and the 30-V source: $V = (1.5 \times 25) + 30 = 67.5$ V. Notice that the parallel 10-Ω resistor does not affect I_2. In general, resistors in parallel with voltage sources that have zero internal resistances (ideal voltage sources) do not affect currents or voltages elsewhere in a circuit. They do, however, cause an increase in current flow in these voltage sources.

4.21 A 90-V source is in series with five resistors having resistances of 4, 5, 6, 7, and 8 Ω. Find the voltage across the 6-Ω resistor. (Here "voltage" refers to the positive voltage, as it will in later problems unless otherwise indicated. The same is true for current.)

By the voltage division formula, the voltage across a resistor in a series circuit equals the resistance of that resistor times the applied voltage divided by the total resistance. So,

$$V_6 = \frac{6}{4+5+6+7+8} \times 90 = 18 \text{ V}$$

4.22 Use voltage division to determine the voltages V_4 and V_5 in the circuit shown in Fig. 4-8.

The total voltage applied across the resistors equals the sum of the voltage rises from the voltage sources, preferably in a clockwise direction: $12 - 5 + 8 = 15$ V. The polarity of this net voltage is such that it produces a clockwise current flow. In this sum the 5 V is negative because it is a drop, and rises are being added. Put another way, the polarity of the 5-V source opposes the polarities of the 12- and 8-V sources. The V_4 voltage division formula should have a positive sign because V_4 is a drop in the clockwise direction—it opposes the polarity of the net applied voltage:

$$V_4 \times \frac{8}{10+15+6+8+11} \times 15 = \frac{8}{50} \times 15 = 2.4 \text{ V}$$

The voltage division formula for V_5 requires a negative sign because both V_5 and the net source voltage are rises in the clockwise direction:

$$V_5 = -\frac{11}{50} \times 15 = -3.3 \text{ V}$$

4.23 Find the voltage V_{ab} across the open circuit in the circuit shown in Fig. 4-10.

There is zero voltage across the 10-Ω resistor since it has zero current flow through it as a result of being in series with an open circuit. Because of this zero voltage, the voltage V_{ab} is equal to the voltage drop, top to bottom, across the 60-Ω resistor. By voltage division,

$$V_{ab} = \frac{60}{60+40} \times 100 = 60 \text{ V}$$

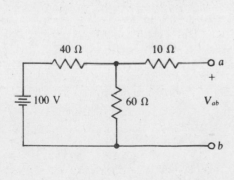

Fig. 4-10

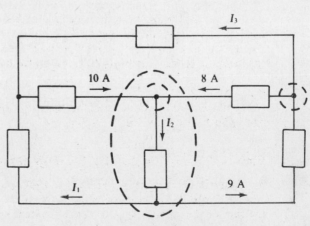

Fig. 4-11

4.24   Find the unknown currents in the circuit shown in Fig. 4-11.   Find I_1 first.

The basic KCL approach is to find closed surfaces such that only one unknown current flows across each surface.   In Fig. 4-11, the large dashed loop represents a closed surface drawn such that I_1 is the only unknown current flowing across it.   Other currents flowing across it are the 10-, 8-, and 9-A currents. I_1 and the 9-A currents leave this closed surface, and the 8- and 10-A currents enter it.   By KCL, the sum of the currents leaving is zero: $I_1 + 9 - 8 - 10 = 0$,   or $I_1 = 9$ A. I_2 is readily found from summing the currents leaving the middle top node: $I_2 - 8 - 10 = 0$,   or $I_2 = 18$ A.   Similarly, at the right top node, $I_3 + 8 - 9 = 0$,   and $I_3 = 1$ A.   Checking at the left top node: $10 - I_1 - I_3 = 10 - 9 - 1 = 0$,   as it should be.

4.25   Find I for the circuit shown in Fig. 4-12.

Since I is the only unknown current flowing across the shown dashed loop, it can be found by setting to zero the sum of the currents leaving this loop: $I - 16 - 8 - 9 + 3 + 2 - 10 = 0$,   from which $I = 38$ A.

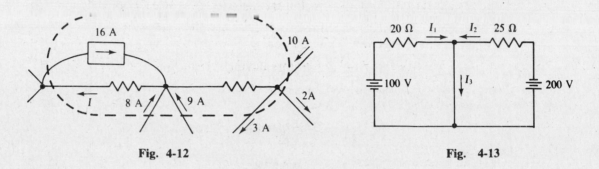

Fig. 4-12   Fig. 4-13

4.26   Find the short-circuit current I_3 for the circuit shown in Fig. 4-13.

The short circuit places the 100 V of the left voltage source across the 20-Ω resistor, and it places the 200 V of the right source across the 25-Ω resistor.   By Ohm's law, $I_1 = 100/20 = 5$ A   and $I_2 = -200/25 = -8$ A.   The negative sign occurs in the I_2 formula because of nonassociated references.   From KCL applied at the top middle node, $I_3 = I_1 + I_2 = 5 - 8 = -3$ A.   Of course the negative sign in the answer means that 3 A actually flows up through the short circuit, opposite the direction of the I_3 current reference arrow.

4.27   Find the total conductance and resistance of four parallel resistors having resistances of 1, 0.5, 0.25, and 0.125 Ω.

The total conductance is the sum of the individual conductances:

$$G_T = \frac{1}{1} + \frac{1}{0.5} + \frac{1}{0.25} + \frac{1}{0.125} = 1 + 2 + 4 + 8 = 15 \text{ S}$$

The total resistance is the inverse of this total conductance: $R_T = 1/G_T = 1/15 = 0.0667 \ \Omega$.

4.28   Find the total resistance of fifty 200-Ω resistors connected in parallel.

The total resistance equals the common resistance divided by the number of resistors: $200/50 = 4 \ \Omega$.

4.29   A resistor is to be connected in parallel with a 10-kΩ resistor and a 20-kΩ resistor to produce a total resistance of 12 kΩ.   What is the resistance of the resistor?

Assuming that the added resistor is a conventional resistor, no added parallel resistor will give a total resistance of 12 kΩ because the total resistance of parallel resistors is always less than the least individual resistance, which is 10 kΩ. With transistors, however, it is possible to make a component that has a negative resistance and that in parallel can cause an increase in total resistance. Generally, however, the term *resistor* means a conventional resistor that has only positive resistance.

4.30   Three parallel resistors have a total conductance of 1.75 S. If two of the resistances are 1 and 2 Ω, what is the third resistance?

The sum of the individual conductances equals the total conductance:

$$\tfrac{1}{1} + \tfrac{1}{2} + G_3 = 1.75 \qquad \text{or} \qquad G_3 = 1.75 - 1.5 = 0.25 \text{ S}$$

The resistance of the third resistor is the inverse of this conductance: $R_3 = 1/G_3 = 1/0.25 = 4 \ \Omega$.

4.31   Without using conductances, find the total resistance of parallel resistors having resistances of 5 and 20 Ω.

The total resistance equals the product of the individual resistances divided by the sum: $R_T = (5 \times 20)/(5 + 20) = 100/25 = 4 \ \Omega$.

4.32   Repeat Prob. 4.31 for three parallel resistors having resistances of 12, 24, and 32 Ω.

One approach is to consider the resistances two at a time. For the 12- and the 24-Ω resistances, the equivalent resistance is

$$\frac{12 \times 24}{12 + 24} = \frac{288}{36} = 8 \ \Omega$$

This combined with the 32-Ω resistance gives a total resistance of

$$R_T = \frac{8 \times 32}{8 + 32} = \frac{256}{40} = 6.4 \ \Omega$$

4.33   Three light bulbs, one with a wattage of 60 W, the second with a wattage of 100 W, and the third with a wattage of 200 W, are connected in parallel across a 120-V line. What is the equivalent hot resistance of this combination?

From $R = V^2/P$, the individual resistances are $120^2/60 = 240 \ \Omega$, $120^2/100 = 144 \ \Omega$, and $120^2/200 = 72 \ \Omega$. The 72- and 144-Ω resistances have an equivalent resistance of $(72 \times 144)/(72 + 144) = 48 \ \Omega$. The equivalent resistance of this and the 240-Ω resistance is the total equivalent hot resistance: $(240 \times 48)/(240 + 48) = 40 \ \Omega$. Alternatively, from the total power of 360 W, $R = V^2/P = 120^2/360 = 40 \ \Omega$.

4.34   Determine R_T in $R_T = (4 + 24\|12)\|6$.

It is essential to start evaluating inside the parentheses, and then work out. By definition, the term $24\|12 = (24 \times 12)/(24 + 12) = 8$. This adds to the 4: $4 + 8 = 12$. The expression reduces to $12\|6$, which is $(12 \times 6)/(12 + 6) = 4$. Thus, $R_T = 4 \ \Omega$.

4.35   Find the total resistance R_T of the resistor ladder network shown in Fig. 4-14.

To find the equivalent resistance of a ladder network by combining resistances, always start at the end opposite the input terminals. At this end, the series 4- and 8-Ω resistors have an equivalent resistance of 12 Ω. This combines in parallel with the 24-Ω resistance: $(24 \times 12)/(24 + 12) = 8 \ \Omega$. This adds to the 3 and the 9 Ω of the series resistors for a sum of $8 + 3 + 9 = 20 \ \Omega$. This combines in parallel with the 5-Ω resistance: $(20 \times 5)/(20 + 5) = 4 \ \Omega$. R_T is the sum of this resistance and the resistances of the series 16- and 14-Ω resistors: $R_T = 4 + 16 + 14 = 34 \ \Omega$.

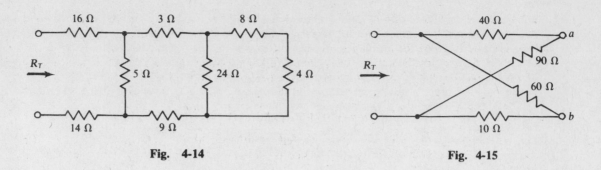

Fig. 4-14   Fig. 4-15

4.36 In the circuit shown in Fig. 4-15 find the total resistance R_T with terminals a and b (a) open-circuited,   and   (b) short-circuited.

(a) With terminals a and b open, the 40- and 90-Ω resistors are in series, as are the 60- and 10-Ω resistors.   And the two series combinations are in parallel.   So,

$$R_1 = \frac{(10 + 90)(60 + 10)}{40 + 90 + 60 + 10} = 45 \text{ } \Omega$$

(b) For terminals a and b short-circuited, the 40- and 60-Ω resistors are in parallel, as are the 90- and 10-Ω resistors.   The two parallel combinations are in series, making

$$R_T = \frac{40 \times 60}{40 + 60} + \frac{90 \times 10}{90 + 10} = 33 \text{ } \Omega$$

4.37 A 90-A current flows into four parallel resistors having resistances of 5, 6, 12, and 20 Ω.   Find the current in each resistor.

The total resistance is

$$R_T = \frac{1}{1/5 + 1/6 + 1/12 + 1/20} = 2 \text{ } \Omega$$

This value times the current gives the voltage across the parallel combination: $2 \times 90 = 180$ V. Then by Ohm's law, $I_5 = 180/5 = 36$ A, $I_6 = 180/6 = 30$ A, $I_{12} = 180/12 = 15$ A,   and $I_{20} = 180/20 = 9$ A.

4.38 Find the voltage and unknown currents in the circuit shown in Fig. 4-16.

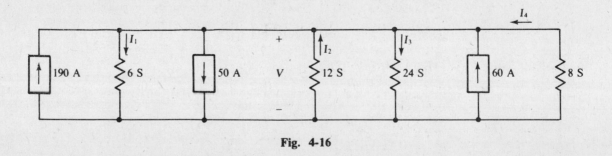

Fig. 4-16

Even though it has several dots, the top line is just a single node because the entire line is at the same potential.   The same is true of the bottom line.   Thus, there are just two nodes and one voltage V.   The total conductance of the parallel-connected resistors is $G = 6 + 12 + 24 + 8 = 50$ S.   Also, the total current entering the top node from current sources is $190 - 50 + 60 = 200$ A.   This conductance and current can be used in the conductance version of Ohm's law, $I = GV$,   to obtain the voltage: $V = I/G = 200/50 = 4$ V.   Since this is the voltage across each resistor, the resistor currents are $I_1 = 6 \times 4 = 24$ A, $I_2 = -12 \times 4 = -48$ A, $I_3 = 24 \times 4 = 96$ A, and $I_4 = -8 \times 4 = -32$ A. The negative signs are the result of nonassociated references. Of course, all the actual resistor currents leave the top node.

Note that the parallel current sources have the same effect as a single current source, the current of which is the algebraic sum of the individual currents from the sources.

4.39 Use current division to find the currents I_2 and I_3 in the circuit shown in Fig. 4-16.

The sum of the currents from current sources into the top node is $190 - 50 + 60 = 200$ A. Also, the sum of the conductances is $6 + 12 + 24 + 8 = 50$ S. By the current division formula,

$$I_2 = -\frac{12}{50} \times 200 = -48 \text{ A} \quad \text{and} \quad I_3 = \frac{24}{50} \times 200 = 96 \text{ A}$$

The formula for I_2 has a negative sign because I_2 has a reference into the top node, and the sum of the currents from current sources is also into the top node. For a positive sign, one current in the formula must be into a node and the other current must be out of the same node.

4.40 A 90-A current flows into two parallel resistors having resistances of 12 and 24 Ω. Find the current in the 24-Ω resistor.

The current in the 24-Ω resistor equals the resistance of the other parallel resistor divided by the sum of the resistances, all times the input current:

$$I_{24} = \frac{12}{12 + 24} \times 90 = 30 \text{ A}$$

As a check, this current results in a voltage of $30 \times 24 = 720$ V, which is also across the 12-Ω resistor. Thus, $I_{12} = 720/12 = 60$ A, and $I_{24} + I_{12} = 30 + 60 = 90$ A, the input current.

4.41 Use voltage division twice to find V_1 in the circuit shown in Fig. 4-17.

Clearly, V_1 can be found from V_2 by voltage division. And V_2 can be found from the source voltage by voltage division used with the equivalent resistance to the right of the 16-Ω resistor. This resistance is

$$\frac{(54 + 18)(36)}{54 + 18 + 36} = 24 \ \Omega$$

By voltage division,

$$V_2 = \frac{24}{16 + 24} \times 80 = 48 \text{ V} \quad \text{and} \quad V_1 = \frac{18}{54 + 18} \times 48 = 12 \text{ V}$$

A common error in finding V_2 is to neglect the loading of the resistors to the right of the V_2 node.

4.42 Use current division twice to find I_1 in the circuit shown in Fig. 4-18.

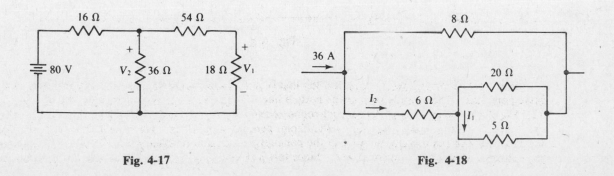

Fig. 4-17   Fig. 4-18

Obviously I_1 can be found from I_2 by current division. And, if the total resistance of the bottom three branches is found, current division can be used to find I_2 from the input current. The needed total resistance is

$$6 + \frac{20 \times 5}{20 + 5} = 10 \ \Omega$$

By the two-resistance form of the current division formula,

$$I_2 = \frac{8}{10 + 8} \times 36 = 16 \text{ A} \quad \text{and} \quad I_1 = \frac{20}{20 + 5} \times 16 = 12.8 \text{ A}$$

Supplementary Problems

4.43 Determine the number of nodes, branches, loops, and meshes in the circuit shown in Fig. 4-19.
Ans. 6 nodes, 8 branches, 7 loops, 3 meshes

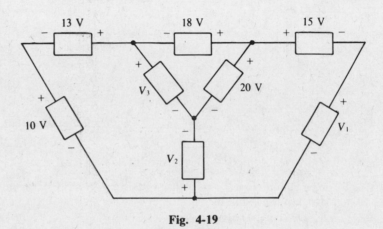

Fig. 4-19

4.44 Find V_1, V_2, and V_3 for the circuit shown in Fig. 4-19.
Ans. $V_1 = 2$ V, $V_2 = -21$ V, $V_3 = 26$ V

4.45 Four resistors in series have a total resistance of 500 Ω. If three of the resistors have resistances of 100, 150, and 200 Ω, what is the resistance of the fourth resistor? *Ans.* 50 Ω

4.46 Find the total conductance of 2-, 4-, 8-, and 10-S resistors connected in series. *Ans.* 1.03 S

4.47 A 60-W, 120-V light bulb is to be connected in series with a resistor across a 277-V line. What is the resistance and minimum power rating of the resistor required if the light bulb is to operate under rated conditions? *Ans.* 314 Ω, 78.5 W

4.48 A series circuit consists of a dc voltage source and 4-, 5-, and 6-Ω resistors. If the current is 7 A, find the source voltage. *Ans.* 105 V

4.49 A 12-V battery with a 0.3-Ω internal resistance is to be charged from a 15-V source. If the charging current should not exceed 2 A, what is the minimum resistance of a series resistor that will limit the current to this safe value? *Ans.* 1.2 Ω

4.50   A resistor in series with a 100-Ω resistor absorbs 80 W when the two are connected across a 240-V line.   Find the unknown resistance.   *Ans.* 20 or 500 Ω

4.51   A series circuit consists of a 4-V source and 2-, 4-, and 6-Ω resistors.   What is the minimum power rating of each resistor if the resistors are available in power ratings of $\frac{1}{2}$ W, 1 W, and 2 W?   *Ans.* $P_2 = \frac{1}{2}$ W, $P_4 = \frac{1}{2}$ W, $P_6 = 1$ W

4.52   Find V_{ab} in the circuit shown in Fig. 4-20.   *Ans.* 20 V

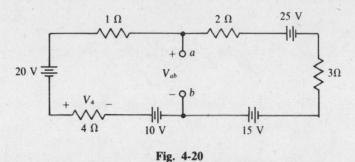

Fig. 4-20

4.53   Use voltage division to find the voltage V_4 in the circuit shown in Fig. 4-20.   *Ans.* -8 V

4.54   A series circuit consists of a 100-V source and 4-, 5-, 6-, 7-, and 8-Ω resistors.   Use voltage division to determine the voltage across the 6-Ω resistor.   *Ans.* 20 V

4.55   Find the indicated unknown currents in the circuits shown in Fig. 4-21.
Ans. $I_1 = 2$ A, $I_2 = -6$ A, $I_3 = -5$ A, $I_4 = 3$ A

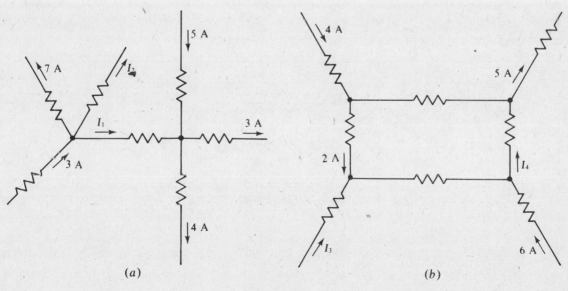

(a)   *(b)*

Fig. 4-21

4.56   Find the short-circuit current I for the circuit shown in Fig. 4-22.   *Ans.* 3 A

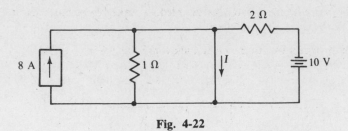

Fig. 4-22

4.57 What are the different resistances that can be obtained with three 4-Ω resistors?
Ans. 1.33, 2, 4, 6, 8, and 12 Ω

4.58 A 100-Ω resistor and another resistor in parallel have an equivalent resistance of 75 Ω.   What is the resistance of the other resistor?   *Ans.* 300 Ω

4.59 Find the equivalent resistance of four parallel resistors having resistances of 2, 4, 6, and 8 Ω.
Ans. 0.96 Ω

4.60 Three parallel resistors have a total conductance of 2 mS.   If two of the resistances are 1 and 5 kΩ, what is the third resistance?   *Ans.* 1.25 kΩ

4.61 The equivalent resistance of three parallel resistors is 10 Ω.   If two of the resistors have resistances of 40 and 60 Ω, what is the resistance of the third resistor?   *Ans.* 17.1 Ω

4.62 Determine R_T in $R_T = (24\|48 + 24)\|10$.   *Ans.* 8 Ω

4.63 Determine R_T in $R_T = (6\|12 + 10\|40)\|(6 + 2)$.   *Ans.* 4.8 Ω

4.64 Find the total resistance R_T of the resistor ladder network shown in Fig. 4-23.   *Ans.* 26.6 kΩ

4.65 Repeat Prob. 4.64 with all resistances doubled.   *Ans.* 53.2 kΩ

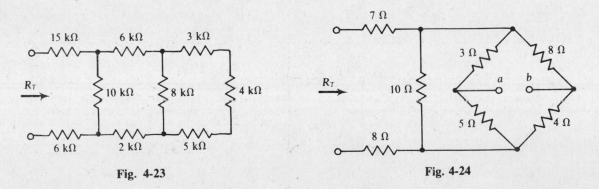

Fig. 4-23   Fig. 4-24

4.66 In the circuit shown in Fig. 4-24, find R_T with terminals a and b (*a*) open-circuited, and (*b*) short-circuited.   *Ans.* (*a*) 18.2 Ω,   (*b*) 18.1 Ω

4.67 A 15-mA current flows into four parallel resistors having resistances of 4, 6, 8, and 12 kΩ.   Find each resistor current.   *Ans.* $I_4 = 6$ mA, $I_6 = 4$ mA, $I_8 = 3$ mA, $I_{12} = 2$ mA

4.68   Repeat Prob. 4.67 with all resistances doubled.   *Ans.* Same currents

4.69   Find the unknown currents in the circuit shown in Fig. 4-25.
Ans. $I_1 = -10$ A, $I_2 = -8$ A, $I_3 = 6$ A, $I_4 = -2$ A, $I_5 = 12$ A

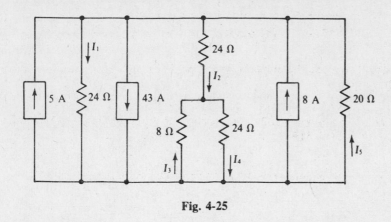

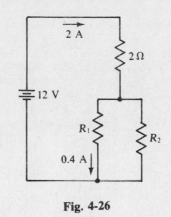

Fig. 4-25   Fig. 4-26

4.70   Find R_1 and R_2 for the circuit shown in Fig. 4-26.   *Ans.* $R_1 = 20$ Ω, $R_2 = 5$ Ω

4.71   In the circuit shown in Fig. 4-26, let $R_1 = 6$ Ω   and $R_2 = 12$ Ω.   Then find the current in the 6-Ω resistor by current division.   *Ans.* 1.33 A

4.72   A 60-A current flows into a resistor network described by $R_T = 40\|(12 + 40\|10)$.   Find the current in the 10-Ω resistor.   *Ans.* 32 A

4.73   A 620-V source connected to a resistor network described by $R_T = 50 + R\|20$ provides 120 V to the 20-Ω resistor.   What is R?   *Ans.* 30 Ω

4.74   Find I for the circuit shown in Fig. 4-27.   *Ans.* 4 A

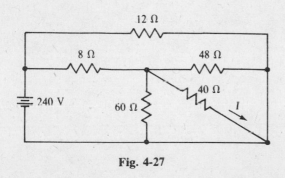

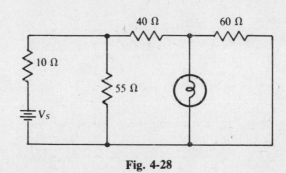

Fig. 4-27   Fig. 4-28

4.75   In the circuit shown in Fig. 4-28 there is a 120-V, 60-W light bulb.   What must be the supply voltage V_S for the light bulb to operate under rated conditions?   *Ans.* 285 V

4.76   Use voltage division twice to find the voltage V in the circuit shown in Fig. 4-29.   *Ans.* 36 V

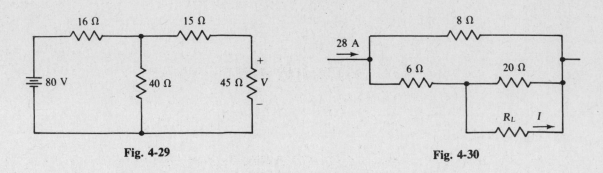

Fig. 4-29   Fig. 4-30

4.77   In the circuit shown in Fig. 4-30, use current division twice to calculate the current I in the load
resistor R_L for   (a) $R_L = 0\,\Omega$,   (b) $R_L = 5\,\Omega$,   and   (c) $R_L = 20\,\Omega$.
Ans.   (a) 16 A,   (b) 9.96 A,   (c) 4.67 A

Chapter 5

DC Circuit Analysis

CRAMER'S RULE

A knowledge of *determinants* is necessary for using *Cramer's rule,* which is a popular method for solving the equations occurring in the analysis of a circuit. A determinant is a square arrangement of numbers between two vertical lines, as follows:

$$\begin{vmatrix} a_{11} & a_{12} & a_{13} \\ a_{21} & a_{22} & a_{23} \\ a_{31} & a_{32} & a_{33} \end{vmatrix}$$

in which each a is a number. The first and second subscripts indicate the row and column, respectively, that each number is in.

A determinant with two rows and columns is a second-order determinant. One with three rows and columns is a third-order determinant, and so on.

Determinants have values. The value of the second-order determinant

$$\begin{vmatrix} a_{11} & a_{12} \\ a_{21} & a_{22} \end{vmatrix}$$

is $a_{11}a_{22} - a_{21}a_{12}$, which is the product of the numbers on the principal diagonal minus the product of the numbers on the other diagonal:

$$a_{11} \quad - \quad a_{12}$$
$$a_{22} \quad a_{21}$$

For example, the value of

$$\begin{vmatrix} 8 & -2 \\ 6 & -4 \end{vmatrix}$$

is $8(-4) - 6(-2) = -32 + 12 = -20$.

A convenient method for evaluating a third-order determinant is to repeat the first two columns to the right of the third column and then take the sum of the products of the diagonal numbers indicated by downward arrows, as follows, and subtract from this the sum of the products of the diagonal numbers indicated by upward arrows. The result is

$$a_{11}a_{22}a_{33} + a_{12}a_{23}a_{31} + a_{13}a_{21}a_{32} - a_{31}a_{22}a_{13} - a_{32}a_{23}a_{11} - a_{33}a_{21}a_{12}$$

$$\begin{vmatrix} a_{11} & a_{12} & a_{13} \\ a_{21} & a_{22} & a_{23} \\ a_{31} & a_{32} & a_{33} \end{vmatrix} \begin{matrix} a_{11} & a_{12} \\ a_{21} & a_{22} \\ a_{31} & a_{32} \end{matrix}$$

For example, the value of

$$\begin{vmatrix} 2 & -3 & 4 \\ 6 & 10 & 8 \\ 7 & -5 & 9 \end{vmatrix}$$

53

from

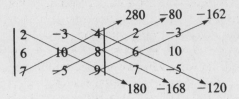

is $180 - 168 - 120 - (280 - 80 - 162) = -146.$

Evaluations of higher-order determinants require other methods that will not be considered here.

Before Cramer's rule can be applied to solve for the unknowns in a set of equations, the equations must be arranged with the unknowns on one side, say the left, of the equal signs and the knowns on the right-hand side.   The unknowns should have the same order in each equation.   For example, I_1 may be the first unknown in each equation, I_2 the second, and so on.   Then, by Cramer's rule, each unknown is the ratio of two determinants.   The denominator determinants are the same, being formed from the coefficients of the unknowns.   Each numerator determinant differs from the denominator determinant in only one column.   For the first unknown, the numerator determinant has a *first* column that is the right-hand side of the equations.   For the second unknown, the numerator determinant has a *second* column that is the right-hand side of the equations, and so on.   As an illustration, for

$$10I_1 - 2I_2 - 4I_3 = 32$$
$$-2I_1 + 12I_2 - 9I_3 = -43$$
$$-4I_1 - 9I_2 + 15I_3 = 13$$

$$I_1 = \frac{\begin{vmatrix} 32 & -2 & -4 \\ -43 & 12 & -9 \\ 13 & -9 & 15 \end{vmatrix}}{\begin{vmatrix} 10 & -2 & -4 \\ -2 & 12 & -9 \\ -4 & -9 & 15 \end{vmatrix}} \qquad I_2 = \frac{\begin{vmatrix} 10 & 32 & -4 \\ -2 & -43 & -9 \\ -4 & 13 & 15 \end{vmatrix}}{\begin{vmatrix} 10 & -2 & -4 \\ -2 & 12 & -9 \\ -4 & -9 & 15 \end{vmatrix}} \qquad I_3 = \frac{\begin{vmatrix} 10 & -2 & 32 \\ -2 & 12 & -43 \\ -4 & -9 & 13 \end{vmatrix}}{\begin{vmatrix} 10 & -2 & -4 \\ -2 & 12 & -9 \\ -4 & -9 & 15 \end{vmatrix}}$$

SOURCE CONVERSIONS

Depending on the type of analysis, a circuit with either no voltage sources or no current sources may be preferable.   Because a circuit may have an undesired type of source, it is convenient to be able to convert voltage sources to equivalent current sources, and current sources to equivalent voltage sources.   For a conversion, each voltage source must have a *series* internal resistance, and each current source a *parallel* internal resistance.

Figure 5-1a shows the conversion from a voltage source to an equivalent current source, and Fig. 5-1b the conversion from a current source to an equivalent voltage source.   This equivalence

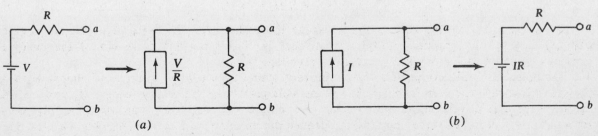

(a)   (b)

Fig. 5-1

applies only to the external circuit connected to these sources. The voltages and currents of this external circuit will be the same with either source. Internally, the sources are *not* equivalent.

As shown, in the conversion of a voltage source to a current source, the same resistor is in parallel with the current source, and the source current equals the original source voltage divided by the resistance of this resistor. The current source arrow is toward the terminal nearest the positive terminal of the voltage source. In the conversion from a current source to a voltage source, the same resistor is in series with the voltage source, and the source voltage equals the original source current times the resistance of this resistor. The positive terminal of the voltage source is nearest the terminal toward which the arrow of the current source is directed.

MESH ANALYSIS

In *mesh analysis*, KVL is applied with mesh currents, which are currents preferably referenced to flow clockwise around the meshes, as shown in Fig. 5-2.

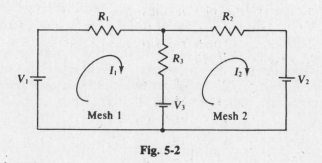

Fig. 5-2

KVL is applied to each mesh, one at a time. The voltage drops across the resistors taken in the direction of the mesh currents are set equal to the voltage rises across the voltage sources. As an illustration, in the circuit shown in Fig. 5-2, around mesh 1 the drops across resistors R_1 and R_3 are $I_1 R_1$ and $(I_1 - I_2)R_3$, respectively, the latter because the current through R_3 in the direction of I_1 is $I_1 - I_2$. The total voltage rise from voltage sources is $V_1 - V_3$, in which V_3 has a negative sign because it is a voltage drop. The result for mesh 1 is

$$I_1 R_1 + (I_1 - I_2)R_3 = V_1 - V_3 \qquad \text{or} \qquad (R_1 + R_3)I_1 - R_3 I_2 = V_1 - V_3$$

Notice that $R_1 + R_3$, the coefficient of I_1, is the sum of the resistances of the resistors in mesh 1. This sum is called the *self-resistance* of mesh 1. Also, $-R_3$, the coefficient of I_2, is the negative of the resistance of the resistor that is common to or mutual to meshes 1 and 2. R_3 is called the *mutual resistance.* In mesh equations, mutual resistance terms always have negative signs because the other mesh currents always flow through the mutual resistors in directions opposite to those of the principal mesh currents.

It is easier to write mesh equations using self-resistances and mutual resistances than it is to directly apply KVL. Doing this for mesh 2 results in

$$-R_3 I_1 + (R_2 + R_3)I_2 = V_3 - V_2$$

In a mesh equation, the voltage for a voltage source has a positive sign if the voltage source aids the flow of the principal mesh current—that is, if this current flows out of the positive terminal—because this aiding is equivalent to a voltage rise.

For mesh analysis, conversion of all current sources to voltage sources is almost always preferable because there is no formula for the voltages across current sources. If, however, a current source is positioned at the exterior of a circuit such that only one mesh current flows through it, that current source can remain because the mesh current through it is known—it is the source current or the negative of it, depending on direction. KVL is not applied to this mesh.

The number of mesh equations equals the number of meshes minus the number of current sources, if there are any.

LOOP ANALYSIS

Loop analysis is similar to mesh analysis, the principal difference being that the current paths selected are loops that are not necessarily meshes. Also, there is no convention on the direction of loop currents; they can be clockwise or counterclockwise. As a result, mutual terms can be positive when KVL is applied to the loops.

For loop analysis, no current source has to be converted to a voltage source. But each current source should have only one loop current flowing through it so that the loop current is known. Also, then KVL need not be applied to this loop, as is desirable because the current source voltage is unknown.

Of course, the loops for the loop currents must be selected such that every component has at least one loop current flowing through it. The number of these loops equals the number of meshes if the circuit is *planar*—that is, if the circuit can be drawn on a flat surface with no wires crossing. In general, the number of loop currents required is $B - N + 1$, where B is the number of branches and N is the number of nodes.

If the current through only one component is desired, the loops should be selected such that only one loop current flows through this component. Then, only one current has to be solved for. In contrast, for mesh analysis, finding the current through an interior component requires solving for two mesh currents.

NODAL ANALYSIS

For *nodal analysis*, preferably all voltage sources are converted to current sources and all resistances are converted to conductances. KCL is applied to all nodes but the ground node, which is often indicated by a ground symbol at the bottom node of the circuit, as shown in Fig. 5-3. As mentioned in Chap. 4, almost always the bottom node is selected as the ground node even though any node can be. Conventionally, voltages on all other nodes are referenced positive with respect to the ground node. As a consequence, showing node voltage polarity signs is not necessary.

In nodal analysis, KCL is applied to each node, one at a time. The currents leaving a node through resistors are set equal to the currents entering the node from current sources. As an illustration, in the circuit shown in Fig. 5-3, the current flowing down through the resistor with conductance G_1 is $G_1 V_1$. The current to the right through the resistor with conductance G_3 is $G_3(V_1 - V_2)$. *This current is the conductance times the voltage at the node at which the current enters the resistor minus the voltage at the node at which the current leaves the resistor.* This voltage is, of course, just the resistor voltage referenced positive at the node at which the current

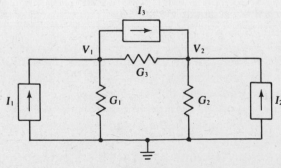

Fig. 5-3

enters the resistor and negative at the node at which the current leaves the resistor, as is required for associated references. The current entering node 1 from current sources is $I_1 - I_3$, in which I_3 has a negative sign because it is actually leaving node 1. The result for node 1 is

$$G_1 V_1 + G_3(V_1 - V_2) = I_1 - I_3 \quad \text{or} \quad (G_1 + G_3)V_1 - G_3 V_2 = I_1 - I_3$$

Notice that the V_1 coefficient of $G_1 + G_3$ is the sum of the conductances of the resistors connected to node 1. This sum is called the *self-conductance* of node 1. The coefficient of V_2 is $-G_3$, the negative of the conductance of the resistor connected between nodes 1 and 2. G_3 is called the *mutual conductance* of nodes 1 and 2. Mutual conductance terms always have negative signs because all nongrounded node voltages have the same reference polarity—all are positive.

It is easier to write nodal equations using self-conductances and mutual conductances than it is to directly apply KCL. Doing this for node 2 results in

$$-G_3 V_1 + (G_2 + G_3)V_2 = I_2 + I_3$$

Conversion of all voltage sources to current sources is not absolutely essential for nodal analysis, but is usually preferable for the shortcut approach with self-conductances and mutual conductances. The problem with voltage sources is that there is no formula for the currents flowing through them. Nodal analysis, though, is fairly easy to use with circuits having grounded voltage sources, each of which has a terminal connected to ground. Such voltage sources give known voltages at their nonground terminal nodes, making it unnecessary to apply KCL at these nodes. Other voltage sources—floating voltage sources—should be converted to current sources.

The number of nodal equations equals the number of nongrounded nodes minus the number of grounded voltage sources.

DEPENDENT SOURCES AND CIRCUIT ANALYSIS

A *dependent source*, or *controlled source*, produces a voltage or current that depends on a voltage or current elsewhere in a circuit. In contrast, *independent sources*, the only ones considered to this point, produce voltages or currents independently of any voltages or currents anywhere in a circuit. A dependent source that produces a current that is dependent on a current is a *current-dependent* current source. There are also *voltage-dependent* current sources. They produce currents that depend on voltages elsewhere in a circuit. Similarly, there are voltage-dependent and current-dependent voltage sources.

As a practical matter, dependent sources are seldom actual physical components. Rather, they appear in circuit diagrams as parts of models that operate similarly to physical transistors and other electronic components.

Mesh, loop, and nodal analyses are about the same for circuits having dependent sources as for circuits having only independent sources. Usually, though, there are a few more equations. Also, positive terms may appear in the circuit equations where only negative mutual resistance or conductance terms appear for circuits having no dependent sources.

Source conversions apply to dependent sources in the same way as to independent sources if the controlling quantities are external to the sources. Otherwise, methods described in the next chapter must be used.

Solved Problems

5.1 Evaluate the following determinants:

$$(a) \quad \begin{vmatrix} 1 & -2 \\ 3 & 4 \end{vmatrix} \qquad (b) \quad \begin{vmatrix} -5 & 6 \\ 7 & -8 \end{vmatrix}$$

(a) The product of the numbers on the principal diagonal is $1 \times 4 = 4$, and for the numbers on the other diagonal is $-2 \times 3 = -6$. The value of the determinant is the first product minus the second product: $4 - (-6) = 10$.

(b) Similarly, the value of the second determinant is $-5(-8) - 7(6) = 40 - 42 = -2$.

5.2   Evaluate the following determinant:

$$\begin{vmatrix} 8 & -9 & 4 \\ 3 & -2 & 1 \\ 6 & 5 & -4 \end{vmatrix}$$

One method of evaluation is to repeat the first two columns to the right of the third column and then find the products of the numbers on the diagonals, as indicated:

The value of the determinant is the sum of the products for the downward-pointing arrows minus the sum of the products for the upward-pointing arrows:

$$(64 - 54 + 60) - (-48 + 40 + 108) = -30$$

5.3   Use Cramer's rule to solve for the unknowns in

$$5V_1 + 4V_2 = 31$$
$$-4V_1 + 8V_2 = 20$$

The first unknown V_1 equals the ratio of two determinants. The denominator determinant has elements that are the coefficients of V_1 and V_2. The numerator determinant differs only in having the first column replaced by the right-hand sides of the equations:

$$V_1 = \frac{\begin{vmatrix} 31 & 4 \\ 20 & 8 \end{vmatrix}}{\begin{vmatrix} 5 & 4 \\ -4 & 8 \end{vmatrix}} = \frac{31(8) - 20(4)}{5(8) - (-4)(4)} = \frac{168}{56} = 3 \text{ V}$$

The denominator determinant for V_2 has the same value of 56. In the numerator determinant the second column, instead of the first, is replaced by the right-hand sides of the equations:

$$V_2 = \frac{\begin{vmatrix} 5 & 31 \\ -4 & 20 \end{vmatrix}}{56} = \frac{5(20) - (-4)(31)}{56} = \frac{224}{56} = 4 \text{ V}$$

5.4   Use Cramer's rule to solve for the unknowns in

$$6I_1 - 3I_2 = -4$$
$$-18I_1 + 9I_2 = 12$$

From the ratio of determinants,

$$I_1 = \frac{\begin{vmatrix} -4 & -3 \\ 12 & 9 \end{vmatrix}}{\begin{vmatrix} 6 & -3 \\ -18 & 9 \end{vmatrix}} = \frac{-4(9) - 12(-3)}{6(9) - (-18)(-3)} = \frac{0}{0} = ?$$

The two determinant values of zero indicate that the specified equations are not *independent*—they are *dependent*, and as such are equivalent to just a single equation. In fact, one equation can be obtained from the other: The second equation is -3 times the first. Another independent equation is needed to solve for a unique I_1 and I_2. When dependent equations occur in an analysis, probably an error has been made in selecting the proper equations to be solved.

5.5 Without using Cramer's rule, solve for the unknowns in

$$4I_1 - 2I_2 = 14$$
$$6I_1 + 5I_2 = -3$$

For solutions to just two equations, the *elimination method* is often easier to use than Cramer's rule. In this method one equation is multiplied by a number selected to make the coefficient of an unknown in one equation the negative of the coefficient for the same unknown in the other equation. Then, adding the two equations results in a single equation with just one unknown. Here, I_2 can be eliminated by multiplying the first equation by 2.5 to get $10I_1 - 5I_2 = 35$, and adding this to the second equation. This results in

$$10I_1 + 6I_1 = 35 - 3 \quad \text{and} \quad I_1 = \frac{32}{16} = 2 \text{ A}$$

Substitution of $I_1 = 2$ A into the first equation gives

$$4(2) - 2I_2 = 14 \quad \text{and} \quad I_2 = \frac{6}{-2} = -3 \text{ A}$$

5.6 Use Cramer's rule to solve for the unknowns in

$$10I_1 - 2I_2 - 4I_3 = 10$$
$$-2I_1 + 12I_2 - 6I_3 = -34$$
$$-4I_1 - 6I_2 + 14I_3 = 40$$

All three unknowns have the same denominator determinant of coefficients, which evaluates to

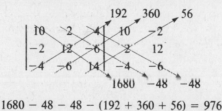

$$1680 - 48 - 48 - (192 + 360 + 56) = 976$$

In the numerator determinants, the right-hand sides of the equations replace the first column for I_1, the second column for I_2, and the third column for I_3:

$$I_1 = \frac{\begin{vmatrix} 10 & -2 & -4 \\ -34 & 12 & -6 \\ 40 & -6 & 14 \end{vmatrix}}{976} \qquad I_2 = \frac{\begin{vmatrix} 10 & 10 & -4 \\ -2 & -34 & -6 \\ -4 & 40 & 14 \end{vmatrix}}{976} \qquad I_3 = \frac{\begin{vmatrix} 10 & -2 & 10 \\ -2 & 12 & -34 \\ -4 & -6 & 40 \end{vmatrix}}{976}$$

$$= \frac{1952}{976} - 2 \text{ A} \qquad \qquad = \frac{-976}{976} = -1 \text{ A} \qquad \qquad = \frac{2928}{976} = 3 \text{ A}$$

5.7 Convert the voltage sources shown in Fig. 5-4 to current sources.

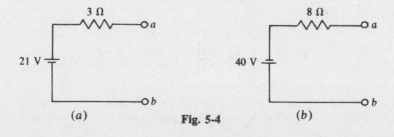

| (a) | **Fig. 5-4** | (b) |

(a) The current of the equivalent current source equals the voltage of the original voltage source divided by the resistance: 21/3 = 7 A. The current direction is toward node *a* because the positive terminal of the voltage source is toward that node. The parallel resistor is the same 3-Ω resistor of the original voltage source. The equivalent current source is shown in Fig. 5-5*a*.

(b) The current of the current source is 40/8 = 5 A. It is directed toward node *b* because the positive terminal of the voltage source is toward that node. The parallel resistor is the same 8-Ω resistor of the voltage source. Figure 5-5*b* shows the equivalent current source.

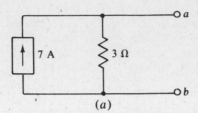

(a)

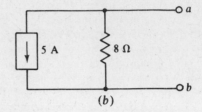

(b)

Fig. 5-5

5.8 Convert the current sources shown in Fig. 5-6 to voltage sources.

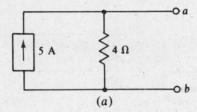

(a)

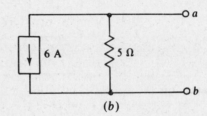

(b)

Fig. 5-6

(a) The voltage of the equivalent voltage source equals the current of the original current source times the resistance: 5 × 4 = 20 V. The positive terminal is toward node *a* because the current of the original current source is toward that node. Of course, the source resistance remains 4 Ω, but is in series instead of in parallel. Figure 5-7*a* shows the equivalent voltage source.

(b) The voltage is 6 × 5 = 30 V, positive toward node *b* because the current of the original current source is toward that node. The source resistance is the same 5 Ω, but is in series. The equivalent voltage source is shown in Fig. 5-7*b*.

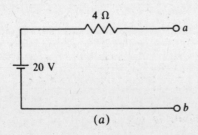

(a)

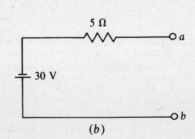

(b)

Fig. 5-7

5.9 Find the currents down through the resistors in the circuit shown in Fig. 5-8. Then convert the current source and 2-Ω resistor to an equivalent voltage source and again find the resistor currents. Compare results.

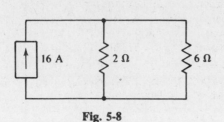

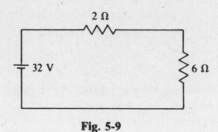

Fig. 5-8   Fig. 5-9

By current division, the current down through the 2-Ω resistor is

$$\frac{6}{2+6} \times 16 = 12 \text{ A}$$

The remainder of the source current (16 − 12 = 4 A) flows down through the 6-Ω resistor.

Conversion of the current source produces a voltage source of $16 \times 2 = 32$ V in series with a 2-Ω resistor, all in series with the 6-Ω resistor, as shown in the circuit of Fig. 5-9. In this circuit, the same current $32/(2+6) = 4$ A flows through both resistors. The 6-Ω resistor current is the same as for the original circuit, but the 2-Ω resistor current is different. This result illustrates the fact that a converted source produces the same voltages and currents only in the circuit exterior to the source. Inside the source, the voltages and currents change.

5.10 Find the mesh currents in the circuit shown in Fig. 5-10.

Using self-resistances and mutual resistances is almost always best for getting mesh equations. The self-resistance of mesh 1 is $5 + 6 = 11 \Omega$, and the resistance mutual with mesh 2 is 6 Ω. The sum of the source voltage rises in the direction of I_1 is $62 - 16 = 46$ V. So, the mesh 1 KVL equation is $11I_1 - 6I_2 = 46$.

No KVL equation is needed for mesh 2 because I_2 is the only current flowing through the 4-A current source, with the result that $I_2 = -4$ A. The current I_2 is negative because its reference direction is down through the current source, but the 4-A source current actually flows up. Incidentally, a KVL equation cannot be written for mesh 2 without introducing a variable for the voltage across the current source because this voltage is unknown.

The substitution of $I_2 = -4$ A into the mesh 1 equation results in

$$11I_1 - 6(-4) = 46 \qquad \text{and} \qquad I_1 = \frac{22}{11} = 2 \text{ A}$$

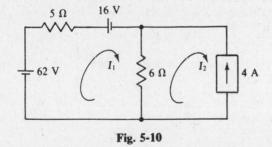

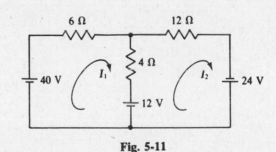

Fig. 5-10   Fig. 5-11

5.11 Solve for the mesh currents in the circuit shown in Fig. 5-11.

The self-resistance of mesh 1 is $6 + 4 = 10 \Omega$, the mutual resistance with mesh 2 is 4 Ω, and the sum of the source voltage rises in the direction of I_1 is $40 - 12 = 28$ V. So, the mesh 1 KVL equation is $10I_1 - 4I_2 = 28$.

Similarly, for mesh 2 the self-resistance is $4 + 12 = 16 \Omega$, the mutual resistance is 4 Ω, and the sum of the voltage rises from voltage sources is $24 + 12 = 36$ V. These give a mesh 2 KVL equation of $-4I_1 + 16I_2 = 36$.

Placing the two mesh equations together shows the symmetry of coefficients (here -4) about the principal diagonal as a result of the common mutual resistance:

$$10I_1 - 4I_2 = 28$$
$$-4I_1 + 16I_2 = 36$$

A good way to solve these two equations is to add four times the first equation to the second equation to eliminate I_2.   The result is

$$40I_1 - 4I_1 = 112 + 36 \qquad \text{from which} \qquad I_1 = \frac{148}{36} = 4.11 \text{ A}$$

This substituted into the second equation gives

$$-4(4.11) + 16I_2 = 36 \qquad \text{and} \qquad I_2 = \frac{52.44}{16} = 3.28 \text{ A}$$

5.12   Find the battery currents flowing from the positive terminals in the circuit shown in Fig. 5-12.

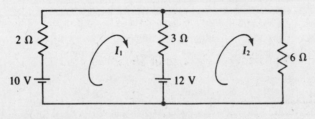

Fig. 5-12

For mesh 1 the self-resistance is $2 + 3 = 5\ \Omega$, the mutual resistance is $3\ \Omega$, and the net aiding source voltage is $10 - 12 = -2$ V. For mesh 2 the self-resistance is $6 + 3 = 9\ \Omega$, the mutual resistance is $3\ \Omega$, and the aiding source voltage is 12 V.   So, the mesh equations are

$$5I_1 - 3I_2 = -2$$
$$-3I_1 + 9I_2 = 12$$

Multiplying the first equation by 3 and adding the result to the second equation eliminates I_2:

$$15I_1 - 3I_1 = -6 + 12 \qquad \text{from which} \qquad I_1 = \frac{6}{12} = 0.5 \text{ A}$$

This substituted into the second equation gives

$$I_2 = \frac{12 + 3(0.5)}{9} = 1.5 \text{ A}$$

The current flowing from the positive terminal of the 10-V battery is $I_1 = 0.5$ A.   The current flowing from the positive terminal of the 12-V source is $I_2 - I_1 = 1.5 - 0.5 = 1$ A.

5.13   Find the mesh currents for the circuit shown in Fig. 5-13.

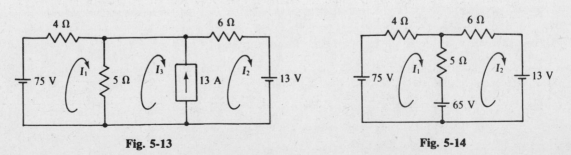

Fig. 5-13   **Fig. 5-14**

The first step is the conversion of 13-A current source and parallel 5-Ω resistor to a voltage source, as shown in the circuit of Fig. 5-14.

The self-resistance of mesh 1 is $4 + 5 = 9\ \Omega$, and that of mesh 2 is $6 + 5 = 11\ \Omega$. The mutual resistance is 5 Ω. The voltage rises from sources are $75 - 65 = 10$ V for mesh 1 and $65 - 13 = 52$ V for mesh 2. The corresponding mesh equations are

$$9I_1 - 5I_2 = 10$$
$$-5I_1 + 11I_2 = 52$$

Multiplying the first equation by 5 and the second by 9 and then adding them eliminates I_1:

$$25I_2 + 99I_2 = 50 + 468 \quad \text{from which} \quad I_2 = \frac{518}{74} = 7 \text{ A}$$

This substituted into the first equation produces

$$9I_1 - 5(7) = 10 \quad \text{or} \quad I_1 = \frac{10 + 35}{9} = 5 \text{ A}$$

From the original circuit shown in Fig. 5-13, the current through the current source is $I_2 - I_3 = 13$ A, and so

$$I_3 = I_2 - 13 = 7 - 13 = -6 \text{ A}$$

5.14 Solve for the mesh currents in the circuit shown in Fig. 5-15.

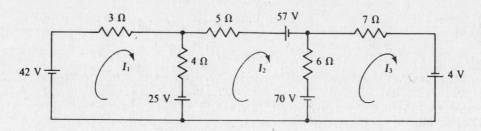

Fig. 5-15

The self-resistances are $3 + 4 = 7\ \Omega$ for mesh 1, $4 + 5 + 6 = 15\ \Omega$ for mesh 2, and $6 + 7 = 13\ \Omega$ for mesh 3. The mutual resistances are 4 Ω for meshes 1 and 2, 6 Ω for meshes 2 and 3, and 0 Ω for meshes 1 and 3. The aiding source voltages are $42 + 25 = 67$ V for mesh 1, $-25 - 57 - 70 = -152$ V for mesh 2, and $70 + 4 = 74$ V for mesh 3. So, the mesh equations are

$$7I_1 - 4I_2 - 0I_3 = 67$$
$$-4I_1 + 15I_2 - 6I_3 = -152$$
$$-0I_1 - 6I_2 + 13I_3 = 74$$

Notice the indicated symmetry of the mutual coefficients about the principal diagonal, shown as a dashed line. Because of the common mutual resistances, this symmetry always occurs—unless a circuit has dependent sources. Also, notice for each mesh that the self-resistance is equal to or greater than the sum of the mutual resistances because the self-resistance includes the mutual resistances.

By Cramer's rule,

$$I_1 = \frac{\begin{vmatrix} 67 & -4 & 0 \\ -152 & 15 & -6 \\ 74 & -6 & 13 \end{vmatrix}}{\begin{vmatrix} 7 & -4 & 0 \\ -4 & 15 & -6 \\ 0 & -6 & 13 \end{vmatrix}} = \frac{4525}{905} = 5 \text{ A}$$

$$I_2 = \frac{\begin{vmatrix} 7 & 67 & 0 \\ -4 & -152 & -6 \\ 0 & 74 & 13 \end{vmatrix}}{905}$$
$$= \frac{-7240}{905} = -8 \text{ A}$$

$$I_3 = \frac{\begin{vmatrix} 7 & -4 & 67 \\ -4 & 15 & -152 \\ 0 & -6 & 74 \end{vmatrix}}{905}$$
$$= \frac{1810}{905} = 2 \text{ A}$$

5.15 Find the mesh currents in the circuit shown in Fig. 5-16.

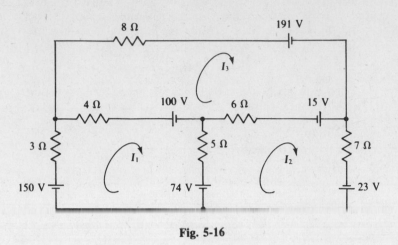

Fig. 5-16

The self-resistances are $3 + 4 + 5 = 12\ \Omega$ for mesh 1, $5 + 6 + 7 = 18\ \Omega$ for mesh 2, and $6 + 4 + 8 = 18\ \Omega$ for mesh 3. The mutual resistances are $5\ \Omega$ for meshes 1 and 2, $6\ \Omega$ for meshes 2 and 3, and $4\ \Omega$ for meshes 1 and 3. The aiding source voltages are $150 - 100 - 74 = -24\ \text{V}$ for mesh 1, $74 + 15 + 23 = 112\ \text{V}$ for mesh 2, and $100 - 191 - 15 = -106\ \text{V}$ for mesh 3. So, the mesh equations are

$$12I_1 - 5I_2 - 4I_3 = -24$$
$$-5I_1 + 18I_2 - 6I_3 = 112$$
$$-4I_1 - 6I_2 + 18I_3 = -106$$

For a check, notice the symmetry of the coefficients about the principal diagonal.
By Cramer's rule,

$$I_1 = \frac{\begin{vmatrix} -24 & -5 & -4 \\ 112 & 18 & -6 \\ -106 & -6 & 18 \end{vmatrix}}{\begin{vmatrix} 12 & -5 & -4 \\ -5 & 18 & -6 \\ -4 & -6 & 18 \end{vmatrix}} = \frac{-4956}{2478} = -2\ \text{A}$$

$$I_2 = \frac{\begin{vmatrix} 12 & -24 & -4 \\ -5 & 112 & -6 \\ -4 & -106 & 18 \end{vmatrix}}{2478}$$
$$= \frac{9912}{2478} = 4\ \text{A}$$

$$I_3 = \frac{\begin{vmatrix} 12 & -5 & -24 \\ -5 & 18 & 112 \\ -4 & -6 & -106 \end{vmatrix}}{2478}$$
$$= \frac{-12\,390}{2478} = -5\ \text{A}$$

5.16 Show a circuit corresponding to the mesh equations

$$8I_1 - 3I_2 = 10$$
$$-3I_1 + 12I_2 = -8$$

Because there are two equations, the circuit has two meshes. The -3 coefficients indicate that meshes 1 and 2 have a 3-Ω mutual resistance. The 8 coefficient of I_1 in the first equation indicates that the resistors in mesh 1 have a total resistance of $8\ \Omega$. Since $3\ \Omega$ of this is in mutual resistance, there must be $8 - 3 = 5\ \Omega$ of resistance in mesh 1 that is not mutual. Similarly, mesh 2 has a separate resistance of $12 - 3 = 9\ \Omega$. The 10 on the right-hand side of the first equation can be the result of a total of 10 V of voltage source rises, or aiding source voltages, in mesh 1. One way to get this is with a single 10-V source that is not in the mutual branch and that is arranged such that I_1 flows out of its positive terminal. Likewise, for the mesh 2 equation the -8 can result from a single 8-V source, not in the mutual branch, arranged such that I_2 flows into its positive terminal. Figure 5-17 shows the corresponding circuit.

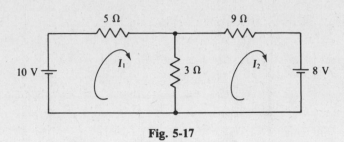

Fig. 5-17

5.17 Show a circuit corresponding to

$$I_2 = \frac{\begin{vmatrix} 12 & 10 & -3 \\ -4 & -8 & -6 \\ -3 & 5 & 15 \end{vmatrix}}{\begin{vmatrix} 12 & -4 & -3 \\ -4 & 14 & -6 \\ -3 & -6 & 15 \end{vmatrix}}$$

Because the determinants are third-order, the circuit has three meshes. From the first row of the bottom determinant, there is 4 Ω mutual to meshes 1 and 2 and 3 Ω mutual to meshes 1 and 3, leaving $12 - 4 - 3 = 5\ \Omega$ in mesh 1 that is not mutual with other meshes. Similarly, from the second row, there is 6 Ω common to meshes 2 and 3 and $14 - 4 - 6 = 4\ \Omega$ in mesh 2 that is not mutual. From row 3, there is $15 - 6 - 3 = 6\ \Omega$ in mesh 3 that is not mutual. The second column in the numerator determinant, the only column that differs from corresponding columns in the denominator determinant, shows that the voltage sources are 10 V aiding in mesh 1, 8 V opposing in mesh 2, and 5 V aiding in mesh 3. Figure 5-18 shows one circuit that satisfies these conditions.

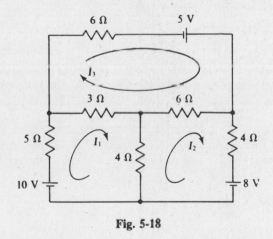

Fig. 5-18

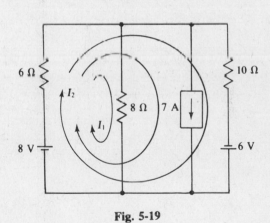

Fig. 5-19

5.18 Use loop analysis to find the current down through the 8-Ω resistor in the circuit shown in Fig. 5-19.

Because the circuit has three meshes, the analysis requires three loop currents. The loops can be selected as shown, with only one current I_1 flowing through the 8-Ω resistor so that only one current needs to be solved for. Also, only one loop current should flow through the 7-A source so that this loop current is known, making it unnecessary to apply KVL to the corresponding loop. There are other ways of selecting the loop current paths to satisfy these conditions.

The self-resistance of the first loop is $6 + 8 = 14\ \Omega$, and the resistance mutual with the second loop is 6 Ω. The 7-A current flowing through the 6-Ω resistor produces a 42-V drop in the first loop. The resulting loop equation is

$$14I_1 + 6I_2 + 42 = 8 \qquad \text{or} \qquad 14I_1 + 6I_2 = -34$$

The 6 coefficient of I_2 is positive because I_2 flows through the 6-Ω resistor in the same direction as I_1.

For the second loop, the self-resistance is $6 + 10 = 16 \ \Omega$, of which 6 Ω is mutual with the first loop. The second loop equation is

$$6I_1 + 16I_2 + 42 = 8 + 6 \qquad \text{or} \qquad 6I_1 + 16I_2 = -28$$

The two loop equations together are

$$14I_1 + \ 6I_2 = -34$$
$$6I_1 + 16I_2 = -28$$

Multiplying the first equation by 8 and the second by -3 and then adding them eliminates I_2:

$$112I_1 - 18I_1 = -272 + 84 \qquad \text{from which} \qquad I_1 = -\frac{118}{94} = -2 \text{ A}$$

5.19 Use loop analysis to find the current flowing to the right through the 5-kΩ resistor in the circuit shown in Fig. 5-20.

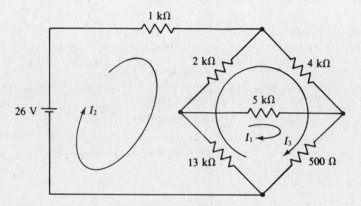

Fig. 5-20

Three loop currents are required because the circuit has three meshes. Only one loop current should flow through the 5-kΩ resistor so that only one current needs to be solved for. The paths for the two other loop currents can be selected as shown, but there are other suitable paths.

Since working with kilohms is inconvenient, a common practice is to drop those units—to divide each resistance by 1000. But then the current answers will be in milliamperes. With this approach, and from self-resistances, mutual resistances, and aiding source voltages, the loop equations are

$$18.5I_1 - 13I_2 + 13.5I_3 = \ 0$$
$$-13I_1 + 16I_2 - \ 15I_3 = 26$$
$$13.5I_1 - 15I_2 + 19.5I_3 = \ 0$$

Notice the symmetry of the I coefficients about the principal diagonal, just as for mesh equations. But there is the difference that some of these coefficients are positive. This is the result of two loop currents flowing through a mutual resistor in the same direction—something that cannot happen in mesh analysis if all mesh currents are selected in the clockwise direction, as is conventional.

From Cramer's rule,

$$I_1 = \frac{\begin{vmatrix} 0 & -13 & 13.5 \\ 26 & 16 & -15 \\ 0 & -15 & 19.5 \end{vmatrix}}{\begin{vmatrix} 18.5 & -13 & 13.5 \\ -13 & 16 & -15 \\ 13.5 & -15 & 19.5 \end{vmatrix}} = \frac{1326}{663} = 2 \text{ mA}$$

5.20 Use loop analysis to solve for the current flowing down through the 5-Ω resistor in the circuit shown in Fig. 5-16.

   The current loops should be selected such that only one loop current flows down through the 5-Ω resistor. The advantage is that only one current needs to be solved for. Of course, three current loops are necessary because the circuit has three meshes. Figure 5-21 shows the circuit with one suitable selection of current loops.

   From self-resistances, mutual resistances, and aiding source voltages, the loop equations are

$$12I_1 + 7I_2 - 4I_3 = -24$$
$$7I_1 + 20I_2 - 10I_3 = 88$$
$$-4I_1 - 10I_2 + 18I_3 = -106$$

The coefficient of I_2 in the first equation and of I_1 in the second equation is 7 because the mutual resistance of loops 1 and 2 is $4 + 3 = 7$ Ω. This coefficient is positive because I_1 and I_2 flow through the 4- and 3-Ω resistors in the same direction. The other terms are obvious.

   By Cramer's rule,

$$I_1 = \frac{\begin{vmatrix} -24 & 7 & -4 \\ 88 & 20 & -10 \\ -106 & -10 & 18 \end{vmatrix}}{\begin{vmatrix} 12 & 7 & -4 \\ 7 & 20 & -10 \\ -4 & -10 & 18 \end{vmatrix}} = \frac{-14\,868}{2478} = -6 \text{ A}$$

This answer agrees with the answers to Prob. 5.15, which deals with the same circuit. In that problem, the current flowing down through the 5-Ω resistor is $I_1 - I_2$, and $I_1 = -2$ A and $I_2 = 4$ A, giving $-2 - 4 = -6$ A, the same answer as here

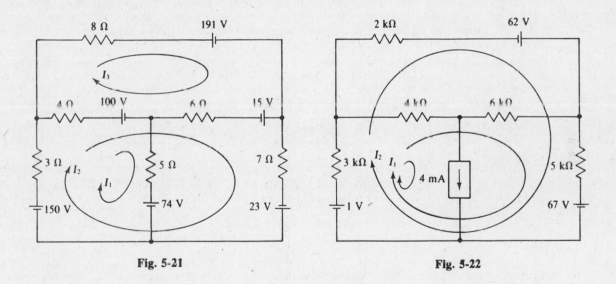

Fig. 5-21   **Fig. 5-22**

5.21 Use loop analysis to find the current flowing to the right in the 6-kΩ resistor in the circuit shown in Fig. 5-22.

   Because the circuit has three meshes, three current loops are required. Of course, the loops should be selected such that only one loop current flows through the current source and only one through the 6-kΩ resistor. Figure 5-22 shows a suitable selection. From the self-resistance and mutual-resistance approach, the loop equations are

$$18I_1 + 8I_2 + 4 \times 7 = 1 - 67$$
$$8I_1 + 10I_2 + 4 \times 3 = 1 + 62 - 67$$

in which the simplifying kilohm-milliampere method is used.   These equations simplify to

$$18I_1 + 8I_2 = -94$$
$$8I_1 + 10I_2 = -16$$

Multiplying the first equation by 5 and the second by -4 and then adding them eliminates I_2, giving

$$90I_1 - 32I_1 = -470 + 64 \quad \text{and} \quad I_1 = -\frac{406}{58} = -7 \text{ mA}$$

5.22 Two 12-V batteries are being charged from a 16-V generator.   The internal resistances are 0.5 and 0.8 Ω for the batteries and 2 Ω for the generator.   Find the currents flowing into the positive battery terminals.

The arrangement is basically parallel, with just two nodes.   If the voltage at the positive node with respect to the negative node is called V, the current flowing away from the positive node through the sources is

$$\frac{V - 12}{0.5} + \frac{V - 12}{0.8} + \frac{V - 16}{2} = 0$$

Multiplying by 4 produces

$$8V - 96 + 5V - 60 + 2V - 32 = 0 \quad \text{or} \quad 15V = 188 \quad \text{and} \quad V = \frac{188}{15} = 12.533 \text{ V}$$

Consequently, the current into the 12-V battery with 0.5-Ω internal resistance is $(12.533 - 12)/0.5 = 1.07$ A,   and the current into the other 12-V battery is $(12.533 - 12)/0.8 = 0.667$ A.

5.23 Solve for the node voltages in the circuit shown in Fig. 5-23.

Using self-conductances and mutual conductances is almost always best for getting the nodal equations.   The self-conductance of node 1 is $5 + 8 = 13$ S, and the mutual conductance is 8 S.   The sum of the currents from current sources into this node is $36 + 48 = 84$ A.   So, the node 1 KCL equation is $13V_1 - 8V_2 = 84$.

No KCL equation is needed for node 2 because a grounded voltage source is connected to it, making $V_2 = -5$ V.   Anyway, a KCL equation cannot be written for this node without introducing a variable for the current through the 5-V source because this current is unknown.

The substitution of $V_2 = -5$ V into the node 1 equation results in

$$13V_1 - 8(-5) = 84 \quad \text{and} \quad V_1 = \frac{44}{13} = 3.38 \text{ V}$$

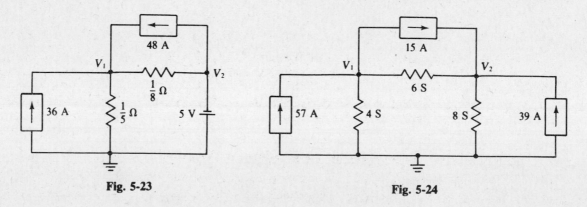

Fig. 5-23   **Fig. 5-24**

5.24 Solve for the node voltages in the circuit shown in Fig. 5-24.

The self-conductance of node 1 is $6 + 4 = 10$ S.   The conductance mutual with node 2 is 6 S, and the sum of the currents into node 1 from current sources is $57 - 15 = 42$ A.   So, the node 1 KCL equation is $10V_1 - 6V_2 = 42$.

Similarly, for node 2 the self-conductance is $6 + 8 = 14$ S, the mutual conductance is 6 S, and the sum of the input currents from current sources is $39 + 15 = 54$ A. These give a node 2 KCL equation of $-6V_1 + 14V_2 = 54$.

Placing the two nodal equations together shows the symmetry of the coefficients (-6 here) about the principal diagonal as a result of the same mutual conductance coefficient in both equations:

$$10V_1 - 6V_2 = 42$$
$$-6V_1 + 14V_2 = 54$$

Three times the first equation added to five times the second eliminates V_1. The result is

$$-18V_2 + 70V_2 = 126 + 270 \quad \text{from which} \quad V_2 = \frac{396}{52} = 7.62 \text{ V}$$

This substituted into the first equation gives

$$10V_1 - 6(7.62) = 42 \quad \text{and} \quad V_1 = \frac{87.7}{10} = 8.77 \text{ V}$$

5.25 Find the node voltages in the circuit shown in Fig. 5-25.

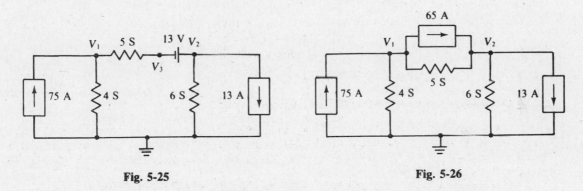

Fig. 5-25   Fig. 5-26

The first step is the conversion of the voltage source and series resistor to a current source and parallel resistor, as shown in the circuit of Fig. 5-26.

The self-conductance of node 1 is $4 + 5 = 9$ S, and that of node 2 is $5 + 6 = 11$ S. The mutual conductance is 5 S. The sum of the currents into node 1 from current sources is $75 - 65 = 10$ A, and that into node 2 is $65 - 13 = 52$ A. Thus, the corresponding nodal equations are

$$9V_1 - 5V_2 = 10$$
$$-5V_1 + 11V_2 = 52$$

Except for V's instead of I's, these are the same equations as for Prob. 5.13. Consequently, the answers are the same: $V_1 = 5$ V and $V_2 = 7$ V. Circuits having such similar equations are called *duals*.

From the original circuit shown in Fig. 5-25, the 13-V source makes V_3 13 V more negative than V_2: $V_3 = V_2 - 13 = 7 - 13 = -6$ V.

5.26 Obtain the nodal equations for the circuit shown in Fig. 5-27.

The self-conductances are $3 + 4 = 7$ S for node 1, $4 + 5 + 6 = 15$ S for node 2, and $6 + 7 = 13$ S for node 3. The mutual conductances are 4 S for nodes 1 and 2, 6 S for nodes 2 and 3, and 0 S for nodes 1 and 3. The currents flowing into the nodes from current sources are $42 + 25 = 67$ A for node 1, $-25 - 57 - 70 = -152$ A for node 2, and $70 + 4 = 74$ A for node 3. So, the nodal equations are

$$\begin{aligned}
7V_1 - 4V_2 - 0V_3 &= 67 \\
-4V_1 + 15V_2 - 6V_3 &= -152 \\
0V_1 - 6V_2 + 13V_3 &= 74
\end{aligned}$$

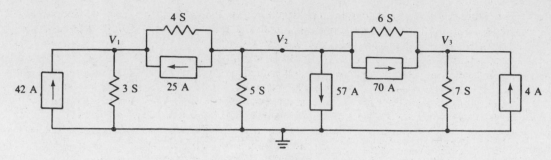

Fig. 5-27

Notice the symmetry of coefficients about the principal diagonal. This symmetry always occurs for circuits that do not have dependent sources.

Since this set of equations is the same as that for Prob. 5.14, except for having V's instead of I's, the answers are the same: $V_1 = 5$ V, $V_2 = -8$ V, and $V_3 = 2$ V

5.27 Obtain the nodal equations for the circuit shown in Fig. 5-28.

The self-conductances are $3 + 4 + 5 = 12$ S for node 1, $5 + 6 + 7 = 18$ S for node 2, and $6 + 4 + 8 = 18$ S for node 3. The mutual conductances are 5 S for nodes 1 and 2, 6 S for nodes 2 and 3, and 4 S for nodes 1 and 3. The currents into the nodes from current sources are $150 - 100 - 74 = -24$ A for node 1, $74 + 15 + 23 = 112$ A for node 2, and $100 - 191 - 15 = -106$ A for node 3. So, the nodal equations are

$$12V_1 - 5V_2 - 4V_3 = -24$$
$$-5V_1 + 18V_2 - 6V_3 = 112$$
$$-4V_1 - 6V_2 + 18V_3 = -106$$

As a check, notice the symmetry of the coefficients about the principal diagonal.

Since these equations are basically the same as those in Prob. 5.15, the answers are the same: $V_1 = -2$ V, $V_2 = 4$ V, and $V_3 = -5$ V.

5.28 Show a circuit corresponding to the nodal equations

$$10V_1 - 4V_2 = -8$$
$$-4V_1 + 12V_2 = 16$$

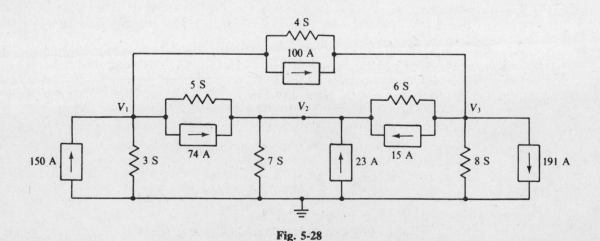

Fig. 5-28

Since there are two equations, the circuit has three nodes, one of which is the ground or reference node. The -4 coefficients indicate that nodes 1 and 2 have a 4-S mutual conductance. The 10 coefficient of V_1 in the first equation indicates that the resistors connected to node 1 have a total conductance of 10 S. Since 4 S of this is in mutual conductance, there must be a $10 - 4 = 6$-S resistor connected between node 1 and ground. Similarly, from the second equation there must be a $12 - 4 = 8$-S resistor connected between node 2 and ground. The -8 on the right-hand side of the first equation is the result of a total current of 8 A *leaving* node 1 through current sources. The negative sign indicates that the current is leaving. Perhaps the easiest way to get this current is with a single 8-A current source connected between node 1 and ground, with the current directed away from node 1. Similarly, for node 2 the 16 can be from a 16-A current source connected between node 2 and ground, with the current directed into node 2. Figure 5-29 shows the circuit.

For a circuit of three resistors, the conductance values are unique. But there are two other source arrangements and values for two current sources, and an infinite number of sets of suitable current source values for three current sources.

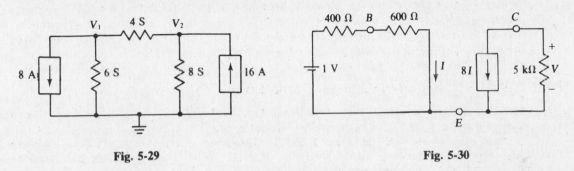

Fig. 5-29　　　　　　　　　　　　　　　　　　　　**Fig. 5-30**

5.29 Find V for the circuit shown in Fig. 5-30.

This circuit has a current source that produces a current that is dependent on a current flowing in another part of the circuit. The portion of this circuit having this source and the 600-Ω resistor is a simple *model* for a transistor. A model is a circuit formed of convenient components that produce approximately the same operation, to the desired degree of accuracy, as the original circuit component, which in this case is a transistor (not shown). The E, B, and C terminals correspond, respectively, to the emitter, base, and collector terminals of the transistor.

From the left-hand portion of the circuit, $I = 1/1000$ A $= 1$ mA. And from the right-hand portion, $V = -5000 \times 8I = -5000 \times 8 \times 10^{-3} = -40$ V. The negative sign is necessary because for the 5-kΩ resistor the references for $8I$ and V are not associated.

Notice that the output voltage is 40 times the input voltage and is negative. If the input voltage had been varying instead of constant, the output voltage would, if plotted, look exactly like the input except it would be 40 times larger and would be inverted. Consequently, this circuit is an amplifier for a varying input voltage.

5.30 Find V for the circuit shown in Fig. 5-31.

Here, the transistor model has two dependent sources, one a voltage-dependent voltage source in the input circuit, and the other a current-dependent current source in the output circuit. In the output circuit, the 60- and 40-kΩ resistors, being in parallel, have a net resistance of $(60 \times 40)/(60 + 40) = 24$ kΩ. Thus, the output voltage is $V = -20I \times 24 \times 10^3 = -480 \times 10^3 I$. At the input, KVL gives $2 = 15.733 \times 10^3 I + 0.005V$. Substituting in for V produces a single equation in I:

$$2 = 15.733 \times 10^3 I + 0.005(-480 \times 10^3 I) = 13\,333I$$

from which

$$I = 2/13\,333 \text{ A} = 0.15 \text{ mA}$$

Substituting this into the output equation gives V:

$$V = -480 \times 10^3 I = (-480 \times 10^3)(0.15 \times 10^{-3}) = -72 \text{ V}$$

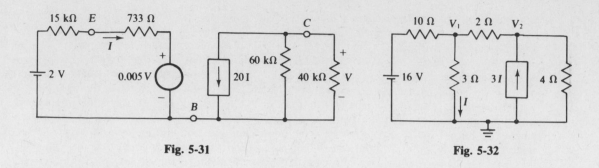

Fig. 5-31　　　　　　　　　　　　　Fig. 5-32

5.31 Solve for V_1 in the circuit shown in Fig. 5-32.

What analysis method is best here? Usually, preference is given to the method that requires fewest equations. Here, both loop and nodal analyses require two equations. Mesh analysis can be used if the 4-Ω resistor and $3I$-A current source are interchanged, but mesh analysis also requires two equations. So, the work is about the same regardless of the method. However, nodal analysis may be slightly preferable because the unknown to be solved for is a voltage, and nodal analysis gives voltages directly.

For nodal analysis, the 16-V source and series 10-Ω resistor are preferably converted to a current source with parallel resistor. The current source has a current of $16/10 = 1.6$ A directed into node 1, and the parallel resistor has a resistance of 10 Ω.

The self-conductances are $\frac{1}{10} + \frac{1}{3} + \frac{1}{2} = \frac{14}{15}$ S for node 1, and $\frac{1}{2} + \frac{1}{4} = \frac{3}{4}$ S for node 2. The mutual conductance is $\frac{1}{2}$ S. The currents into the nodes from current sources are 1.6 A for node 1, and $3I$ for node 2. Using this $3I$ in the nodal equations is inconvenient. Since, by Ohm's law, $I = V_1/3$, the dependent source current can be expressed as $3(V_1/3) = V_1$. Then, the nodal equations are

$$\frac{14}{15} V_1 - \frac{1}{2} V_2 = 1.6$$
$$-\frac{1}{2} V_1 + \frac{3}{4} V_2 = V_1$$

Multiplying the first equation by 30 and the second by 4, and subtracting $4V_1$ from both sides of the second equation, simplifies them to

$$28V_1 - 15V_2 = 48$$
$$-6V_1 + 3V_2 = 0$$

Adding the first equation to five times the second eliminates V_2, giving

$$28V_1 - 30V_1 = 48 \quad \text{from which} \quad V_1 = \frac{48}{-2} = -24 \text{ V}$$

In the solution, notice that the dependent source variable was expressed as a function of the variable being solved for. Often this is a desirable early step in the analysis of a circuit having dependent sources.

5.32 Find I_1 for the circuit shown in Fig. 5-33.

Loop analysis is a good choice here since the desired quantity is a current that flows in an interior branch. For the current loop selection of the left-hand mesh for I_1 and the outside loop for I_2, the self-resistances are $4 + 2 + 3 = 9$ Ω for loop 1 and $4 + 2 = 6$ Ω for loop 2. The mutual resistance is 4 Ω. Also, the mutual terms are positive because both currents flow through the 4-Ω resistor in the same direction. The aiding voltages from voltage sources are 6 V for loop 1, and $6 - 3V$ for loop 2. The V for the dependent source can be expressed as a function of I_1, the variable being solved for. Since $V = 3I_1$ from the center branch, the aiding voltage for loop 2 can be written as $6 - 9I_1$. The resulting loop equations are

$$9I_1 + 4I_2 = 6$$
$$4I_1 + 6I_2 = 6 - 9I_1$$

which simplify to

$$9I_1 + 4I_2 = 6$$
$$13I_1 + 6I_2 = 6$$

Notice the lack of symmetry about the principal diagonal caused by the circuit having a dependent source.

I_2 can be eliminated by adding the second equation to -1.5 times the first. The result is

$$13I_1 - 13.5I_1 = 6 - 9 \qquad \text{from which} \qquad I_1 = \frac{-3}{-0.5} = 6 \text{ A}$$

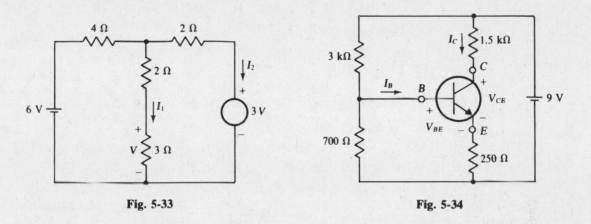

Fig. 5-33　　　　　　　　　　　　　　　　　　　Fig. 5-34

5.33 Figure 5-34 shows a transistor with a bias circuit. If $I_C = 50I_B$ and if $V_{BE} = 0.7$ V, find V_{CE}.

Perhaps the best way to find V_{CE} is to first find I_B and I_C, and from them the voltage drops across the 1.5-kΩ and 250-Ω resistors. Then, use KVL on the right-hand mesh and obtain V_{CE} from 9 V minus these two drops.

I_B can be found from the two left-hand meshes. The current through the 250-Ω resistor is $I_C + I_B = 50I_B + I_B = 51I_B$, giving a voltage drop of $(51I_B)(250)$. This drop added to V_{BE} is the drop across the 700-Ω resistor. Thus, the current through this resistor is $[0.7 + (51I_B)(250)]/700$. From KCL applied at the left-hand node, this current plus I_B is the total current flowing through the 3-kΩ resistor. The voltage drop across this resistor added to the drop across the 700-Ω resistor equals 9 V, as is evident from the outside loop:

$$\left[\frac{0.7 + (51I_B)(250)}{700} + I_B\right](3000) + 0.7 + (51I_B)(250) = 9$$

From this, $I_B = 75.3 \ \mu\text{A}$. So, $I_C = 50I_B = 3.76$ mA and

$$V_{CE} = 9 - 1500I_C - 250(I_C + I_B) = 2.39 \text{ V}$$

Supplementary Problems

5.34 Evaluate the following determinants:

$$(a) \quad \begin{vmatrix} 4 & 3 \\ -2 & -6 \end{vmatrix} \qquad (b) \quad \begin{vmatrix} 8 & -30 \\ 42 & 56 \end{vmatrix}$$

Ans. (a) -18, (b) 1708

5.35   Evaluate the following determinants:

$$(a) \begin{vmatrix} 16 & 0 & -25 \\ -32 & 15 & -19 \\ 13 & 21 & -18 \end{vmatrix} \qquad (b) \begin{vmatrix} -27 & 33 & -45 \\ -52 & 64 & -73 \\ 18 & -92 & 46 \end{vmatrix}$$

Ans.   (a) 23 739,   (b) −26 022

5.36   Use Cramer's rule to solve for the unknowns in

$$(a) \begin{array}{l} 26V_1 - 18V_2 = -124 \\ -18V_1 + 30V_2 = 156 \end{array} \qquad (b) \begin{array}{l} 16I_1 - 12I_2 = 560 \\ -12I_1 + 21I_2 = -708 \end{array}$$

Ans.   (a) $V_1 = -2$ V, $V_2 = 4$ V;   (b) $I_1 = 17$ A, $I_2 = -24$ A

5.37   Without using Cramer's rule, solve for the unknowns in

$$(a) \begin{array}{l} 11I_1 - 20I_2 = -704 \\ -28I_1 + 37I_2 = 659 \end{array} \qquad (b) \begin{array}{l} 62V_1 - 42V_2 = 694 \\ -42V_1 + 77V_2 = 161 \end{array}$$

Ans.   (a) $I_1 = -9$ A, $I_2 = 11$ A;   (b) $V_1 = 20$ V, $V_2 = 13$ V

5.38   Use Cramer's rule to solve for the unknowns in

$$\begin{array}{l} 26V_1 - 11V_2 - 9V_3 = -166 \\ -11V_1 + 45V_2 - 23V_3 = 1963 \\ -9V_1 - 23V_2 + 56V_3 = -2568 \end{array}$$

Ans. $V_1 = -11$ V, $V_2 = 21$ V, $V_3 = -39$ V

5.39   What is the current source equivalent of a 12-V battery with a 0.5-Ω internal resistance?
Ans. $I = 24$ A, $R = 0.5$ Ω

5.40   What is the voltage-source equivalent of a 3-A current source in parallel with a 2-kΩ resistor?   *Ans.* $V = 6$ kV, $R = 2$ kΩ

5.41   Solve for the mesh currents in the circuit shown in Fig. 5-35.   *Ans.* $I_1 = 3$ A, $I_2 = -8$ A, $I_3 = 7$ A

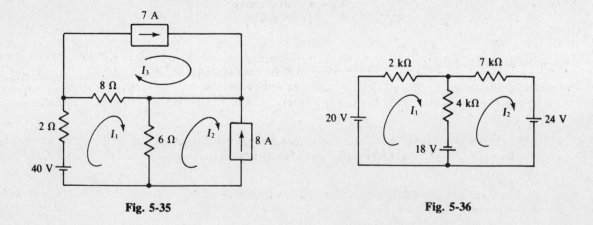

Fig. 5-35   **Fig. 5-36**

5.42   Solve for the mesh currents in the circuit shown in Fig. 5-36.   *Ans.* $I_1 = 5$ mA, $I_2 = -2$ mA

5.43   Repeat Prob. 5.42 with the 24-V source changed to −1 V.   *Ans.* $I_1 = 7$ mA, $I_2 = 1$ mA

5.44   Two 12-V batteries in parallel provide current to a light bulb that has a hot resistance of 0.5 Ω.   If the battery internal resistances are 0.1 and 0.2 Ω, find the power consumed by the light bulb.
Ans. 224 W

5.45   Find the mesh currents in the circuit shown in Fig. 5-37.
  Ans. $I_1 = -2$ mA, $I_2 = 6$ mA, $I_3 = 4$ mA

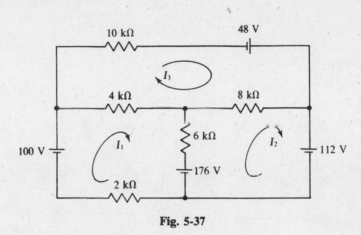

Fig. 5-37

5.46   Double the voltages of the voltage sources in the circuit shown in Fig. 5-37 and redetermine the mesh currents.   Compare them with the original mesh currents.
  Ans. $I_1 = -4$ mA, $I_2 = 12$ mA, $I_3 = 8$ mA,   double

5.47   Double the resistances of the resistors in the circuit shown in Fig. 5-37 and redetermine the mesh currents.   Compare them with the original mesh currents.
  Ans. $I_1 = -1$ mA, $I_2 = 3$ mA, $I_3 = 2$ mA,   half

5.48   Repeat Prob. 5.45 with the three voltage-source changes of 176 to 108 V, 112 to 110 V, and 48 to 66 V.   *Ans.* $I_1 = 3$ mA, $I_2 = 4$ mA, $I_3 = 5$ mA

5.49   For a certain three-mesh circuit, the self-resistances are 20, 25, and 32 Ω for meshes 1, 2, and 3, respectively.   The mutual resistances are 10 Ω for meshes 1 and 2, 12 Ω for meshes 2 and 3, and 6 Ω for meshes 1 and 3.   The aiding voltages from voltage sources are −74, 227, and −234 V for meshes 1, 2, and 3, respectively.   Find the mesh currents.   *Ans.* $I_1 = -3$ A, $I_2 = 5$ A, $I_3 = -6$ A

5.50   Repeat Prob. 5.49 for the same self-resistances and mutual resistances, but for aiding source voltages of 146, −273, and 182 V for meshes 1, 2, and 3, respectively.   *Ans.* $I_1 = 5$ A, $I_2 = -7$ A, $I_3 = 4$ A

5.51   Use loop analysis to find the current flowing down through the 6-Ω resistor in the circuit shown in Fig. 5-35.   *Ans.* 11 A

5.52   Use loop analysis to find the current flowing to the right through the 8-kΩ resistor in the circuit shown in Fig. 5-37.   *Ans.* 2 mA

5.53   Use loop analysis to find the current I in the circuit shown in Fig. 5-38.   *Ans.* 0.375 A

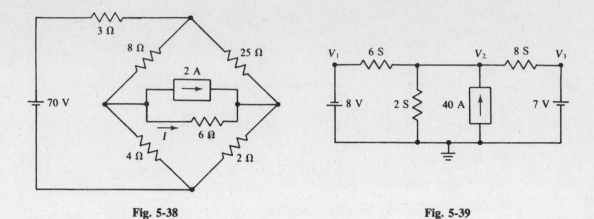

Fig. 5-38   **Fig. 5-39**

5.54   Solve for the node voltages in the circuit shown in Fig. 5-39.
          Ans. $V_1 = -8$ V, $V_2 = 3$ V, $V_3 = 7$ V

5.55   Solve for the node voltages in the circuit shown in Fig. 5-40.   *Ans.* $V_1 = 5$ V, $V_2 = -2$ V

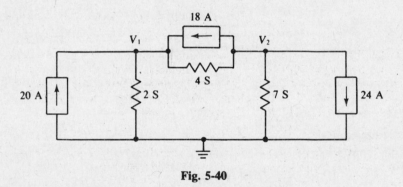

Fig. 5-40

5.56   Double the currents from the current sources in the circuit shown in Fig. 5-40 and redetermine the node
          voltages.   Compare them with the original node voltages.
          Ans. $V_1 = 10$ V, $V_2 = -4$ V,   double

5.57   Double the conductances of the resistors in the circuit shown in Fig. 5-40 and redetermine the node
          voltages.   Compare them with the original node voltages.   *Ans.* $V_1 = 2.5$ V, $V_2 = -1$ V,   half

5.58   Repeat Prob. 5.55 with the 24-A source changed to -1 A.   *Ans.* $V_1 = 7$ V, $V_2 = 1$ V

5.59   Find the voltages V_1, V_2, and V_3 in the circuit shown in Fig. 5-41.
          Ans. $V_1 = 5$ V, $V_2 = -2$ V, $V_3 = 3$ V

5.60   Find the node voltages in the circuit shown in Fig. 5-42.   *Ans.* $V_1 = -2$ V, $V_2 = 6$ V, $V_3 = 4$ V

5.61   Repeat Prob. 5.60 with the three current-source changes of 176 to 108 A, 112 to 110 A, and 48 to
          66 A.   *Ans.* $V_1 = 3$ V, $V_2 = 4$ V, $V_3 = 5$ V

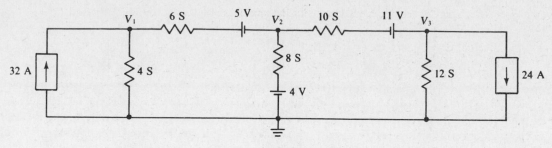

Fig. 5-41

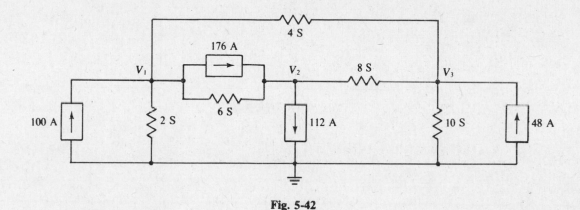

Fig. 5-42

5.62   For a certain four-node circuit, including a ground node, the self-conductances are 40, 50, and 64 S for nodes 1, 2, and 3, respectively. The mutual conductances are 20 S for nodes 1 and 2, 24 S for nodes 2 and 3, and 12 S for nodes 1 and 3. Currents flowing in current sources connected to these nodes are 74 A away from node 1, 227 A into node 2, and 234 A away from node 3. Find the node voltages.   *Ans.* $V_1 = -1.5$ V, $V_2 = 2.5$ V, $V_3 = -3$ V

5.63   Repeat Prob. 5.62 for the same self-conductances and mutual conductances, but for source currents of 292 A into node 1, 546 A away from node 2, and 364 A into node 3.
        Ans $V_1 = 5$ V, $V_2 = -7$ V, $V_3 = 4$ V

5.64   Find V for the circuit shown in Fig. 5-43.   *Ans.* -50 V

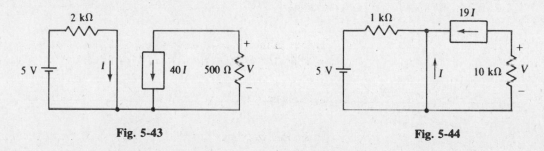

Fig. 5-43   **Fig. 5-44**

5.65   Find V for the circuit shown in Fig. 5-44.   *Ans.* 47.5 V

5.66   Find V for the circuit shown in Fig. 5-45.   *Ans.* -1000 V

5.67   Solve for V in the circuit shown in Fig. 5-46.   *Ans.* 180 V

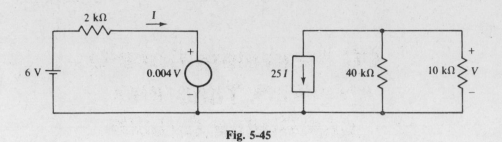

Fig. 5-45

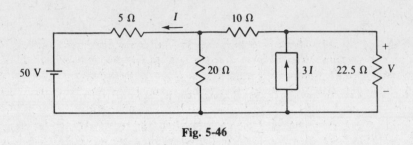

Fig. 5-46

5.68 In the circuit shown in Fig. 5-47, find V_{CE} if $I_C = 30I_B$ and $V_{BE} = 0.7$ V.   *Ans.* 3.68 V

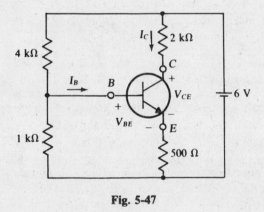

Fig. 5-47

5.69 Repeat Prob. 5.68 with the dc voltage source changed to 9 V and the collector resistor changed from 2 to 2.5 kΩ.   *Ans.* 2.89 V

Chapter 6

DC Equivalent Circuits, Network Theorems, and Bridge Circuits

INTRODUCTION

Network theorems are often important aids for network analyses. Some theorems apply only to linear, bilateral circuits, or portions of them. A *linear* electric circuit is constructed of linear electric elements as well as of independent sources. A linear electric element has an excitation-response relation such that doubling the excitation doubles the response, tripling the excitation triples the response, and so on. A *bilateral* circuit is constructed of bilateral elements as well as of independent sources. A bilateral element operates the same upon reversal of the excitation, except that the response also reverses. Resistors are both linear and bilateral if they have voltage-current relations that obey Ohm's law. On the other hand, a diode, which is a common electronic component, is neither linear nor bilateral.

Some of the theorems require "killing" of independent sources. The term "killing" refers to replacing all *independent* sources by their internal resistances. In other words, all ideal voltage sources are replaced by short circuits, and all ideal current sources by open circuits. Internal resistances are not affected, as neither are *dependent* sources.

THEVENIN'S AND NORTON'S THEOREMS

Thevenin's and *Norton's theorems* are probably the most important network theorems. For the application of either of them, a network is divided into two parts, A and B, as shown in Fig. 6-1a, with two joining wires. One part must be linear and bilateral, but the other part can be anything.

Thevenin's theorem specifies that the linear, bilateral part, say part A, can be replaced by a *Thevenin equivalent circuit* consisting of a voltage source and a resistor in series, as shown in Fig. 6-1b, without any changes in voltages or currents in part B. The voltage V_{Th} of the voltage source is called the *Thevenin voltage*, and the resistance R_{Th} of the resistor is called the *Thevenin resistance*.

As should be apparent from Fig. 6-1b, V_{Th} is the voltage across terminals a and b if part B is replaced by an open circuit. So, if the wires are cut at terminals a and b in either circuit shown in Fig. 6-1, and if a voltmeter is connected to measure the voltage across these terminals, the voltmeter reading is V_{Th}. This voltage is almost always different from the voltage across terminals a and b with part B connected. The Thevenin or open-circuit voltage V_{Th} is sometimes designated by V_{OC}.

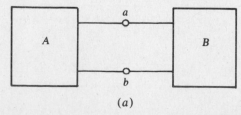
(a)

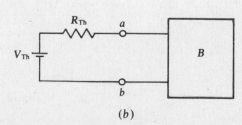
(b)

Fig. 6-1

79

With the joining wires cut, R_{Th} is the resistance of part A with all independent sources killed.   In other words, if all independent sources in part A are replaced by their internal resistances, an ohmmeter connected to terminals a and b reads Thevenin's resistance.   For some applications in which part B is just a load, R_{Th} is called the *output resistance* of part A.

Thevenin's theorem guarantees only that the voltages and currents in part B do not change when part A is replaced by its Thevenin equivalent circuit.   The voltages and currents in the Thevenin circuit itself are almost always different from those in the original part A, except at terminals a and b where they are the same, of course.

Although R_{Th} is most often determined by finding the resistance at terminals a and b with the connecting wires cut and the independent sources killed, it can also be found from the current I_{SC} that flows in a short circuit placed across terminals a and b, as shown in Fig. 6-2a.   As is apparent from Fig. 6-2b, this short-circuit current from terminal a to b is related to the Thevenin voltage and resistance.   Specifically,

$$R_{Th} = \frac{V_{Th}}{I_{SC}}$$

So, R_{Th} is equal to the ratio of the open-circuit voltage at terminals a and b, and the short-circuit current between them.   With this approach to determining R_{Th}, no sources are killed.   Of course, the short circuit is placed across terminals a and b of the original circuit A, and not across terminals a and b of the Thevenin equivalent circuit.

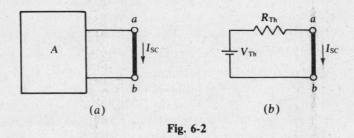

Fig. 6-2

The *Norton equivalent circuit* can be derived by applying a source conversion to the Thevenin equivalent circuit, as illustrated in Fig. 6-3a.   The Norton equivalent circuit is sometimes illustrated as in Fig. 6-3b, in which $I_N = V_{Th}/R_{Th}$ and $R_N = R_{Th}$.   Notice that, if a short circuit is placed across terminals a and b in the circuit shown in Fig. 6-3b, the short-circuit current I_{SC} from terminal a to b is equal to the Norton current I_N.   Often in circuit diagrams, the notation I_{SC} is used for the source current instead of I_N.   Also, often R_{Th} is used for the resistance instead of R_N.

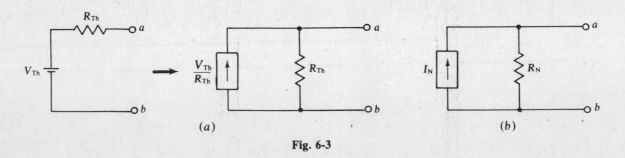

Fig. 6-3

MAXIMUM POWER TRANSFER THEOREM

The *maximum power transfer theorem* specifies that a resistive load receives maximum power from a linear, bilateral dc circuit if the load resistance equals the Thevenin resistance of the circuit as "seen" by the load.   The proof is based on calculus.   Selecting the load resistance to be equal to the circuit Thevenin resistance is called *matching* the resistances.   With matching, the load voltage is $V_{Th}/2$, and so the power consumed by the load is $(V_{Th}/2)^2/R_{Th} = V_{Th}^2/4R_{Th}$.

SUPERPOSITION THEOREM

The *superposition theorem* specifies that, in a linear circuit containing several independent sources, the current or voltage of a circuit element equals the *algebraic sum* of the component voltages or currents produced by the independent sources acting alone.   Put another way, the voltage or current contribution from each independent source can be found separately, and then all the contributions algebraically added to get the actual voltage or current with all independent sources in the circuit.

This theorem applies only to independent sources—not to dependent ones.   Also, it applies only to finding voltages and currents.   In particular, it cannot be used to find power in dc circuits.   Additionally, the theorem applies to each independent source acting alone, which means that the other independent sources must be killed.   In practice, though, it is not essential that the independent sources be considered one at a time; any number can be considered simultaneously.

Because applying the superposition theorem requires several analyses, more work may be done than with a single mesh, loop, or nodal analysis with all sources present.

MILLMAN'S THEOREM

Millman's theorem is a method for reducing a circuit by combining parallel voltage sources into a single voltage source.   It is just a special case of the application of Thevenin's theorem.

Figure 6-4 illustrates the theorem for only three parallel voltage sources, but the theorem applies to any number of such sources.   The derivation of Millman's theorem is simple.   If the voltage sources shown in Fig. 6-4a are converted to current sources (Fig. 6-4b) and the currents added, and if the conductances are added, the result is a single current source of $G_1V_1 + G_2V_2 + G_3V_3$ in parallel with a resistor having a conductance of $G_1 + G_2 + G_3$ (Fig. 6-4c).   Then, the conversion of this current source to a voltage source gives the final result indicated in Fig. 6-4d.   In general, for N parallel voltage sources the Millman voltage source has a voltage of

$$V_M = \frac{G_1V_1 + G_2V_2 + \cdots + G_NV_N}{G_1 + G_2 + \cdots + G_N}$$

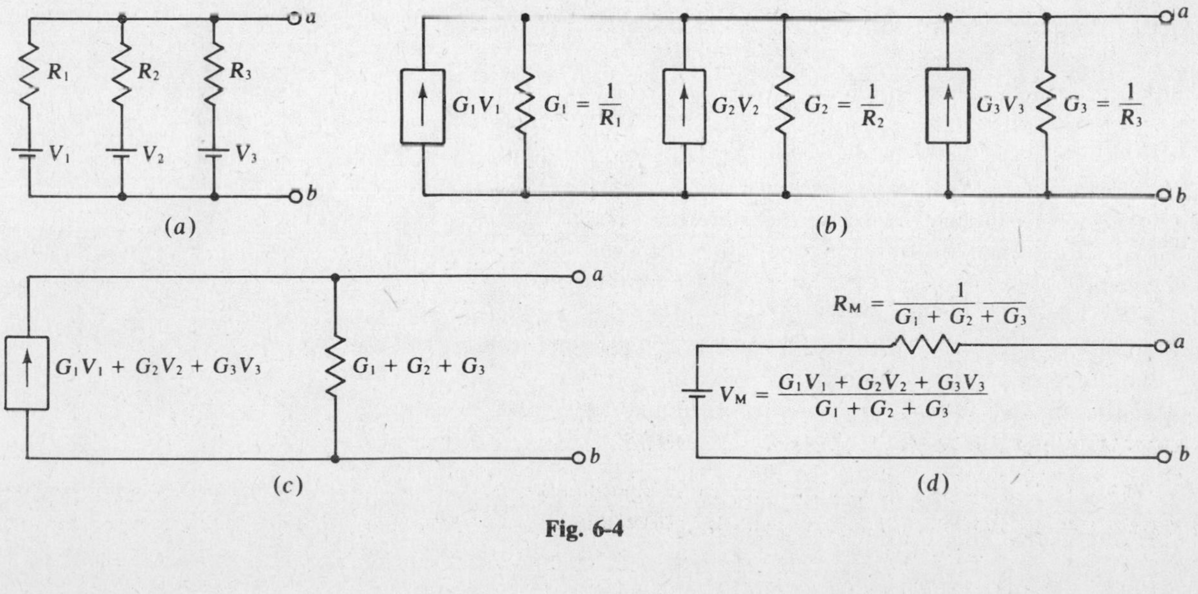

Fig. 6-4

and the Millman series resistor has a resistance of

$$R_M = \frac{1}{G_1 + G_2 + \cdots + G_N}$$

Note from the voltage source formula that, if all the sources have the same voltage, this is also the Millman source voltage.

Y-Δ AND Δ-Y CONVERSIONS

Figure 6-5a shows a Y (wye) resistor circuit and Fig. 6-5b a Δ (delta) resistor circuit. There are other names. If the Y circuit is drawn in the shape of a T, it is also called a T (tee) circuit. And, if the Δ circuit is drawn in the shape of a Π, it is also called a Π (pi) circuit.

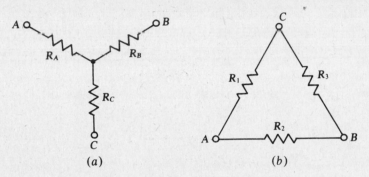

(a)   (b)

Fig. 6-5

Conversion is possible from a Y to an equivalent Δ and also from a Δ to an equivalent Y. The corresponding circuits are equivalent only for voltages and currents *external* to the Y and Δ circuits. Internally, the voltages and currents are different.

Conversion formulas can be found from equating resistances between two lines to a Δ and a Y when the third line to each is open. This equating is done three times, with a different line open each time. Some algebraic manipulation of the results produces the following Δ-to-Y conversion formulas:

$$R_A = \frac{R_1 R_2}{R_1 + R_2 + R_3} \qquad R_B = \frac{R_2 R_3}{R_1 + R_2 + R_3} \qquad R_C = \frac{R_1 R_3}{R_1 + R_2 + R_3}$$

Also produced are the following Y-to-Δ conversion formulas:

$$R_1 = \frac{R_A R_B + R_A R_C + R_B R_C}{R_B} \qquad R_2 = \frac{R_A R_B + R_A R_C + R_B R_C}{R_C} \qquad R_3 = \frac{R_A R_B + R_A R_C + R_B R_C}{R_A}$$

Notice in the Δ-to-Y conversion formulas that the denominators are the same: $R_1 + R_2 + R_3$, the sum of the Δ resistances. In the Y-to-Δ conversion formulas, the numerators are the same: $R_A R_B + R_A R_C + R_B R_C$, the sum of the different products of the Y resistances taken two at a time.

Drawing the Y inside the Δ, as in Fig. 6-6, is a good aid for remembering the numerators of the Δ-to-Y conversion formulas and the denominators of the Y-to-Δ conversion formulas. For each Y resistor in the Δ-to-Y conversion formulas, the two resistances in each numerator product are those of the two Δ resistors adjacent to the Y resistor being found. In the Y-to-Δ conversion formulas, the single Y resistance in each denominator is that of the Y resistor opposite the Δ resistor being found.

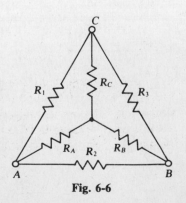

Fig. 6-6

If it happens that each Y resistor has the same value R, then each resistance of the corresponding Δ is $3R$, as the formulas give.   And if each Δ resistance is R, each resistance of the corresponding Y is $R/3$.

BRIDGE CIRCUITS

As illustrated in Fig. 6-7a, a *bridge* resistor circuit has two joined Δ's or, depending on the point of view, two joined Y's with a shared branch.   Although the circuit usually appears in this form, the forms shown in Fig. 6-7b and c are also common.   The circuit illustrated in Fig. 6-7c is often called a *lattice*.   If a Δ part of a bridge is converted to a Y, or a Y part converted to a Δ, the circuit becomes series-parallel.   Then the resistances can be easily combined, and the circuit reduced.

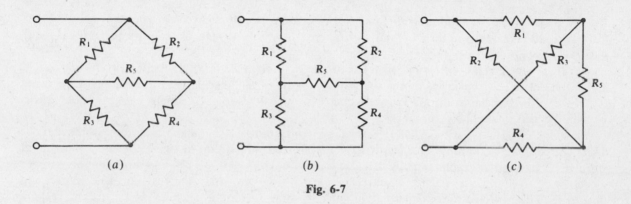

| (a) | (b) | (c) |

Fig. 6-7

A bridge circuit can be used for precision resistance measurements.   A *Wheatstone bridge* has a center branch that is a sensitive current indicator such as a galvanometer, as shown in Fig. 6-8.   Three of the other branches are precision resistors, one of which is variable as indicated.   The fourth branch is the resistor with the unknown resistance R_X that is to be measured.

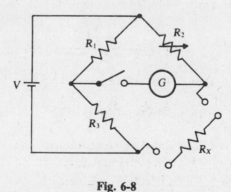

Fig. 6-8

For a resistance measurement, the resistance R_2 of the variable resistor is adjusted until the galvanometer needle does not deflect when the switch in the center branch is closed.   This lack of deflection is the result of zero voltage across the galvanometer, and this means that, even with the switch open, the voltage across R_1 equals that across R_2, and the voltage across R_3 equals that across R_X.   In this condition the bridge is said to be *balanced*.   By voltage division,

$$\frac{R_1 V}{R_1 + R_3} = \frac{R_2 V}{R_2 + R_X} \quad \text{and} \quad \frac{R_3 V}{R_1 + R_3} = \frac{R_X V}{R_2 + R_X}$$

Taking the ratio of the two equations produces the *bridge balance equation*:

$$R_X = \frac{R_2 R_3}{R_1}$$

Presumably, R_1 and R_3 are known standard resistances and a dial connected to R_2 gives this resistance so that R_X can be solved for. Of course, a commercial Wheatstone bridge has dials that directly indicate R_X upon balance.

A good way to remember the bridge balance equation is to equate products of the resistances of opposite branch arms: $R_1 R_X = R_2 R_3$. Another way is to equate the ratio of the top and bottom resistances of one side to that of the other: $R_1/R_3 = R_2/R_X$.

INPUT AND OUTPUT RESISTANCES

The resistance at the input terminals of an electronic circuit is the *input resistance* R_{in}. For a purely resistive circuit, the terms "total resistance," "equivalent resistance," and just "resistance" are far more popular.

An electronic circuit with a load also has an *output resistance* R_{out}. If the load is disconnected, and if the source at the input of the electronic circuit is replaced by its internal resistance, the output resistance of the electronic circuit is the resistance at its load terminals. As mentioned, it is the same as the Thevenin resistance.

Since electronic circuits usually have dependent sources, the input and output resistances cannot be found by combining resistances. For an electronic circuit having either no independent sources or having all such sources killed, a general approach for finding an input or output resistance is to apply a 1-A current source to the terminals at which the resistance is desired and then to find the voltage across these terminals by some technique such as nodal, mesh, or loop analysis. The desired resistance is equal to this voltage divided by the applied current. Since the applied current is 1 A, the value of the resistance equals the value of the voltage. Alternatively, a voltage source can be applied and the input current found. The desired resistance is equal to the applied voltage divided by the input current. Actually, it does not matter whether a voltage or current source is applied, or what the value of the applied voltage or current is. Applying a 1-A current source is, however, usually the simplest approach.

This method for finding input and output resistances is also convenient for finding the total resistance of a purely resistive circuit in which the resistors are not in a series-parallel configuration.

Solved Problems

6.1 A car battery has an open-circuit terminal voltage of 12.6 V. The terminal voltage drops to 10.8 V when the battery supplies 40 A to a starter motor. What is the Thevenin equivalent circuit for this battery?

The Thevenin voltage is the 12.6-V open-circuit voltage ($V_{Th} = 12.6$ V). The voltage drop when the battery supplies 40 A is the same drop that would occur across the Thevenin resistor in the Thevenin equivalent circuit because this resistor is in series with the Thevenin voltage source. From this drop,

$$R_{Th} = \frac{12.6 - 10.8}{40} = 0.045 \ \Omega$$

6.2 Find the Thevenin equivalent circuit for a dc power supply that has a 30-V terminal voltage when delivering 400 mA and a 27-V terminal voltage when delivering 600 mA.

For the Thevenin equivalent circuit, the terminal voltage is the Thevenin voltage minus the drop across the Thevenin resistor.   Consequently, from the two specified conditions of operation,

$$V_{Th} - (400 \times 10^{-3})R_{Th} = 30$$
$$V_{Th} - (600 \times 10^{-3})R_{Th} = 27$$

Subtracting,

$$-(400 \times 10^{-3})R_{Th} + (600 \times 10^{-3})R_{Th} = 30 - 27$$

From which

$$R_{Th} = \frac{3}{200 \times 10^{-3}} = 15 \ \Omega$$

This value of R_{Th} substituted into the first equation gives

$$V_{Th} - (400 \times 10^{-3})(15) = 30 \qquad \text{which solves to} \qquad V_{Th} = 36 \text{ V}$$

6.3   Find the Thevenin equivalent circuit for a battery box containing four batteries with their positive terminals connected together and their negative terminals connected together.   The open-circuit voltages and internal resistances of the batteries are 12.2 V and 0.5 Ω, 12.1 V and 0.1 Ω, 12.4 V and 0.16 Ω, and 12.4 V and 0.2 Ω.

The first step is to convert each voltage source to a current source.   The result is four ideal current sources and four resistors, all in parallel.   The next step is to add the currents from the current sources and also to add the conductances of the resistors, the effect of which is to combine the current sources into a single current source and the resistors into a single resistor.   The final step is to convert this source and resistor to a voltage source in series with a resistor to get the Thevenin equivalent circuit.

The currents of the equivalent sources are

$$\frac{12.2}{0.5} = 24.4 \text{ A} \qquad \frac{12.1}{0.1} = 121 \text{ A} \qquad \frac{12.4}{0.16} = 77.5 \text{ A} \qquad \frac{12.4}{0.2} = 62 \text{ A}$$

which add to

$$24.4 + 121 + 77.5 + 62 = 284.9 \text{ A}$$

The conductances add to

$$\frac{1}{0.5} + \frac{1}{0.1} + \frac{1}{0.16} + \frac{1}{0.2} = 23.25 \text{ S}$$

From this current and conductance, the Thevenin voltage and resistance are

$$V_{Th} = \frac{I}{G} = \frac{284.9}{23.25} = 12.3 \text{ V} \qquad R_{Th} = \frac{1}{23.25} = 0.043 \ \Omega$$

6.4   Find the Norton equivalent circuit for the power supply of Prob. 6.2 if the terminal voltage is 28 V instead of 27 V when the power supply delivers 600 mA.

For the Norton equivalent circuit, the load current is the Norton current minus the loss of current through the Norton resistor.   Consequently, from the two specified conditions of operation,

$$I_N - \frac{30}{R_N} = 400 \times 10^{-3}$$

$$I_N - \frac{28}{R_N} = 600 \times 10^{-3}$$

Subtracting,

$$-\frac{30}{R_N} + \frac{28}{R_N} = 400 \times 10^{-3} - 600 \times 10^{-3}$$

or

$$-\frac{2}{R_N} = -200 \times 10^{-3} \quad \text{from which} \quad R_N = \frac{2}{200 \times 10^{-3}} = 10 \ \Omega$$

Substituting this into the first equation gives

$$I_N - \frac{30}{10} = 400 \times 10^{-3} \quad \text{which solves to} \quad I_N = 3.4 \ \text{A}$$

6.5   What resistor draws a current of 5 A when connected across terminals a and b of the circuit shown in Fig. 6-9?

A good approach is to use Thevenin's theorem to simplify the circuit to the Thevenin equivalent of a V_{Th} voltage source in series with an R_{Th} resistor.   Then the load resistor R is in series with these, and Ohm's law can be used to find R:

$$5 = \frac{V_{Th}}{R_{Th} + R} \quad \text{from which} \quad R = \frac{V_{Th}}{5} - R_{Th}$$

The open-circuit voltage at terminals a and b is the voltage across the 20-Ω resistor since there is 0 V across the 6-Ω resistor because no current flows through it.   By voltage division this voltage is

$$V_{Th} = \frac{20}{20 + 5} \times 100 = 80 \ \text{V}$$

R_{Th} is the resistance at terminals a and b with the 100-V source replaced by a short circuit.   This short circuit places the 5- and 20-Ω resistors in parallel for a net resistance of $5 \| 20 = 4 \ \Omega$.   So, $R_{Th} = 6 + 4 = 10 \ \Omega$.

With V_{Th} and R_{Th} known, the load resistance R for a 5-A current can be found from the previously derived equation:

$$R = \frac{V_{Th}}{5} - R_{Th} = \frac{80}{5} - 10 = 6 \ \Omega$$

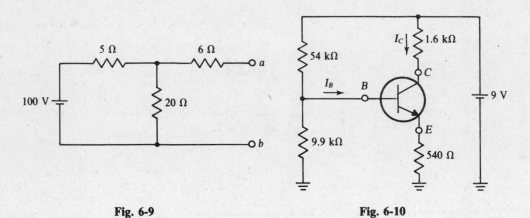

Fig. 6-9   **Fig. 6-10**

6.6   In the circuit shown in Fig. 6-10, find the base current I_B if $I_C = 30I_B$.   The base current is provided by a bias circuit consisting of 54- and 9.9-kΩ resistors and a 9-V source.   There is a 0.7-V drop from base to emitter.

One way to find the base current is to break the circuit at the base lead and determine the Thevenin equivalent of the bias circuit.   For this approach it helps to consider the 9-V source to be two 9-V sources, one of which is connected to the 1.6-kΩ collector resistor and the other of which is connected to the 54-kΩ bias resistor.   Then the bias circuit appears as illustrated in Fig. 6-11a.   From it, the voltage V_{Th} is, by voltage division,

$$V_{Th} = \frac{9.9}{9.9 + 54} \times 9 = 1.394 \ \text{V}$$

Replacing the 9-V source by a short circuit places the 54- and 9.9-kΩ resistors in parallel for an R_{Th} of

$$R_{Th} = \frac{9.9 \times 54}{9.9 + 54} = 8.37 \text{ k}\Omega$$

and the circuit simplifies to that shown in Fig. 6-11b.

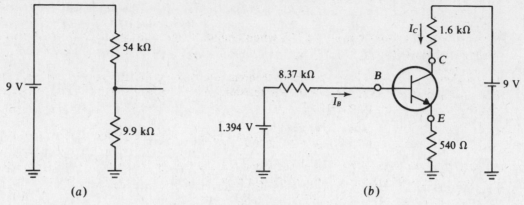

(a)   *(b)*

Fig. 6-11

From KVL applied to the base loop, and from the fact that $I_C + I_B = 31I_B$ flows through the 540-Ω emitter resistor,

$$1.394 = 8.37I_B + 0.7 + 0.54 \times 31I_B$$

from which

$$I_B = \frac{0.694}{25.1} = 0.0277 \text{ mA} = 27.7 \ \mu\text{A}$$

Of course, the simplifying kilohm-milliampere method was used in some of the calculations.

6.7   Find the Thevenin equivalent circuit at terminals a and b of the circuit with transistor model shown in Fig. 6-12.

The open-circuit voltage is $500 \times 30I_B = 15\,000I_B$,   positive at terminal b.   From the base circuit, $I_B = 10/1000 \text{ A} = 10 \text{ mA}$.   Substituting in for I_B gives

$$V_{Th} = 15\,000(10 \times 10^{-3}) = 150 \text{ V}$$

The best way to find R_{Th} is to kill the independent 10-V source and determine the resistance at terminals a and b.   With this source killed, $I_B = 0$ A,   and so $30I_B = 0$ A,   which means that the dependent current source acts like an open circuit—it produces zero current regardless of the voltage across it.   The result is that the resistance at terminals a and b is just the shown 500 Ω.

The Thevenin equivalent circuit is a 500-Ω resistor in series with a 150-V source that has its positive terminal toward terminal b, as shown in Fig. 6-13.

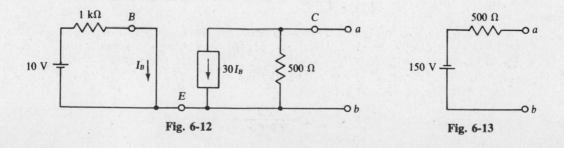

Fig. 6-12   **Fig. 6-13**

6.8   What is the Norton equivalent circuit for the transistor circuit shown in Fig. 6-14?

A good approach is to first find I_{SC}, which is the Norton current I_N; next find V_{OC}, which is the Thevenin voltage V_{Th}; and then take their ratio to get the Norton resistance R_N, which is the same as R_{Th}.

Placing a short circuit across terminals a and b makes $V_C = 0$ V, which in turn causes the dependent voltage source in the base circuit to be a short circuit. As a result, $I_B = 6/2000$ A = 3 mA. This short circuit also places 0 V across the 40-kΩ resistor, preventing any current flow through it. So, all the $25I_B = 25 \times 3 = 75$-mA current from the dependent current source flows through the short circuit in a direction from terminal b to terminal a: $I_{SC} = I_N = 75$ mA.

The open-circuit voltage is more difficult to find. From the collector circuit, $V_C = (-25I_B)(40\,000) = -10^6 I_B$. This substituted into the KVL equation for the base circuit produces an equation in which I_B is the only unknown:

$$6 = 2000I_B + 0.0004V_C = 2000I_B + 0.0004(-10^6 I_B) = 1600I_B$$

So, $I_B = 6/1600$ A $= 3.75$ mA, and $V_C = -10^6 I_B = -10^6(3.75 \times 10^{-3}) = -3750$ V. The result is that $V_{OC} = 3750$ V, positive at terminal b.

In the calculation of R_N, signs are important when, as here, a circuit has dependent sources that can cause R_N to be negative. From Fig. 6-2b, $R_{Th} = R_N$ is the ratio of the open-circuit voltage referenced positive at terminal a and the short-circuit current referenced from terminal a to terminal b. Alternatively, both references can be reversed, which is convenient here. So,

$$R_N = \frac{V_{OC}}{I_{SC}} = \frac{3750}{75 \times 10^{-3}} \Omega = 50 \text{ k}\Omega$$

The Norton equivalent circuit is a 50-kΩ resistor in parallel with a 75-mA current source that is directed toward terminal b, as shown in Fig. 6-15.

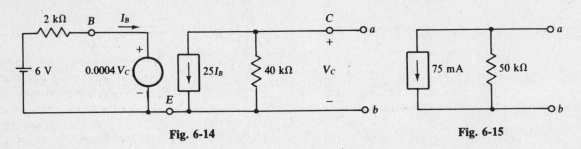

Fig. 6-14   **Fig. 6-15**

6.9   Find the Thevenin equivalent of the circuit shown in Fig. 6-16.

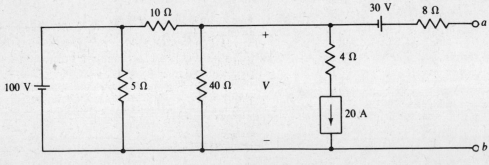

Fig. 6-16

The Thevenin or open-circuit voltage, positive at terminal a, is the indicated V plus the 30 V of the 30-V source. The 8-Ω resistor has no effect on this voltage because there is zero current flow through it as a result of the open circuit. With zero current there is zero voltage. V can be found from a single nodal equation:

$$\frac{V - 100}{10} + \frac{V}{40} + 20 = 0$$

Multiplying by 40 and simplifying produces

$$5V = 400 - 800 \qquad \text{from which } V = -80 \text{ V}$$

So, $V_{Th} = -80 + 30 = -50$ V.   Notice that the 5- and 4-Ω resistors have no effect on V_{Th}.

Figure 6-17a shows the circuit with the voltage sources replaced by short circuits and the current source by an open circuit.   Notice that the 5-Ω resistor has no effect on R_{Th} because it is shorted, and neither does the 4-Ω resistor because it is in series with an open circuit.   Since the resistor arrangement in Fig. 6-17a is series-parallel, R_{Th} is easy to calculate by combining resistances: $R_{Th} = 8 + 40\|10 = 16 \ \Omega$.

Figure 6-17b shows the Thevenin equivalent circuit.

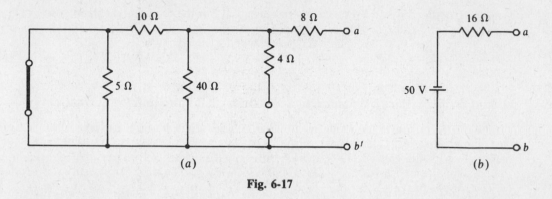

(a)   (b)

Fig. 6-17

The fact that neither the 5- nor the 4-Ω resistor has an effect on V_{Th} and R_{Th} leads to the generalization that resistors in parallel with ideal voltage sources, and resistors in series with ideal current sources, have no effect on voltages and currents elsewhere in a circuit.

6.10 What is the maximum power that can be drawn from a 12-V battery that has an internal resistance of 0.25 Ω?

A resistive load of 0.25 Ω draws maximum power because it has the same resistance as the Thevenin or internal resistance of the source.   For this load, half the source voltage drops across the load, making the power $6^2/0.25 = 144$ W.

6.11 What is the maximum power that can be drawn by a resistor connected to terminals a and b of the circuit shown in Fig. 6-14?

In the solution to Prob. 6.8, the Thevenin resistance of the circuit shown in Fig. 6-14 was found to be 50 kΩ and the Norton current was found to be 75 mA.   So, a load resistor of 50 kΩ absorbs maximum power.   By current division, half the Norton current flows through it, producing a power of

$$\left(\frac{75}{2} \times 10^{-3}\right)^2 (50 \times 10^3) = 70.3 \text{ W}$$

6.12 Use superposition to find the power absorbed by the 12-Ω resistor in the circuit shown in Fig. 6-18.

Superposition cannot be used to find power in a dc circuit because the method applies only to linear quantities—power has a squared voltage or current relation instead of a linear one.   To illustrate, the current through the 12-Ω resistor from the 100-V source is, with the 6-A source replaced by an open circuit, $100/(12 + 6) = 5.556$ A.   The corresponding power is $5.556^2 \times 12 = 370$ W.   With the voltage source replaced by a short circuit, the current through the 12-Ω resistor from the 6-A current source is, by current division, $[6/(12 + 6)](6) = 2$ A.   The corresponding power is $2^2 \times 12 = 48$ W.   So, if superposition could be applied to power, the result would be $370 + 48 = 418$ W   for the power dissipated in the 12-Ω resistor.

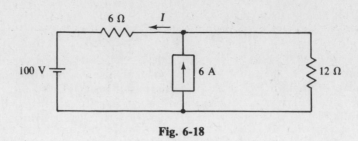

Fig. 6-18

Superposition does, however, apply to currents. So, the total current through the 12-Ω resistor is 5.556 + 2 = 7.556 A, and the power consumed is $7.556^2 \times 12 = 685$ W, which is much different than the 418 W found by erroneously applying superposition to power.

6.13 In the circuit shown in Fig. 6-18, change the 100-V source to a 360-V source, and the 6-A current source to an 18-A source, and use superposition to find the current I.

Figure 6-19a shows the circuit with the current source replaced by an open circuit. Obviously, the component I_V of I from the voltage source is $I_V = -360/(6 + 12) = -20$ A. Figure 6-19b shows the circuit with the voltage source replaced by a short circuit. By current division, I_C, the current-source component of I, is $I_C = [12/(12 + 6)](18) = 12$ A. The total current is the algebraic sum of the current components: $I = I_V + I_C = -20 + 12 = -8$ A.

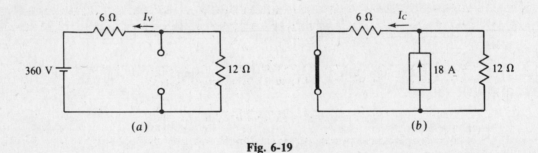

(a)   (b)

Fig. 6-19

6.14 For the circuit shown in Fig. 6-16, use superposition to find V_{Th} referenced positive on terminal a.

Clearly, the 30-V source contributes 30 V to V_{Th} because this source, being in series with an open circuit, cannot cause any currents to flow. Zero currents mean zero resistor voltage drops, and so the only voltage in the circuit is that of the source.

Figure 6-20a shows the circuit with all independent sources killed except the 100-V source. Notice that the voltage across the 40-Ω resistor appears across terminals a and b because there is a zero voltage drop across the 8-Ω resistor. By voltage division this component of V_{Th} is

$$V_{Thv} = \frac{40}{40 + 10} \times 100 = 80 \text{ V}$$

Figure 6-20b shows the circuit with the current source as the only independent source. The voltage across the 40-Ω resistor is the open-circuit voltage since there is a zero voltage drop across the 8-Ω resistor. Note that the short circuit replacing the 100-V source prevents the 5-Ω resistor from having an effect, and also it places the 40- and 10-Ω resistors in parallel for a net resistance of 40∥10 = 8 Ω. So, the component of V_{Th} from the current source is $V_{ThC} = -20 \times 8 = -160$ V.

V_{Th} is the algebraic sum of the three components of voltage:

$$V_{Th} = 30 + 80 - 160 = -50 \text{ V}$$

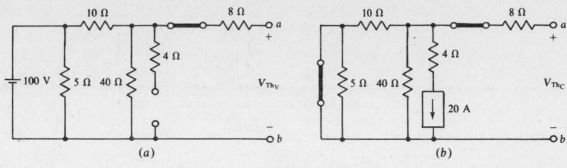

Fig. 6-20

Notice that finding V_{Th} by superposition requires more work than finding it by nodal analysis, as was done in the solution to Prob. 6.9.

6.15 Use superposition to find V_{Th} for the circuit shown in Fig. 6-14.

Although this circuit has three sources, superposition cannot be used since two of the sources are dependent. Only one source is independent. The superposition theorem does not apply to dependent sources.

6.16 Use Millman's theorem to find the current flowing to a $0.2\text{-}\Omega$ resistor from four batteries operating in parallel. Each battery has a 12.8-V open-circuit voltage. The internal resistances are 0.1, 0.12, 0.2, and 0.25 Ω.

Because the battery voltages are the same, being 12.8 V, the Millman voltage is $V_M = 12.8$ V. The Millman resistance is the inverse of the sum of the conductances:

$$R_M = \frac{1}{1/0.1 + 1/0.12 + 1/0.2 + 1/0.25}\,\Omega = 36.6\text{ m}\Omega$$

Of course, the resistor current equals the Millman voltage divided by the sum of the Millman and load resistances:

$$I = \frac{V_M}{R_M + R} = \frac{12.8}{0.2 + 0.0366} = 54.1\text{ A}$$

6.17 Use Millman's theorem to find the current drawn by a $5\text{-}\Omega$ resistor from four batteries operating in parallel. The battery open-circuit voltages and internal resistances are 18 V and 1 Ω, 20 V and 2 Ω, 22 V and 5 Ω, and 24 V and 4 Ω.

The Millman voltage and resistance are

$$V_M = \frac{(1)(18) + (1/2)(20) + (1/5)(22) + (1/4)(24)}{1 + 1/2 + 1/5 + 1/4} = 19.7\text{ V}$$

$$R_M = \frac{1}{1 + 1/2 + 1/5 + 1/4} = 0.513\ \Omega$$

The current is, of course, the Millman voltage divided by the sum of the Millman and load resistances:

$$I = \frac{V_M}{R_M + R} = \frac{19.7}{0.513 + 5} = 3.57\text{ A}$$

6.18 Use Millman's theorem to find I for the circuit shown in Fig. 6-21.

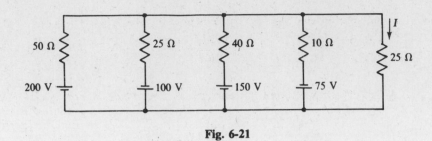

Fig. 6-21

The Millman voltage and resistance are

$$V_M = \frac{(1/50)(200) + (1/25)(-100) + (1/40)(150) + (1/10)(-75)}{1/50 + 1/25 + 1/40 + 1/10} = -20.27 \text{ V}$$

$$R_M = \frac{1}{1/50 + 1/25 + 1/40 + 1/10} = 5.41 \text{ } \Omega$$

And so

$$I = \frac{V_M}{R_M + R} = \frac{-20.27}{5.41 + 25} = -0.667 \text{ A}$$

6.19 Convert the Δ shown in Fig. 6-22a to the Y shown in Fig. 6-22b for (a) $R_1 = R_2 = R_3 = 36 \text{ } \Omega$, and (b) $R_1 = 20 \text{ } \Omega$, $R_2 = 30 \text{ } \Omega$, and $R_3 = 50 \text{ } \Omega$.

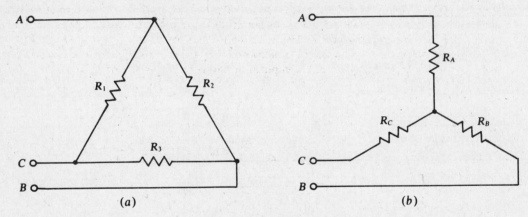

Fig. 6-22

(a) For Δ resistances of the same value, $R_Y = R_\Delta/3$. So, here, $R_A = R_B = R_C = 36/3 = 12 \text{ } \Omega$.

(b) The denominators of the R_Y formulas are the same: $R_1 + R_2 + R_3 = 20 + 30 + 50 = 100 \text{ } \Omega$. The numerators are products of the adjacent resistor resistances if the Y is placed inside the Δ:

$$R_A = \frac{R_1 R_2}{100} = \frac{20 \times 30}{100} = 6 \text{ } \Omega \qquad R_B = \frac{R_2 R_3}{100} = \frac{30 \times 50}{100} = 15 \text{ } \Omega \qquad R_C = \frac{R_1 R_3}{100} = \frac{20 \times 50}{100} = 10 \text{ } \Omega$$

6.20 Convert the Y shown in Fig. 6-22b to the Δ shown in Fig. 6-22a for (a) $R_A = R_B = R_C = 5 \text{ } \Omega$, and (b) $R_A = 10 \text{ } \Omega$, $R_B = 5 \text{ } \Omega$, $R_C = 20 \text{ } \Omega$.

(a) For Y resistances of the same value, $R_\Delta = 3R_Y$. So, here, $R_1 = R_2 = R_3 = 3 \times 5 = 15 \text{ } \Omega$.

(b) The numerators of the R_Δ formulas are the same: $R_A R_B + R_A R_C + R_B R_C = 10 \times 5 + 10 \times 20 + 5 \times 20 = 350$. The denominators of the R_Δ formulas are the resistances of the Y arms opposite the Δ arms if the Y is placed inside the Δ.   Thus,

$$R_1 = \frac{350}{R_B} = \frac{350}{5} = 70\ \Omega \qquad R_2 = \frac{350}{R_C} = \frac{350}{20} = 17.5\ \Omega \qquad R_3 = \frac{350}{R_A} = \frac{350}{10} = 35\ \Omega$$

6.21   Use a Δ-to-Y conversion to find the currents I_1, I_2, and I_3 for the circuit shown in Fig. 6-23.

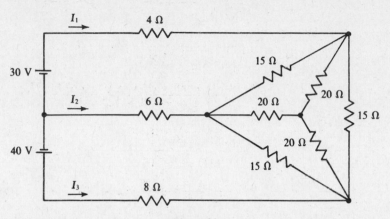

Fig. 6-23

The Δ of 15-Ω resistors converts to a Y of $15/3 = 5$-Ω resistors that are in parallel with the Y of 20-Ω resistors. It is not obvious that they are in parallel, and in fact they would not be if the resistances for each Y were not all the same value. When, as here, they are the same value, an analysis would show that the middle nodes are at the same potential, just as if a wire were connected between them. So, corresponding resistors of the two Y's are in parallel, as shown in Fig. 6-24a. The two Y's can be reduced to the single Y shown in Fig. 6-24b, in which each Y resistance is $5\|20 = 4\ \Omega$. With this Y replacing the Δ-Y combination, the circuit is as shown in Fig. 6-24c.

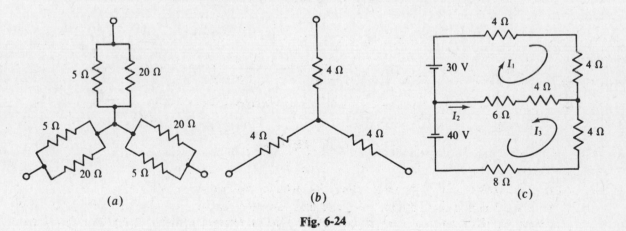

(a) (b) (c)

Fig. 6-24

With the consideration of I_1 and I_3 as loop currents, the corresponding KVL equations are

$$30 = 18I_1 + 10I_3$$
$$40 = 10I_1 + 22I_3$$

which solve to $I_1 = 0.88\ \text{A}$ and $I_3 = 1.42\ \text{A}$.   Then, from KCL applied at the right-hand node, $I_2 = -I_1 - I_3 = -2.3\ \text{A}$.

6.22   Using a Y-to-Δ conversion, find the total resistance R_T of the circuit shown in Fig. 6-25, which has a bridged-T attenuator.

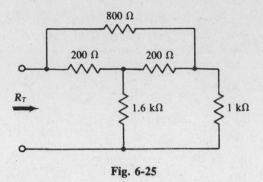

Fig. 6-25

Figure 6-26a shows the T part of the circuit inside a Δ as an aid in finding the Δ resistances.   From the Δ-to-Y conversion formulas,

$$R_1 = R_2 = \frac{200(200) + 200(1600) + 200(1600)}{200} = \frac{680\,000}{200}\,\Omega = 3.4\text{ k}\Omega$$

$$R_2 = \frac{680\,000}{1600} = 425\ \Omega$$

As a result of this conversion, the circuit becomes series-parallel as in Fig. 6-26b, and the total resistance is easy to find:

$$R_T = 3400\|(800\|425 + 3400\|1000) = 3400\|1050 = 802\ \Omega$$

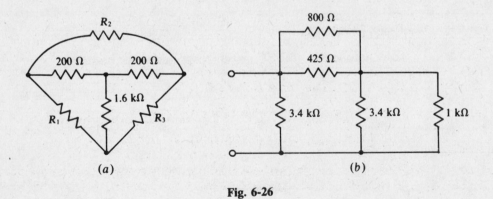

Fig. 6-26

6.23   Find *I* for the circuit shown in Fig. 6-27 by using a Δ-Y conversion.

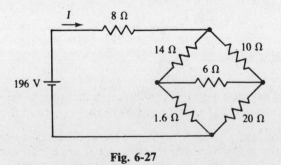

Fig. 6-27

The bridge simplifies to a series-parallel configuration from a conversion of either the top or bottom Δ to a Y, or the left- or right-hand Y to a Δ.   Perhaps the most common approach is to convert one of the Δ's to a Y, although the work required is about the same for any type of conversion used.   Figure 6-28a shows the top Δ enclosing a Y as a memory aid for the conversion of this Δ to a Y.   All three Y formulas have the same denominator: $14 + 10 + 6 = 30$.   The numerators, though, are the products of the resistances of the adjacent Δ resistors:

$$R_A = \frac{10 \times 14}{30} = 4.67 \ \Omega \qquad R_B = \frac{14 \times 6}{30} = 2.8 \ \Omega \qquad R_C = \frac{6 \times 10}{30} = 2 \ \Omega$$

With this conversion the circuit simplifies to that shown in Fig. 6-28b in which all the resistors are in series-parallel.   From it,

$$I = \frac{196}{8 + 4.67 + (2.8 + 1.6)\|(2 + 20)} = 12 \text{ A}$$

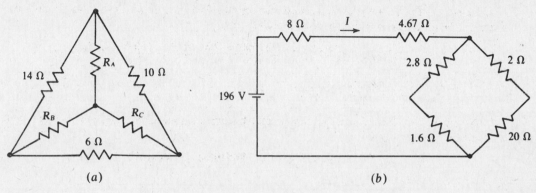

(a)   (b)

Fig. 6-28

6.24   In the circuit shown in Fig. 6-27, what resistor R replacing the 20-Ω resistor causes the bridge to be balanced?   Also, what is I then?

For balance, the product of the resistances of opposite bridge arms are equal:

$$R \times 14 = 1.6 \times 10 \qquad \text{from which} \qquad R = \frac{16}{14} = 1.14 \ \Omega$$

With the bridge in balance, the center arm can be considered as an open circuit because it carries no current.   This being the case, and because the bridge is a series-parallel arrangement, the current I is

$$I = \frac{196}{8 + (14 + 1.6)\|(10 + 1.14)} = 13.5 \text{ A}$$

Alternatively, the center arm can be considered to be a short circuit because both ends of it are at the same potential.   From this point of view,

$$I = \frac{196}{8 + 14\|10 + 1.6\|1.14} = 13.5 \text{ A}$$

which is the same, of course.

6.25   The slide-wire bridge shown in Fig. 6-29 has a uniform resistance wire that is 1 m long.   If balance occurs with the slider at 24 cm from the top, what is the resistance of R_X?

Let R_W be the total resistance of the resistance wire.   Then the resistance from the top of the wire to the slider is $(24/100)R_W = 0.24R_W$.   That from the slider to the bottom of the wire is $(76/100)R_W = 0.76R_W$.   So, the bridge resistances are $0.24R_W$, $0.76R_W$, $30 \ \Omega$, and R_X.   These inserted into the bridge balance equation give

$$R_X = \frac{0.76R_W}{0.24R_W} \times 30 = 95 \ \Omega$$

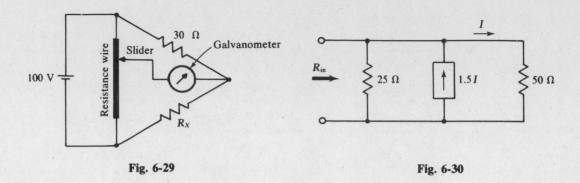

Fig. 6-29   **Fig. 6-30**

6.26   Find the input resistance R_{in} of the circuit shown in Fig. 6-30.

Since this circuit has a dependent source but no independent sources, the approach to finding the input resistance is to apply a source at the input. Then the input resistance is equal to the input voltage divided by the input current. Probably the best source to apply is a 1-A current, as shown in Fig. 6-31.

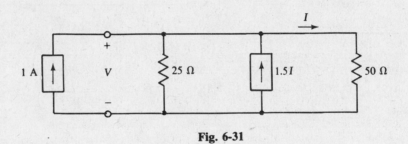

Fig. 6-31

By nodal analysis,

$$\frac{V}{25} - 1.5I + \frac{V}{50} = 1$$

But from the right-hand branch, $I = V/50$. With this substitution the equation becomes

$$\frac{V}{25} - 1.5\frac{V}{50} + \frac{V}{50} = 1$$

which solves to $V = 33.3$ V. The input resistance is

$$R_{in} = \frac{V}{1} = \frac{33.3}{1} = 33.3 \ \Omega$$

Of course, for an input current of 1 A, the input resistance and voltage have the same numerical value, although different units.

6.27   Find the input resistance of the circuit shown in Fig. 6-30 if the dependent current source has a current of $5I$ instead of $1.5I$.

For a 1-A current source applied at the input terminals, the nodal equation at the top node is

$$\frac{V}{25} - 5I + \frac{V}{50} = 1$$

But, from the right-hand branch, $I = V/50$.   With this substitution the equation is

$$\frac{V}{25} - 5\frac{V}{50} + \frac{V}{50} = 1$$

from which $V = -25$ V.   The input resistance is $R_{in} = -25/1 = -25\ \Omega$.

A negative resistance may be somewhat disturbing to the mind when first encountered, but it is physically real even though it takes a transistor circuit, an operational amplifier, or the like to get it.   Physically, a negative input resistance means that the circuit supplies power to whatever source is applied at the input, with the dependent source being the source of power.

6.28   Directly find the output resistance of the circuit shown in Fig. 6-14.

Figure 6-32 shows the circuit with the 6-V independent source killed and a 1-A current source applied at the output a and b terminals.   From Ohm's law applied to the base circuit,

$$I_B = -\frac{0.0004V_C}{2000} = -2 \times 10^{-7}V_C$$

Nodal analysis applied to the top node of the collector circuit gives

$$\frac{V_C}{40\,000} + 25I_B = 1 \quad \text{or} \quad \frac{V_C}{40\,000} + 25(-2 \times 10^{-7}V_C) = 1$$

upon substitution for I_B.   This equation solves to $V_C = 50\,000$ V, and so $R_{out} = R_{Th} = 50\,k\Omega$. This checks with the $R_N = R_{Th}$ answer from the Prob. 6.8 solution in which the $R_N = R_{Th} = V_{OC}/I_{SC}$ approach was used.

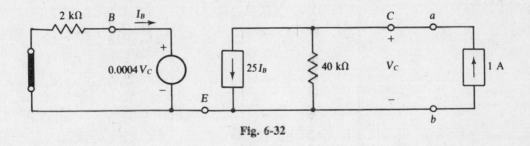

Fig. 6-32

6.29   Figure 6-33 shows an emitter-follower circuit for getting a low output resistance for resistance matching.   Find R_{out}.

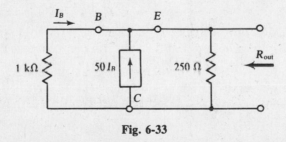

Fig. 6-33

Because the circuit has a dependent source but no independent sources, R_{out} must be found by applying a source at the output terminals, preferably a 1-A current source as shown in Fig. 6-34.

From KCL applied at the top node,

$$\frac{V}{1000} - 50I_B + \frac{V}{250} = 1$$

But from Ohm's law applied to the 1-kΩ resistor, $I_B = -V/1000$. With this substitution the equation becomes

$$\frac{V}{1000} - 50\left(-\frac{V}{1000}\right) + \frac{V}{250} = 1$$

from which $V = 18.2$ V. Since $R_{out} = V/1 = 18.2 \; \Omega$, the output resistance is much smaller than the resistance of either resistor in the circuit.

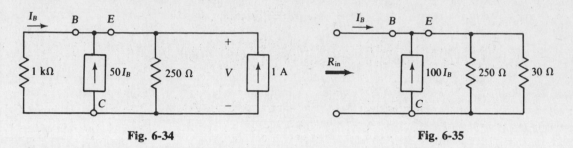

Fig. 6-34   Fig. 6-35

6.30   Figure 6-35 shows an emitter-follower circuit for getting a large input resistance for resistance matching. The load is a 30-Ω resistor, as shown. Find the input resistance R_{in}.

Because the circuit has a dependent source and no independent sources, the way to find R_{in} is, preferably, from the input voltage when a 1-A current source is applied, as shown in Fig. 6-36. Here, $I_B = 1$ A, and so the total current to the parallel resistors is $I_B + 100I_B = 101I_B = 101$ A, and the voltage V is

$$V = 101(250\|30) \; V = 2.7 \; kV$$

The input resistance is $R_{in} = V/1 = 2.7 \; k\Omega$, which is much greater than the 30 Ω of the load.

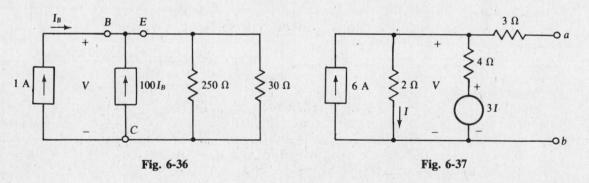

Fig. 6-36   Fig. 6-37

6.31   Use two methods to find the resistance of the resistor that will consume maximum power when connected to terminals a and b of the circuit shown in Fig. 6-37.

The desired resistance is, of course, the Thevenin resistance. One way to find it is from the quotient of the open-circuit voltage and the short-circuit current. The open-circuit voltage is the voltage V across the three parallel branches since there is a 0-V drop across the 3-Ω resistor. By KCL,

$$\frac{V}{2} + \frac{V - 3I}{4} = 6$$

From the 2-Ω branch, $I = V/2$. With this substitution the equation becomes

$$\frac{V}{2} + \frac{V - 1.5V}{4} = 6 \qquad \text{from which} \qquad V = 16 \; V$$

One way of obtaining the short-circuit current is from V when there is a short circuit across terminals a and b. Then, the short-circuit current, which also flows through the 3-Ω resistor, is

$V/3$. With a short circuit between terminals a and b, the KCL equation differs from the open-circuit KCL equation only by the addition of a $V/3$ term to account for the current that flows through the 3-Ω resistor:

$$\frac{V}{2} + \frac{V - 1.5V}{4} + \frac{V}{3} = 6 \qquad \text{from which} \qquad V = 8.47 \text{ V}$$

So, the short-circuit current is $I_{SC} = V/3 = 8.47/3 = 2.82$ A, and

$$R_{Th} = \frac{V_{OC}}{I_{SC}} = \frac{16}{2.82} = 5.67 \ \Omega$$

The other way of finding R_{Th} is from the voltage-current ratio of a source, preferably a 1-A current source, applied at terminals a and b with the 6-A current source killed. Figure 6-38 shows this circuit.

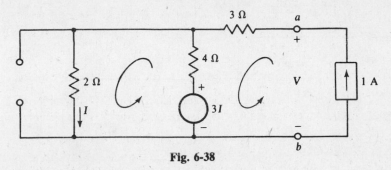

Fig. 6-38

KVL applied to the I loop results in

$$(2 + 4)I - 3I = 4 \times 1 \qquad \text{from which} \qquad I = 1.33 \text{ A}$$

The source voltage equals the drop across the 3-Ω resistor plus the drop across the 2-Ω resistor:

$$V = (3 \times 1) + (2 \times 1.33) = 5.67 \text{ V}$$

So, $R_{Th} = V/1 = 5.67/1 = 5.67 \ \Omega$, the same as for the V_{OC}/I_{SC} approach.

Supplementary Problems

6.32   A car battery has a 12.1-V terminal voltage when supplying 10 A to the car lights. When the starter motor is turned over, the extra 40 A drawn drops the battery terminal voltage to 11.1 V. What is the Thevenin equivalent circuit of this battery?   *Ans.* $R_{Th} = 25$ mΩ, $V_{Th} = 12.35$ V

6.33   In full sunlight a 2- by 2-cm solar cell has a short-circuit current of 80 mA, and the current is 75 mA for a terminal voltage of 0.6 V. What is the Norton equivalent circuit?
Ans. $R_N = 120 \ \Omega$, $I_N = 80$ mA

6.34   Find the Thevenin equivalent of the circuit shown in Fig. 6-39. Reference V_{Th} positive toward terminal a.   *Ans.* $R_{Th} = 12 \ \Omega$, $V_{Th} = 12$ V

6.35   In the circuit shown in Fig. 6-39, change the 5-A current source to a 7-A current source, the 12-Ω resistor to an 18-Ω resistor, and the 48-V source to a 96-V source. Then find the Norton equivalent circuit with the current arrow directed toward terminal a.   *Ans.* $R_N = 12.5 \ \Omega$, $I_N = 3.24$ A

6.36   Find the Thevenin equivalent of the grounded-base transistor circuit shown in Fig. 6-40. Reference V_{Th} positive toward terminal a.   *Ans.* $R_{Th} = 4$ kΩ, $V_{Th} = 3.9$ V

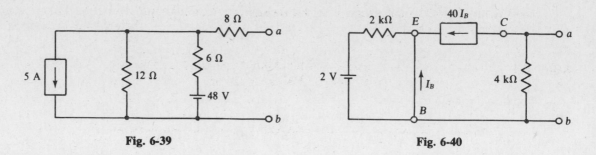

Fig. 6-39   Fig. 6-40

6.37   In the transistor circuit shown in Fig. 6-41, find the base current I_B if $I_C = 40I_B$.   There is a 0.7-V drop from base to emitter.   *Ans.*   90.1 μA

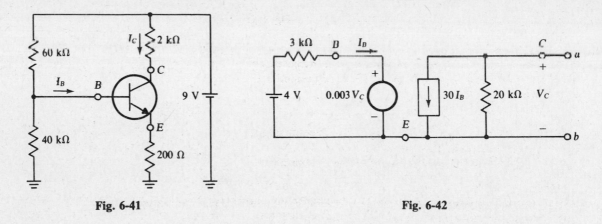

Fig. 6-41   Fig. 6-42

6.38   Find the Thevenin equivalent of the transistor circuit shown in Fig. 6-42.   Reference V_{Th} positive toward terminal a.   *Ans.* $R_{Th} = 50$ kΩ, $V_{Th} = -2000$ V

6.39   A standard method of analyzing a circuit containing a nonlinear element, as in the circuit shown in Fig. 6-43, is to find the Thevenin equivalent of the linear part of the circuit at the terminals of the nonlinear element.   This simplifies the circuit to three elements in series.   Then some numerical scheme is applied to the simplified circuit.   Find I in the circuit shown in Fig. 6-43, which contains a nonlinear element having a V-I relation of $V = 3I^2$.   Use Thevenin's theorem and the quadratic formula.   *Ans.* $I = 2$ A

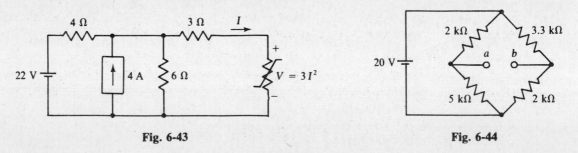

Fig. 6-43   Fig. 6-44

6.40   What resistor connected between terminals a and b in the bridge circuit shown in Fig. 6-44 absorbs maximum power and what is this power?   *Ans.* 2.67 kΩ, 4.25 mW

6.41   What will be the reading of a zero-resistance ammeter connected across terminals a and b of the bridge circuit shown in Fig. 6-44?   Assume that the ammeter is connected to have an upscale

reading.   What will be the reading if a 1-kΩ resistor is in series with the ammeter?
Ans. 2.52 mA, 1.83 mA

6.42   For the circuit shown in Fig. 6-45, find the Norton equivalent with I_N referenced positive toward terminal *a*.   *Ans.* $R_N = 4 \ \Omega$, $I_N = -3$ A

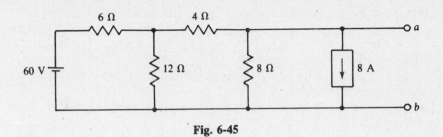

Fig. 6-45

6.43   Some solar cells are interconnected for increased power output.   Each has the specifications given in Prob. 6.33.   What area of solar cells is required for a power output of 1 W?   Assume a matching load.   *Ans.* 20.8 cm^2

6.44   For the circuit shown in Fig. 6-39, use superposition to find the contribution of each source to V_{Th} if it is referenced positive toward terminal *a*.
Ans. 32 V from the 48-V source, -20 V from the 5-A source

6.45   In the circuit shown in Fig. 6-43, replace the nonlinear resistor with an open circuit and use superposition to find the contribution of each source to the open-circuit voltage referenced positive at the top.   *Ans.* 13.2 V from the 22-V source, 9.6 V from the 4-A source

6.46   For the circuit shown in Fig. 6-45, use superposition to find the contribution of each source to the current in a short circuit connected between terminals *a* and *b*.   The short-circuit current reference is from terminal *a* to terminal *b*.   *Ans.* 5 A from the 60-V source, -8 A from the 8-A source

6.47   An automobile generator operating in parallel with a battery energizes a 0.8-Ω load.   The open-circuit voltages and internal resistances are 14.8 V and 0.4 Ω for the generator, and 12.8 V and 0.5 Ω for the battery.   Use Millman's theorem to find the load current.   *Ans.* 13.6 A

6.48   For the automobile circuit of Prob. 6.47 use superposition to find the load current contribution from each source.   *Ans.* 8.04 A from the generator, 5.57 A from the battery

6.49   Convert the Δ shown in Fig. 6-46*a* to the Y in Fig. 6-46*b* for $R_1 = 2$ kΩ, $R_2 = 4$ kΩ,   and $R_3 = 6$ kΩ.   *Ans.* $R_A = 667 \ \Omega$, $R_B = 2$ kΩ, $R_C = 1$ kΩ

6.50   Repeat Prob. 6.49 for $R_1 = 8 \ \Omega$, $R_2 = 5 \ \Omega$,   and $R_3 = 7 \ \Omega$.
Ans. $R_A = 2 \ \Omega$, $R_B = 1.75 \ \Omega$, $R_C = 2.8 \ \Omega$

6.51   Convert the Y shown in Fig. 6-46*b* to the Δ in Fig. 6-46*a* for $R_A = 12 \ \Omega$, $R_B = 15 \ \Omega$,   and $R_C = 18 \ \Omega$.   *Ans.* $R_1 = 44.4 \ \Omega$, $R_2 = 37 \ \Omega$, $R_3 = 55.5 \ \Omega$

6.52   Repeat Prob. 6.51 for $R_A = 10$ kΩ, $R_B = 18$ kΩ,   and $R_C = 12$ kΩ.
Ans. $R_1 = 28.7$ kΩ, $R_2 = 43$ kΩ, $R_3 = 51.6$ kΩ

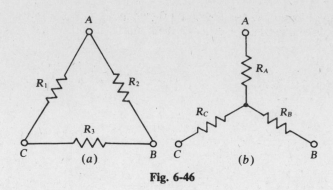

Fig. 6-46

6.53   For the lattice circuit shown in Fig. 6-47, use a Δ-Y conversion to find the V that makes I = 3 A.   *Ans.* 177 V

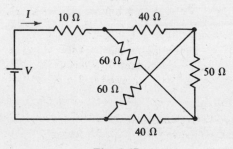

Fig. 6-47

6.54   Use a Δ-Y conversion to find the currents in the circuit shown in Fig. 6-48.
Ans. $I_1 = 7.72$ A, $I_2 = -0.36$ A, $I_3 = -7.36$ A

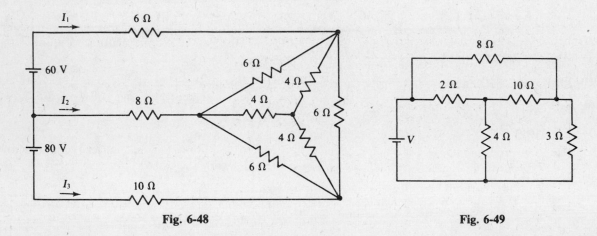

Fig. 6-48   **Fig. 6-49**

6.55   Use a Δ-to-Y conversion in finding the voltage V that causes 2 A to flow down through the 3-Ω resistor in the circuit shown in Fig. 6-49.   *Ans.* 17.8 V

6.56   In the lattice circuit shown in Fig. 6-47, what resistor substituted for the top 40-Ω resistor causes zero current flow in the 50-Ω resistor?   *Ans.* 90 Ω

6.57   If in the slide-wire bridge shown in Fig. 6-29, balance occurs with the slider at 67 cm from the top, what is the resistance R_x ?   *Ans.* 14.8 Ω

6.58   Use a Δ-Y conversion to find I in the circuit shown in Fig. 6-50.   Remember that, for a Δ-Y conversion, only the voltages and currents external to the Δ and Y do not change.   *Ans.* 0.334 A

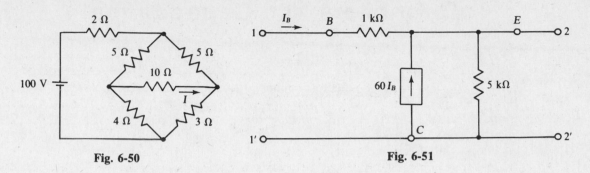

Fig. 6-50   **Fig. 6-51**

6.59   Find the input resistance at terminals 1 and 1' of the transistor circuit shown in Fig. 6-51 if a 2-kΩ resistor is connected across terminals 2 and 2'.   *Ans.* 88.1 kΩ

6.60   Find the output resistance at terminals 2 and 2' of the transistor circuit shown in Fig. 6-51 if a source with a 1-kΩ internal resistance is connected to terminals 1 and 1'.   Remember in finding the output resistance to replace the source by its internal resistance.   *Ans.* 32.6 Ω

6.61   Find the input resistance at terminals 1 and 1' of the transistor circuit shown in Fig. 6-52 if a 5-kΩ load resistor is connected between terminals 2 and 2', from collector to emitter.   *Ans.* 760 Ω

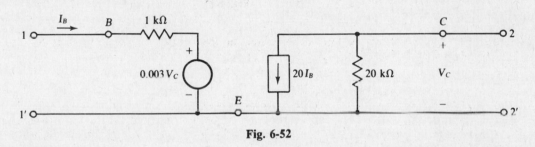

Fig. 6-52

6.62   Find the output resistance at terminals 2 and 2' of the transistor circuit shown in Fig. 6-52 if a source with a 500-Ω internal resistance is connected to terminals 1 and 1'.   *Ans.* 100 kΩ

Chapter 7

Capacitors and Capacitance

INTRODUCTION

A *capacitor* consists of two conductors separated by an insulator. The chief feature of a capacitor is its ability to store electric charge, with negative charge on one of its two conductors and positive charge on the other. Accompanying this charge is energy, which a capacitor can release. Figure 7-1 shows the circuit symbol for a capacitor.

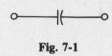

Fig. 7-1

CAPACITANCE

Capacitance, the electrical property of capacitors, is a measure of the ability of a capacitor to store charge on its two conductors. Specifically, if the potential difference between the two conductors is V volts when there is a positive charge of Q coulombs on one conductor and a negative charge of the same amount on the other, the capacitor has a capacitance of

$$C = \frac{Q}{V}$$

where C is the quantity symbol of capacitance.

The SI unit of capacitance is the *farad*, with symbol F. Unfortunately, the farad is much too large a unit for practical applications. The microfarad (μF) and picofarad (pF) are much more common.

CAPACITOR CONSTRUCTION

One common type of capacitor is the parallel-plate capacitor of Fig. 7-2a. This capacitor has two spaced conducting plates that can be rectangular, as shown, but that often are circular. The insulator between the plates is called a *dielectric*. The dielectric is air in Fig. 7-2a, and is a slab of solid insulator in Fig. 7-2b.

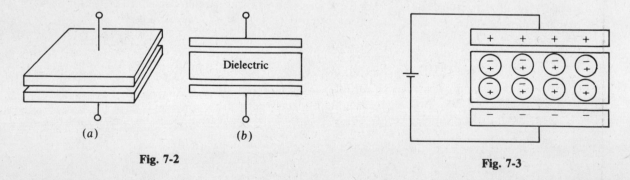

Dielectric

(a)   (b)

Fig. 7-2

Fig. 7-3

A voltage source connected to a capacitor, as shown in Fig. 7-3, causes the capacitor to become charged. Electrons from the top plate are attracted to the positive terminal of the source, and they pass through the source to the negative terminal where they are repelled to the bottom plate. Because each electron lost by the top plate is gained by the bottom plate, the magnitude of charge Q is the same on both plates. Of course, the voltage across the capacitor from this charge exactly equals the source voltage.   The voltage source did work on the electrons in moving them to the bottom plate, which work becomes energy stored in the capacitor.

For the parallel-plate capacitor, the capacitance in farads is

$$C = \epsilon \frac{A}{d}$$

where A is the area of either plate in square meters, d is the separation in meters, and ϵ is the *permittivity* in farads per meter (F/m) of the dielectric.   The larger the plate area or the smaller the plate separation, or the greater the dielectric permittivity, the greater the capacitance.

The permittivity ϵ relates to atomic effects in the dielectric.   As shown in Fig. 7-3, the charges on the capacitor plates distort the dielectric atoms, with the result that there is a net negative charge on the top dielectric surface and a net positive charge on the bottom dielectric surface.   This dielectric charge partially neutralizes the effects of the stored charge to permit an increase in charge for the same voltage.

The permittivity of vacuum, designated by ϵ_0, is 8.85 pF/m.   Permittivities of other dielectrics are related to that of vacuum by a factor called the *dielectric constant* or *relative permittivity*, designated by ϵ_r. The relation is $\epsilon = \epsilon_r \epsilon_0$.   The dielectric constants of some common dielectrics are 1.0006 for air, 2.5 for paraffined paper, 5 for mica, 7.5 for glass, and 7500 for ceramic.

TOTAL CAPACITANCE

The total or equivalent capacitance (C_T or C_{eq}) of parallel capacitors, as seen in Fig. 7-4a, can be found from the total stored charge and the $Q = CV$ formula.   The total stored charge Q_T equals the sum of the individual stored charges: $Q_T = Q_1 + Q_2 + Q_3$.   With the substitution of the appropriate $Q = CV$ for each Q, this equation becomes $C_T V = C_1 V + C_2 V + C_3 V$. Upon division by V, it reduces to $C_T = C_1 + C_2 + C_3$.   Because the number of capacitors is not significant in this derivation, this result can be generalized to any number of parallel capacitors:

$$C_T = C_1 + C_2 + C_3 + C_4 + \cdots$$

So, the total or equivalent capacitance of parallel capacitors is the sum of the individual capacitances.

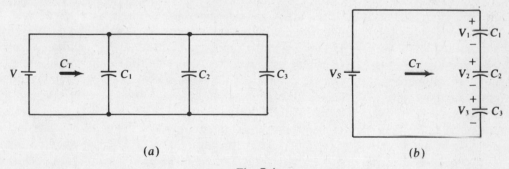

(a)   (b)

Fig. 7-4

For series capacitors, as shown in Fig. 7-4b, the formula for the total capacitance is derived by substituting Q/C for each V in the KVL equation.   The Q in each term is the same.   This is because the charge gained by a plate of any capacitor must have come from a plate of an adjacent capacitor.   The KVL equation for the circuit shown in Fig. 7-4b is $V_S = V_1 + V_2 + V_3$.   With the substitution of the appropriate Q/C for each V, this equation becomes

$$\frac{Q}{C_T} = \frac{Q}{C_1} + \frac{Q}{C_2} + \frac{Q}{C_3} \quad \text{or} \quad \frac{1}{C_T} = \frac{1}{C_1} + \frac{1}{C_2} + \frac{1}{C_3}$$

upon division by Q.   This can also be written as

$$C_T = \frac{1}{1/C_1 + 1/C_2 + 1/C_3}$$

Generalizing,

$$C_T = \frac{1}{1/C_1 + 1/C_2 + 1/C_3 + 1/C_4 + \cdots}$$

which specifies that the total capacitance of series capacitors equals the reciprocal of the sum of the reciprocals of the individual capacitances.   Notice that the total capacitance of series capacitors is found in the same way as the total resistance of parallel resistors.

For the special case of N series capacitors having the same capacitance C, this formula simplifies to $C_T = C/N$.   And for two capacitors in series it is $C_T = C_1 C_2/(C_1 + C_2)$.

ENERGY STORAGE

As can be shown using calculus, the energy stored in a capacitor is

$$W_C = \tfrac{1}{2}CV^2$$

where W_C is in joules, C is in farads, and V is in volts.   Notice that this stored energy does not depend on the capacitor current.

TIME-VARYING VOLTAGES AND CURRENTS

In dc resistor circuits, the currents and voltages are constant—never varying.   Even if switches are included, a switching operation can, at most, cause a voltage or current to jump from one constant level to another.   (The term "jump" means a change from one value to another in zero time.)   When capacitors are included, though, almost never does a voltage or a current jump from one constant level to another when switches open or close.   Some voltages or currents may initially jump at switching, but the jumps are almost never to final values.   Instead, they are to values from which the voltages or currents change *exponentially* to their final values.   These voltages and currents vary with time—they are *time-varying*.

Quantity symbols for time-varying quantities are distinguished from those for constant quantities by the use of lowercase letters instead of uppercase letters.   For example, v and i are the quantity symbols for time-varying voltages and currents.   Sometimes, the lowercase t, for time, is shown as an argument with lowercase quantity symbols as in $v(t)$ and $i(t)$.   Numerical values of v and i are called *instantaneous values*, or *instantaneous voltages* and *currents*, because these values depend on (vary with) exact instants of time.

As explained in Chap. 2, a constant current is the quotient of the charge Q passing a point in a wire and the time T required for this charge to pass: $I = Q/T$.   The specific time T is not important because the charge in a resistive dc circuit flows at a steady rate.   This means that doubling the time T doubles the charge Q, tripling the time triples the charge, and so on, keeping I the same.

For a time-varying current, though, the value of i usually changes from instant to instant.   So, finding the current at any particular time requires using a very short time interval Δt.   If Δq is the small charge that flows during this time interval, then the current is approximately $\Delta q/\Delta t$.   For an exact value of current, this quotient must be found in the limit as Δt approaches zero ($\Delta t \to 0$):

$$i = \lim_{\Delta t \to 0} \frac{\Delta q}{\Delta t} = \frac{dq}{dt}$$

This limit, designated by dq/dt, is called the *derivative* of charge with respect to time.

CAPACITOR CURRENT

An equation for capacitor current can be found by substituting $q = Cv$ into $i = dq/dt$:

$$i = \frac{dq}{dt} = \frac{d}{dt}(Cv)$$

But C is a constant, and a constant can be factored from a derivative. The result is

$$i = C\frac{dv}{dt}$$

with associated references assumed. If the references are not associated, a negative sign must be included. This equation specifies that the capacitor current at any time equals the product of the capacitance and the time rate of change of voltage at that time. But the current does *not* depend on the value of voltage at that time.

If a capacitor voltage is constant, then the voltage is not changing and so dv/dt is zero, making the capacitor current zero. Of course, from physical considerations, if a capacitor voltage is constant, no charge can be entering or leaving the capacitor, which means that the capacitor current is zero. With a voltage across it and zero current flow through it, the capacitor acts like an open circuit: *a capacitor is an open circuit to dc.* Remember, though, it is only after a capacitor voltage becomes constant that the capacitor acts like an open circuit. Capacitors are often used in electronic circuits to block dc currents and voltages.

Another important fact from $i = C\,dv/dt$ or $i \simeq C\,\Delta v/\Delta t$ is that *a capacitor voltage cannot jump*. If, for example, a capacitor voltage could jump from 3 to 5 V or, in other words, change by 2 V in zero time, then Δv would be 2 and Δt would be 0, with the result that the capacitor current would be infinite. An infinite current is impossible because no source can deliver this current. Further, such a current flowing through a resistor would produce an infinite power loss, and there are no sources of infinite power and no resistors that can absorb such power. Capacitor current has no similar restriction. It can jump or even change directions, instantaneously. Capacitor voltage not jumping means that capacitor voltages right after a switching operation are the same as right before the operation. This is an important fact for resistor-capacitor (RC) circuit analysis.

SINGLE-CAPACITOR DC-EXCITED CIRCUITS

When switches open or close in a dc RC circuit with a single capacitor, all voltages and currents that change do so exponentially from their initial values to their final constant values, as can be shown from differential equations. The exponential terms in a voltage or current expression are called *transient terms* because they eventually become zero in practical circuits.

Figure 7-5 shows these exponential changes for a switching operation at $t = 0$ s. In Fig. 7-5a the initial value is greater than the final value, and in Fig. 7-5b the final value is greater. Although both initial and final values are shown as positive, both can be negative or one can be positive and the other negative.

The voltages and currents approach their final values asymptotically, graphically speaking, which means that they never actually reach them. As a practical matter, however, after five time constants (described next) they are close enough to their final values to be considered to be at them.

Time constant, with symbol τ, is a measure of the time required for certain changes in voltages and currents. For a single-capacitor RC circuit, the time constant of the circuit is the product of the capacitance and the Thevenin's resistance as "seen" by the capacitor:

$$RC \text{ time constant} = \tau = R_{\text{Th}}C$$

The expressions for the voltages and currents shown in Fig. 7-5 are

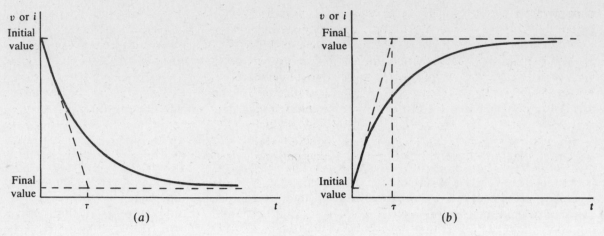

Fig. 7-5

$$v(t) = v(\infty) + [v(0+) - v(\infty)]e^{-t/\tau} \text{ V}$$

$$i(t) = i(\infty) + [i(0+) - i(\infty)]e^{-t/\tau} \text{ A}$$

for all time greater than zero ($t > 0$ s). In these equations, $v(0+)$ and $i(0+)$ are initial values immediately after switching; $v(\infty)$ and $i(\infty)$ are final values; $e = 2.718$, the base of natural logarithms; and τ is the time constant of the circuit of interest. These equations apply to all voltages and currents in a linear, RC, single-capacitor circuit having either only dc independent sources or no independent sources.

By letting $t = \tau$ in these equations, it is easy to see that, in a time equal to one time constant, the voltages and currents change by 63.2 percent of their total change of $v(\infty) - v(0+)$ or $i(\infty) - i(0+)$. And by letting $t = 5\tau$, it is easy to see that, after five time constants, the voltages and currents change by 99.3 percent of their total change, and so can be considered to be at their final values for most practical purposes.

RC TIMERS AND OSCILLATORS

An important use for capacitors is in circuits for measuring time—*timers*. A simple timer consists of a switch, capacitor, resistor, and dc voltage source, all in series. At the beginning of a time interval to be measured, the switch is closed to cause the capacitor to start charging. At the end of the time interval, the switch is opened to stop the charging and "trap" the capacitor charge. The corresponding capacitor voltage is a measure of the time interval. A voltmeter connected across the capacitor can have a scale calibrated in time to give a direct time measurement.

As indicated in Fig. 7-5, for times much less than one time constant, the capacitor voltage changes almost linearly. Further, the capacitor voltage would get to its final value in one time constant if the rate of change were constant at its initial value. This linear change approximation is valid if the time to be measured is one-tenth or less of a time constant, or, what amounts to the same thing, if the voltage change during the time interval is one-tenth or less of the difference between the initial and final voltages.

A timing circuit can be used with a gas tube to make an *oscillator*—a circuit that produces a repeating waveform. A gas tube has a very large resistance—approximately an open circuit—for small voltages. But at a certain voltage it will fire or, in other words, conduct and have a very low resistance—approximately a short circuit for some purposes. After beginning to conduct, it will continue to conduct even if its voltage drops, provided that this voltage does not drop below a certain low voltage at which the tube stops firing (extinguishes) and becomes an open circuit again.

The circuit illustrated in Fig. 7-6a is an oscillator for producing a sawtooth capacitor voltage as shown in Fig. 7-6b. If the firing voltage V_F of the gas tube is one-tenth or less of the source voltage V_S, the capacitor voltage increases almost linearly, as shown in Fig. 7-6b, to the voltage V_F, at which time T the gas tube fires. If the resistance of the conducting gas tube is small and much less than that of the resistor R, the capacitor rapidly discharges through the tube until the capacitor voltage drops to V_E, the extinguishing voltage, which is not great enough to keep the tube conducting. Then the tube cuts off, the capacitor starts charging again, and the process keeps repeating indefinitely. The time T for one charging and discharging cycle is called a *period*.

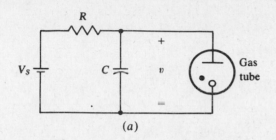   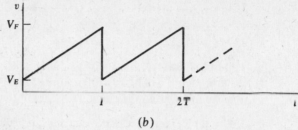

(a)   (b)

Fig. 7-6

Solved Problems

7.1   Find the capacitance of an initially uncharged capacitor for which the movement of 3×10^{15} electrons from one capacitor plate to another produces a 200-V capacitor voltage.

From the basic capacitor formula $C = Q/V$, in which Q is in coulombs,

$$C = \frac{-3 \times 10^5 \text{ electrons}}{200 \text{ V}} \times \frac{-1 \text{ C}}{6.241 \times 10^{18} \text{ electrons}} = 2.4 \times 10^{-6} \text{ F} = 2.4 \ \mu\text{F}$$

7.2   What is the charge stored on a 2-μF capacitor with 10 V across it?

From $C = Q/V$,

$$Q = CV = (2 \times 10^{-6})(10) \text{ C} = 20 \ \mu\text{C}$$

7.3   What is the change of voltage produced by 8×10^9 electrons moving from one plate to the other of an initially charged 10-pF capacitor?

Since $C = Q/V$ is a linear relation, C also relates changes in charge and voltage: $C = \Delta Q/\Delta V$. In this equation, ΔQ is the change in stored charge and ΔV is the accompanying change in voltage. From this,

$$\Delta V = \frac{\Delta Q}{C} = \frac{-8 \times 10^9 \text{ electrons}}{10 \times 10^{-12} \text{ F}} \times \frac{-1 \text{ C}}{6.241 \times 10^{18} \text{ electrons}} = 128 \text{ V}$$

7.4   Find the capacitance of a parallel-plate capacitor if the dimensions of each rectangular-shaped plate is 1 by 0.5 cm and if the distance between plates is 0.1 mm. The dielectric is air. Also, find the capacitance if the dielectric is mica instead of air.

The dielectric constant of air is so close to 1 that the permittivity of vacuum can be used for that of air in the parallel-plate capacitor formula:

$$C = \epsilon \frac{A}{d} = \frac{(8.85 \times 10^{-12})(10^{-2})(0.5 \times 10^{-2})}{0.1 \times 10^{-3}} \text{ F} = 4.43 \text{ pF}$$

Because the dielectric constant of mica is 5, a mica dielectric increases the capacitance by a factor of 5: $C = 5 \times 4.43 = 22.1$ pF.

7.5   Find the distance between the plates of a 0.01-μF parallel-plate capacitor if the area of each plate is 0.07 m^2 and the dielectric is glass.

From rearranging $C = \epsilon A/d$ and using 7.5 for the dielectric constant of glass,

$$d = \frac{\epsilon A}{C} = \frac{7.5(8.85 \times 10^{-12})(0.07)}{0.01 \times 10^{-6}} \text{ m} = 0.465 \text{ mm}$$

7.6   A capacitor has a disk-shaped dielectric of ceramic that has a 0.5-cm diameter and is 0.521 mm thick.   The disk is coated on both sides with silver, this coating being the plates.   Find the capacitance.

With the ceramic dielectric constant of 7500 in the parallel-plate capacitor formula,

$$C = \epsilon \frac{A}{d} = \frac{7500(8.85 \times 10^{-12})[\pi \times (0.25 \times 10^{-2})^2]}{0.521 \times 10^{-3}} \text{ F} = 2500 \text{ pF}$$

7.7   A 1-F parallel-plate capacitor has a ceramic dielectric 1 mm thick.   If the plates are square, find the length of a side of a plate.

Because each plate is square, a length l of a side is $l = \sqrt{A}$.   From this and $C = \epsilon A/d$,

$$l = \sqrt{\frac{dC}{\epsilon}} = \sqrt{\frac{10^{-3} \times 1}{7500(8.85 \times 10^{-12})}} = 123 \text{ m}$$

Each side is 123 m long or, approximately, 1.3 times the length of a football field.   This problem demonstrates that the farad is an extremely large unit.

7.8   What are the different capacitances that can be obtained with a 1- and a 3-μF capacitor?

The capacitors can produce 1 and 3 μF individually, also $1 + 3 = 4$ μF   in parallel, and $(1 \times 3)/(1 + 3) = 0.75$ μF   in series.

7.9   Find the total capacitance C_T of the circuit shown in Fig. 7-7.

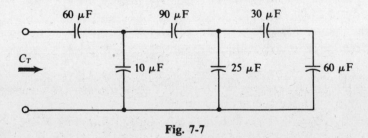

Fig. 7-7

At the end opposite the input, the series 30- and 60-μF capacitors have a total capacitance of $30 \times 60/(30 + 60) = 20$ μF.   This adds to the capacitance of the parallel 25-μF capacitor for a total of 45 μF to the right of the 90-μF capacitor.   The 45- and 90-μF capacitances combine to $45 \times 90/(45 + 90) = 30$ μF.   This adds to the capacitance of the parallel 10-μF capacitor for a total of $30 + 10 = 40 \ \mu\text{F}$ to the right of the 60-μF capacitor. Finally,

$$C_T = \frac{60 \times 40}{60 + 40} = 24 \ \mu\text{F}$$

7.10 A 4-μF capacitor, a 6-μF capacitor, and an 8-μF capacitor are in parallel across a 300-V source. Find (a) the total capacitance, (b) the magnitude of charge stored by each capacitor, and (c) the total stored energy.

(a) Because the capacitors are in parallel, the total or equivalent capacitance is the sum of the individual capacitances: $C_T = 4 + 6 + 8 = 18 \ \mu\text{F}$.

(b) The three charges are, from $Q = CV$, $(4 \times 10^{-6})(300) \text{ C} = 1.2 \text{ mC}$, $(6 \times 10^{-6})(300) \text{ C} = 1.8 \text{ mC}$, and $(8 \times 10^{-6})(300) \text{ C} = 2.4 \text{ mC}$ for the 4-, 6-, and 8-μF capacitors, respectively.

(c) The total capacitance can be used to get the total stored energy:

$$W = \tfrac{1}{2}C_T V^2 = 0.5(18 \times 10^{-6})(300)^2 = 0.81 \text{ J}$$

7.11 Repeat Prob. 7.10 for the capacitors in series instead of in parallel, but find each capacitor voltage instead of each charge stored.

(a) Because the capacitors are in series, the total capacitance is the reciprocal of the sum of the reciprocals of the individual capacitances:

$$C_T = \frac{1}{1/4 + 1/6 + 1/8} = 1.846 \ \mu\text{F}$$

(b) The voltage across each capacitor depends on the charge stored, which is the same for each capacitor. This charge can be gotten from the total capacitance and the applied voltage:

$$Q = C_T V = (1.846 \times 10^{-6})(300) \text{ C} = 554 \ \mu\text{C}$$

From $V = Q/C$, the individual capacitor voltages are

$$\frac{554 \times 10^{-6}}{4 \times 10^{-6}} = 138.5 \text{ V} \qquad \frac{554 \times 10^{-6}}{6 \times 10^{-6}} = 92.3 \text{ V} \qquad \frac{554 \times 10^{-6}}{8 \times 10^{-6}} = 69.2 \text{ V}$$

for the 4-, 6-, and 8-μF capacitors, respectively.

(c) The total stored energy is

$$W = \tfrac{1}{2}C_T V^2 = 0.5(1.846 \times 10^{-6})(300)^2 \text{ J} = 83.1 \text{ mJ}$$

7.12 A 24-V source and two capacitors are connected in series. If one capacitor has 20 μF of capacitance and has 16 V across it, what is the capacitance of the other capacitor?

By KVL, the other capacitor has $24 - 16 = 8 \text{ V}$ across it. Also, the charge on it is the same as that on the other capacitor: $Q = CV = (20 \times 10^{-6})(16) \text{ C} = 320 \ \mu\text{C}$. So, $C = Q/V = 320 \times 10^{-6}/8 \text{ F} = 40 \ \mu\text{F}$

7.13 Find each capacitor voltage in the circuit shown in Fig. 7-8.

The approach is to find the equivalent capacitance, use it to find the charge, and then use this charge to find the voltages across the 6- and 12-μF capacitors, which have this same charge because they are in series with the source.

At the end opposite the source, the two parallel capacitors have an equivalent capacitance of $5 + 1 = 6 \ \mu\text{F}$. With this reduction, the capacitors are in series, making

$$C_T = \frac{1}{1/6 + 1/12 + 1/6} = 2.4 \ \mu\text{F}$$

The desired charge is

$$Q = CV = (2.4 \times 10^{-6})(100) \text{ C} = 240 \ \mu\text{C}$$

which is the charge on the 6-μF capacitor as well as on the 12-μF capacitor.   From $V = Q/C$,

$$V_1 = \frac{240 \times 10^{-6}}{6 \times 10^{-6}} = 40 \text{ V} \qquad V_2 = \frac{240 \times 10^{-6}}{12 \times 10^{-6}} = 20 \text{ V}$$

and, by KVL, $V_3 = 100 - V_1 - V_2 = 40$ V.

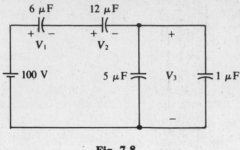 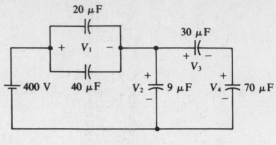

Fig. 7-8   **Fig. 7-9**

7.14 Find each capacitor voltage in the circuit shown in Fig. 7-9.

A good analysis method is to reduce the circuit to a series circuit with two capacitors and the voltage source, find the charge on each reduced capacitor, and from it find the voltages across these capacitors.   Then the process can be partially repeated to find all the capacitor voltages in the original circuit.

The parallel 20- and 40-μF capacitors reduce to a single 60-μF capacitor.   The 30- and 70-μF capacitors reduce to a $30 \times 70/(30 + 70) = 21$-$\mu$F   capacitor in parallel with the 9-μF capacitor.   So, all three of these capacitors reduce to a $21 + 9 = 30$-μF   capacitor that is in series with the reduced 60-μF capacitor, and the total capacitance at the source terminals is $30 \times 60/(30 + 60) = 20 \ \mu$F.   The desired charge is

$$Q = C_T V = (20 \times 10^{-6})(400) \text{ C} = 8 \text{ mC}$$

This charge can be used to get V_1 and V_2:

$$V_1 = \frac{8 \times 10^{-3}}{60 \times 10^{-6}} = 133 \text{ V} \qquad \text{and} \qquad V_2 = \frac{8 \times 10^{-3}}{30 \times 10^{-6}} = 267 \text{ V}$$

Alternatively, $V_2 = 400 - V_1 = 400 - 133 = 267$ V.

The charge on the 30-μF capacitor and also on the series 70-μF capacitor is the 8 mC minus the charge on the 9-μF capacitor:

$$8 \times 10^{-3} - (9 \times 10^{-6})(267) \text{ C} = 5.6 \text{ mC}$$

Consequently, from $V = Q/C$,

$$V_3 = \frac{5.6 \times 10^{-3}}{30 \times 10^{-6}} = 187 \text{ V} \qquad \text{and} \qquad V_4 = \frac{5.6 \times 10^{-3}}{70 \times 10^{-6}} = 80 \text{ V}$$

As a check, $V_3 + V_4 = 187 + 80 = 267 \text{ V} = V_2$.

7.15 A 3-μF capacitor charged to 100 V is connected across an uncharged 6-μF capacitor.   Find the voltage and also the initial and final stored energies.

The charge and capacitance are needed to find the voltage from $V = Q/C$.   Initially, the charge on the 3-μF capacitor is $Q = CV = (3 \times 10^{-6})(100) \text{ C} = 0.3 \text{ mC}$.   When the capacitors are connected together, this charge distributes over the two capacitors, but does not change.   Since the same voltage is across both capacitors, they are in parallel.   So, $C_T = 3 + 6 = 9 \ \mu$F,   and

$$V = \frac{Q}{C_T} = \frac{0.3 \times 10^{-3}}{9 \times 10^{-6}} = 33.3 \text{ V}$$

The initial energy is all stored by the 3-μF capacitor: $\frac{1}{2}CV^2 = 0.5(3 \times 10^{-6})(100)^2 \text{ J} = 15 \text{ mJ}$.   The final energy is stored by both capacitors: $0.5(9 \times 10^{-6})(33.3)^2 \text{ J} = 5 \text{ mJ}$.

7.16 Repeat Prob. 7.15 for an added 2-kΩ series resistor in the circuit.

The resistor has no effect on the final voltage, which is 33.3 V, because this voltage depends only on the equivalent capacitance and the charge stored, neither of which are affected by the presence of the resistor. Since the final voltage is the same, the final energy storage is the same: 5 mJ. Of course, the resistor has no effect on the initial 15 mJ stored. The resistor will, however, slow the time taken for the voltage to reach its final value, which time is five time constants after the switching. This time is zero if the resistance is zero. The presence of the resistor also makes it easier to account for the 10 mJ decrease in stored energy—it is dissipated in the resistor.

7.17 A 2-μF capacitor charged to 150 V and a 1-μF capacitor charged to 50 V are connected together with plates of opposite polarity joined. Find the voltage and the initial and final stored energies.

Because of the opposite polarity connection, some of the charge on one capacitor cancels that on the other. The initial charges are $(2 \times 10^{-6})(150)$ C $= 300$ μC for the 2-μF capacitor and $(1 \times 10^{-6})(50) = 50$ μC for the 1-μF capacitor. The final charge distributed over both capacitors is the difference of these two charges: $300 - 50 - 250$ μC. It produces a voltage of

$$V = \frac{Q}{C_T} = \frac{250 \times 10^{-6}}{2 \times 10^{-6} + 1 \times 10^{-6}} = 83.3 \text{ V}$$

The initial stored energy is the sum of the energies stored by both capacitors:

$$0.5(2 \times 10^{-6})(150)^2 + 0.5(1 \times 10^{-6})(50)^2 \text{ J} = 23.8 \text{ mJ}$$

The final stored energy is

$$\tfrac{1}{2}C_T V_F^2 = 0.5(3 \times 10^{-6})(83.3)^2 \text{ J} = 10.4 \text{ mJ}$$

7.18 What is the current flowing through a 2-μF capacitor when the capacitor voltage is 10 V?

There is not enough information to find the capacitor current. That current depends on the rate of change of capacitor voltage and *not* the voltage, and this rate is not given.

7.19 If the voltage across a 0.1-μF capacitor is 3000t V, find the capacitor current.

The capacitor current equals the product of the capacitance and the time derivative of the voltage. Since the time derivative of 3000t is 3000,

$$i = C\frac{dv}{dt} = (0.1 \times 10^{-6})(3000) \text{ A} = 0.3 \text{ mA}$$

which is a constant value.

The capacitor current can also be found from $i = C \, \Delta v/\Delta t$ because the voltage is increasing linearly. If Δt is taken as, say, 2 s, from 0 to 2 s, the corresponding Δv is $3000 \, \Delta t = 3000(2 - 0) = 6000$ V. So,

$$i = C\frac{\Delta v}{\Delta t} = \frac{(0.1 \times 10^{-6})(6000)}{2} \text{ A} = 0.3 \text{ mA}$$

7.20 Sketch the waveform of the current that flows through a 2-μF capacitor when the capacitor voltage is as shown in Fig. 7-10. As always, assume associated references because there is no statement to the contrary.

Graphically, the dv/dt in $i = C \, dv/dt$ is the *slope* of the voltage graph. For straight lines this slope is the same as $\Delta v/\Delta t$. For this voltage graph, the straight line for the interval of $t = 0$ s to $t = 1$ μs has a slope of $(20 - 0)/(1 \times 10^{-6} - 0)$ V/s $= 20$ MV/s, which is the voltage at $t = 1$ μs minus the voltage at $t = 0$ s, divided by the time at $t = 1$ μs minus the time at $t = 0$ s. As a result, during this time interval the current is $i = C \, dv/dt = (2 \times 10^{-6})(20 \times 10^{6}) = 40$ A.

From $t = 1\ \mu s$ to $t = 4\ \mu s$, the voltage graph is horizontal, which means that the slope and, consequently, the current are zero: $i = 0$ A.

For the time interval from $t = 4\ \mu s$ to $t = 6\ \mu s$, the straight line has a slope of $(-20 - 20)/(6 \times 10^{-6} - 4 \times 10^{-6})$ V/s $= -20$ MV/s. This change in voltage produces a current of $i = C\ dv/dt = (2 \times 10^{-6})(-20 \times 10^{6}) = -40$ A.

Finally, from $t = 6\ \mu s$ to $t = 8\ \mu s$, the slope of the straight line is $[0 - (-20)]/(8 \times 10^{-6} - 6 \times 10^{-6})$ V/s $= 10$ MV/s and the capacitor current is $i = C\ dv/dt = (2 \times 10^{-6})(10 \times 10^{6}) = 20$ A.

Figure 7-11 is a graph of the capacitor current. Notice that, unlike capacitor voltage, capacitor current can jump, as it does at 1, 4, and 6 μs. In fact, at 6-μs the current reverses direction instantaneously.

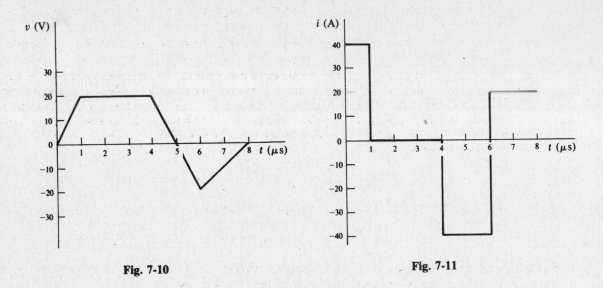

Fig. 7-10   Fig. 7-11

7.21   Find the time constant of the circuit shown in Fig. 7-12.

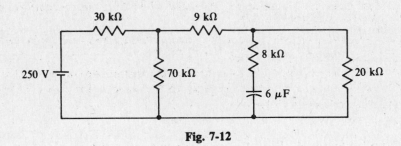

Fig. 7-12

The time constant is $\tau = R_{Th}C$, where R_{Th} is the Thevenin resistance at the capacitor terminals. Here,

$$R_{Th} = 8 + 20\|(9 + 70\|30) = 8 + 20\|30 = 20\ k\Omega$$

and so the time constant is $\tau = R_{Th}C = (20 \times 10^{3})(6 \times 10^{-6}) = 0.12$ s.

7.22   How long does a 20-μF capacitor charged to 150 V take to discharge through a 3-MΩ resistor? Also, at what time does the maximum discharge current occur and what is its value?

The discharge is considered to be completed after five time constants:

$$5\tau = 5RC = 5(3 \times 10^{6})(20 \times 10^{-6}) = 300\ s$$

Since the current decreases as the capacitor discharges, it has a graph as shown in Fig. 7-5a with a maximum value at the time of switching, $t = 0$ s here. In this circuit the current has an initial value of $150/(3 \times 10^6)$ A $= 50\ \mu$A because initially the capacitor voltage of 150 V, which cannot jump, is across the 3-MΩ resistor.

7.23 At $t = 0$ s, a 100-V source is switched in series with a 1-kΩ resistor and an uncharged 2-μF capacitor. What are (a) the initial capacitor voltage, (b) the initial current, (c) the initial rate of capacitor voltage increase, and (d) the time required for the capacitor voltage to reach its maximum value?

(a) Since the capacitor voltage is zero before the switching, it is also zero immediately after the switching—a capacitor voltage cannot jump: $v(0+) = 0$ V.

(b) By KVL, at $t = 0+$ s the 100 V of the source is all across the 1-kΩ resistor because the capacitor voltage is 0 V. Consequently, $i(0+) = 100/10^3$ A $= 100$ mA.

(c) As can be seen from Fig. 7-5b, the initial rate of capacitor voltage increase equals the total change in capacitor voltage divided by the circuit time constant. In this circuit the capacitor voltage eventually equals the 100 V of the source. Of course, the initial value is 0 V. Also, the time constant is $\tau = RC = 10^3(2 \times 10^{-6})$ s $= 2$ ms. So, the initial rate of capacitor voltage increase is $100/(2 \times 10^{-3}) = 50\,000$ V/s. This initial rate can also be found from $i = C\, dv/dt$ evaluated at $t = 0+$ s:

$$\frac{dv}{dt}(0+) = \frac{i(0+)}{C} = \frac{100 \times 10^{-3}}{2 \times 10^{-6}} = 50\,000\ \text{V/s}$$

(d) It takes five time constants, $5 \times 2 = 10$ ms, for the capacitor voltage to reach its final value of 100 V.

7.24 Repeat Prob. 7.23 for an initial capacitor charge of 50 μC. The positive plate of the capacitor is toward the positive terminal of the 100-V source.

(a) The initial capacitor voltage is $V = Q/C = (50 \times 10^{-6})/(2 \times 10^{-6}) = 25$ V.

(b) At $t = 0+$ s, the voltage across the resistor is, by KVL, the source voltage minus the initial capacitor voltage. This voltage difference divided by the resistance is the initial current: $i(0+) = (100 - 25)/10^3$ A $= 75$ mA.

(c) The initial rate of capacitor voltage increase equals the total change in capacitor voltage divided by the time constant: $75/(2 \times 10^{-3}) = 37\,500$ V/s.

(d) The initial capacitor voltage has no effect on the circuit time constant and so also not on the time required for the capacitor voltage to reach its final value. This time is 10 ms, the same as for the circuit discussed in Prob. 7.23.

7.25 In the circuit shown in Fig. 7-13, find the indicated voltages and currents at $t = 0+$ s, immediately after the switch closes. The capacitors are initially uncharged. Also, find these voltages and currents a long time after the switch closes.

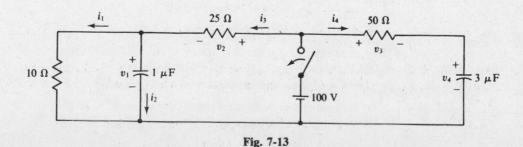

Fig. 7-13

At $t = 0+$ s, the capacitors have 0 V across them because the capacitor voltages cannot jump from the 0-V values that they have at $t = 0-$ s, immediately before the switching: $v_1(0+) = 0$ V and $v_4(0+) = 0$ V. Further, with 0 V across them, the capacitors act like short circuits at $t = 0+$ s, with the result that the 100 V of the source is across both the 25- and 50-Ω resistors: $v_2(0+) = v_3(0+) = 100$ V. Three of the initial currents can be found from these voltages:

$$i_1(0+) = \frac{0}{10} = 0 \text{ A} \qquad i_3(0+) = \frac{100}{25} = 4 \text{ A} \qquad i_4(0+) = \frac{100}{50} = 2 \text{ A}$$

The remaining initial current, $i_2(0+)$, can be found by applying KCL at the node at the top of the 1-μF capacitor:

$$i_2(0+) = i_3(0+) - i_1(0+) = 4 - 0 = 4 \text{ A}$$

A "long time" after the switch closes means more than five time constants later. At this time the capacitor voltages are constant, and so the capacitors act like open circuits, blocking i_2 and i_4: $i_2(\infty) = i_4(\infty) = 0$ A. With the 1-μF capacitor acting like an open circuit, the 10- and 25-Ω resistors are in series across the 100-V source, and so $i_1(\infty) = i_3(\infty) = 100/35 = 2.86$ A. From the resistances and the calculated currents, $v_1(\infty) = 10 \times 2.86 = 28.6$ V, $v_2(\infty) = 25 \times 2.86 = 71.4$ V, and $v_3(\infty) = 0 \times 50 = 0$ V. Finally, from the right-hand mesh,

$$v_4(\infty) = 100 - v_3(\infty) = 100 - 0 = 100 \text{ V}$$

7.26 A 2-μF capacitor, initially charged to 300 V, is discharged through a 270-kΩ resistor. What is the capacitor voltage at 0.25 s after the capacitor starts to discharge?

The voltage formula is $v = v(\infty) + [v(0+) - v(\infty)]e^{-t/\tau}$. Since the time constant is $\tau = RC = (270 \times 10^3)(2 \times 10^{-6}) = 0.54$ s, the initial capacitor voltage is $v(0+) = 300$ V and the final capacitor voltage is $v(\infty) = 0$ V, it follows that the equation for the capacitor voltage is

$$v(t) = 0 + (300 - 0)e^{-t/0.54} = 300e^{-1.85t} \text{ V} \qquad \text{for} \qquad t > 0 \text{ s}$$

From this, $v(0.25) = 300e^{-1.85(0.25)} = 189$ V.

7.27 Closing a switch connects in series a 200-V source, a 2-MΩ resistor, and an uncharged 0.1-μF capacitor. Find the capacitor voltage and current at 0.1 s after the switch closes.

The voltage formula is $v = v(\infty) + [v(0+) - v(\infty)]e^{-t/\tau}$. Here, $v(\infty) = 200$ V, $v(0+) = 0$ V, and $\tau = (2 \times 10^6)(0.1 \times 10^{-6}) = 0.2$ s. So,

$$v(t) = 200 + [0 - 200]e^{-t/0.2} = 200 - 200e^{-5t} \text{ V} \qquad \text{for} \qquad t > 0 \text{ s}$$

Substitution of 0.1 for t gives $v(0.1)$:

$$v(0.1) = 200 - 200e^{-0.5} = 78.7 \text{ V}$$

Similarly, $i = i(\infty) + [i(0+) - i(\infty)]e^{-t/\tau}$, in which $i(0+) = 200/(2 \times 10^6)$ A $= 0.1$ mA, $i(\infty) = 0$ A, and of course $\tau = 0.2$ s. With these values inserted,

$$i(t) = 0 + (0.1 - 0)e^{-5t} = 0.1e^{-5t} \text{ mA}$$

From this, $i(0.1) = 0.1e^{-0.5}$ mA $= 60.7$ μA. This current can also be found by using the voltage across the resistor at $t = 0.1$ s: $i(0.1) = (200 - 78.7)/(2 \times 10^6)$ A $= 60.7$ μA.

7.28 For the circuit used in Prob. 7.27, find the time required for the capacitor voltage to reach 50 V. Then find the time required for the capacitor voltage to increase another 50 V, from 50 to 100 V. Compare times.

From the solution to Prob. 7.27, $v(t) = 200 - 200e^{-5t}$ V. To find the time at which the voltage is 50 V, it is only necessary to substitute 50 for $v(t)$ and solve for t: $50 = 200 - 200e^{-5t}$ or $e^{-5t} = 150/200 = 0.75$. The exponential can be eliminated by taking the natural logarithm of both sides:

$$\ln e^{-5t} = \ln 0.75 \qquad \text{from which} \qquad -5t = -0.288 \qquad \text{and} \qquad t = 0.288/5 \text{ s} = 57.5 \text{ ms}$$

The same procedure can be used to find the time at which the capacitor voltage is 100 V: $100 = 200 - 200e^{-5t}$ or $e^{-5t} = 100/200 = 0.5$. Further,

$$\ln e^{-5t} = \ln 0.5 \quad \text{from which} \quad -5t = -0.693 \quad \text{and} \quad t = 0.693/5 \text{ s} = 138.6 \text{ ms}$$

The voltage required 57.5 ms to reach 50 V, and $138.6 - 57.5 = 81.1$ ms to increase another 50 V, which verifies the fact that the rate of increase becomes less and less as time increases.

7.29 In the circuit shown in Fig. 7-14, the switch closes at $t = 0$ s. Find v_C and i for $t > 0$ s if $v_C(0) = 100$ V.

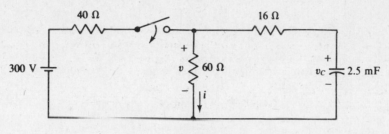

Fig. 7-14

All that are needed for the v and i formulas are $v_C(0+)$, $v_C(\infty)$, $i(0+)$, $i(\infty)$, and $\tau = R_{Th}C$. Of course, $v_C(0+) = 100$ V because the capacitor voltage cannot jump. The voltage $v_C(\infty)$ is the same as the voltage across the 60-Ω resistor a long time after the switch closes, because at this time the capacitor acts like an open circuit.   So, by voltage division

$$v_C(\infty) = \frac{60}{60 + 40} \times 300 = 180 \text{ V}$$

Also, $i(\infty) = v_C(\infty)/60 = 180/60 = 3$ A. It is easy to obtain $i(0+)$ from $v(0+)$, which can be solved for using a nodal equation at the middle top node for the time $t = 0+$ s:

$$\frac{v(0+) - 300}{40} + \frac{v(0+)}{60} + \frac{v(0+) - 100}{16} = 0$$

from which $v(0+) = 132$ V. So, $i(0+) = 132/60 = 2.2$ A. Since the Thevenin resistance at the capacitor terminals is $16 + 60\|40 = 40 \ \Omega$, the time constant is $\tau = RC = 40(2.5 \times 10^{-3}) = 0.1$ s.
With these quantities substituted into the v and i formulas,

$$v_C(t) = v_C(\infty) + [v_C(0+) - v_C(\infty)]e^{-t/\tau} = 180 + (100 - 180)e^{-10t} = 180 - 80e^{-10t} \text{ V} \quad \text{for} \quad t > 0 \text{ s}$$

$$i(t) = i(\infty) + [i(0+) - i(\infty)]e^{-t/\tau} = 3 + (2.2 - 3)e^{-10t} = 3 - 0.8e^{-10t} \text{ A} \quad \text{for} \quad t > 0 \text{ s}$$

7.30 The switch is closed at $t = 0$ s in the circuit shown in Fig. 7-15. Find i for $t > 0$ s. The capacitor is initially uncharged.

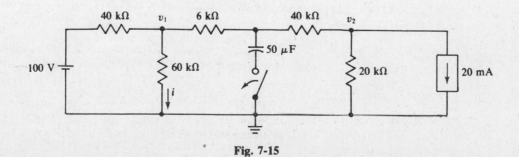

Fig. 7-15

The quantities $i(0+)$, $i(\infty)$, and τ are needed for the current formula $i = i(\infty) + [i(0+) - i(\infty)]e^{-t/\tau}$.

At $t = 0+$ s, the short-circuiting action of the capacitor prevents the 20-mA current source from affecting $i(0+)$. Also, it places the 6-kΩ resistor in parallel with the 60-kΩ resistor. Consequently, by current division,

$$i(0+) = \left(\frac{6}{60 + 6}\right)\left(\frac{100}{40 + 6\|60}\right) = 0.2 \text{ mA}$$

in which the simplifying kilohm-milliampere method is used.

After five time constants the capacitor no longer conducts current and can be considered to be an open circuit and so neglected in the calculations. By nodal analysis,

$$\left(\frac{1}{40} + \frac{1}{60} + \frac{1}{46}\right)v_1(\infty) - \frac{1}{46} v_2(\infty) = \frac{100}{40} \qquad - \frac{1}{46} v_1(\infty) + \left(\frac{1}{46} + \frac{1}{20}\right)v_2(\infty) = -20$$

from which $v_1(\infty) = -62.67$ V. So, $i(\infty) = -62.67/(60 \times 10^3)$ A $= -1.04$ mA.

The Thevenin resistance at the capacitor terminals is $(6 + 40\|60)\|(40 + 20) = 20$ kΩ. This can be used to find the time constant:

$$\tau = R_{Th}C = (20 \times 10^3)(50 \times 10^{-6}) = 1 \text{ s}$$

Now that $i(0+)$, $i(\infty)$, and τ are known, the current i can be found:

$$i = -1.04 + [0.2 - (-1.04)]e^{-t} = -1.04 + 1.24e^{-t} \text{ mA} \qquad \text{for} \qquad t > 0 \text{ s}$$

7.31 After a long time in position 1, the switch in the circuit shown Fig. 7-16 is thrown to position 2 at $t = 0$ s for a duration of 30 s and then returned to position 1. (a) Find the equations for v for $t > 0$ s. (b) Find v at $t = 5$ s and at $t = 40$ s. (c) Make a sketch of v for $0 \text{ s} < t < 80 \text{ s}$.

(a) At the time that the switch is thrown to position 2, the initial capacitor voltage is 20 V, the same as immediately before the switching; the final capacitor voltage is 70 V, the voltage of the source in the circuit; and the time constant is $(20 \times 10^6)(2 \times 10^{-6}) = 40$ s. Consequently, while the switch is in position 2,

$$v = 70 + (20 - 70)e^{-t/40} = 70 - 50e^{-0.025t} \text{ V}$$

Of course, the capacitor voltage never reaches the "final voltage" because a switching operation interrupts the charging, but the circuit does not "know" this ahead of time.

When the switch is returned to position 1, the circuit changes, and so the equation for v changes. The initial voltage at this $t = 30$-s switching can be found by substituting 30 for t in the equation for v that was just calculated: $v(30) = 70 - 50e^{-0.025(30)} = 46.4$ V. The final capacitor voltage is 20 V, and the time constant is $(5 \times 10^6)(2 \times 10^{-6}) = 10$ s. For these values, the basic voltage formula must be modified since the switching occurs at $t = 30$ s instead of

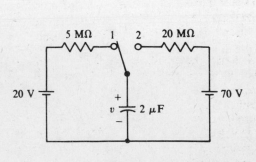

Fig. 7-16

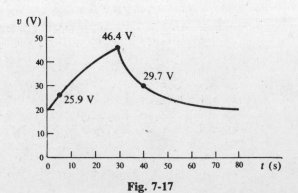

Fig. 7-17

at $t = 0+$ s. The modified formula is

$$v(t) = v(\infty) + [v(30+) - v(\infty)]e^{-(t-30)/\tau} \text{ V} \qquad \text{for} \qquad t > 30 \text{ s}$$

The $t - 30$ is necessary in the exponent to account for the time shift. With the values inserted into this formula, the capacitor voltage is

$$v(t) = 20 + [46.4 - 20]e^{-0.1(t-30)} = 20 + 26.4e^{-0.1(t-30)} \text{ V} \qquad \text{for} \qquad t > 30 \text{ s}$$

(b) For v at $t = 5$ s, the first voltage equation must be used because it is the one that is valid for the first 30 s: $v(5) = 70 - 50e^{-0.025(5)} = 25.9$ V. For v at $t = 40$ s, the second equation must be used because it is the one that is valid after 30 s: $v(40) = 20 + 26.4e^{-0.1(40-30)} = 29.7$ V.

(c) Figure 7-17 shows the voltage graph which is based on the two voltage equations. The voltage rises exponentially to 46.4 V at $t = 30$ s, heading toward 70 V. After 30 s, the voltage decays exponentially to the final value of 20 V, reaching it at 80 s, five time constants after the switch returns to position 1.

7.32 A simple *RC* timer has a switch that when closed connects in series a 300-V source, a 16-MΩ resistor, and an uncharged 10-μF capacitor. Find the time between the closing and opening of the switch if the capacitor charges to 10 V during this time.

Since 10 V is less than one-tenth of the final voltage of 300 V, a linear approximation can be used. In this approximation the rate of voltage change is considered to be constant at its initial value. Although not needed, this rate is the quotient of the possible total voltage change of 300 V and the time constant of $RC = (16 \times 10^6)(10 \times 10^{-6}) = 160$ s. Since the voltage that the capacitor charges to is 1/30th of the possible total voltage change, the time required for this charging is approximately 1/30th of the time constant: $t \simeq 160/30 = 5.33$ s.

This time can be found more accurately, but with more effort, from the voltage formula. For it, $v(0+) = 0$ V, $v(\infty) = 300$ V, and $\tau = 160$ s. With these values inserted, the capacitor voltage equation is $v = 300 - 300e^{-t/160}$. For $v = 10$ V, it becomes $10 = 300 - 300e^{-t/160}$, from which $t = 160 \ln(300/290) = 5.42$ s. The approximation of 5.33 s is within 2 percent of this formula value of 5.42 s.

7.33 Repeat Prob. 7.32 for a capacitor voltage of 250 V.

The approximation cannot be used since 250 V is more than one-tenth of 300 V. The exact formula must be used. From the solution to Prob. 7.32, $v = 300 - 300e^{-t/160}$. For $v = 250$ V, it becomes $250 = 300 - 300e^{-t/160}$, which simplifies to $t = 160 \ln(300/50) = 287$ s. By comparison, the linear approximation gives $t = (250/300)(160) = 133$ s, which is considerably in error.

7.34 For the oscillator circuit shown in Fig. 7-18, find the period of oscillation if the gas tube fires at 90 V and extinguishes at 10 V. The gas tube has a 50-Ω resistance when firing and a 10^{10}-Ω resistance when extinguished.

When extinguished, the gas tube has such a large resistance (10^{10} Ω) compared to the 1-MΩ resistance of the resistor that it can be considered to be an open circuit and neglected during the charging time of the capacitor. During this time, the capacitor charges from an initial 10 V toward the 1000 V of the source, but stops charging when its voltage reaches 90 V, at which time the tube fires. Although this voltage change is $90 - 10 = 80$ V, the initial circuit action is as if the total voltage change will be $1000 - 10 = 990$ V. Since 80 V is less than one-tenth of 990 V, a linear approximation can be used to find the proportion that the charging time is of the time constant of $10^6(2 \times 10^{-6}) = 2$ s. The proportionality is $t/2 = 80/990$, from which $t = 160/990 = 0.162$ s. If an exact analysis is made, the result is 0.168 52 s.

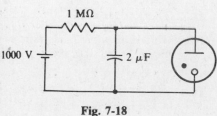

Fig. 7-18

When the tube fires, its 50-Ω resistance is so small compared to the 1-MΩ resistance of the resistor that the resistor can be considered to be an open circuit and neglected along with the voltage source. So, the discharging circuit is essentially an initially charged 2-μF capacitor and a 50-Ω resistor, until the voltage drops from the 90-V initial voltage to the 10-V extinguishing voltage. The time constant of this circuit is just $(2 \times 10^{-6})(50)$ s $= 0.1$ ms. This is so short compared to the charging time that the discharging time can usually be neglected even if five time constants are used for the discharge time. If an exact analysis is made, the result is a time of 0.22 ms for the capacitor to discharge from 90 to 10 V.

In summary, by approximations the period is $T = 0.162 + 0 = 0.162$ s, as compared to the exact method result of $T = 0.168\,52 + 0.000\,22 = 0.168\,74$ s or 0.169 s to three significant digits. Note that the approximate result is within about 4 percent of the actual result. This is usually good enough, especially in view of the fact that in the actual circuit the component values probably differ from the specified values by more than this.

7.35 Repeat Prob. 7.34 with the source voltage changed from 1000 to 100 V.

During the charge cycle the capacitor charges toward 100 V from an initial 10 V, the same as if the total voltage change will be $100 - 10 = 90$ V. Since the actual voltage change of $90 - 10 = 80$ V is considerably more than one-tenth of 90 V, a linear approximation is not valid. The exact method must be used. For this, $v(\infty) = 100$ V, $v(0+) = 10$ V, and $\tau = 2$ s. The corresponding voltage formula is

$$v = 100 + (10 - 100)e^{-t/2} = 100 - 90e^{-t/2} \text{ V}$$

The desired time is found by letting $v = 90$ V, and solving for t: $90 = 100 - 90e^{-t/2}$, which simplifies to $t = 2 \ln (90/10) = 4.39$ s. This is the period because the discharge time, which is the same as that found in the solution to Prob. 7.34, is negligible compared to this time.

Supplementary Problems

7.36 What electron movement between the plates of a 0.1-μF capacitor produces a 110-V change of voltage?   *Ans.* 6.87×10^{13} electrons

7.37 If the movement of 4.68×10^{14} electrons between the plates of a capacitor produces a 150-V change in capacitor voltage, find the capacitance.   *Ans.* 0.5 μF

7.38 What change in voltage of a 20-μF capacitor is produced by a movement of 9×10^{14} electrons between plates?   *Ans.* 7.21 V

7.39 A tubular capacitor consists of two sheets of aluminum foil 3 cm wide and 1 m long, rolled into a tube with separating sheets of waxed paper of the same size. What is the capacitance if the paper is 0.1 mm thick and has a dielectric constant of 3.5?   *Ans.* 9.29 nF

7.40 Find the area for each plate of a 10-μF parallel-plate capacitor that has a ceramic dielectric 0.5 mm thick.   *Ans.* 0.0753 m^2

7.41 Find the thickness of the mica dielectric of a 10-pF parallel-plate capacitor if the area of each plate is 10^{-4} m^2.   *Ans.* 0.443 mm

7.42 Find the diameter of a disk-shaped 0.001-μF capacitor that has a ceramic dielectric 1 mm thick.   *Ans.* 4.38 mm

7.43 What are the different capacitances that can be found with a 1-μF capacitor, a 2-μF capacitor, and a 3-μF capacitor?   *Ans.* 0.545 μF, 0.667 μF, 0.75 μF, 1 μF, 1.2 μF, 2 μF, 3 μF, 4 μF, 5 μF, 6 μF

7.44 Find the total capacitance C_T of the circuit shown in Fig. 7-19.   *Ans.* 2.48 μF

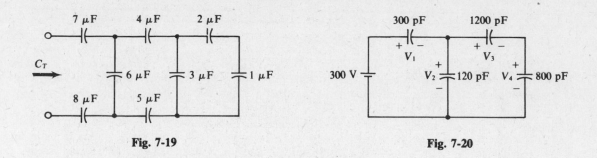

Fig. 7-19   **Fig. 7-20**

7.45 A 5-, a 7-, and a 9-μF capacitor are in parallel across a 200-V source.   Find the magnitude of charge stored by each capacitor and the total energy stored.
Ans. $Q_5 = 1$ mC, $Q_7 = 1.4$ mC, $Q_9 = 1.8$ mC,   0.42 J

7.46 A 6-, a 16-, and a 48-μF capacitor are in series with a 180-V source.   Find the voltage across each capacitor and the total energy stored.   *Ans.* $V_6 = 120$ V, $V_{16} = 45$ V, $V_{48} = 15$ V,   64.8 mJ

7.47 Two capacitors are in series across a 50-V source.   If one is a 1-μF capacitor with 16 V across it, what is the capacitance of the other?   *Ans.* 0.471 μF

7.48 Find each capacitor voltage in the circuit shown in Fig. 7-20.
Ans. $V_1 = 200$ V, $V_2 = 100$ V, $V_3 - 40$ V, $V_4 = 60$ V

7.49 A 0.1-μF capacitor charged to 100 V and a 0.2-μF capacitor charged to 60 V are connected together with plates of the same polarity joined.   Find the voltage and the initial and final stored energies.   *Ans.* 73.3 V, 860 μJ, 807 μJ

7.50 Repeat Prob. 7.49 for plates of opposite polarity joined.   *Ans.* 6.67 V, 860 μJ, 6.67 μJ

7.51 Find the voltage across a 0.1-μF capacitor when the capacitor current is 0.5 mA.
Ans. It can be any value.   There is not enough information to determine a unique value.

7.52 Repeat Prob. 7.51 if the capacitor voltage is 6 V at $t - 0$ s   and if the 0.5-mA capacitor current is constant.   Of course, assume associated references.   *Ans.* $6 + 5000t$ V

7.53 If the voltage across a 2-μF capacitor is $200t$ V for $t < 1$ s,   is 200 V for 1 s $< t < 5$ s,   and is $3200 - 600t$ V   for $t > 5$ s,   find the capacitor current.
Ans.   0.4 mA for $t < 1$ s,   0 A for 1 s $< t < 5$ s, -1.2 mA for $t > 5$ s

7.54 Find the time constant of the circuit shown in Fig. 7-21.   *Ans.* 60 μs

7.55 Find the time constant of the circuit shown in Fig. 7-22.   *Ans.* 66.3 ms

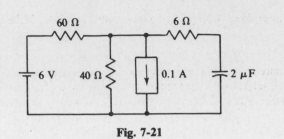

Fig. 7-21

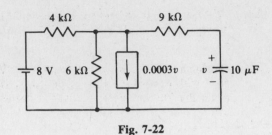

Fig. 7-22

7.56 How long does it take a 10-μF capacitor charged to 200 V to discharge through a 160-kΩ resistor, and what is the total energy dissipated in the resistor?   *Ans.* 8 s, 0.2 J

7.57 At $t = 0$ s, the closing of a switch connects in series a 150-V source, a 1.6-kΩ resistor, and the parallel combination of a 1-kΩ resistor and an uncharged 0.2-μF capacitor. Find (*a*) the initial capacitor current, (*b*) the initial and final 1-kΩ resistor currents, (*c*) the final capacitor voltage,   and (*d*) the time required for the capacitor voltage to reach its final value.
Ans. (*a*) 93.8 mA,   (*b*) 0 A and 57.7 mA,   (*c*) 57.7 V,   (*d*) 0.615 ms

7.58 Repeat Prob. 7.57 for a 200-V source and an initial capacitor voltage of 50 V opposed in polarity to that of the source.   *Ans.*   (*a*) 43.8 mA,   (*b*) 50 mA and 76.9 mA,   (*c*) 76.9 V,   (*d*) 0.615 ms

7.59 In the circuit shown in Fig. 7-23, find the indicated voltages and currents at $t = 0+$ s, immediately after the switch closes. Notice that the current source is active in the circuit before the switch closes.
Ans. $v_1(0+) = v_2(0+) = 20$ V $i_3(0+) = -0.106$ A
$i_1(0+) = 1$ A $i_4(0+) = 0.17$ A
$i_2(0+) = 0.106$ A $i_5(0+) = 63.8$ mA

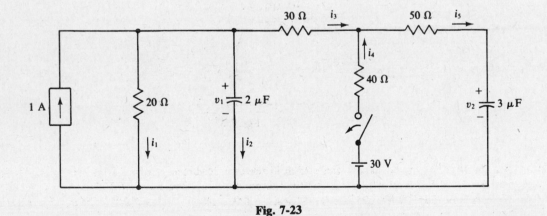

Fig. 7-23

7.60 In the circuit shown in Fig. 7-23, find the indicated voltages and currents a long time after the switch closes.
Ans. $v_1(\infty) = 22.2$ V $i_1(\infty) = 1.11$ A $i_3(\infty) = -0.111$ A $i_5(\infty) = 0$ A
$v_2(\infty) = 25.6$ V $i_2(\infty) = 0$ A $i_4(\infty) = 0.111$ A

7.61 A 0.1-μF capacitor, initially charged to 230 V, is discharged through a 3-MΩ resistor. Find the capacitor voltage 0.2 s after the capacitor starts to discharge.   *Ans.* 118 V

7.62 For the circuit discussed in Prob. 7.61, how long does it take the capacitor to discharge to 40 V?   *Ans.* 0.525 s

7.63 Closing a switch connects in series a 300-V source, a 2.7-MΩ resistor, and a 2-μF capacitor charged to 50 V with its positive plate toward the positive terminal of the source. Find the capacitor current 3 s after the switch closes. Also, find the time required for the capacitor voltage to increase to 250 V.   *Ans.* 53.1 μA,   8.69 s

7.64 The switch is closed at $t = 0$ s in the circuit shown in Fig. 7-24. Find v and i for $t > 0$ s. The capacitor is initially uncharged.   *Ans.* $60(1 - e^{-2t})$ V, $1 - 0.4e^{-2t}$ mA

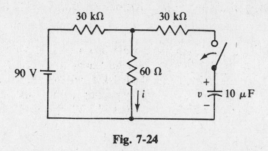

Fig. 7-24

7.65 Repeat Prob. 7.64 for $v(0+) = 20$ V and for the 60-kΩ resistor replaced by a 70-kΩ resistor.
Ans. $63 - 43e^{-1.96t}$ V, $0.9 - 0.253e^{-1.96t}$ mA

7.66 After a long time in position 1, the switch in the circuit shown in Fig. 7-25 is thrown to position 2 for 2 s, after which it is returned to position 1. Find v for $t > 0$ s.
Ans. $-200 + 300e^{-0.1t}$ V for 0 s $< t < 2$ s, $100 - 54.4e^{-0.2(t-2)} = 100 - 81.1e^{-0.2t}$ V for $t > 2$ s

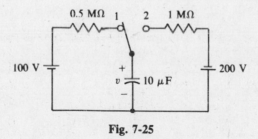

Fig. 7-25

7.67 After a long time in position 2, the switch in the circuit shown in Fig. 7-25 is thrown at $t = 0$ s to position 1 for 4 s, after which it is returned to position 2. Find v for $t > 0$ s.
Ans. $100 - 300e^{-0.2t}$ V for 0 s $< t < 4$ s, $-200 + 165e^{-0.1(t-4)} - -200 + 246e^{-0.1t}$ V for $t > 4$ s

7.68 A simple *RC* timer has a 50-V source, a switch, an uncharged 1-μF capacitor, and a resistor, all in series. Closing the switch and then opening it 5 s later produces a capacitor voltage of 3 V. Find the resistance of the resistor.   *Ans.* 83.3 MΩ approximately, 80.8 MΩ more exactly

7.69 Repeat Prob. 7.68 for a capacitor voltage of 40 V.   *Ans.* 3.11 MΩ

7.70 In the oscillator circuit shown in Fig. 7-18, replace the 1-MΩ resistor with a 4.3-MΩ resistor and the 1000-V source with a 150-V source and find the period of oscillation.   *Ans.* 7.29 s

Chapter 8

Inductors and Inductance

INTRODUCTION

The following material on inductors and inductance is similar to that on capacitors and capacitance presented in Chap. 7.   The reason for this similarity is that, mathematically speaking, the capacitor and inductor formulas are the same.   Only the symbols differ.   Where one has v, the other has i, and vice versa; where one has the capacitance quantity symbol C, the other has the inductance quantity symbol L; and where one has R, the other has G.   It follows then that the basic inductor voltage-current formula is $v = L\, di/dt$ in place of $i = C\, dv/dt$,   that the energy stored is $\frac{1}{2}Li^2$ instead of $\frac{1}{2}Cv^2$, that inductor currents cannot jump instead of capacitor voltages, that inductors are short circuits to dc instead of open circuits, and that the time constant is $LG = L/R$ instead of CR.   Although it is possible to approach the study of inductor action on the basis of this duality, the standard approach is to use magnetic flux.

MAGNETIC FLUX

Magnetic phenomena are explained using *magnetic flux*, or just flux, which relates to magnetic lines of force that, for a magnet, extend in continuous lines from the magnetic north pole to the south pole outside the magnet and from the south pole to the north pole inside the magnet; this is shown in Fig. 8-1a.   The SI unit of flux is the *weber*, with unit symbol Wb.   The quantity symbol is Φ for a constant flux and ϕ for a time-varying flux.

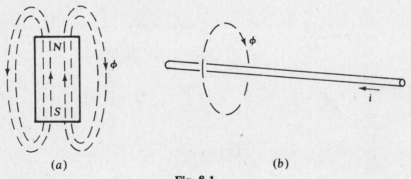

| (a) | (b) |

Fig. 8-1

Current flowing in a wire also produces flux, as shown in Fig. 8-1b.   The relation between the direction of flux and the direction of current can be remembered from one version of the *right-hand rule*.   If the thumb of the right hand is placed along the wire in the direction of the current flow, the four fingers of the right hand curl in the direction of the flux about the wire.   Coiling the wire enhances the flux, as does placing certain material, called *ferromagnetic material*, in and around the coil.   For example, a current flowing in a coil wound on an iron cylindrical core produces more flux than the same current flowing in an identical coil wound on a plastic cylinder.

Permeability, with quantity symbol μ, is a measure of this flux-enhancing property.   It has an SI unit of *henry per meter* and a unit symbol of H/m.   (The henry, with unit symbol H, is the SI unit of inductance.)   The permeability of vacuum, designated by μ_0, is $0.4\pi\ \mu$H/m.   Permeabilities of other materials are related to that of vacuum by a factor called the *relative permeability*, with symbol μ_r.   The relation is $\mu = \mu_r\mu_0$.   Most materials have relative permeabilities close to 1, but pure iron has them in the range of 6000 to 8000, and nickel has them in the range of 400 to 1000. Permalloy, an alloy of 78.5 percent nickel and 21.5 percent iron, has a relative permeability of over 80 000.

If a coil of N turns is linked by a ϕ amount of flux, this coil has a flux linkage of $N\phi$. Any change in flux linkages induces a voltage in the coil of

$$v = \lim_{\Delta t \to 0} \frac{\Delta N\phi}{\Delta t} = \frac{d}{dt}(N\phi) = N\frac{d\phi}{dt}$$

This is known as *Faraday's law*. The voltage polarity is such that any current resulting from this voltage produces a flux that opposes the original change in flux.

INDUCTANCE AND INDUCTOR CONSTRUCTION

For most coils, a current i produces flux linkages $N\phi$ that are proportional to i. The equation relating $N\phi$ and i has a constant of proportionality L that is the quantity symbol for the *inductance* of the coil. Specifically, $Li = N\phi$ and $L = N\phi/i$. The SI unit of inductance is the *henry*, with unit symbol H. A component designed to be used for its inductance property is called an *inductor*. The terms "coil" and "choke" are also used. Figure 8-2 shows the circuit symbol for an inductor.

The inductance of a coil depends on the shape of the coil, the permeability of the surrounding material, the number of turns, the spacing of the turns, and other factors. For the single-layer coil shown in Fig. 8-3, the inductance is approximately $L = N^2\mu A/l$, where N is the number of turns of wire, A is the core cross-sectional area in square meters, l is the coil length in meters, and μ is the core permeability. The greater the length to diameter, the more accurate the formula. For a length of 10 times the diameter, the actual inductance is 4 percent less than the value given by the formula.

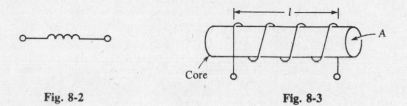

Fig. 8-2   Fig. 8-3

INDUCTOR VOLTAGE AND CURRENT RELATION

Inductance instead of flux is used in analyzing circuits containing inductors. The equation relating inductor voltage, current, and inductance can be found from substituting $N\phi = Li$ into $v = d(N\phi)/dt$. The result is $v = L\,di/dt$, with associated references assumed. If the voltage and current references are not associated, a negative sign must be included. Notice that the voltage at any instant depends on the rate of change of inductor current at that instant, but not at all on the value of current then.

One important fact from $v = L\,di/dt$ is that if an inductor current is constant, not changing, then the inductor voltage is zero because $di/dt = 0$. With a current flowing through it, but zero voltage across it, an inductor acts like a short circuit: *An inductor is a short circuit to dc.* Remember, though, that it is only after an inductor current becomes constant that an inductor acts like a short circuit.

The relation $v = L\,di/dt \simeq L\Delta i/\Delta t$ also means that *an inductor current cannot jump.* For a jump to occur, Δi would be nonzero while Δt was zero, with the result that $\Delta i/\Delta t$ would be infinite, making the inductor voltage infinite. In other words, a jump in inductor current requires an infinite inductor voltage. But, of course, there are no sources of infinite voltage. Inductor voltage has no similar restriction. It can jump or even change polarity instantaneously. Inductor currents not jumping means that inductor currents immediately after a switching operation are the same as immediately before the operation. This is an important fact for *RL* (resistor-inductor) circuit analysis.

TOTAL INDUCTANCE

The total or equivalent inductance (L_T or L_{eq}) of inductors connected in series, as in the circuit shown in Fig. 8-4a, can be found from KVL: $v_S = v_1 + v_2 + v_3$. Substituting from $v = L\ di/dt$ results in

$$L_T \frac{di}{dt} = L_1 \frac{di}{dt} + L_2 \frac{di}{dt} + L_3 \frac{di}{dt}$$

which upon division by di/dt reduces to $L_T = L_1 + L_2 + L_3$. Since the number of series inductors is not significant in this derivation, the result can be generalized to any number of series inductors:

$$L_T = L_1 + L_2 + L_3 + L_4 + \cdots$$

which specifies that the total or equivalent inductance of series inductors is equal to the sum of the individual inductances.

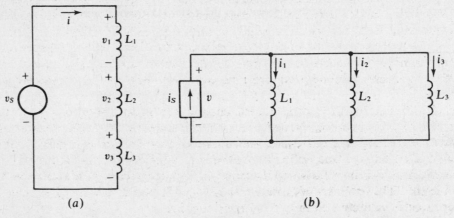

Fig. 8-4

The total inductance of inductors connected in parallel, as in the circuit shown in Fig. 8-4b, can be found starting with the voltage-current equation at the source terminals: $v = L_T \cdot di_S/dt$, and substituting in $i_S = i_1 + i_2 + i_3$:

$$v = L_T \frac{d}{dt}(i_1 + i_2 + i_3) = L_T\left(\frac{di_1}{dt} + \frac{di_2}{dt} + \frac{di_3}{dt}\right)$$

Each derivative can be eliminated using the appropriate $di/dt = v/L$:

$$v = L_T\left(\frac{v}{L_1} + \frac{v}{L_2} + \frac{v}{L_3}\right) \quad \text{or} \quad \frac{1}{L_T} = \frac{1}{L_1} + \frac{1}{L_2} + \frac{1}{L_3}$$

which can also be written as

$$L_T = \frac{1}{1/L_1 + 1/L_2 + 1/L_3}$$

Generalizing,

$$L_T = \frac{1}{1/L_1 + 1/L_2 + 1/L_3 + 1/L_4 + \cdots}$$

which specifies that the total inductance of parallel inductors equals the reciprocal of the sum of the reciprocals of the individual inductances. For the special case of N parallel inductors having the same inductance L, this formula simplifies to $L_T = L/N$. And for two parallel inductors it is $L_T = L_1 L_2/(L_1 + L_2)$. Notice that the formulas for finding total inductances are the same as those for finding total resistances.

ENERGY STORAGE

As can be shown by using calculus, the energy stored in an inductor is

$$w_L = \tfrac{1}{2}Li^2$$

in which w_L is in joules, L is in henries, and i is in amperes. This energy is considered to be stored in the magnetic field surrounding the inductor.

SINGLE-INDUCTOR DC-EXCITED CIRCUITS

When switches open or close in an RL dc-excited circuit with a single inductor, all voltages and currents that are not constant change exponentially from their initial values to their final constant values, as can be proved from differential equations. These exponential changes are the same as those illustrated in Fig. 7-5 for capacitors. Consequently, the voltage and current equations are the same: $v = v(\infty) + [v(0+) - v(\infty)]e^{-t/\tau}$ V and $i = i(\infty) + [i(0+) - i(\infty)]e^{-t/\tau}$ A. The time constant τ, though, is different. It is $\tau = L/R_{Th}$, in which R_{Th} is the circuit Thevenin resistance at the inductor terminals. Of course, in one time constant the voltages and currents change by 63.2 percent of their total changes, and after five time constants they can be considered to be at their final values.

Because of the similarity of the RL and RC equations, it is possible to make RL timers. But, practically speaking, RC timers are much better. One reason is that inductors are not nearly as ideal as capacitors because the coils have resistances that are seldom negligible. Also, inductors are relatively bulky, heavy, and difficult to fabricate using integrated-circuit techniques. Additionally, the magnetic fields extending out from the inductors can induce unwanted voltages in other components. The problems with inductors are significant enough that designers of electronic circuits often exclude inductors entirely from their circuits.

Solved Problems

8.1 Find the voltage induced in a 50-turn coil from a constant flux of 10^4 Wb, and also from a changing flux of 3 Wb/s.

A constant flux linking a coil does not induce any voltage—only a changing flux does. A changing flux of 3 Wb/s induces a voltage of $v = N\,d\phi/dt = 50 \times 3 = 150$ V.

8.2 What is the rate of change of flux linking a 200-turn coil when 50 V is across the coil?

This rate of change is the $d\phi/dt$ in $v = N\,d\phi/dt$:

$$\frac{d\phi}{dt} = \frac{v}{N} = \frac{50}{200} = 0.25 \text{ Wb/s}$$

8.3   Find the number of turns of a coil for which a change of 0.4 Wb/s of flux linking the coil induces a coil voltage of 20 V.

This number of turns is the N in $v = N \, d\phi/dt$:

$$N = \frac{v}{d\phi/dt} = \frac{20}{0.4} = 50 \text{ turns}$$

8.4   Find the inductance of a 100-turn coil that is linked by 3×10^{-4} Wb when a 20-mA current flows through it.

The pertinent formula is $Li = N\phi$.   Thus we have

$$L = \frac{N\phi}{i} = \frac{100(3 \times 10^{-4})}{20 \times 10^{-3}} = 1.5 \text{ H}$$

8.5   Find the approximate inductance of a single-layer coil that has 300 turns wound on a plastic cylinder 12 cm long and 0.5 cm in diameter.

The relative permeability of plastic is so nearly 1 that the permeability of vacuum can be used in the inductance formula for a single-layer cylindrical coil:

$$L = \frac{N^2 \mu A}{l} = \frac{300^2(0.4\pi \times 10^{-6})[\pi \times (0.25 \times 10^{-2})^2]}{12 \times 10^{-2}} \text{ H} = 18.5 \ \mu H$$

8.6   Find the approximate inductance of a single-layer 50-turn coil that is wound on a ferromagnetic cylinder 1.5 cm long and 1.5 mm in diameter.   The ferromagnetic material has a relative permeability of 7000.

$$L = \frac{N^2 \mu A}{l} = \frac{50^2(7000 \times 0.4\pi \times 10^{-6})[\pi \times (0.75 \times 10^{-3})^2]}{1.5 \times 10^{-2}} \text{ H} = 2.59 \text{ mH}$$

8.7   A 3-H inductor has 2000 turns.   How many turns must be added to increase the inductance to 5 H?

In general, inductance is proportional to the square of the number of turns.   By this proportionality,

$$\frac{5}{3} = \frac{N^2}{2000^2} \qquad \text{or} \qquad N = 2000\sqrt{\frac{5}{3}} = 2582 \text{ turns}$$

So, $2582 - 2000 = 582$ turns must be added without making any other changes.

8.8   Find the voltage induced in a 150-mH coil when the current is constant at 4 A.   Also, find the voltage when the current is changing at a rate of 4 A/s.

If the current is constant, $di/dt = 0$ and so the coil voltage is zero.   For a rate of change of $di/dt = 4$ A/s,

$$v = L \frac{di}{dt} = (150 \times 10^{-3})(4) = 0.6 \text{ V}$$

8.9   Find the voltage induced in a 200-mH coil at $t = 3$ ms   if the current increases uniformly from 30 mA at $t = 2$ ms   to 90 mA at $t = 5$ ms.

Because the current increases uniformly, the induced voltage is constant over the time interval.   The rate of increase is $\Delta i/\Delta t$, where Δi is the current at the end of the time interval minus the current at the beginning of the time interval: $90 - 30 = 60$ mA.   Of course, Δt is the time interval: $5 - 2 = 3$ ms. The voltage is

$$v = L \frac{\Delta i}{\Delta t} = \frac{(200 \times 10^{-3})(60 \times 10^{-3})}{3 \times 10^{-3}} = 4 \text{ V} \qquad \text{for} \qquad 2 \text{ ms} < t < 5 \text{ ms}$$

8.10 What is the inductance of a coil for which a changing current increasing uniformly from 30 to 80 mA in 100 μs induces 50 mV in the coil?

Because the increase is uniform (linear), the time derivative of the current equals the quotient of the current change and the time interval:

$$\frac{di}{dt} = \frac{\Delta i}{\Delta t} = \frac{80 \times 10^{-3} - 30 \times 10^{-3}}{100 \times 10^{-6}} = 500 \text{ A/s}$$

Then, from $v = L \, di/dt$,

$$L = \frac{v}{di/dt} = \frac{50 \times 10^{-3}}{500} \text{ H} = 100 \ \mu\text{H}$$

8.11 Find the voltage induced in a 400-mH coil from 0 s to 8 ms when the current shown in Fig. 8-5 flows through the coil.

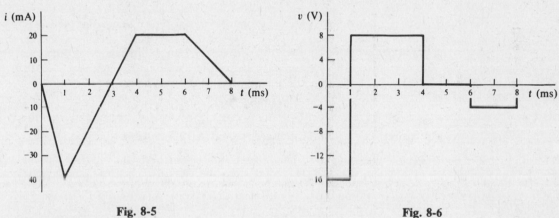

Fig. 8-5   Fig. 8-6

The approach is to find di/dt, the slope, from the graph and insert it into $v = L \, di/dt$ for the various time intervals. For the first millisecond, the current decreases uniformly from 0 A to -40 mA. So, the slope is $(-40 \times 10^{-3} - 0)/(1 \times 10^{-3}) = -40$ A/s, which is the change in current divided by the corresponding change in time. The resulting voltage is $v = L \, di/dt = (400 \times 10^{-3})(-40) = -16$ V. For the next three milliseconds, the slope is $[20 \times 10^{-3} - (-40 \times 10^{-3})]/(3 \times 10^{-3}) = 20$ A/s and the voltage is $v = (400 \times 10^{-3})(20) = 8$ V. For the next two milliseconds, the current graph is horizontal, which means that the slope is zero. Consequently, the voltage is zero: $v = 0$ V. For the last two milliseconds, the slope is $(0 - 20 \times 10^{-3})/(2 \times 10^{-3}) = -10$ A/s and $v = (400 \times 10^{-3})(-10) = -4$ V.

Figure 8-6 shows the graph of voltage. Notice that the inductor voltage can jump and can even instantaneously change polarity.

8.12 Find the total inductance of three parallel inductors having inductances of 45, 60, and 75 mH.

$$L_T = \frac{1}{1/45 + 1/60 + 1/75} = 19.1 \text{ mH}$$

8.13 Find the inductance of the inductor that when connected in parallel with a 40-mH inductor produces a total inductance of 10 mH.

As has been derived, the reciprocal of the total inductance equals the sum of the reciprocals of the inductances of the individual parallel inductors:

$$\frac{1}{10} = \frac{1}{40} + \frac{1}{L} \quad \text{from which} \quad \frac{1}{L} = 0.075 \quad \text{and} \quad L = 13.3 \text{ mH}$$

8.14 Find the total inductance L_T of the circuit shown in Fig. 8-7.

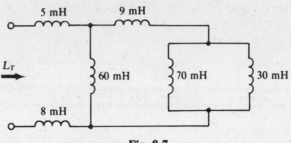

Fig. 8-7

The approach, of course, is to combine inductances starting with inductors at the end opposite the terminals at which L_T is to be found. There, the parallel 70- and 30-mH inductors have a total inductance of $70(30)/(70 + 30) = 21$ mH. This adds to the inductance of the 9-mH series inductor: $21 + 9 = 30$ mH. This combines with the inductance of the parallel 60-mH inductor: $60(30)/(60 + 30) = 20$ mH. And, finally, this adds with the inductances of the series 5- and 8-mH inductors: $L_T = 20 + 5 + 8 = 33$ mH.

8.15 Find the energy stored in a 200-mH inductor that has 10 V across it.

Not enough information is given to determine the stored energy. The inductor current is needed, not the voltage, and there is no way of finding this current from the specified voltage.

8.16 A current $i = 0.32t$ A flows through a 150-mH inductor. Find the energy stored at $t = 4$ s.

At $t = 4$ s the inductor current is $i = 0.32 \times 4 = 1.28$ A, and so the stored energy is

$$w = \tfrac{1}{2}Li^2 = 0.5(150 \times 10^{-3})(1.28)^2 = 0.123 \text{ J}$$

8.17 Find the time constant of the circuit shown in Fig. 8-8.

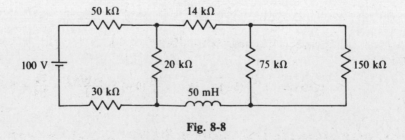

Fig. 8-8

The time constant is L/R_{Th}, where R_{Th} is the Thevenin resistance of the circuit at the inductor terminals.   For this circuit,

$$R_{Th} = (50 + 30)\|20 + 14 + 75\|150 = 80 \text{ k}\Omega$$

and $\tau = (50 \times 10^{-3})/(80 \times 10^3) \text{ s} = 0.625 \text{ } \mu\text{s}.$

8.18   What is the energy stored in the inductor of the circuit shown in Fig. 8-8?

The inductor current is needed.   Presumably, the circuit has been constructed long enough ($5\tau = 5 \times 0.625 = 3.13 \text{ } \mu\text{s}$) for the inductor current to become constant and so for the inductor to be a short circuit.   The current in this short circuit can be found from Thevenin's resistance and voltage.   The Thevenin resistance is 80 kΩ, as found in the solution to Prob. 8.17.   The Thevenin voltage is the voltage across the 20-kΩ resistor if the inductor is replaced by an open circuit.   This voltage will appear across the open circuit since the 14-, 75-, and 150-kΩ resistors will not carry any current. By voltage division, this voltage is

$$V_{Th} = \frac{20}{20 + 50 + 30} \times 100 = 20 \text{ V}$$

Because of the short-circuit inductor load, the inductor current is $V_{Th}/(R_{Th} + 0) = 20/80 - 0.25$ mA,   and the stored energy is $0.5(50 \times 10^{-3})(0.25 \times 10^{-3})^2 \text{ J} = 1.56 \text{ nJ}.$

8.19   Closing a switch connects in series a 20-V source, a 2-Ω resistor, and a 3.6-H inductor.   How long does it take the current to get to its maximum value, and what is this value?

The current reaches its maximum value five time constants after the switch closes: $5L/R = 5(3.6)/2 = 9$ s.   Since the inductor acts like a short circuit at that time, only the resistance limits the current: $i(\infty) = 20/2 = 10$ A.

8.20   Closing a switch connects in series a 21-V source, a 3-Ω resistor, and a 2.4-H inductor.   Find   (a) the initial and final currents,   (b) the initial and final inductor voltages,   and   (c) the initial rate of current increase.

(a)   Immediately after the switch closes, the inductor current is 0 A because it was 0 A immediately before the switch closed, and an inductor current cannot jump.   The current increases from 0 A until it reaches its maximum value five time constants ($5 \times 2.4/3 = 4$ s) after the switch closes.   Then, because the current is constant, the inductor becomes a short circuit, and so $i(\infty) = V/R = 21/3 = 7$ A.

(b)   Since the current is zero immediately after the switch closes, the resistor voltage is 0 V, which means, by KVL, that all the source voltage is across the inductor: The initial inductor voltage is 21 V.   Of course, the final inductor voltage is zero because the inductor is a short circuit to dc after five time constants.

(c)   As can be seen from Fig. 7-5b, the current initially increases at a rate such that the final current value would be reached in one time constant if the rate did not change.   This initial rate is

$$\frac{i(\infty) - i(0+)}{\tau} = \frac{7 - 0}{0.8} = 8.75 \text{ A/s}$$

Another way of finding this initial rate, which is di/dt at $t = 0+$,   is from the initial inductor voltage:

$$v_L(0+) = L \frac{di}{dt}(0+) \quad \text{or} \quad \frac{di}{dt}(0+) = \frac{v_L(0+)}{L} = \frac{21}{2.4} = 8.75 \text{ A/s}$$

8.21   A closed switch connects a 120-V source to the field coils of a dc motor.   These coils have 6 H of inductance and 30 Ω of resistance.   A discharge resistor in parallel with the coil limits the maximum coil and switch voltages at the instants at which the switch is opened.   Find the maximum value of the discharge resistor that will prevent the coil voltage from exceeding 300 V.

With the switch closed, the current in the coils is   120/30 = 4 A   because the inductor part of the coils is a short circuit.   Immediately after the switch is opened, the current must still be 4 A because an inductor current cannot jump—the magnetic field about the coil will change to produce whatever coil voltage is necessary to maintain this 4 A.   In fact, if the discharge resistor were not present, this voltage would become great enough—thousands of volts—to produce arcing at the switch contacts to provide a current path to enable the current to decrease continuously.   Such a large voltage might be destructive to the switch contacts and to the coil insulation.   The discharge resistor provides an alternative path for the inductor current, which has a maximum value of 4 A.   To limit the coil voltage to 300 V, the maximum value of discharge resistance is   300/4 = 75 Ω.   Of course, any value less than 75 Ω will limit the voltage to less than 300 V, but a smaller resistance will result in more power dissipation when the switch is closed.

8.22   In the circuit shown in Fig. 8-9, find the indicated currents a long time after the switch has been in position 1.

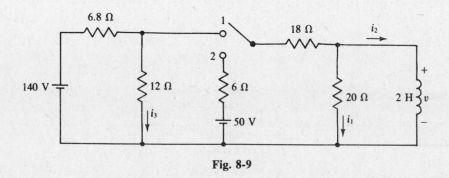

Fig. 8-9

Of course, the inductor is a short circuit, shorting out the 20-Ω resistor.   As a result, $i_1 = 0$ A.   This short circuit also places the 18-Ω resistor in parallel with the 12-Ω resistor.   Together they have a total resistance of   18(12)/(18 + 12) = 7.2 Ω.   This adds to the resistance of the series 6.8-Ω resistor to produce   7.2 + 6.8 = 14 Ω   at the source terminals.   So, the source current is   140/14 = 10 A.   By current division,

$$i_2 = \frac{12}{12 + 18} \times 10 = 4 \text{ A} \qquad \text{and} \qquad i_3 = \frac{18}{12 + 18} \times 10 = 6 \text{ A}$$

8.23   For the circuit shown in Fig. 8-9, find the indicated voltage and currents immediately after the switch is thrown to position 2 from position 1, where it has been a long time.

As soon as the switch leaves position 1, the left-hand side of the circuit is isolated, becoming a series circuit in which $i_3 = 140/(6.8 + 12) = 7.45$ A.   In the other part of the circuit, the inductor current cannot jump, and is 4 A as was found in the solution to Prob. 8.22: $i_2 = 4$ A.   Since this is a known current, it can be considered to be from a current source, as shown in Fig. 8-10.   Remember, though, that this circuit is valid only for the one instant of time immediately after the switch is thrown to position 2.   By nodal analysis,

$$\frac{v}{20} + \frac{v - 50}{6 + 18} + 4 = 0 \qquad \text{from which} \qquad v = -20.9 \text{ V}$$

And $i_1 = v/20 = -20.9/20 = -1.05$ A.

This technique of replacing inductors in a circuit by current sources is completely general for an analysis at an instant of time immediately after a switching operation.   (Similarly, capacitors can be replaced by voltage sources.)   Of course, if an inductor current is zero, then the current source carries 0 A and so is equivalent to an open circuit.

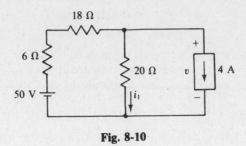

Fig. 8-10

8.24 A short is placed across a coil that at the time is carrying 0.5 A. If the coil has an inductance of 0.5 H and a resistance of 2 Ω, what is the coil current 0.1 s after the short is applied?

The current equation is needed. For the basic formula $i = i(\infty) + [i(0+) - i(\infty)]e^{-t/\tau}$, the initial current is $i(0+) = 0.5$ A because the inductor current cannot jump, the final current is $i(\infty) = 0$ A because the current will decay to zero after all the initially stored energy is dissipated in the resistance, and the time constant is $\tau = L/R = 0.5/2 = 0.25$ s. So,

$$i(t) = 0 + (0.5 - 0)e^{-t/0.25} = 0.53^{-4t} \text{ A}$$

and $i(0.1) = 0.5e^{-4(0.1)} = 0.335$ A.

8.25 A coil for a relay has a resistance of 30 Ω and an inductance of 2 H. If the relay requires 250 mA to operate, how soon will it operate after 12 V is applied to the coil?

For the current formula, $i(0+) = 0$ A, $i(\infty) = 12/30 = 0.4$ A, and $\tau = 2/30 = 1/15$ s. So,

$$i = 0.4 + (0 - 0.4)e^{-15t} = 0.4(1 - e^{-15t}) \text{ A}$$

The time at which the current is 250 mA $= 0.25$ A can be found by substituting 0.25 for i and solving for t:

$$0.25 = 0.4(1 - e^{-15t}) \qquad \text{or} \qquad e^{-15t} = 0.375$$

Taking the natural logarithm of both sides results in

$$\ln e^{-15t} = \ln 0.375 \qquad \text{from which} \qquad -15t = -0.9809 \qquad \text{and} \qquad t = 65.4 \text{ ms}$$

8.26 For the circuit shown in Fig. 8-11, find v and i for $t > 0$ s if at $t = 0$ s the switch is thrown to position 2 after having been in position 1 for a long time.

The shown switch is a make-before-break switch that makes contact at the beginning of position 2 before breaking contact at position 1. This temporary double contacting provides a path for the inductor current during switching and prevents arcing at the switch contacts. To find the voltage and current, it is only necessary to get their initial and final values, along with the time constant, and insert these into the voltage and current formulas. The initial current $i(0+)$ is the same as the inductor current immediately before the switching operation, with the switch in position 1: $i(0+) = 50/(4 + 6) = 5$ A. When the switch is in position 2, this current produces initial voltage drops of $5 \times 6 = 30$ V and $14 \times 5 = 70$ V across the 6- and 14-Ω resistors, respectively. By KVL, $30 + 70 + v(0+) = 20$, from which $v(0+) = -80$ V. For the final values, clearly $v(\infty) = 0$ V and $i(\infty) = 20/(14 + 6) = 1$ A. The time constant is $4/20 = 0.2$ s. With these values inserted, the voltage and current formulas are

$$v = 0 + (-80 - 0)e^{-t/0.2} = -80e^{-5t} \text{ V} \qquad \text{for} \qquad t > 0 \text{ s}$$
$$i = 1 + (5 - 1)e^{-t/0.2} = 1 + 4e^{-5t} \text{ A} \qquad \text{for} \qquad t > 0 \text{ s}$$

8.27 For the circuit shown in Fig. 8-12, find i for $t > 0$ s if the switch is closed at $t = 0$ s after being open for a long time.

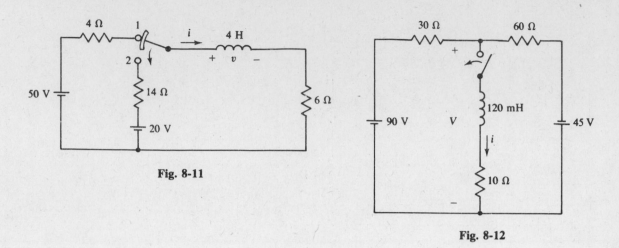

Fig. 8-11

Fig. 8-12

A good approach is to use the Thevenin equivalent circuit at the inductor terminals. The Thevenin resistance is easy to find because the resistors are in series-parallel when the sources are killed: $R_{Th} = 10 + 30\|60 = 30 \ \Omega$. The Thevenin voltage is the indicated V with the center branch removed because replacing the inductor by an open circuit prevents the center branch from affecting this voltage. By nodal analysis,

$$\frac{V - 90}{30} + \frac{V - (-45)}{60} = 0 \qquad \text{from which} \qquad V = 45 \ \text{V}$$

So, the Thevenin equivalent circuit is a 30-Ω resistor in series with a 45-V source, and the polarity of the source is such as to produce a positive current i. With the Thevenin circuit connected to the inductor, it should be obvious that $i(0+) = 0 \ \text{A}$, $i(\infty) = 45/30 = 1.5 \ \text{A}$, $\tau = (120 \times 10^{-3})/30 = 4 \times 10^{-3} \ \text{s}$, and $1/\tau = 250$. These values inserted into the current formula result in $i = 1.5 - 1.5e^{-250t} \ \text{A}$ for $t > 0 \ \text{s}$.

8.28 In the circuit shown in Fig. 8-13, switch S_1 is closed at $t = 0 \ \text{s}$, and switch S_2 is opened at $t = 3 \ \text{s}$. Find $i(2)$ and $i(4)$, and make a sketch of i for $t > 0 \ \text{s}$.

Two equations for i are needed: one with both switches closed, and the other with switch S_1 closed and switch S_2 open. At the time that S_1 is closed, $i(0+) = 0 \ \text{A}$, and i starts increasing toward a final value of $i(\infty) = 6/(0.1 + 0.2) = 20 \ \text{A}$. The time constant is $1.2/(0.1 + 0.2) = 4 \ \text{s}$. The 1.2-$\Omega$ resistor does not affect the current or time constant because this resistor is shorted by switch S_2. So, for the first three seconds, $i = 20 - 20e^{-t/4} \ \text{A}$, and from this, $i(2) = 20 - 20e^{-2/4} = 7.87 \ \text{A}$.

After switch S_2 opens at $t = 3 \ \text{s}$, the equation for i must change because the circuit changes as a result of the insertion of the 1.2-Ω resistor. With the switching occurring at $t = 3 \ \text{s}$ instead of at $t = 0 \ \text{s}$, the basic formula for i is $i = i(\infty) + [i(3+) - i(\infty)]e^{-(t-3)/\tau} \ \text{A}$. The current $i(3+)$ can be calculated from the first i equation since the current cannot jump at $t = 3 \ \text{s}$: $i(3+) = 20 - 20e^{-3/4} = 10.55 \ \text{A}$. Of course, $i(\infty) = 6/(0.1 + 1.2 + 0.2) = 4 \ \text{A}$ and $\tau = 1.2/1.5 = 0.8 \ \text{s}$. With these values inserted, the current formula is

$$i = 4 + (10.55 - 4)e^{-(t-3)/0.8} = 4 + 6.55e^{-1.25(t-3)} \ \text{A} \qquad \text{for} \qquad t > 3 \ \text{s}$$

from which $i(4) = 4 + 6.55e^{-1.25(4-3)} = 5.88 \ \text{A}$.

Figure 8-14 shows the graph of current based on the two current equations.

8.29 In the circuit shown in Fig. 8-15, the switch is closed at $t = 0 \ \text{s}$ after being open a long time. Find the indicated voltages and currents both immediately after and a long time after the switch closes.

Immediately after the switch closes, the inductors are open circuits because they carried zero current before the switch closed, and inductor currents cannot jump. With the inductors replaced by

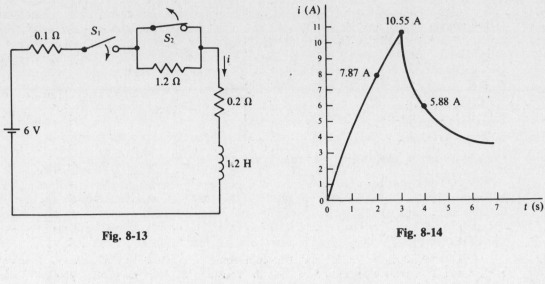

Fig. 8-13   Fig. 8-14

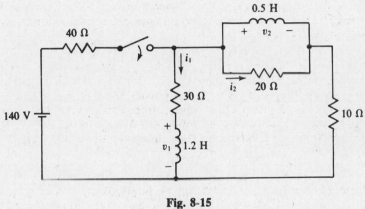

Fig. 8-15

open circuits, the circuit is, as shown in Fig. 8-16a, a single loop—the outside loop. From this circuit, the currents are $i_1(0+) = 0$ A and $i_2(0+) = 140/(40 + 20 + 10) = 2$ A. The voltage $v_1(0+)$ is the source voltage minus the drop across the 40-Ω resistor: $v_1(0+) = 140 - 2 \times 40 = 60$ V. Of course, $v_2(0+) = 20i_2(0+) = 20 \times 2 = 40$ V.

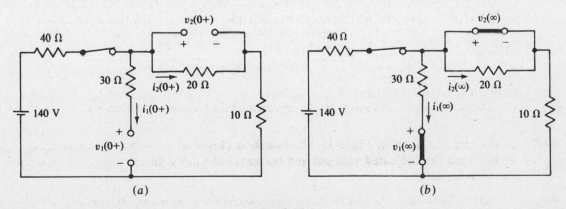

Fig. 8-16

A long time after the switch closes, the inductors are short circuits, as shown in Fig. 8-16b.   As a result, $v_1(\infty) = 0$ V   and $v_2(\infty) = 0$ V.   Also, $i_2(\infty) = 0$ A   because the 20-Ω resistor is short-circuited.   The short circuits also place the 10-Ω resistor in parallel with the 30-Ω resistor for a total resistance of $30(10)/(30 + 10) = 7.5$ Ω.   This resistance added to that of the 40-Ω series resistor is the resistance at the source terminals: $40 + 7.5 = 47.5$ Ω.   Consequently, the source current is $140/47.5 = 2.947$ A.   From it, by current division,

$$i_1(\infty) = \frac{10}{10 + 30} \times 2.947 = 0.737 \text{ A}$$

8.30   For the circuit shown in Fig. 8-15, find i_1 for $t > 0$ s.

Even though the initial and final values of i_1 are known from the solution to Prob. 8.29, i_1 cannot be found from $i = i(\infty) + [i(0+) - i(\infty)]e^{-t/\tau}$.   The reason is that this formula applies only to a dc-excited or initial-condition-excited circuit with a single inductor (or a single capacitor), but the circuit in Fig. 8-15 has two inductors that cannot be combined into a single equivalent inductor.   Advanced mathematical techniques are required to find i_1 for $t > 0$ s.

Supplementary Problems

8.31   Find the voltage induced in a 500-turn coil when the flux changes uniformly by 16×10^{-5} Wb in 2 ms.   *Ans.* 40 V

8.32   Find the change in flux linking an 800-turn coil when 3.2 V is induced for 6 ms.   *Ans.* 24 μWb

8.33   What is the number of turns of a coil for which a flux change of 40×10^{-6} Wb in 0.4 ms induces 70 V in the coil?   *Ans.* 700 turns

8.34   Find the flux linking a 500-turn, 0.1-H coil carrying a 2-mA current.   *Ans.* 0.4 μWb

8.35   Find the approximate inductance of a single-layer, 300-turn air-core coil that is 3 in long and 0.25 in in diameter.   *Ans.* 47 μH

8.36   Find the approximate inductance of a single-layer 500-turn coil that is wound on a ferromagnetic cylinder that is 1 in long and 0.1 in in diameter.   The ferromagnetic material has a relative permeability of 8000.   *Ans.* 0.501 H

8.37   A 250-mH inductor has 500 turns.   How may turns must be added to increase the inductance to 400 mH?   *Ans.* 132 turns

8.38   The current in a 300-mH inductor increases uniformly from 0.2 to 1 A in 0.5 s.   What is the inductor voltage for this time?   *Ans.* 0.48 V

8.39   If a change in current in a 0.2-H inductor produces a constant 5-V inductor voltage, how long does the current take to increase from 30 to 200 mA?   *Ans.* 6.8 ms

8.40   What is the inductance of a coil for which a changing current increasing uniformly from 150 to 275 mA in 300 μs induces 75 mV in the coil?   *Ans.* 180 μH

8.41 Find the voltage induced in a 200-mH coil from 0 to 5 ms when a current i described as follows flows through the coil: $i = 250t$ A for $0 < t < 1$ ms, $i = 250$ mA for 1 ms $< t < 2$ ms, and $i = 416 - 83\,000t$ mA   for 2 ms $< t < 5$ ms.
Ans. $v = 50$ V for $0 < t < 1$ ms,
$\quad\quad v = 0$ V   for 1 ms $< t < 2$ ms,
$\quad\quad v = -16.6$ V   for 2 ms $< t < 5$ ms

8.42 Find the total inductance of four parallel inductors having inductances of 80, 125, 200, and 350 mH.   *Ans.* 35.3 mH

8.43 Find the total inductance of a 40-mH inductor in series with the parallel combination of a 60-mH inductor, an 80-mH inductor, and a 100-mH inductor.   *Ans.* 65.5 mH

8.44 A 2-H inductor, a 430-Ω resistor, and a 50-V source have been connected in series for a long time.   What is the energy stored in the inductor?   *Ans.* 13.5 mJ

8.45 A current $i = 0.56t$ A   flows through a 0.5-H inductor.   Find the energy stored at $t = 6$ s.
Ans. 2.82 J

8.46 What is the energy stored by the inductor in the circuit shown in Fig. 8.17 if $R = 20\ \Omega$?
Ans. 667 mJ

8.47 Find the time constant of the circuit shown in Fig. 8-17 for $R = 90\ \Omega$.   *Ans.* 4.21 ms

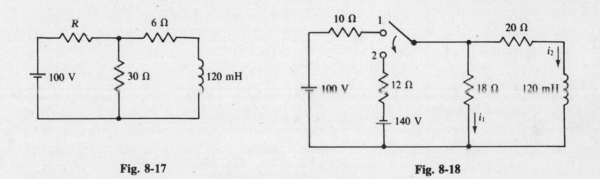

Fig. 8-17   **Fig. 8-18**

8.48 How long after a short circuit is placed across a coil carrying a current of 2 A does the current go to zero if the coil has 1.2 H of inductance and 40 Ω of resistance?   Also, how much energy is dissipated?   *Ans.* 0.15 s, 2.4 J

8.49 A switch closing connects in series a 10-V source, an 8.2-Ω resistor, and a 1.2-H inductor.   How long does the current take to reach its maximum value, and what is this value?   *Ans.* 732 ms, 1.22 A

8.50 In closing, a switch connects a 100-V source with 5 Ω of internal resistance across the parallel combination of a 20-Ω resistor and a 0.4-H inductor.   What are the initial and final source currents, and what is the initial rate of inductor current increase?   *Ans.* 4 A, 20 A, 200 A/s

8.51 In the circuit shown in Fig. 8-18, the switch is thrown at $t = 0$ s   from an open position to position 1.   Find the indicated currents at $t = 0+$ s   and also at a long time later.
Ans. $i_1(0+) = 3.57$ A, $i_2(0+) = 0$ A, $i_1(\infty) = 2.7$ A, $i_2(\infty) = 2.43$ A

8.52   In the circuit shown in Fig. 8-18, the switch is thrown at $t = 0$ s   to position 2 from position 1 where
it has been a long time.   Find the indicated currents at $t = 0+$ s   and also at a long time later.
Ans. $i_1(0+) = -5.64$ A, $i_2(0+) = 2.43$ A, $i_1(\infty) = -3.43$ A, $i_2(\infty) = -3.09$ A

8.53   A switch closing at $t = 0$ s   connects a 20-mH inductor to a 40-V source that has 10 Ω of internal
resistance.   Find the inductor voltage and current for $t > 0$ s.
Ans. $v = 40e^{-500t}$ V, $i = 4(1 - e^{-500t})$ A

8.54   A switch closing at $t = 0$ s   connects a 100-V source with a 15-Ω internal resistance to a coil that has
200 mH of inductance and 5 Ω of resistance.   Find the coil voltage for $t > 0$ s.
Ans. $25 + 75e^{-100t}$ V

8.55   A coil for a relay has a resistance of 20 Ω and an inductance of 1.2 H.   The relay requires 300 mA to
operate.   How soon will the relay operate after a 20-V source with 5 Ω of internal resistance is applied
to the coil?   *Ans.* 22.6 ms

8.56   For the circuit shown in Fig. 8-19, find i as a function of time after the switch closes at $t = 0$ s.   *Ans.* $0.04(1 - e^{-500t})$ A

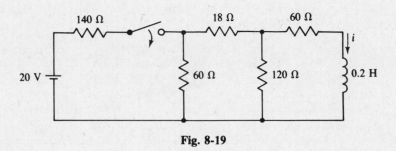

Fig. 8-19

8.57   Assume that the switch in the circuit shown in Fig. 8-19 has been closed a long time.   Find i as a
function of time after the switch opens at $t = 0$ s.   *Ans.* $0.04e^{-536t}$ A

8.58   In the circuit shown in Fig. 8-20, the switch is thrown to position 1 at $t = 0$ s   after being open a long
time.   Then it is thrown to position 2 at $t = 2.5$ s.   Find i for $t > 0$ s.
Ans. $50(1 - e^{-0.1t})$ A   for 0 s $< t < 2.5$ s, $-20 + 31.1e^{-0.05(t-2.5)}$ A   for $t > 2.5$ s

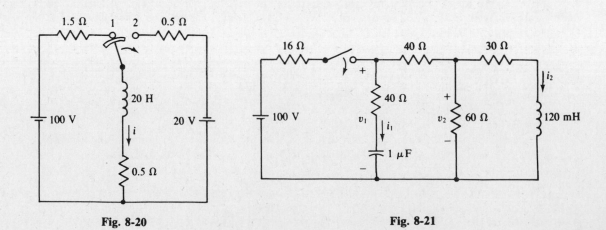

Fig. 8-20   **Fig. 8-21**

8.59   In the circuit shown in Fig. 8-21, the switch is closed at $t = 0$ s   after being open a long time.   Find the indicated voltages and currents both immediately after and a long time after the switch is closed.

   Ans. $i_1(0+) = 1.6$ A $i_1(\infty) = 0$ A
            $i_2(0+) = 0$ A $i_2(\infty) = 0.877$ A
            $v_1(0+) = 64.1$ V $v_1(\infty) = 78.9$ V
            $v_2(0+) = 38.5$ V $v_2(\infty) = 26.3$ V

8.60   In the circuit shown in Fig. 8-21, the switch is opened at $t = 0$ s   after being closed a long time.   Find the indicated voltages and currents immediately after the switch is opened.

   Ans. $i_1(0+) = -0.94$ A $v_1(0+) = 41.4$ V
            $i_2(0+) = 0.877$ A $v_2(0+) = 3.76$ V

Chapter 9

Sinusoidal Alternating Voltage
and Current

INTRODUCTION

In the circuits considered so far, the independent sources have all been dc. From this point on, though, the circuits have *alternating-current* (*ac*) sources.

An ac voltage (or ac current) varies *sinusoidally* with time, as shown in Fig. 9-1a. This is a *periodic* voltage since it varies with time such that it continually repeats. The smallest nonrepeatable portion of a periodic waveform is a *cycle*, and the duration of a cycle is the *period T* of the wave. The reciprocal of the period, and the number of cycles in a period, is the *frequency*, which has a quantity symbol f:

$$f = \frac{1}{T}$$

The SI unit of frequency is the *hertz*, with unit symbol Hz.

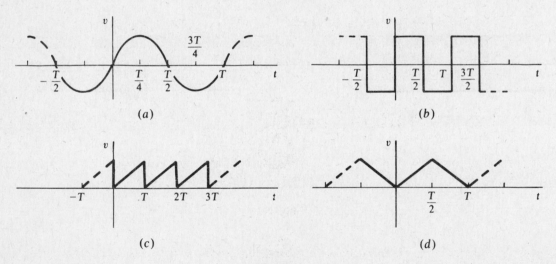

Fig. 9-1

In these definitions, notice the terms *wave* and *waveform*. They do not refer to the same thing. A wave is a varying voltage or current, but a waveform is a graph of such a voltage or current. Often, however, these terms are used interchangeably.

Although the *sine wave* of Fig. 9-1a is by far the most common periodic wave, there are other common ones: Figure 9-1b shows a square wave, Fig. 9-1c a sawtooth wave, and Fig. 9-1d a triangular wave. The dashed lines at both ends indicate that the waves have no beginnings and no ends, as is strictly required for periodic waves. But, of course, all practical voltages and currents have beginnings and ends. When a wave is obviously periodic, these dashed lines are often omitted.

The voltage waveforms shown in Fig. 9-1a and b are negative or below the time axis for part of each period. During these times, the corresponding voltages have polarities opposite the reference polarities. Of course, when the waveforms are above the time axis, these voltages have the same polarities as the references. For similar graphs of currents, the currents flow in the current reference directions when the waveforms are above the time axis, and in opposite directions when the waveforms are below that axis.

SINE AND COSINE WAVES

Figure 9-2 shows the basics of an ac generator or alternator for generating a sinusoidal voltage. The conductor, which in practice is a coil of wire, is rotated by a steam turbine or by some other source of mechanical energy. This rotation causes a continuous change of magnetic flux linking the conductor, thereby inducing a sine wave voltage in the conductor. This change of flux, and so the induced voltage, varies from zero when the conductor is horizontal to a maximum when the conductor is vertical. If $t = 0$ s corresponds to a time when the conductor is horizontal and the induced voltage is increasing, the induced voltage is $v = V_m \sin \omega t$, where V_m is the peak value or *amplitude*, sin is the operation designator for a sine wave, ωt is the *argument*, and ω is the quantity symbol for the *radian frequency* of the voltage. (Some authors use the terms "angular velocity" or "angular frequency" instead of radian frequency.) The SI unit of radian frequency is *radian per second*, and the unit symbol is rad/s. The frequency f and the radian frequency ω are related by

$$\omega = 2\pi f$$

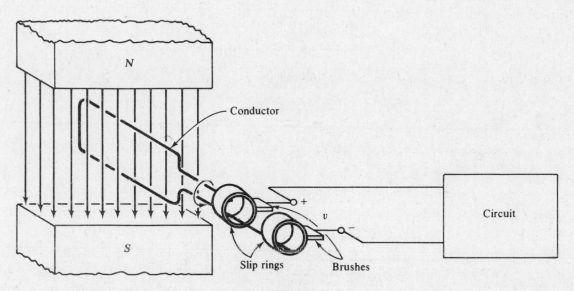

Fig. 9-2

The *radian* in radian per second is an SI angular unit, with symbol rad, and it is an alternative to degrees. A radian is the angle subtended by an arc on the circumference of a circle if the arc has a length equal to the radius. Since the circumference of a circle equals $2\pi r$, where r is the radius, it follows that 2π rad equals 360° or

$$1 \text{ rad} = \frac{360°}{2\pi} = \frac{180°}{\pi} = 57.3°$$

This relation is useful for converting from degrees to radians and from radians to degrees. Specifically,

$$\text{Angle in radians} = \frac{\pi}{180°} \times \text{angle in degrees}$$

and

$$\text{Angle in degrees} = \frac{180°}{\pi} \times \text{angle in radians}$$

The waveform of sin ωt has the shape shown in Fig. 9-1a. In each cycle it varies from 0 to a positive peak or maximum of 1, back to 0, then to a negative peak or minimum of -1, and back to 0 again.   For any value of the argument ωt, sin ωt can be evaluated with a calculator operated in the radians mode.   Alternatively, the argument can be converted to degrees and the calculator operated in the more popular decimal degrees mode.   For example,   sin $(\pi/6)$ = sin 30° = 0.5.

The abscissa of a graph of a sine wave can be expressed in radians, degrees, or time.   Sometimes, when time is used, it is in fractions of the period T, as in Fig. 9-1a.   Usually, determining what the fractions should be is obvious from the corresponding proportions of a cycle.

Consider the graphing of one cycle of a specific ac voltage: v_1 = 20 sin 377t V.   The peak value or amplitude is 20 V because sin 377t has a maximum value of 1.   The radian frequency is ω = 377 rad/s, which corresponds to $f = \omega/2\pi$ = 60 Hz,   the frequency of the electrical power systems in the United States.   The period is T = 1/60 = 16.7 ms.   A cycle of this voltage can be

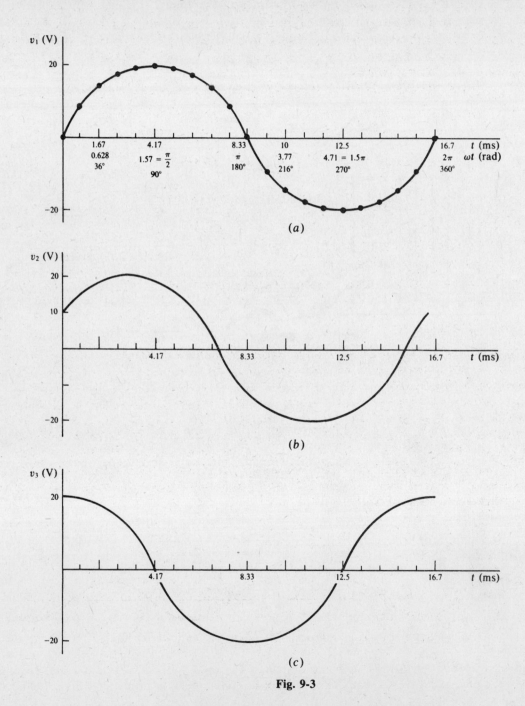

Fig. 9-3

plotted by substituting into $20 \sin 377t$ different times for t from the time interval of $t = 0$ s to $t = 16.7$ ms. Figure 9-3a shows the results of evaluating this sine wave at 21 different times and drawing a smooth curve through the plotted points. For comparison purposes, all three abscissa units—seconds, radians, and degrees—are shown.

Figure 9-3b shows a graph of one cycle of $v_2 = 20 \sin (377t + 30°)$ V. Notice that the argument $377t + 30°$ is the sum of two terms, the first of which is in radians and the second of which is in degrees. Showing such an addition is common despite the fact that before the terms can be added, either the first term must be converted to degrees, or the second term must be converted to radians. The $30°$ in the argument is called the *phase angle*.

The *cosine wave*, with designator cos, is as important as the sine wave. Its waveform has the same shape as the sine waveform, but is shifted $90°$—a fourth of a period—ahead of it. Sine and cosine waves are so similar that the same term "sinusoid" is applied to both as well as to phase-shifted sine and cosine waves. Figure 9-3c is a graph of $v_3 = 20 \sin (377t + 90°) = 20 \cos 377t$ V. Notice that the values of the cosine wave v_3 occur one-fourth period earlier than corresponding ones for the sine wave v_1.

Some sine and cosine identities are important in the study of ac circuit analysis:

$$\sin (-x) = -\sin x \qquad \cos (-x) = \cos x \qquad \sin (x + 90°) = \cos x$$

$$\sin (x - 90°) = -\cos x \qquad \cos (x + 90°) = -\sin x \qquad \cos (x - 90°) = \sin x$$

$$\sin (x \pm 180°) = -\sin x \qquad \cos (x \pm 180°) = -\cos x \qquad \sin^2 x = \frac{1 - \cos 2x}{2}$$

$$\cos^2 x = \frac{1 + \cos 2x}{2} \qquad \sin (x + y) = \sin x \cos y + \sin y \cos x$$

$$\sin (x - y) = \sin x \cos y - \sin y \cos x$$

$$\cos (x + y) = \cos x \cos y - \sin x \sin y$$

$$\cos (x - y) = \cos x \cos y + \sin x \sin y$$

$$\sin x = \sin (x \pm N \times 360°) \qquad \text{and} \qquad \cos x = \cos (x \pm N \times 360°) \qquad \text{for any integer } N$$

PHASE RELATIONS

Sinusoids of the *same frequency* have *phase relations* that have to do with the angular difference of the sinusoidal arguments. For example, because of the added $30°$ in its argument, $v_2 = 20 \sin (377t + 30°)$ V of the last section *leads* $v_1 = 20 \sin 377t$ V by $30°$. Alternatively, v_1 *lags* v_2 by $30°$. This means that the peaks, zeros, and other values of v_2 occur earlier than those of v_1 by a time corresponding to $30°$. Another but less specific way of expressing this phase relation is to say that v_1 and v_2 have a $30°$ *phase difference* or that they are $30°$ *out of phase*. Similarly, the cosine wave v_3 leads the sine wave v_1 by $90°$ or v_1 lags v_3 by $90°$. They have a phase difference of $90°$; they are $90°$ out of phase. Sinusoids that have a $0°$ phase difference are said to be *in phase*. Figure 9-4a shows sinusoids that are in phase, and Fig. 9-4b shows sinusoids that are $180°$ out of phase.

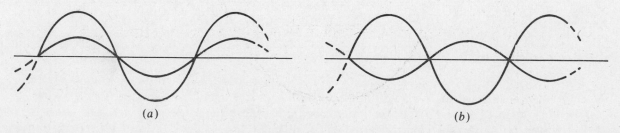

(a)   (b)

Fig. 9-4

The phase difference between two sinusoids can be found by subtracting the phase angle of one from that of the other, provided that both sinusoids have either the sine form or the cosine form, and that the amplitudes have the same sign—both positive or both negative. Additionally, of course, the two sinusoids must have the same frequency.

AVERAGE VALUE

The average value of a periodic wave is a quotient of area and time—the area being that between the corresponding waveform and the time axis for one period, and the time being one period. Areas above the time axis are positive, and areas below are negative. The areas must be algebraically added (signs must be included) to get the total area between the waveform and time axis for one period.

The average value of a sinusoid is zero because over one period the positive and negative areas cancel in the sum of the two areas. For some purposes, though, a nonzero "average" is used. By definition, it is the average of a positive half-cycle. From calculus, this average is $2/\pi = 0.637$ of the peak value.

RESISTOR SINUSOIDAL RESPONSE

If a resistor of R ohms has a voltage $v = V_m \sin(\omega t + \theta)$ across it, the current is, by Ohm's law, $i = v/R = (V_m/R)\sin(\omega t + \theta)$. The multiplier V_m/R is the current peak I_m: $I_m = V_m/R$. Notice that the current is in phase with the voltage. To repeat, *a resistor current and voltage are in phase.*

Instantaneous resistor power dissipation varies with time because the instantaneous voltage and current vary with time, and the power is the product of the two. Specifically,

$$p = vi = [V_m \sin(\omega t + \theta)][I_m \sin(\omega t + \theta)] = V_m I_m \sin^2(\omega t + \theta)$$

which shows that the peak power is $P_m = V_m I_m$, and it occurs each time that $\sin(\omega t + \theta) = \pm 1$. From the identity $\sin^2 x = (1 - \cos 2x)/2$,

$$p = \frac{V_m I_m}{2} - \frac{V_m I_m}{2} \cos(2\omega t + 2\theta)$$

which is a constant plus a sinusoid of twice the frequency of the voltage and current. This instantaneous power is zero each time that the voltage and current are zero, but it is never negative because the positive first term is always equal to or greater than the second term, which is negative half the time. The power never being negative means that a resistor never delivers power to a circuit. Rather, it dissipates as heat all the energy it receives.

The average power supplied to a resistor is just the first term: $P_{av} = V_m I_m/2$, because the average value of the second term is zero. From $V_m = I_m R$,

$$P_{av} = \frac{V_m I_m}{2} = \frac{V_m^2}{2R} = \frac{I_m^2 R}{2}$$

These formulas differ from the corresponding dc formulas by a factor of $\frac{1}{2}$.

EFFECTIVE OR RMS VALUES

Although periodic voltages and currents vary with time, it is convenient to associate with them specific values called *effective values*. Effective voltages are used, for example, in the rating of electrical appliances. The 120-V rating of an electric hair dryer and the 240-V rating of an electric clothes dryer are effective values. Also, most ac ammeters and voltmeters give readings in effective values.

By definition, the effective value of a periodic voltage or current (V_{eff} or I_{eff}) is the *positive* dc voltage or current that produces the same average power loss in a resistor: $P_{av} = V_{\text{eff}}^2/R$ and $P_{av} = I_{\text{eff}}^2 R$. Since for a sinusoidal voltage the average power loss is $P_{av} = V_m^2/2R$,

$$\frac{V_{rms}^2}{R} = P_{av} = \frac{V_{\text{eff}}^2}{R} = \frac{V_m^2}{2R} \quad \text{from which} \quad V_{\text{eff}} = \frac{V_m}{\sqrt{2}} = 0.707 V_m = V_{rms} =$$

Similarly, $I_{\text{eff}} = I_m/\sqrt{2} = 0.707 I_m$. So, *the effective value of a sinusoidal voltage or current equals the peak value divided by* $\sqrt{2}$.

Another name for effective value is *root mean square*, or more commonly *rms*, the abbreviation. The corresponding voltage and current notations are V_{rms} and I_{rms}, which are the same as V_{eff} and I_{eff}. This name stems from a procedure for finding the effective or rms value of any periodic voltage or current—not just sinusoids. As can be derived using calculus, this procedure is to

1. *Square* the periodic voltage or current.
2. Find the average of this squared wave over one period. Another name for this average is the *mean*.
3. Find the positive square *root* of this average.

Unfortunately, except for square-type waves, finding the area in step 2 requires calculus. Incidentally, if this procedure is applied to a sawtooth and a triangular wave, the result is the same effective value—the peak value divided by $\sqrt{3}$.

INDUCTOR SINUSOIDAL RESPONSE

If an inductor of L henries has a current $i = I_m \sin(\omega t + \theta)$ flowing through it, the voltage across the inductor is

$$v = L \frac{di}{dt} = L \frac{d}{dt} [I_m \sin(\omega t + \theta)] = \omega L I_m \cos(\omega t + \theta)$$

The multiplier $\omega L I_m$ is the peak voltage V_m: $V_m = \omega L I_m$ and $I_m = V_m/\omega L$. From a comparison of $I_m = V_m/\omega L$ and $I_m = V_m/R$, clearly ωL has a current-limiting action similar to that of R.

The quantity ωL is called the *inductive reactance* of the inductor. Its quantity symbol is X_L:

$$X_L = \omega L$$

It has the same ohm unit as resistance. Unlike resistance, though, inductive reactance depends on frequency—the greater the frequency the greater its value and so the greater its current-limiting action. For sinusoids of very low frequency, approaching 0 Hz or dc, an inductive reactance is almost zero, which means that an inductor is almost a short circuit to such sinusoids, in agreement with dc results. At the other frequency extreme, for sinusoids of very high frequencies, approaching infinity, an inductive reactance approaches infinity, which means that an inductor is almost an open circuit to such sinusoids.

From a comparison of the inductor current and voltage sinusoids, it can be seen that *the inductor voltage leads the inductor current by 90° or the inductor current lags the inductor voltage by 90°*.

The instantaneous power absorbed by an inductor is

$$p = vi = [V_m \cos(\omega t + \theta)][I_m \sin(\omega t + \theta)] = V_m I_m \cos(\omega t + \theta) \sin(\omega t + \theta)$$

which from sine and cosine identities reduces to

$$p = \frac{V_m I_m}{2} \sin(2\omega t + 2\theta) = V_{\text{eff}} I_{\text{eff}} \sin(2\omega t + 2\theta)$$

This power is sinusoidal at twice the voltage and current frequency.   Being sinusoidal, its average value is zero—*a sinusoidally excited inductor absorbs zero average power.*   In terms of energy, at the times when p is positive, an inductor absorbs energy.   And at the times when p is negative, an inductor returns energy to the circuit and acts like a source.   Over a period, it delivers just as much energy as it receives.

CAPACITOR SINUSOIDAL RESPONSE

If a capacitor of C farads has a voltage $v = V_m \sin(\omega t + \theta)$ across it, the capacitor current is

$$i = C\frac{dv}{dt} = C\frac{d}{dt}[V_m \sin(\omega t + \theta)] = \omega C V_m \cos(\omega t + \theta)$$

The multiplier $\omega C V_m$ is the peak current I_m: $I_m = \omega C V_m$ and $V_m/I_m = 1/\omega C$.   So, a capacitor has a current-limiting action similar to that of a resistor with $1/\omega C$ corresponding to R.   Because of this action, some electric circuits books have *capacitive reactance* defined as $1/\omega C$.   But, almost all electrical engineering circuits books include a negative sign and have capacitive reactance defined as

$$X_C = -\frac{1}{\omega C}$$

The negative sign relates to phase shift, as will be explained in Chap. 11.   Of course, the quantity symbol for capacitive reactance is X_C and the unit is the ohm.

Because $1/\omega C$ is inversely proportional to frequency, the greater the frequency the greater the current for the same voltage peak.   For high-frequency sinusoids, a capacitor is almost a short circuit, and for low-frequency sinusoids approaching 0 Hz or dc, a capacitor is almost an open circuit.

From a comparison of the capacitor voltage and current sinusoids, it can be seen that *the capacitor current leads the capacitor voltage by 90°, or the capacitor voltage lags the capacitor current by 90°.*   This is the opposite of the inductor voltage and current phase relation.

The instantaneous power absorbed by a capacitor is

$$p = vi = [V_m \sin(\omega t + \theta)][I_m \cos(\omega t + \theta)] = \frac{V_m I_m}{2}\sin(2\omega t + 2\theta)$$

the same as for an inductor.   The instantaneous power absorbed is sinusoidal at twice the voltage and current frequency and has a zero average value.   So, *a capacitor absorbs zero average power.*   Over a period a capacitor delivers just as much energy as it absorbs.

Solved Problems

9.1   Find the periods of periodic voltages that have frequencies of   (*a*) 0.2 Hz,   (*b*) 12 kHz, and   (*c*) 4.2 MHz.

(*a*)   From $T = 1/f$, $T = 1/0.2 = 5$ s

(*b*)   Similarly, $T = 1/(12 \times 10^3)$ s $= 83.3$ μs

(*c*) $T = 1/(4.2 \times 10^6)$ s $= 238$ ns

9.2   Find the frequencies of periodic currents that have periods of   (*a*) 50 μs,   (*b*) 42 ms, and   (*c*) 1 h.

(a) From $f = 1/T$, $f = 1/(50 \times 10^{-6})$ Hz = 20 kHz

(b) Similarly, $f = 1/(42 \times 10^{-3}) = 23.8$ Hz

(c) $f = \dfrac{1}{1\,\cancel{h}} \times \dfrac{1\,\cancel{h}}{3600\text{ s}} = 2.78 \times 10^{-4}$ Hz = 0.278 mHz

9.3   What are the period and frequency of a periodic voltage that has 12 cycles in 46 ms?

The period is the time taken for one cycle, which can be found by dividing the 12 cycles into the time that it takes for them to occur (46 ms): $T = 46/12 = 3.83$ ms.   Of course, the frequency is the reciprocal of the period: $f = 1/(3.83 \times 10^{-3}) = 261$ Hz.   Alternatively, but what amounts to the same thing, the frequency is the number of cycles that occur in 1 s: $f = 12/(46 \times 10^{-3}) = 261$ Hz.

9.4   Find the period, the frequency, and the number of cycles shown for the periodic wave illustrated in Fig. 9-5.

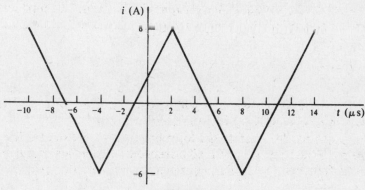

Fig. 9-5

The wave has a positive peak at 2 μs and another positive peak at 14 μs, between which times there is one cycle. So, the period is $T = 14 - 2 = 12$ μs,   and the frequency is $f = 1/T = 1/(12 \times 10^{-6})$ Hz = 83.3 kHz.   There is one other cycle shown—from −10 to 2 μs.

9.5   Convert the following angles in degrees to angles in radians: (a) 49°,   (b) −130°,   and   (c) 435°.

(a) $49° \times \dfrac{\pi}{180°} = 0.855$ rad

(b) $-130° \times \dfrac{\pi}{180°} = -2.27$ rad

(c) $435° \times \dfrac{\pi}{180°} = 7.59$ rad

9.6   Convert the following angles in radians to angles in degrees: (a) π/18 rad,   (b) −0.562 rad,   and   (c) 4 rad.

(a) $\dfrac{\pi}{18} \times \dfrac{180°}{\pi} = 10°$

(b) $-0.562 \times \dfrac{180°}{\pi} = -32.2°$

(c) $4 \times \dfrac{180°}{\pi} = 229°$

9.7 Find the periods and frequencies of sinusoidal currents that have radian frequencies of (a) 9π rad/s, (b) 0.042 rad/s, and (c) 13 Mrad/s.

From $f = \omega/2\pi$ and $T = 1/f$,

(a) $f = 9\pi/2\pi = 4.5$ Hz, $T = 1/4.5 = 0.222$ s

(b) $f = 0.042/2\pi$ Hz $= 6.68$ mHz, $T = 1/(6.68 \times 10^{-3}) = 150$ s

(c) $f = 13 \times 10^6/2\pi$ Hz $= 2.07$ MHz, $T = 1/(2.07 \times 10^6)$ s $= 0.483$ μs

9.8 Find the radian frequencies of sinusoidal voltages that have periods of (a) 4 s, (b) 6.3 ms, and (c) 7.9 μs.

From $\omega = 2\pi f = 2\pi/T$,

(a) $\omega = 2\pi/4 = 1.57$ rad/s

(b) $\omega = 2\pi/(6.3 \times 10^{-3}) = 997$ rad/s

(c) $\omega = 2\pi/(7.9 \times 10^{-6})$ rad/s $= 0.795$ Mrad/s

9.9 Find the amplitudes and frequencies of
(a) $42.1 \sin (377t + 30°)$ and (b) $-6.39 \cos (10^5 t - 20°)$.

(a) The amplitude is the *magnitude* of the multiplier: $|42.1| = 42.1$. Note the vertical lines about 42.1 for designating the magnitude operation, which removes a negative sign, if there is one. The radian frequency is the multiplier of t: 377 rad/s. From it, and $f = \omega/2\pi$, the frequency is $f = 377/2\pi = 60$ Hz.

(b) Similarly, the amplitude is $|-6.39| = 6.39$. The radian frequency is 10^5, from which $f = \omega/2\pi = 10^5/2\pi$ Hz $= 15.9$ kHz.

9.10 Find the instantaneous value of $v = 70 \sin 400\pi t$ V at $t = 3$ ms.

Substituting for t: $v(3 \text{ ms}) = 70 \sin (400\pi \times 3 \times 10^{-3}) = 70 \sin 1.2\pi$ V. Since the 1.2π sinusoidal argument is in radians, a calculator must be operated in the radians mode for this evaluation. The result is -41.1 V. Alternatively, the angle can be converted to degrees, $1.2\pi \times 180°/\pi = 216°$, and a calculator operated in the more popular decimal degrees mode: $v(3 \text{ ms}) = 70 \sin 216° = -41.1$ V.

9.11 A current sine wave has a peak of 58 mA and a radian frequency of 90 rad/s. Find the instantaneous current at $t = 23$ ms.

From the specified peak current and frequency, the expression for the current is $i = 58 \sin 90t$ mA. For $t = 23$ ms, this evaluates to

$$i(23 \text{ ms}) = 58 \sin (90 \times 23 \times 10^{-3}) = 58 \sin 2.07 = 50.9 \text{ mA}$$

Of course, the 2.07 in radians could have been converted to degrees; $2.07 \times 180°/\pi = 118.6°$, and then $58 \sin 118.6°$ evaluated.

9.12 Evaluate (a) $v = 200 \sin (3393t + \pi/7)$ V and (b) $i = 67 \cos (3016t - 42°)$ mA at $t = 1.1$ ms.

From substituting 1.1×10^{-3} for t,

(a) $v(1.1 \text{ ms}) = 200 \sin (3393 \times 1.1 \times 10^{-3} + \pi/7) = 200 \sin 4.18 = -172$ V

Operating a calculator in the radians mode is convenient for this calculation because both parts of the sinusoidal argument are in radians.

(b)	$i(1.1 \text{ ms}) = 67 \cos (3016 \times 1.1 \times 10^{-3} - 42°) = 67 \cos (190° - 42°) = -60.9$ mA

Note that the first term was converted from radians to degrees so that it could be added to the second term. Alternatively, the second term could have been converted to radians.

9.13 Find expressions for the sinusoids shown in Fig. 9-6.

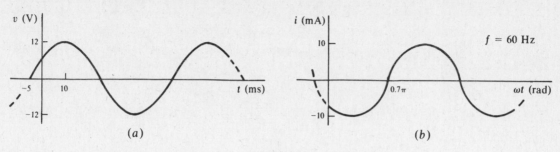

Fig. 9-6

The sinusoid shown in Fig. 9-6a can be considered to be either a phase-shifted sine wave or a phase-shifted cosine wave—it does not make any difference. For the selection of a phase-shifted sine wave, the general expression is $v = 12 \sin (\omega t + \theta)$, since the peak value is shown as 12. The radian frequency ω can be found from the period. One-fourth of a period occurs in the 15-ms time interval from -5 to 10 ms, which means that $T = 4 \times 15 = 60$ ms, and $\omega = 2\pi/T = 2\pi/(60 \times 10^{-3}) = 104.7$ rad/s. From the zero value at $t = -5$ ms and the fact that the waveform is going from negative to positive then just as a sine wave does for a zero argument, the argument can be zero at this time: $104.7(-5 \times 10^{-3}) + \theta = 0$, from which $\theta = 0.524$ rad $= 30°$. The result is $v = 12 \sin (105t + 0.524) = 12 \sin (105t + 30°)$ V.

Now consider the equation for the current shown in Fig. 9-6b. From $\omega = 2\pi f = 2\pi (60) = 377$ rad/s and the peak value of 10 mA, $i = 10 \cos (377t + \theta)$ mA, with the arbitrary selection of a phase-shifted cosine wave. The angle θ can be determined from the zero value at $\omega t = 0.7\pi$ from the graph. For this value of ωt, the phase-shifted cosine argument can be 1.5π rad because at 1.5π rad $= 270°$ a cosine waveform is zero and going from negative to positive, as can be seen from Fig. 9-3c. So, for $\omega t = 0.7\pi$, the argument can be $\omega t + \theta = 0.7\pi + \theta = 1.5\pi$, from which $\theta = 0.8\pi$ rad $= 144°$. The result is $i = 10 \cos (377t + 0.8\pi) = 10 \cos (377t + 144°)$ mA.

9.14 Sketch a cycle of $v = 30 \sin (754t + 60°)$ V for the period beginning at $t = 0$ s. Have all three abscissa units of time, radians, and degrees.

A fairly accurate sketch can be made from the initial value, the peaks of 30 and -30 V, and the times at which the waveform is zero and at its peaks. Also needed is the period, which is $T = 2\pi/\omega = 2\pi/754 = 8.33$ ms. The initial value can be found by substituting 0 for t in the argument. The result is $v = 30 \sin 60° = 26$ V. The waveform is zero for the first time when the argument is π radians since $\sin \pi = 0$. This time can be found from the argument with the 60° converted to $\pi/3$ radians: $754t + \pi/3 = \pi$, from which $t = 2.78$ ms. The next zero is half a period later: $2.78 + 8.33/2 = 6.94$ ms. The positive peak for this cycle occurs at a time when the sinusoidal argument is $\pi/2$: $754t + \pi/3 = \pi/2$, from which $t = 0.694$ ms. The negative peak is half a period later: $t = 0.694 + 8.33/2 = 4.86$ ms. The radian units for these times can be found from $\omega t = 754t = 240\pi t$. Of course, the corresponding degree units can be found by converting from radians to degrees. Figure 9-7 shows the sinusoid.

9.15 What is the shortest time required for a 2.1-krad/s sinusoid to increase from zero to four-fifths of its peak value?

The expression for the sinusoid can be considered to be $V_m \sin (2.1 \times 10^3 t)$, for convenience. The time required for this wave to equal $0.8V_m$ can be found from $V_m \sin (2.1 \times 10^3 t) =$

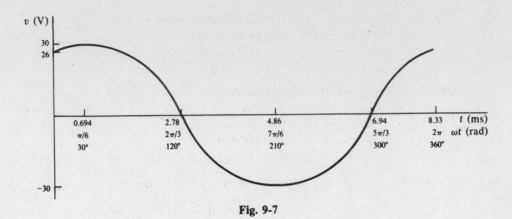

Fig. 9-7

$0.8V_m$, which simplifies to $\sin (2.1 \times 10^3 t) = 0.8$. This can be evaluated for t by taking the inverse sine, called the *arcsine*, of both sides. This operation causes the sin operation to be cancelled, leaving the argument. On a calculator, the arcsine is designated by "$\sin^{-1}$." Taking the arcsine of both sides produces

$$\sin^{-1}[\sin (2.1 \times 10^3 t)] = \sin^{-1} 0.8$$

which simplifies to $2.1 \times 10^3 t = \sin^{-1} 0.8$, from which

$$t = \frac{\sin^{-1} 0.8}{2.1 \times 10^3} = \frac{0.9273}{2.1 \times 10^3} \text{ s} = 0.442 \text{ ms}$$

The 0.9273 is, of course, in radians.

9.16 If 50 V is the peak voltage induced in the conductor of the alternator shown in Fig. 9-2, find the voltage induced after the conductor has rotated through an angle of 35° from its vertical position.

When the conductor is in a vertical position, the induced voltage is a maximum in magnitude, but can be either positive or negative. The vertical position can, for convenience, be considered to correspond to 0°. Then, since the induced voltage is sinusoidal, and since the cosine wave has a peak at 0°, the voltage can be considered to be $v = \pm 50 \cos \theta$, in which θ is the angle of the conductor from the vertical. So, with the conductor at 35° from the vertical, the induced voltage is $v = \pm 50 \cos 35° = \pm 41$ V.

9.17 If the conductor in the alternator shown in Fig. 9-2 is rotating at 60 Hz, and if the induced voltage has a peak of 20 V, find the induced voltage 20 ms after the conductor passes through a horizontal position if the voltage is increasing then.

The simplest expression for the induced voltage is $v = 20 \sin 377t$ V if $t = 0$ s corresponds to the time at which the conductor is in the specified horizontal position. This is the voltage expression because the induced voltage is sinusoidal, 20 V is specified as the peak, 377 rad/s corresponds to 60 Hz, and the sin ωt is zero at $t = 0$ s and is increasing. So,

$$v(20 \times 10^{-3}) = 20 \sin (377 \times 20 \times 10^{-3}) = 20 \sin 7.54 = 20 \sin 432° = 19 \text{ V}$$

9.18 Find the periods of (a) $7 - 4 \cos (400t + 30°)$, (b) $3 \sin^2 4t$, and (c) $4 \cos 3t \sin 3t$.

(a) The expression $7 - 4 \cos (400t + 30°)$ is a sinusoid of $-4 \cos (400t + 30°)$ "riding" on a constant 7. Since only the sinusoid contributes to the variations of the wave, only it determines the period: $T = 2\pi/\omega = 2\pi/400$ s $= 15.7$ ms.

(b) Because of the square, it is not immediately obvious what the period is. The identity $\sin^2 x = (1 - \cos 2x)/2$ can be used to eliminate the square:

$$3 \sin^2 4t = 3\left[\frac{1 - \cos (2 \times 4t)}{2}\right] = 1.5(1 - \cos 8t)$$

From the cosine wave portion, the period is $T = 2\pi/\omega = 2\pi/8 = 0.785$ s.

(c) Because of the product of the sinusoids in $4 \cos 3t \sin 3t$, some simplification must be done before the period can be determined. The identity $\sin (x + y) = \sin x \cos y + \sin y \cos x$ can be used for this by setting $y = x$. The result is

$$\sin (x + x) = \sin x \cos x + \sin x \cos x \qquad \text{or} \qquad \sin 2x = 2 \sin x \cos x$$

from which $\sin x \cos x = (\sin 2x)/2$. Here, $x = 3t$,   and so

$$4 \cos 3t \sin 3t = 4\left[\frac{\sin (2 \times 3t)}{2}\right] = 2 \sin 6t$$

From this, the period is $T = 2\pi/\omega = 2\pi/6 = 1.05$ s.

9.19 Find the phase relations for the following pairs of sinusoids:

(a) $v = 60 \sin (377t + 50°)$ V, $i = 3 \sin (754t - 10°)$ A

(b) $v_1 = 6.4 \sin (7.1\pi t + 30°)$ V, $v_2 = 7.3 \sin (7.1\pi t - 10°)$ V

(c) $v = 42.3 \sin (400t + 60°)$ V, $i = -4.1 \sin (400t - 50°)$ A

(a) There is no phase relation because the sinusoids have different frequencies.

(b) The angle by which v_1 leads v_2 is the phase angle of v_1 minus the phase angle of v_2:   ang v_1 − ang $v_2 = 30° - (-10°) = 40°$. Alternatively, v_2 lags v_1 by 40°.

(c) The amplitudes must have the same sign before a phase comparison can be made. The negative sign of i can be eliminated by using the identity $-\sin x = \sin (x \pm 180°)$. The positive sign in $\pm$ is more convenient because, as will be seen, it leads to a phase difference of the smallest angle, as is generally preferable. The result is

$$i = -4.1 \sin (400t - 50°) = 4.1 \sin (400t - 50° + 180°) = 4.1 \sin (400t + 130°) \text{ A}$$

The angle by which v leads i is the phase angle of v minus the phase angle of i: ang v − ang $i = 60° - 130° = -70°$. The negative sign indicates that v lags, instead of leads, i by 70°. Alternatively, i leads v by 70°. If the negative sign in $\pm$ had been used, the result would have been that v leads i by 290°, which is equivalent to $-70°$ because 360° can be subtracted from (or added to) a sinusoidal angle without affecting the value of the sinusoid.

9.20 Find the angle by which $i_1 = 3.1 \sin (754t - 20°)$ mA leads $i_2 = -2.4 \cos (754t + 30°)$ mA.

Before a phase comparison can be made, both amplitudes must have the same sign, and both sinusoids must be of the same form: either phase-shifted sine waves or phase-shifted cosine waves. The negative sign of i_2 can be eliminated by using the identity $-\cos x = \cos (x \pm 180°)$. At this point it is not clear whether the positive or negative sign is preferable, and so both will be kept:

$$i_2 = 2.4 \cos (754t + 210°) = 2.4 \cos (754t - 150°) \text{ A}$$

Both of these phase-shifted cosine waves can be converted to phase-shifted sine waves by using the identity $\cos x = \sin (x + 90°)$:

$$i_2 = 2.4 \sin (754t + 300°) = 2.4 \sin (754t - 60°) \text{ A}$$

Now a phase angle comparison can be made: i_1 leads i_2 by $-20° - 300° = -320°$ from the first i_2 expression, or by $-20° - (-60°) = 40°$ from the second i_2 expression. Being smaller in magnitude, the 40° lead is preferable to a −320° lead. But both are equivalent.

9.21 Find the average values of the periodic waveforms shown in Fig. 9-8.

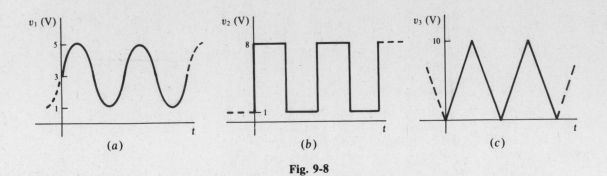

Fig. 9-8

The waveform shown in Fig. 9-8a is a sinusoid "riding" on top of a constant 3 V. Since the average value of the sinusoid is zero, the average value of the waveform equals the constant 3 V.

The average value of the waveform shown in Fig. 9-8b, and of any waveform, is the area under the waveform for one period, divided by the period. Since for the cycle beginning at $t = 0$ s, the waveform is at 8 V for half a period and is at 1 V for the other half-period, the area underneath the curve for this one cycle is $8 \times T/2 + 1 \times T/2 = 4.5T$ from the height-times-base formula for a rectangular area. So, the average value is $4.5T/T = 4.5$ V. Note that the average value does not depend on the period. This is generally true.

The cycle of the waveform shown in Fig. 9-8c beginning at $t = 0$ s is a triangle with a height of 10 and a base of T. Thus, the area under the curve for this one cycle is $0.5 \times 10 \times T = 5T$ from the triangular area formula of one-half the height times the base. And, the average value is $5T/T = 5$ V.

9.22 What are the average values of the periodic waveforms shown in Fig. 9-9?

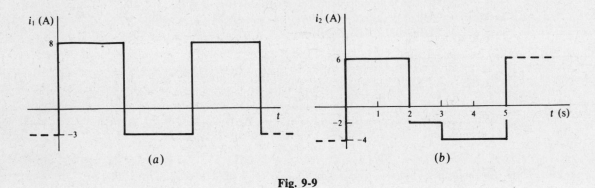

Fig. 9-9

For the cycle starting at $t = 0$ s, the i_1 waveform shown in Fig. 9-9a is at 8 A for half a period and is at -3 A for the next half-period. So, the area for this cycle is $8(T/2) + (-3)(T/2) = 2.5T$, and the average value is $2.5T/T = 2.5$ A.

The i_2 waveform shown in Fig. 9-9b has a complete cycle from $t = 0$ s to $t = 5$ s. For the first 2 s the area under the curve is $6 \times 2 = 12$. For the next second it is $-2 \times 1 = -2$. And for the last 2 s it is $-4 \times 2 = -8$. The algebraic sum of these areas is $12 - 2 - 8 = 2$, which divided by the period of 5 results in an average value of $2/5 = 0.4$ A.

9.23 What is the average power absorbed by a circuit component that has a voltage $v = 6 \sin (377t + 10°)$ V across it when a current $i = 0.3 \sin (377t - 20°)$ A flows through it?

The average power is, of course, the average value of the instantaneous power p:

$$p = vi = [6 \sin (377t + 10°)][0.3 \sin (377t - 20°)] = 1.8 \sin (377t + 10°) \sin (377t - 20°) \text{ W}$$

This can be simplified using a sine-cosine identity derived by subtracting $\cos(x + y) = \cos x \cos y - \sin x \sin y$ from $\cos(x - y) = \cos x \cos y + \sin x \sin y$. The result is the identity $\sin x \sin y = 0.5[\cos(x - y) - \cos(x + y)]$. Here, $x = 377t + 10°$ and $y = 377t - 20°$. So,
So,

$$p = 0.5[1.8 \cos(377t + 10° - 377t + 20°) - 1.8 \cos(377t + 10° + 377t - 20°)]$$
$$= 0.9 \cos 30° - 0.9 \cos(754t - 10°) \text{ W}$$

Since the second term is a sinusoid, and so has an average value of zero, the average power equals the first term:

$$P_{av} = 0.9 \cos 30° = 0.779 \text{ W}$$

Note in particular that the average power is *not* equal to the product of the average voltage (0 V) and the average current (0 A). Nor is it equal to the product of the effective value of voltage $(6/\sqrt{2})$ and the effective value of current $(0.3/\sqrt{2})$.

9.24 If the voltage across a single circuit component is $v = 40 \sin(400t + 10°)$ V for a current through it of $i = 34.1 \sin(400t + 10°)$ mA, and if the references are associated, as should be assumed, what is the component?

Since the voltage and current are in phase, the component is a resistor. The resistance is $R = V_m/I_m = 40/(34.1 \times 10^{-3}) \ \Omega = 1.17 \text{ k}\Omega$.

9.25 The voltage across a 62-Ω resistor is $v = 30 \sin(200\pi t + 30°)$ V. Find the resistor current and plot one cycle of the voltage and current waveforms on the same graph.

From $i = v/R$, $i = [30 \sin(200\pi t + 30°)]/62 = 0.484 \sin(200\pi t + 30°)$ A. Of course, the period is $T = 2\pi/\omega = 2\pi/200\pi$ s $= 10$ ms. For both waves, the curves will be plotted from the initial, peak, and zero values and the times at which they occur. At $t = 0$ s, $v = 30 \sin 30° = 15$ V and $i = 0.484 \sin 30° = 0.242$ A. The positive peaks of 30 V and 0.484 A occur at a time t_p corresponding to 60° since the sinusoidal arguments are 90° then. From the proportionality $t_p/T = 60°/360°$, the peak time is $t_p = 10/6 = 1.67$ ms. Of course, the negative peaks occur at a half-period later, at $1.67 + 5 = 6.67$ ms. The first zero values occur at a time corresponding to 150° because the sinusoidal arguments are 180° then. Using a proportionality again, this time is $(150/360)(10) = 4.17$ ms. The next zeros occur a half a period later at $4.17 + 5 = 9.17$ ms. The voltage and current waveforms are shown in Fig. 9-10. Do not be concerned about the relative heights of the voltage and current peaks. There does not have to be any relation between them because they are in different units.

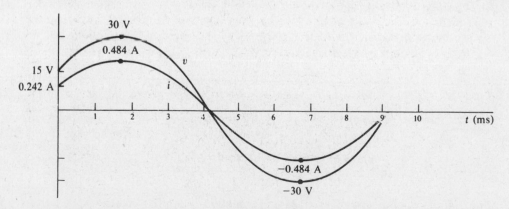

Fig. 9-10

9.26   A 30-Ω resistor has a voltage of $v = 170 \sin (377t + 30°)$ V across it. What is the average power dissipation of the resistor?

$$P_{av} = \frac{V_m^2}{2R} = \frac{170^2}{2 \times 30} = 482 \text{ W}$$

9.27   Find the average power absorbed by a 2.7-Ω resistor when the current $i = 1.2 \sin (377t + 30°)$ A flows through it.

$$P_{av} = \tfrac{1}{2}I_m^2 R = 0.5(1.2)^2(2.7) = 1.94 \text{ W}$$

9.28   What is the peak voltage at a 120-V electric outlet?

The 120 V is the effective value of the sinusoidal voltage at the outlet. Since for a sinusoid the peak is $\sqrt{2}$ times the effective value, the peak voltage at the outlet is $\sqrt{2} \times 120 = 170$ V.

9.29   What is the reading of an ac voltmeter connected across a 680-Ω resistor that has a current of $i = 6.2 \cos (377t - 20°)$ mA flowing through it?

The voltmeter reads the effective value of the resistor voltage, which can be found from I_{eff} and R. Since $V_m = I_m R$, then $V_m/\sqrt{2} = (I_m/\sqrt{2})(R)$ or $V_{eff} = I_{eff}R$. So,

$$V_{eff} = [(6.2 \times 10^{-3})/\sqrt{2}](680) = 2.98 \text{ V}$$

9.30   What is the reading of an ac voltmeter connected across a 10-Ω resistor that has a peak power dissipation of 40 W?

The peak voltage V_m can be found from the peak power: $P_m = V_m I_m = V_m^2/R$, from which $V_m = \sqrt{P_m R} = \sqrt{40(10)} = 20$ V. The effective or rms voltage, which is the voltmeter reading, is $V_m/\sqrt{2} = 20/\sqrt{2} = 14.1$ V.

9.31   What is the expression for a 240-Hz sine wave of voltage that has an rms value of 120 V?

Since the peak voltage is $120 \times \sqrt{2} = 170$ V and the radian frequency is $2\pi \times 240 = 1508$ rad/s, the sine wave is $v = 170 \sin 1508t$ V.

9.32   Find the effective value of a periodic voltage that has a value of 20 V for one half-period and -10 V for the other half-period.

The first step is to square the wave. The result is 400 for the first half-period and $(-10)^2 = 100$ for the second half-period. The next step is to find the average of the squares from the area divided by the period: $(400 \times T/2 + 100 \times T/2)/T = 250$. The last step is to find the square root of this average: $V_{eff} = \sqrt{250} = 15.8$ V.

9.33   Find the effective value of the periodic current shown in Fig. 9-11a.

The first step is to square the wave, which has a period of 8 s. The squared wave is shown in Fig. 9-11b. The next step is to find the average of the squared wave, which can be found by dividing the area by the period: $[16(3) + 9(6 - 4)]/8 = 8.25$. The last step is to find the square root of this average: $I_{eff} = \sqrt{8.25} = 2.87$ A.

9.34   Find the reactances of a 120-mH inductor at (a) 0 Hz (dc), (b) 40 rad/s, (c) 60 Hz, and (d) 30 kHz.

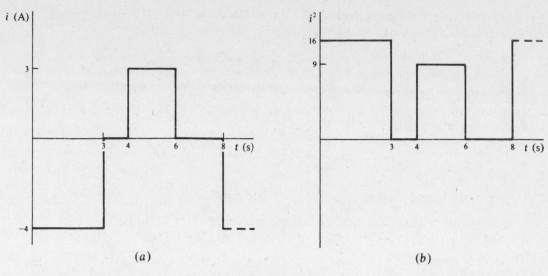

(a)   (b)

Fig. 9-11

From $X_L = \omega L = 2\pi f L,$

(a) $X_L = 2\pi(0)(120 \times 10^{-3}) = 0\ \Omega$

(b) $X_L = 40(120 \times 10^{-3}) = 4.8\ \Omega$

(c) $X_L = 2\pi(60)(120 \times 10^{-3}) = 45.2\ \Omega$

(d) $X_L = 2\pi(30 \times 10^3)(120 \times 10^{-3})\ \Omega = 22.6\ k\Omega$

9.35   Find the inductances of the inductors that have reactances of   (a) 5 Ω at 377 rad/s,   (b) 1.2 kΩ at 30 kHz,   and   (c) 1.6 MΩ at 22.5 Mhz.

Solving for L in $X_L = \omega L$ results in $L = X_L/\omega = X_L/2\pi f$.   So,

(a) $L = 5/377\ H = 13.3\ mH$

(b) $L = (1.2 \times 10^3)/(2\pi \times 30 \times 10^3)\ H = 6.37\ mH$

(c) $L = (1.6 \times 10^6)/(2\pi \times 22.5 \times 10^6)\ H = 11.3\ mH$

9.36   Find the frequencies at which a 250-mH inductor has reactances of 30 Ω and 50 kΩ.

From $X_L = \omega L = 2\pi f L,$ the frequency is $f = X_L/2\pi L,$ and so

$f_1 = 30/(2\pi \times 250 \times 10^{-3}) = 19.1\ Hz$ and $f_2 = (50 \times 10^3)/(2\pi \times 250 \times 10^{-3})\ Hz = 31.8\ kHz$

9.37   What is the voltage across a 30-mH inductor that has a 40-mA, 60-Hz current flowing through it?

The specified current is, of course, the effective value, and the desired voltage is the effective value of voltage, although not specifically stated.   In general, given ac current and voltage values are effective   values   unless   otherwise   specified.   Because $X_L = V_m/I_m$, it   follows   that $X_L = (V_m/\sqrt{2})/(I_m/\sqrt{2}) = V_{\text{eff}}/I_{\text{eff}}$.   So, here, $V_{\text{eff}} = I_{\text{eff}}X_L = (40 \times 10^{-3})(2\pi \times 60)(30 \times 10^{-3}) = 0.452\ V.$

9.38   The voltage $v = 30 \sin(200\pi t + 30°)\ V$ is across an inductor that has a reactance of 62 Ω.   Find the inductor current and plot one cycle of the voltage and current on the same graph.

The current peak equals the voltage peak divided by the reactance: $I_m = 30/62 = 0.484\ A$.   And, since the current lags the voltage by 90°,

$$i = 0.484 \sin (200\pi t + 30° - 90°) = 0.484 \sin (200\pi t - 60°) \text{ A}$$

The voltage graph is the same as that shown in Fig. 9-10. The current graph for these values, though, differs from that in Fig. 9-10 by a shift left by a time corresponding to 90°, which time is one-fourth of a period: 10/4 = 2.5 ms. The waveforms are shown in Fig. 9-12.

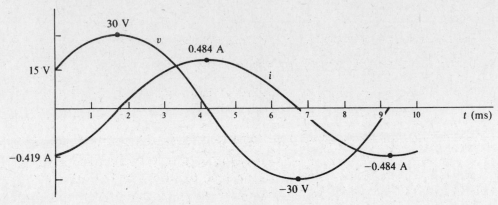

Fig. 9-12

9.39 Find the voltages across a 2-H inductor for currents of (*a*) 10 A, (*b*) 10 sin (377*t* + 10°) A, and (*c*) 10 cos (10⁴*t* − 20°) A. As always, assume associated references because there is no statement to the contrary.

(*a*) The inductor voltage is zero because the current is a constant and the time derivative of a constant is zero: $v = 2 \, d(10)/dt = 0$ V. From another point of view, the reactance is 0 Ω because the frequency is 0 Hz, and so $V_m = I_m X_L = 10(0) = 0$ V.

(*b*) The voltage peak equals the current peak times the reactance of $377 \times 2 = 754$ Ω: $V_m = I_m X_L = 10 \times 754$ V = 7.54 kV. Since the voltage leads the current by 90° and since sin (*x* + 90°) = cos *x*,

$$v = 7.54 \sin (377t + 10° + 90°) = 7.54 \cos (377t + 10°) \text{ kV}$$

(*c*) Similarly, $V_m = I_m X_L = 10(10^4 \times 2)$ V = 0.2 MV, and

$$v = 0.2 \cos (10^4 t - 20° + 90°) = 0.2 \cos (10^4 t + 70°) \text{ MV}$$

9.40 Find the reactances of a 0.1-μF capacitor at (*a*) 0 Hz (dc), (*b*) 377 rad/s, (*c*) 30 kHz, and (*d*) 100 MHz.

From $X_C = -1/\omega C = -1/2\pi f C$,

(*a*) $X_C = \lim_{\omega \to 0} \dfrac{-1}{\omega(0.1 \times 10^{-6})} \ \Omega \to -\infty \ \Omega$ (an open circuit)

(*b*) $X_C = \dfrac{-1}{377(0.1 \times 10^{-6})} \ \Omega = -26.5 \text{ k}\Omega$

(*c*) $X_C = \dfrac{-1}{2\pi(30 \times 10^3)(0.1 \times 10^{-6})} \ \Omega = -53.1 \ \Omega$

(*d*) $X_C = \dfrac{-1}{2\pi(100 \times 10^6)(0.1 \times 10^{-6})} \ \Omega = -15.9 \text{ m}\Omega$

9.41 Find the capacitances of capacitors that have a reactance of −500 Ω at (*a*) 377 rad/s, (*b*) 10 kHz, and (*c*) 22.5 MHz.

Solving for C in $X_C = -1/\omega C$ results in $C = -1/\omega X_C = -1/(2\pi f \times X_C)$. So,

(a) $\quad C = \dfrac{-1}{377(-500)}$ F $= 5.31 \ \mu$F

(b) $\quad C = \dfrac{-1}{2\pi(10 \times 10^3)(-500)}$ F $= 0.0318 \ \mu$F

(c) $\quad C = \dfrac{-1}{2\pi(22.5 \times 10^6)(-500)}$ F $= 14.1$ pF

9.42 Find the frequencies at which a 2-μF capacitor has reactances of -0.1 and $-2500 \ \Omega$.

From $X_C = -1/\omega C = -1/2\pi fC$, the frequency is $f = -1/(X_C \times 2\pi C)$. So,

$$f_1 = \frac{-1}{-0.1 \times 2\pi \times 2 \times 10^{-6}} \text{ Hz} = 796 \text{ kHz} \quad \text{and} \quad f_2 = \frac{-1}{-2500 \times 2\pi \times 2 \times 10^{-6}} = 31.8 \text{ Hz}$$

9.43 What current flows through a 0.1-μF capacitor that has 200 V at 400 Hz across it?

Although not specifically stated, it should be understood that the effective capacitor voltage is specified and the effective capacitor current is to be found. If both sides of $I_m = \omega C V_m$ are divided by $\sqrt{2}$, the result is $I_m/\sqrt{2} = \omega C V_m/\sqrt{2}$ or $I_{\text{eff}} = \omega C V_{\text{eff}}$. So,

$$I_{\text{eff}} = 2\pi(400)(0.1 \times 10^{-6})(200) \text{ A} = 50.3 \text{ mA}$$

9.44 What is the voltage across a capacitor that carries a 120-mA current if the capacitive reactance is $-230 \ \Omega$?

From the solution to Prob. 9.43, $I_{\text{eff}} = \omega C V_{\text{eff}}$ or $V_{\text{eff}} = I_{\text{eff}}(1/\omega C)$. Since $1/\omega C$ is the *magnitude* of capacitive reactance, the effective voltage and current of a capacitor have a relation of $V_{\text{eff}} = I_{\text{eff}}|X_C|$. Consequently, here, $V_{\text{eff}} = (120 \times 10^{-3})|-230| = 27.6$ V.

9.45 The voltage $v = 30 \sin(200\pi t + 30°)$ V is across a capacitor that has a reactance of $-62 \ \Omega$. Find the capacitor current and plot one cycle of the voltage and current on the same graph.

From $V_m/I_m = 1/\omega C = |X_C|$, the current peak equals the voltage peak divided by the magnitude of capacitive reactance: $I_m = 30/|-62| = 0.484$ A. And, since the current leads the voltage by 90°,

$$i = 0.484 \sin(200\pi t + 30° + 90°) = 0.484 \cos(200\pi t + 30°) \text{ A}$$

Notice that the current sinusoid has the same phase angle as the voltage sinusoid, but, because of the 90° lead, is a phase-shifted cosine wave instead of the phase-shifted sine wave of the voltage.

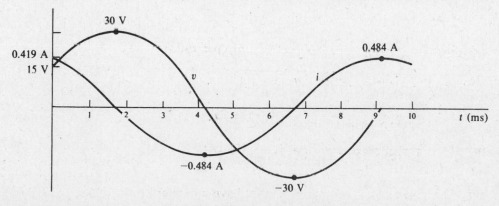

Fig. 9-13

The voltage graph is the same as that in Fig. 9-10.   The current graph differs from that in Fig. 9-10 by a shift right by a time corresponding to 90°, which time is one-fourth of a period: 10/4 = 2.5 ms.   The waveforms are shown in Fig. 9-13.

9.46   What currents flow through a 2-μF capacitor for voltages of   (a) $v = 5 \sin (377t + 10°)$ V   and   (b) $v = 12 \cos (10^4 t - 20°)$ V?

(a)   The current peak equals ωC times the voltage peak:

$$I_m = \omega C V_m = 377(2 \times 10^{-6})(5) \text{ A} = 3.77 \text{ mA}$$

Also, because the capacitor current leads the capacitor voltage by 90° and the voltage is a phase-shifted sine wave, the current can be expressed as a phase-shifted cosine wave with the same phase angle: $i = 3.77 \cos (377t + 10°)$ mA.

(b   The current peak is

$$I_m = \omega C V_m = 10^4 (2 \times 10^{-6})(12) = 0.24 \text{ A}$$

Also, the current leads the voltage by 90°.   As a result,

$$i = 0.24 \cos (10^4 t - 20° + 90°) = 0.24 \cos (10^4 t + 70°) \text{ A}$$

Supplementary Problems

9.47   Find the periods of periodic currents that have frequencies of   (a) 1.2 mHz,   (b) 2.31 kHz,   and   (c) 16.7 MHz.   *Ans.*   (a) 833 s,   (b) 433 μs,   (c) 59.9 ns

9.48   What are the frequencies of periodic voltages that have periods of   (a) 18.3 ps,   (b) 42.3 s,   and   (c) 1 d?   *Ans.*   (a) 546 THz (terahertz—i.e., 10^{12} Hz),   (b) 23.6 mHz,   (c) 11.6 μHz

9.49   What are the period and frequency of a periodic current for which 423 cycles occur in 6.19 ms? *Ans.* 146 μs, 6.83 kHz

9.50   Convert the following angles in degrees to angles in radians: (a) $-40°$,   (b) $-1123°$,   and   (c) 78°.   *Ans.*   (a) -0.698 rad,   (b) -19.6 rad,   (c) 1.36 rad

9.51   Convert the following angles in radians to angles in degrees: (a) 13.4 rad,   (b) 0.675 rad,   and   (c) -11.7 rad.   *Ans.*   (a) 768°,   (b) 38.7°,   (c) $-670°$

9.52   Find the periods of sinusoidal voltages that have radian frequencies of   (a) 120π rad/s,   (b) 0.625 rad/s,   and   (c) 62.1 krad/s.   *Ans.*   (a) 16.7 ms,   (b) 10.1 s,   (c) 101 μs

9.53   Find the radian frequencies of sinusoidal currents that have periods of   (a) 17.6 μs,   (b) 4.12 ms,   and   (c) 1 d.   *Ans.*   357 krad/s,   (b) 1.53 krad/s,   (c) 72.7 μrad/s

9.54   What are the amplitudes and frequencies of   (a) $-63.7 \cos (754t - 50°)$ and   (b) $429 \sin (4000t + 15)°$?   *Ans.*   (a) 63.7, 120 Hz;   (b) 429, 637 Hz

9.55   Find the instantaneous value of $i = 80 \sin 500t$ mA   at   (a) $t = 4$ ms   and   (b) $t = 2.1$ s. *Ans.*   (a) 72.7 mA,   (b) 52 mA

9.56   What is the frequency of a sine wave of voltage that has a 45-V peak and that continuously increases from 0 V at $t = 0$ s to 24 V at $t = 46.2$ ms?   *Ans.* 1.94 Hz

9.57   If a voltage cosine wave has a peak value of 20 V at $t = 0$ s, and if it takes a minimum of 0.123 s for this voltage to decrease from 20 to 17 V, find the voltage at $t = 4.12$ s.   *Ans.* 19.3 V.

9.58   What is the instantaneous value of $i = 13.2 \cos (377t + 50°)$ mA at (*a*) $t = -42.1$ ms and (*b*) $t = 6.3$ s?   *Ans.* (*a*) −10 mA, (*b*) 7.91 mA

9.59   Find an expression for a 400-Hz sinusoidal current that has a 2.3-A positive peak at $t = -0.45$ ms.   *Ans.* $i = 2.3 \cos (800\pi t + 64.8°)$ A

9.60   Find an expression for a sinusoidal voltage that is 0 V at $t = -8.13$ ms after which it increases to a peak of 15 V at $t = 6.78$ ms.   *Ans.* $v = 15 \sin (105t + 49.1°)$ V

9.61   What is the shortest time required for a 4.3-krad/s sinusoid to increase from two-fifths to four-fifths of its peak value?   *Ans.* 120 μs

9.62   If 43.7 V is the peak voltage induced in the conductor of the alternator shown in Fig. 9-2, find the voltage induced after the conductor has rotated through an angle of 43° from its horizontal position.
Ans. ±29.8 V

9.63   If the conductor of the alternator in Fig. 9-2 is rotating at 400 Hz, and if the induced voltage has a 23-V peak, find the induced voltage 0.23 ms after the conductor passes through its vertical position.
Ans. ± 19.3 V

9.64   Find the periods of (*a*) $4 + 3 \sin (800\pi t - 15°)$, (*b*) $8.1 \cos^2 9\pi t$, and (*c*) $8 \sin 16t \cos 16t$.
Ans. (*a*) 2.5 ms, (*b*) 111 ms, (*c*) 196 ms

9.65   Find the phase relations for the following pairs of sinusoids:

   (*a*) $v = 6 \sin (30t - 40°)$ V, $i = 10 \sin (30t - \pi/3)$ mA
   (*b*) $v_1 = -8 \sin (40t - 80°)$ V, $v_2 = -10 \sin (40t - 50°)$ V
   (*c*) $i_1 = 4 \cos (70t - 40°)$ mA, $i_2 = -6 \cos (70t + 80°)$ mA
   (*d*) $v = -4 \sin (45t + 5°)$ V, $i = 7 \cos (45t + 80°)$ mA

   Ans. (*a*) v leads i by 20°, (*b*) v_1 lags v_2 by 30°, (*c*) i_1 leads i_2 by 60°, (*d*) v leads i by 15°

9.66   Find the average value of a half-wave rectified sinusoidal voltage that has a peak of 12 V. This wave consists only of the positive half-cycles of the sinusoidal voltage. It is zero during the times that the sinusoid is negative.   *Ans.* 3.82 V

9.67   Find the average values of the periodic waveforms shown in Fig. 9-14.
Ans. (*a*) 3.5, (*b*) 4, (*c*) 15

9.68   What is the average power absorbed by a circuit component that has a voltage $v = 10$ V across it when a current $i = 5 + 6 \cos 33t$ A flows through it? *Ans.* 50 W

9.69   Find the average power absorbed by a circuit component that has a voltage $v =$

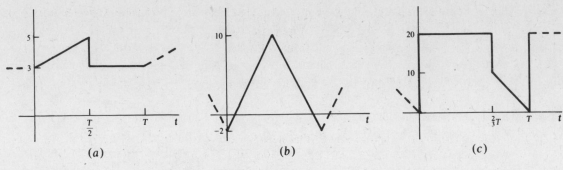

Fig. 9-14

20.3 cos (754t − 10°) V across it when a current i = 15.6 cos (754t − 30°) mA flows through it.
Ans. 149 mW

9.70 What is the conductance of a resistor that has a voltage v = 50.1 sin (200πt + 30°) V across it when a current i = 6.78 sin (200πt + 30°) mA flows through it?   *Ans.* 135 μS

9.71 If the voltage v = 150 cos (377t + 45°) V is across a 33-kΩ resistor, what is the resistor current? *Ans.* i = 4.55 cos (377t + 45°) mA

9.72 What is the average power absorbed by an 82-Ω resistor that has a voltage v = 311 cos (377t − 45°) V across it?   *Ans.* 590 W

9.73 What is the average power absorbed by a 910-Ω resistor that has a current i = 9.76 sin (754t − 36°) mA flowing through it?   *Ans.* 43.3 mW

9.74 Find the average power absorbed by a resistor that has a voltage v = 87.7 cos (400πt − 15°) V across it and a current i = 2.72 cos (400πt − 15°) mA flowing through it. *Ans.* 119 mW

9.75 What is the reading of an ac ammeter that is in series with a 470-Ω resistor that has a voltage v = 150 cos (377t + 30°) V across it?   *Ans.* 226 mA

9.76 What is the reading of an ac ammeter that is in series with a 270-Ω resistor that has a peak power dissipation of 10 W?   *Ans.* 192 mA

9.77 What is the expression for a 400-Hz current cosine wave that has an effective value of 13.2 mA?   *Ans.* i = 18.7 cos 800πt mA

9.78 Find the effective value of v = 3 + 2 sin 4t V. (*Hint:* Use a sinusoidal identity in finding the average value of the squared voltage.)   *Ans.* 3.32 V

9.79 Find the effective value of a periodic current that has a value of 40 mA for two-thirds of a period and 25 mA for the remaining one-third of the period. Would the effective value be different if the current were −25 mA instead of 25 mA for the one-third period?   *Ans.* 35.7 mA, no

9.80   Find the effective value of a periodic current that in a 20-ms period has a value of 0.761 A for 4 ms, 0 A for 2 ms, −0.925 A for 8 ms, and 1.23 A for the remaining 6 ms.   Would the effective value be different if the time segments were in seconds instead of in milliseconds?   Ans. 0.955 A, no

9.81   Find the reactances of a 180-mH inductor at   (a) 754 rad/s,   (b) 400 Hz,   and   (c) 250 kHz.
Ans.   (a) 136 Ω,   (b) 452 Ω,   (c) 283 kΩ

9.82   Find the inductances of the inductors that have reactances of   (a) 72.1 Ω at 754 rad/s,   (b) 11.9 Ω at 12 kHz,   and   (c) 42.1 kΩ at 2.1 MHz.   Ans.   (a) 95.6 mH,   (b) 158 μH,   (c) 3.19 mH

9.83   What are the frequencies at which a 120-mH inductor has reactances of   (a) 45 Ω   and   (b) 97.1 kΩ   Ans.   (a) 59.7 Hz,   (b) 129 kHz

9.84   What current flows through an 80-mH inductor that has 120 V at 60 Hz across it?   Ans. 3.98 A

9.85   What is the inductance of the inductor that will draw a current of 250 mA when connected to a 120-V, 60-Hz voltage source?   Ans. 1.27 H

9.86   What are the currents that flow in a 500-mH inductor for inductor voltages of   (a) $v = 170 \sin (400t + \pi/6)$ V   and   (b) $v = 156 \cos (1000t + 10°)$ V?
Ans.   (a) $i = 0.85 \sin (400t − 60°)$ A,   (b) $i = 0.312 \sin (1000t + 10°)$ A

9.87   Find the reactances of a 0.25-μF capacitor at   (a) 754 rad/s,   (b) 400 Hz,   and   (c) 2 MHz.
Ans.   (a) −5.31 kΩ,   (b) −1.59 kΩ,   (c) −0.318 Ω

9.88   Find the capacitances of the capacitors that have reactances of   (a) −700 Ω at 377 rad/s,   (b) −450 Ω at 400 Hz,   and   (c) −1.23 kΩ at 25 kHz.   Ans.   (a) 3.79 μF,   (b) 0.884 μF,   (c) 5.18 nF

9.89   Find the frequency at which a 0.1-μF capacitor and a 120-mH inductor have the same magnitude of reactance.   Ans. 1.45 kHz

9.90   What is the capacitance of a capacitor that draws 150 mA when connected to a 100-V, 400-Hz voltage source?   Ans. 0.597 μF

9.91   What are the currents that flow in a 0.5-μF capacitor for capacitor voltages of   (a) $v = 190 \sin (377t + 15°)$ V   and   (b) $v = 200 \cos (1000t − 40°)$ V?
Ans   (a) $i = 35.8 \cos (377t + 15°)$ mA,   (b) $i = 0.1 \cos (1000t + 50°)$ A

9.92   What are the voltages across a 2-μF capacitor for currents of   (a) $i = 7 \sin (754t + 15°)$ mA   and (b) $i = 250 \cos (10^3 t − 30°)$ mA?
Ans.   (a) $v = 4.64 \sin (754t − 75°)$ V,   (b) $v = 125 \sin (10^3 t − 30°)$ V

Chapter 10

Complex Algebra and Phasors

INTRODUCTION

Almost every ac circuit is easiest to analyze using *complex algebra*. Complex algebra is an extension of the algebra of real numbers—the common algebra. Complex numbers are included along with their own special rules for adding, subtracting, multiplying, and dividing. As is explained in Chaps. 11 and 12, in ac circuit analysis, sinusoids are *transformed* into complex numbers called *phasors*; also, resistances, inductances, and capacitances are transformed into complex numbers called *impedances*; and then complex algebra is applied in much the same way that ordinary algebra is applied in the analysis of a dc circuit.

IMAGINARY NUMBERS

The common numbers that everyone uses are *real numbers*. But these are not the only kind of numbers. There are also *imaginary numbers*. The name "imaginary" is misleading because it suggests that these numbers are only in the imagination, when actually they are just as much numbers as the common real numbers are. Imaginary numbers were invented when it became necessary to have numbers that are square roots of negative numbers (no real numbers are). This inventing of numbers was not new since it had been preceded by the inventions of noninteger real numbers and negative real numbers.

Imaginary numbers need to be distinguished from real numbers because different rules must be applied in the mathematical operations involving them. There is no one universally accepted way of representing imaginary numbers. In the electrical field, however, it is standard to use the letter j, as in $j2$, $j0.01$, and $-j5.6$.

The rules for adding and subtracting imaginary numbers are the same as those for adding and subtracting real numbers except that the sums and differences are imaginary. To illustrate,

$$j3 + j9 = j12 \qquad j12.5 - j3.4 = j9.1 \qquad j6.25 - j8.4 = -j2.15$$

The multiplication rule, though, is different. The product of two imaginary numbers is a *real* number that is the *negative* of the product that would be found if the numbers were real numbers instead. For example,

$$j2(j6) = -12 \qquad j4(-j3) = 12 \qquad -j5(-j4) = -20$$

Also, $j1(j1) = -1$, from which $j1 = \sqrt{-1}$. Likewise, $j2 = \sqrt{-4}$, $j3 = \sqrt{-9}$, and so forth.

Sometimes powers of $j1$ appear in calculations. These can have values of 1, -1, $j1$, and $-j1$, as can be shown by starting with $j1^2 = j1(j1) = -1$ and then progressively multiplying by $j1$ and evaluating. As an illustration, $j1^3 = j1(j1)^2 = j1(-1) = -j1$ and $j1^4 = j1(j1)^3 = j1(-j1) = 1$.

The product of a real number and an imaginary number is an imaginary number that, except for being imaginary, is the same as if the numbers were both real. For example, $3(j5) = j15$ and $-j5.1(4) = -j20.4$.

In the division of two imaginary numbers, the quotient is real and the same as if the numbers were real. As an illustration,

$$\frac{j8}{j4} = 2 \quad \text{and} \quad \frac{j20}{-j100} = -0.2$$

A convenient memory aid for division is to treat the j's as if they are numbers and to divide them out as in

$$\frac{j16}{j2} = 8$$

This should be viewed as a memory aid only, because j just designates a number as being imaginary and is not a number itself. However, treating j as a number in division, as well as in the other mathematical operations, is often done because of convenience and the fact that it does give correct answers.

If an imaginary number is divided by a real number, the quotient is imaginary but otherwise the same as for real numbers. For example,

$$\frac{j16}{4} = j4 \qquad \text{and} \qquad \frac{j2.4}{-0.6} = -j4.$$

The only difference if the denominator is imaginary and the numerator is real is that the quotient is the negative of the above. To illustrate,

$$\frac{1}{j1} = -j1 \qquad \text{and} \qquad \frac{-100}{j20} = j5$$

The basis for this rule can be shown by multiplying a numerator and denominator by $j1$, as in

$$\frac{225}{j5} = \frac{225 \times j1}{j5 \times j1} = \frac{j225}{-5} = -j45$$

Multiplying to make the denominator real, as here, is called *rationalizing*.

COMPLEX NUMBERS AND THE RECTANGULAR FORM

If a real number and an imaginary number are added, as in $3 + j4$, or subtracted, as in $6 - j8$, the result is considered to be a single *complex number* in *rectangular form*. Other forms of complex numbers are introduced in the next section.

A complex number can be represented by a point on the *complex plane* shown in Fig. 10 1. The horizontal axis, called the *real axis*, and the vertical axis, called the *imaginary axis*, divide the complex plane into four quadrants, as labeled. Both axes must have the same scale. The points for real numbers are on the real axis because a real number can be considered to be a complex number with a zero imaginary part. Figure 10 1 has four of these points: -5, -1, 2, and 4. The points for imaginary numbers are on the imaginary axis because an imaginary number can be considered to be a complex number with a zero real part. Figure 10-1 has four of these points: $j3$, $j1$, $-j2$, and $-j4$. Other complex numbers have nonzero real and imaginary parts, and so correspond to points off the axes. The real part of each number gives the position to the right

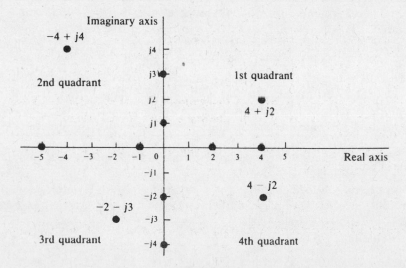

Fig. 10-1

or to the left of the vertical axis, and the imaginary part gives the position above or below the horizontal axis.   Figure 10-1 has four of these numbers, one in each quadrant.

In Fig. 10-1 the complex numbers $4 + j2$ and $4 - j2$ have the same real part, and they also have the same imaginary part—except for sign.   A pair of complex numbers having this relation are said to be *conjugates*: $4 + j2$ is the conjugate of $4 - j2$, and also $4 - j2$ is the conjugate of $4 + j2$.   Points for conjugate numbers have the same horizontal position but opposite vertical positions, being equidistant on opposite sides of the real axis.   If lines are drawn from the origin to these points, both lines will have the same length, and, except for sign, the same angle from the positive real axis.   (Angles are positive if measured in a counterclockwise direction from this axis, and negative if measured in a clockwise direction.)   These graphical relations of conjugates are important for the polar form of complex numbers presented in the next section.

The rectangular form is the only practical form for addition and subtraction.   These operations are applied separately to the real and imaginary parts.   As an illustration, $(3 + j4) + (2 + j6) = 5 + j10$ and $(6 - j7) - (4 - j2) = 2 - j5$.

In the multiplication of complex numbers in the rectangular form, the ordinary rules of algebra are used along with the rules for imaginary numbers.   For example,

$$(2 + j4)(3 + j5) = 2(3) + 2(j5) + j4(3) + j4(j5) = 6 + j10 + j12 - 20 = -14 + j22$$

It follows from this multiplication rule that, if a complex number is multiplied by its conjugate, the product is real and is the sum of the real part squared and the imaginary part squared.   To illustrate,

$$(3 + j4)(3 - j4) = 3(3) + 3(-j4) + j4(3) + j4(-j4) = 9 - j12 + j12 + 16 = 9 + 16 = 3^2 + 4^2 = 25$$

In the division of complex numbers in rectangular form, the numerator and denominator are multiplied by the conjugate of the denominator in order to make the denominator real so that the division will be straightforward.   This process of making the denominator real is the same rationalizing mentioned in the section on imaginary numbers.   As an example of this operation, consider

$$\frac{10 + j24}{6 + j4} = \frac{(10 + j24)(6 - j4)}{(6 + j4)(6 - j4)} = \frac{156 + j104}{6^2 + 4^2} = \frac{156 + j104}{52} = 3 + j2$$

POLAR FORM

The *polar form* of a complex number is a shorthand for the *exponential form*.   Usually, these are the best forms for multiplying and dividing, but are not useful for adding and subtracting unless done graphically.   The exponential form is $Ae^{j\theta}$, where A is the *magnitude* and θ is the *angle* of the complex number.   Also, $e = 2.718\ldots$ is the base of the natural logarithm.   The polar shorthand for $Ae^{j\theta}$ is $A\underline{/\theta}$ as in $4e^{j45°} = 4\underline{/45°}$ and in $-8e^{j60°} = -8\underline{/60°}$.   Although both forms are equivalent, the polar form is much more popular because it is easier to write.

That a number such as $5e^{j60°}$ is a complex number is evident from *Euler's identity*: $e^{j\theta} = \cos\theta + j\sin\theta$.   As an illustration, $7e^{j30°} = 7\underline{/30°} = 7\cos 30° + j7\sin 30° = 6.06 + j3.5$.   This use of Euler's identity not only shows that a number such as $Ae^{j\theta} = A\underline{/\theta}$ is a complex number, but also gives a method for converting a number from exponential or polar form to rectangular form.

Another use of Euler's identity is for deriving formulas for converting a complex number from rectangular form to the exponential and polar forms.   Suppose that x and y are known in $x + jy$, and that A and θ are to be found such that $x + jy = Ae^{j\theta} = A\underline{/\theta}$.   By Euler's identity, $x + jy = A\cos\theta + jA\sin\theta$.   Since two complex numbers are equal only if the real parts are equal and if the imaginary parts are equal, it follows that $x = A\cos\theta$ and $y = A\sin\theta$.   Taking the ratio of these equations eliminates A:

$$\frac{A\sin\theta}{A\cos\theta} = \tan\theta = \frac{y}{x} \qquad \text{from which} \qquad \theta = \tan^{-1}\frac{y}{x}$$

So, θ can be found from the arctangent of the ratio of the imaginary part to the real part. With θ known, A can be found by substituting θ into either $x = A \cos \theta$ or into $y = A \sin \theta$. Perhaps a more popular way of finding A is from a formula based on squaring both sides of $A \cos \theta = x$ and of $A \sin \theta = y$ and adding:

$$A^2 \cos^2 \theta + A^2 \sin^2 \theta = A^2(\cos^2 \theta + \sin^2 \theta) = x^2 + y^2$$

But since, from trigonometry, $\cos^2 \theta + \sin^2 \theta = 1$, it follows that $A^2 = x^2 + y^2$ and $A = \sqrt{x^2 + y^2}$. So, the magnitude of a complex number equals the square root of the sum of the squares of the real and imaginary parts. Of course, many pocket calculators have a built-in feature for converting between the rectangular and polar forms.

This conversion can also be understood from a graphical consideration. Figure 10-2a shows a directed line from the origin to the point for the complex number $x + jy$. As shown in Fig. 10-2b, this line forms a right triangle with its horizontal and vertical projections. From elementary trigonometry, $x = A \cos \theta$, $y = A \sin \theta$, and $A = \sqrt{x^2 + y^2}$, in agreement with the results from Euler's identity. Often this line, instead of the point, is considered to correspond to a complex number because its length and angle are the amplitude and angle of the complex number in polar form.

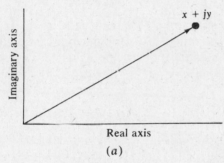
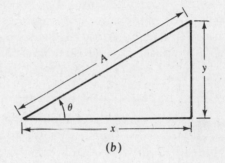

(a)   (b)

Fig. 10-2

As has been mentioned, the conjugate of a complex number in rectangular form differs only in the sign of the imaginary part. In polar form this difference appears as a difference in sign of the angle, as can be shown by converting any two conjugates to polar form. For example, $6 + j5 = 7.81\underline{/39.8°}$ and its conjugate $6 - j5 = 7.81\underline{/-39.8°}$.

As mentioned, the rectangular form is best for adding and subtracting, and the polar form is often best for multiplying and dividing. The multiplication and division formulas for complex numbers in polar form are easy to derive from the corresponding exponential numbers and the law of exponents. The product of the complex numbers $Ae^{j\theta}$ and $Be^{j\phi}$ is $(Ae^{j\theta})(Be^{j\phi}) = ABe^{j(\theta+\phi)}$, which has a magnitude AB that is the product of the individual magnitudes and an angle $\theta + \phi$ that, by the law of exponents, is the sum of the individual angles. In polar form this is $A\underline{/\theta} \times B\underline{/\phi} = AB\underline{/\theta + \phi}$.

For division the result is

$$\frac{Ae^{j\theta}}{Be^{j\phi}} = \frac{A}{B} e^{j(\theta-\phi)} \qquad \text{which in polar form is} \qquad \frac{A\underline{/\theta}}{B\underline{/\phi}} = \frac{A}{B} \underline{/\theta - \phi}$$

So, the magnitude of the quotient is the quotient A/B of the magnitudes, and the angle of the quotient is, by the law of exponents, the difference $\theta - \phi$ of the numerator angle minus the denominator angle.

PHASORS

By definition, a *phasor* is a complex number associated with a phase-shifted sine wave such that, if the phasor is in polar form, its magnitude is the effective (rms) value of the voltage or current and its angle is the phase angle of the phase-shifted sine wave. For example, $\mathbf{V} = 3\underline{/45^\circ}$ V is the phasor for $v = 3\sqrt{2}\sin(377t + 45^\circ)$ V and $\mathbf{I} = 0.439\underline{/-27^\circ}$ A is the phasor for $i = 0.621\sin(754t - 27^\circ)$ A. Of course, $0.621 = \sqrt{2}(0.439)$.

Note the use of the boldface letters $\mathbf{V}$ and $\mathbf{I}$ for the phasor voltage and current quantity symbols. It is conventional to use boldface letter symbols for *all* complex quantities. Also, a superscript asterisk is used to designate a conjugate. As an illustration, if $\mathbf{V} = -6 + j10 = 11.7\underline{/121^\circ}$, then $\mathbf{V}^* = -6 - j10 = 11.7\underline{/-121^\circ}$ V. The magnitude of a phasor variable is indicated by using lightface, and the magnitude of a complex number is indicated by using parallel lines. For example, if $\mathbf{I} = 3 + j4 = 5\underline{/53.1^\circ}$ A, then $I = |3 + j4| = |5\underline{/53.1^\circ}| = 5$ A.

A common error is to equate a phasor and its corresponding sinusoid. They cannot be equal because the phasor is a complex constant, but the sinusoid is a real function of time. In short, it is *wrong* to write something like $3\underline{/30^\circ} = 3\sqrt{2}\sin(\omega t + 30^\circ)$.

Phasors are usually shown in the polar form for convenience. But the rectangular form is just as correct because, being a complex number, a phasor can be expressed in any of the complex number forms. Not all complex numbers, though, are phasors—just those corresponding to sinusoids.

There is no complete agreement on the definition of a phasor. Many electrical engineers use the sinusoidal peak value instead of the effective value. Also, many use the angle from the phase-shifted cosine wave instead of the sine wave.

One use of phasors is for summing sinusoids of the *same* frequency. If each sinusoid is transformed into a phasor and the phasors added and then reduced to a single complex number, this number is the phasor for the sum sinusoid. As an illustration, the single sinusoid corresponding to $v = 3\sin(2t + 30^\circ) + 2\sin(2t - 15^\circ)$ V can be found by adding the corresponding phasors,

$$\mathbf{V} = \frac{3}{\sqrt{2}}\underline{/30^\circ} + \frac{2}{\sqrt{2}}\underline{/-15^\circ} = \frac{4.64}{\sqrt{2}}\underline{/12.2^\circ}\text{ V}$$

and then transforming the sum phasor to a sinusoid. The result is $v = 4.64\sin(2t + 12.2^\circ)$ V. This procedure works for any number of sinusoids being added and subtracted, provided that all have the same frequency. A common error is to forget that they must have the same frequency.

Notice that using $\sqrt{2}$ did not contribute anything to the final result. The $\sqrt{2}$ was introduced in finding the phasors, and then deleted in transforming the sum phasor to a sinusoid. When the problem statement is in sinusoids and the answer is to be a sinusoid, it is easier to forget about the $\sqrt{2}$ and use phasors that are based on peak values instead of rms values.

Phasors are sometimes shown on a complex plane in a diagram called a *phasor diagram*. The phasors are shown as arrows directed out from the origin with lengths corresponding to the phasor magnitudes, and arranged at angles that are the corresponding phasor angles. Such diagrams are convenient for showing the angular relations among voltages and currents of the same frequency. Sometimes they are also used for adding and subtracting, but the accuracy is much less than that obtained by using complex algebra.

Another diagram, called a *funicular diagram*, is more convenient for graphical addition and subtraction. In this type of diagram the adding and subtracting are the same as for vectors. For adding, the arrows of the phasors are placed end to end and the sum phasor is found by drawing an arrow from the tail of the first arrow to the tip of the last. If a phasor is to be subtracted, its arrow is rotated 180° (reversed) and then added.

Solved Problems

10.1 Perform the following operations:

   (a) $j2 + j3 - j6 - j8$ (b) $j2(-j3)(j4)(-j6)$ (c) $\dfrac{1}{j0.25}$ (d) $\dfrac{j100}{j8}$

(a) The rules for adding and subtracting imaginary numbers are the same as for adding and subtracting real numbers, except that the result is imaginary.   So,

$$j2 + j3 - j6 - j8 = j5 - j14 = -j9$$

(b) The numbers can be multiplied two at a time, with the result of

$$[j2(-j3)][j4(-j6)] = 6(24) = 144$$

Or, $j1$ can be factored from each factor and a power of $j1$ found times a product of real numbers:

$$j2(-j3)(j4)(-j6) = j1^4[2(-3)(4)(-6)] = 1(144) = 144$$

(c) The denominator can be made real by multiplying the numerator and denominator by $j1$, and then division performed as if the numbers were real—except that the quotient is imaginary:

$$\frac{1}{j0.25} = \frac{1(j1)}{j0.25(j1)} = \frac{j1}{-0.25} = -j4$$

Alternatively, since $1/j1 = -j1$,

$$\frac{1}{j0.25} = \frac{1}{j1}\left(\frac{1}{0.25}\right) = -j1(4) = -j4$$

(d) For convenience, the j's can be considered to be numbers and divided out:

$$\frac{j100}{j8} = \frac{\cancel{j}100}{\cancel{j}8} = 12.5$$

10.2 Add or subtract as indicated, and express the results in rectangular form:

(a) $(6.21 + j3.24) + (4.13 - j9.47)$

(b) $(7.34 - j1.29) - (5.62 + j8.92)$

(c) $(-24 + j12) - (-36 - j16) - (17 - j24)$

The real and imaginary parts are separately added or subtracted:

(a) $(6.21 + j3.24) + (4.13 - j9.47) = (6.21 + 4.13) + j(3.24 - 9.47) = 10.34 - j6.23$

(b) $(7.34 - j1.29) - (5.62 + j8.92) = (7.34 - 5.62) - j(1.29 + 8.92) = 1.72 - j10.21$

(c) $(-24 + j12) - (-36 - j16) - (17 - j24) = (-24 + 36 - 17) + j(12 + 16 + 24) = -5 + j52$

10.3 Add or subtract as indicated:

(a) $(6 + j10) + (6 - j10)$ (b) $(6 + j10) - (6 - j10)$

(a) $(6 + j10) + (6 - j10) = (6 + 6) + j(10 - 10) = 12$

This result shows that the sum of conjugates is a real number that is twice the real part of either number.

(b) $(6 + j10) - (6 - j10) = (6 - 6) + j(10 + 10) = j20$

This result shows that the difference of conjugates is an imaginary number that is twice the imaginary part of one of the numbers.

10.4 Find the following products and express them in rectangular form:

(a) $(4 + j2)(3 + j4)$ (b) $(6 + j2)(3 - j5)(2 - j3)$

In the multiplication of complex numbers in rectangular form, the ordinary rules of algebra are used along with the rules for imaginary numbers:

(a) $(4 + j2)(3 + j4) = 4(3) + 4(j4) + j2(3) + j2(j4) = 12 + j16 + j6 - 8 = 4 + j22$

(b) It is best to multiply two numbers at a time:

$$(6 + j2)(3 - j5)(2 - j3) = [6(3) + 6(-j5) + j2(3) + j2(-j5)](2 - j3) = (18 - j30 + j6 + 10)(2 - j3)$$
$$= (28 - j24)(2 - j3) = 28(2) + 28(-j3) + (-j24)(2) + (-j24)(-j3)$$
$$= 56 - j84 - j48 - 72 = -16 - j132$$

Multiplying three or more complex numbers in rectangular form usually requires more work than does converting them to polar form and multiplying.

10.5 Evaluate

$$\begin{vmatrix} 4 + j3 & -j2 \\ -j2 & 5 - j6 \end{vmatrix}$$

The value of this second-order determinant equals the product of the elements on the principal diagonal minus the product of the elements on the other diagonal, the same as for one with real elements:

$$\begin{vmatrix} 4 + j3 & -j2 \\ -j2 & 5 - j6 \end{vmatrix} = (4 + j3)(5 - j6) - (-j2)(-j2) = 20 - j24 + j15 + 18 + 4 = 42 - j9$$

10.6 Evaluate

$$\begin{vmatrix} 4 + j6 & -j4 & -2 \\ -j4 & 6 + j10 & 3 \\ -2 & -3 & 2 + j1 \end{vmatrix}$$

The evaluation of a third-order determinant with complex elements is the same as for one with real elements:

$$= (4 + j6)(6 + j10)(2 + j1) + (-j4)(-3)(-2) + (-2)(-j4)(-3)$$

$$- (-2)(6 + j10)(-2) - (-3)(-3)(4 + j6) - (2 + j1)(-j4)(-j4)$$

$$= -148 + j116 - j24 - j24 - 24 - j40 - 36 - j54 + 32 + j16 = -176 - j10$$

10.7 Find the following quotients in rectangular form:

(a) $\dfrac{1}{0.2 + j0.5}$ (b) $\dfrac{14 + j5}{4 - j1}$

For division in the rectangular form, the numerator and denominator should be multiplied by the conjugate of the denominator to make the denominator real. Then the division is straightforward. Doing this results in

(a) $\dfrac{1}{0.2 + j0.5} \times \dfrac{0.2 - j0.5}{0.2 - j0.5} = \dfrac{0.2 - j0.5}{0.2^2 + 0.5^2} = \dfrac{0.2 - j0.5}{0.29} = \dfrac{0.2}{0.29} - j\dfrac{0.5}{0.29} = 0.69 - j1.72$

(b) $\dfrac{14 + j5}{4 - j1} \times \dfrac{4 + j1}{4 + j1} = \dfrac{51 + j34}{17} = 3 + j2$

As easy as these simple divisions are to do with the numbers in rectangular form, they are even easier to do in polar form with the aid of a calculator that does conversions between the two forms.

10.8 Express each of the following as a ratio of two complex numbers in rectangular form:

(a) $\dfrac{6 + j4}{2 - j5} + \dfrac{5 - j4}{3 - j6}$ (b) $\dfrac{7 - j4}{-5 + j2} + \dfrac{4 - j3}{-6 - j4}$

The approach is to form a common denominator that is the product of the individual denominators:

(a) $\dfrac{6 + j4}{2 - j5} + \dfrac{5 - j4}{3 - j6} = \dfrac{(6 + j4)(3 - j6) + (5 - j4)(2 - j5)}{(2 - j5)(3 - j6)} = \dfrac{32 - j57}{-24 - j27}$

(b) $\dfrac{7 - j4}{-5 + j2} + \dfrac{4 - j3}{-6 - j4} = \dfrac{(7 - j4)(-6 - j4) + (4 - j3)(-5 + j2)}{(-5 + j2)(-6 - j4)} = \dfrac{-72 + j19}{38 + j8}$

10.9 Convert the following numbers to polar form:

(a) $6 + j9$ (b) $-21.4 + j33.3$ (c) $-0.521 - j1.42$ (d) $4.23 + j4.23$

If a calculator is used that does not have a rectangular-to-polar conversion feature, then a complex number $x + jy$ can be converted to its equivalent $A\underline{/\theta}$ with the formulas $A = \sqrt{x^2 + y^2}$ and $\theta = \tan^{-1}(y/x)$. With this approach

(a) $6 + j9 = \sqrt{6^2 + 9^2}\,\underline{/\tan^{-1}(9/6)} = 10.8\underline{/56.3°}$

(b) $-21.4 + j33.3 = \sqrt{(-21.4)^2 + 33.3^2}\,\underline{/\tan^{-1}[33.3/(-21.4)]} = 39.6\underline{/122.7°}$

Some calculators will give $\tan^{-1}(-33.3/21.4) = -57.3°$ for the angle, which differs from the correct angle by 180°. For such calculators, this error of 180° always occurs in a rectangular-to-polar form conversion whenever the real part of the complex number is negative. The solution, of course, is to change the calculator angle by either positive or negative 180°, whichever is more convenient.

(c) $-0.521 - j1.42 = \sqrt{(-0.521)^2 + (-1.42)^2}\,\underline{/\tan^{-1}[-1.42/(-0.521)]} = 1.51\underline{/-110°}$

Again, because the real part is negative, some calculators will not give an angle of $-110°$, but instead will give $\tan^{-1}(1.42/0.521) = 70°$.

(d) $4.23 + j4.23 = \sqrt{4.23^2 + 4.23^2}\,\underline{/\tan^{-1}(4.23/4.23)} = \sqrt{2}(4.23)\underline{/\tan^{-1}1} = 5.98\underline{/45°}$

As can be generalized from this result, when the magnitudes of the real and imaginary parts are equal, the polar magnitude is $\sqrt{2}$ times this magnitude. Also, the angle is 45° if the number is in the first quadrant of the complex plane, 135° if it is in the second, $-135°$ if it is in the third, and $-45°$ if it is in the fourth.

10.10 Convert the following numbers to polar form: (a) $1000 + j2$ and   (b) $4 - j963$.

(a) $1000 + j2 = \sqrt{1000^2 + 2^2}\,\underline{/\tan^{-1}(2/1000)} = 1000\underline{/0.115°} \simeq 1000$

(b) $4 - j963 = \sqrt{4^2 + (-963)^2}\,\underline{/\tan^{-1}(-963/4)} = 963\underline{/-89.8°} \simeq 963\underline{/-90°}$
$$= 963\cos(-90°) + j963\sin(-90°) = -j963$$

This problem illustrates the fact that, for a complex number in rectangular form, the smaller part can often be neglected if it is much smaller than the larger. Also, in general from the result of part (b), $A\underline{/-90°} = -jA$. Similarly, $A\underline{/90°} = jA$.

10.11 Convert the following numbers to rectangular form:

(a) $10.2\underline{/20°}$ (b) $6.41\underline{/-30°}$ (c) $-142\underline{/-80.3°}$ (d) $142\underline{/-260.3°}$ (e) $-142\underline{/-440.3°}$

If a calculator is used that does not have a polar-to-rectangular conversion feature, then Euler's identity can be used: $A\underline{/\theta} = A\cos\theta + jA\sin\theta$. With this approach

(a) $10.2\underline{/20°} = 10.2\cos 20° + j10.2\sin 20° = 9.58 + j3.49$

(b) $6.41\underline{/-30°} = 6.41\cos(-30°) + j6.41\sin(-30°) = 5.55 - j3.21$

(c) $-142\underline{/-80.3°} = -142\cos(-80.3°) - j142\sin(-80.3°) = -23.9 + j140$

(d) $142\underline{/-260.3°} = 142\cos(-260.3°) + j142\sin(-260.3°) = -23.9 + j140$

(e) $-142\underline{/-440.3°} = -142\cos(-440.3°) - j142\sin(-440.3°) = -23.9 + j140$

Parts (c) and (d) show that an angular difference of 180° corresponds to multiplying by -1. And parts (c) and (e) show that an angular difference of 360° has no effect. So, in general, $A\underline{/\theta \pm 180°} = -A\underline{/\theta}$ and $A\underline{/\theta \pm 360°} = A\underline{/\theta}$.

10.12 Perform the following operations and express the results in polar form:

(a) $3 + j4 + 9.1\underline{/63°} - 7.2\underline{/-40°}$ (b) $20.1\underline{/135°} - 46.7\underline{/-142°} + 35.2\underline{/64.1°}$

The numbers in polar form must be converted to rectangular form before being added:

(a) $3 + j4 + 9.1\underline{/63°} - 7.2\underline{/-40°} = 3 + j4 + 4.131 + j8.108 - 5.516 + j4.628$
$$= 1.62 + j16.7 = 16.8\underline{/84.5°}$$

(b) $20.1\underline{/135°} - 46.7\underline{/-142°} + 35.2\underline{/64.1°} = -14.2 + j14.2 + 36.8 + j28.7 + 15.4 + j31.7$

$$= 38 + j74.6 = 83.7\underline{/63°}$$

Some pocket calculators, which can convert between rectangular and polar forms, have at least two memories, one in which real parts can be summed and the other in which imaginary parts can be summed. Adding with one of these calculators does not require any writing.

10.13 Find the following products in polar form:

(a) $(3\underline{/25°})(4\underline{/-60°})(-5\underline{/120°})(-6\underline{/-210°}$ (b) $(0.3 + j0.4)(-5 + j6)(7\underline{/35°})(-8 - j9)$

(a) When all the factors are in polar form, the magnitude of the product is the product of the individual magnitudes along with negative signs, if any, and the angle of the product is the sum of the individual angles. So,

$$(3\underline{/25°})(4\underline{/-60°})(-5\underline{/120°})(-6\underline{/-210°}) = 3(4)(-5)(-6)\underline{/25° - 60° + 120° - 210°} = 360\underline{/-125°}$$

(b) The numbers in rectangular form must be converted to polar form before being multiplied:

$$(0.3 + j0.4)(-5 + j6)(7\underline{/35°})(-8 - j9) = (0.5\underline{/53.1°})(7.81\underline{/129.8°})(7\underline{/35°})(12.04\underline{/-131.6°})$$

$$= 0.5(7.81)(7)(12.04)\underline{/53.1° + 129.8° + 35° - 131.6°} = 329\underline{/86.3°}$$

10.14 Find the quotients in polar form for (a) $(81\underline{/45°})/(3\underline{/16°})$ and (b) $(-9.1\underline{/20°})/(-4 + j7)$.

(a) When the numerator and denominator are in polar form, the magnitude of the quotient is the quotient of the magnitudes, and the angle of the quotient is the angle of the numerator minus the angle of the denominator. So,

$$\frac{81\underline{/45°}}{3\underline{/16°}} = \frac{81}{3}\underline{/45° - 16°} = 27\underline{/29°}$$

(b) The denominator should be converted to polar form as a first step:

$$\frac{-9.1\underline{/20°}}{-4 + j7} = \frac{-9.1\underline{/20°}}{8.06\underline{/119.7°}} = -\frac{9.1}{8.06}\underline{/20° - 119.7°} = -1.13\underline{/-99.7°} = 1.13\underline{/-99.7° + 180°} = 1.13\underline{/80.3°}$$

10.15 Find the following quotients in polar form:

(a) $\dfrac{(4 + j6)(5\underline{/20°})(-8 + j9)}{(9.1\underline{/-30°})(7.6\underline{/72°})(-3 + j5)}$ (b) $\dfrac{6 + j7 - 4 + j6 - 7\underline{/20°}}{20 + j17 - 8\underline{/-60°} + 14\underline{/20°}}$

Initially, each numerator and denominator should be made a single complex number, preferably in polar form, to make each division the same as that for two complex numbers. Then

(a) $\dfrac{(4 + j6)(5\underline{/20°})(-8 + j9)}{(9.1\underline{/-30°})(7.6\underline{/72°})(-3 + j5)} = \dfrac{(7.211\underline{/56.3°})(5\underline{/20°})(12.04\underline{/131.6°})}{(9.1\underline{/-30°})(7.6\underline{/72°})(5.83\underline{/121°})} = \dfrac{434\underline{/208°}}{403\underline{/163°}} = 1.08\underline{/45°}$

(b) $\dfrac{6 + j7 - 4 + j6 - 7\underline{/20°}}{20 + j17 - 8\underline{/-60°} + 14\underline{/20°}} = \dfrac{6 + j7 - 4 + j6 - 6.78 - j2.394}{20 + j17 - 4 + j6.928 + 13.16 + j4.788} = \dfrac{-4.578 + j10.61}{29.16 + j28.72}$

$$= \dfrac{11.56\underline{/113.33°}}{40.93\underline{/44.56°}} = 0.282\underline{/68.8°}$$

10.16 For $\mathbf{Z}_1 = 20\underline{/30°}$ and $\mathbf{Z}_2 = 16\underline{/-45°}$, find $\mathbf{Z}_3 = \mathbf{Z}_1\mathbf{Z}_2/(\mathbf{Z}_1 + \mathbf{Z}_2)$.

Substituting for $\mathbf{Z}_1$ and $\mathbf{Z}_2$ results in

$$\mathbf{Z}_3 = \frac{\mathbf{Z}_1\mathbf{Z}_2}{\mathbf{Z}_1 + \mathbf{Z}_2} = \frac{(20\underline{/30°})(16\underline{/-45°})}{20\underline{/30°} + 16\underline{/-45°}} = \frac{320\underline{/-15°}}{17.32 + j10 + 11.31 - j11.31}$$

$$= \frac{320\underline{/-15°}}{28.63 - j1.31} = \frac{320\underline{/-15°}}{28.66\underline{/-2.63°}} = 11.2\underline{/-12.4°}$$

The **Z**'s are not phasor symbols. Rather, as presented in Chap. 11, they are quantity symbols for impedances, which are to ac circuits what resistances are to dc circuits.

10.17 Find the following quotient:

$$\frac{(1.2\underline{/35°})^3(4.2\underline{/-20°})^6}{(2.1\underline{/-10°})^4(-3 + j6)^5}$$

Since each exponent of a number indicates how many times the number is to be multiplied by itself, the effect of an exponent is to raise the number magnitude to this exponent and to multiply the number angle by this exponent. Thus,

$$\frac{(1.2\underline{/35°})^3(4.2\underline{/-20°})^6}{(2.1\underline{/-10°})^4(-3 + j6)^5} = \frac{(1.2\underline{/35°})^3(4.2\underline{/-20°})^6}{(2.1\underline{/-10°})^4(6.71\underline{/117°})^5} = \frac{1.2^3(4.2)^6\underline{/3(35°) - 6(20°)}}{2.1^4(6.71)^5\underline{/4(-10°) + 5(117°)}}$$

$$= \frac{1.73(5489)\underline{/-15°}}{19.4(13\,584)\underline{/543°}} = \frac{9.49 \times 10^3\underline{/-15°}}{2.64 \times 10^3\underline{/543°}} = 0.0359\underline{/-558°} = 0.0359\underline{/-198°} = -0.0359\underline{/-18°}$$

10.18 Find the corresponding phasor voltages and currents for the following:

(a) $v = \sqrt{2}(50) \sin (377t - 35°)$ V   (c) $v = 83.6 \cos (400t - 15°)$ V

(b) $i = \sqrt{2}(90.4) \sin (754t + 48°)$ mA   (d) $i = 3.46 \cos (815t + 30°)$ A

A phasor in polar form has a magnitude that is the effective value of the corresponding sinusoidal voltage or current, and an angle that is the phase angle of the sinusoid if it is in phase-shifted sine-wave form. So,

(a) $v = \sqrt{2}(50) \sin (377t - 35°)$ V   →   **V** = $50\underline{/-35°}$ V

(b) $i = \sqrt{2}(90.4) \sin (754t + 48°)$ mA   →   **I** = $90.4\underline{/48°}$ mA

(c) $v = 83.6 \cos (400t - 15°) = 83.6 \sin (400t - 15° + 90°) = 83.6 \sin (400t + 75°)$ V
       →   **V** = $(83.6/\sqrt{2})\underline{/75°} = 59.1\underline{/75°}$ V

(d) $i = 3.46 \cos (815t + 30°) = 3.46 \sin (815t + 30° + 90°) = 3.46 \sin (815t + 120°)$ A
       →   **I** = $(3.46/\sqrt{2})\underline{/120°} = 2.45\underline{/120°}$ A

10.19 Find the voltages and currents corresponding to the following phasor voltages and currents (each sinusoid has a radian frequency of 377 rad/s):

(a)   **V** = $20\underline{/35°}$ V   (b)   **I** = $10.2\underline{/-41°}$ mA   (c)   **V** = $4 - j6$ V   (d)   **I** = $-3 + j1$ A

If a phasor is in polar form, the corresponding voltage or current is a phase-shifted sine wave that has a phase angle that is the phasor angle, and a peak value that is the $\sqrt{2}$ times the phasor magnitude. Thus,

(a)   **V** = $20\underline{/35°}$ V   → $v = 20\sqrt{2} \sin (377t + 35°) = 28.3 \sin (377t + 35°)$ V

(b)   **I** = $10.2\underline{/-41°}$ mA   → $i = \sqrt{2}(10.2) \sin (377t - 41°) = 14.4 \sin (377t - 41°)$ mA

(c)   **V** = $4 - j6 = 7.21\underline{/-56.3°}$ V   → $v = \sqrt{2}(7.21) \sin (377t - 56.3°) = 10.2 \sin (377t - 56.3°)$ V

(d)   **I** = $-3 + j1 = 3.16\underline{/161.6°}$ A   → $i = \sqrt{2}(3.16) \sin (377t + 161.6°) = 4.47 \sin (377t + 161.6°)$ A

10.20 Find a single sinusoid that is the equivalent of each of the following:

(a) $6.23 \sin \omega t + 9.34 \cos \omega t$

(b) $5 \sin (4t - 20°) + 6 \sin (4t + 45°) - 7 \cos (4t - 60°) + 8 \cos (4t + 30°)$

(c) $5 \sin 377t + 6 \cos 754t$

A phasor approach can be used since the terms are sinusoids. The procedure is to find the phasor corresponding to each sinusoid, add the phasors to get a single complex number, and then find the sinusoid corresponding to this number. Preferably the phasors are based on peak values because there is no advantage in introducing a factor of $\sqrt{2}$ since the problems statements are in sinusoids and the answers are to be in sinusoids.   Thus,

(a) $6.23 \sin \omega t + 9.34 \cos \omega t \quad \rightarrow \quad 6.23\underline{/0°} + 9.34\underline{/90°} = 11.2\underline{/56.3°} \quad \rightarrow \quad 11.2 \sin (\omega t + 56.3°)$

(b) $5 \sin (4t - 20°) + 6 \sin (4t + 45°) - 7 \cos (4t - 60°) + 8 \cos (4t + 30°)$
$\quad \rightarrow \quad 5\underline{/-20°} + 6\underline{/45°} - 7\underline{/30°} + 8\underline{/120°} = 6.07\underline{/100.7°} = -6.07\underline{/-79.3°} \quad \rightarrow \quad -6.07 \sin (4t - 79.3°)$

(c)   The sinusoids cannot be combined because they have different frequencies.

10.21   For the circuit shown in Fig. 10-3, find v_S if $v_1 = 10.2 \sin (754t + 30°)$ V, $v_2 = 14.9 \sin (754t - 10°)$ V, and $v_3 = 16.1 \cos (754t - 25°)$ V.

By KVL, $v_S = v_1 - v_2 + v_3 = 10.2 \sin (754t + 30°) - 14.9 \sin (754t - 10°) + 16.1 \cos (754t - 25°)$ V.   The sum sinusoid can be found by using phasors:

$$\mathbf{V}_S = \mathbf{V}_1 - \mathbf{V}_2 + \mathbf{V}_3 = \frac{10.2}{\sqrt{2}}\underline{/30°} - \frac{14.9}{\sqrt{2}}\underline{/-10°} + \frac{16.1}{\sqrt{2}}\underline{/65°} = \frac{22.3}{\sqrt{2}}\underline{/87.5°} \text{ V}$$
$$\rightarrow \quad v_S = 22.3 \sin (754t + 87.5°) \text{ V}$$

Since the problem statement is in sinusoids and the final result is a sinusoid, finding the solution would have been slightly easier using phasors based on peak rather than rms values.

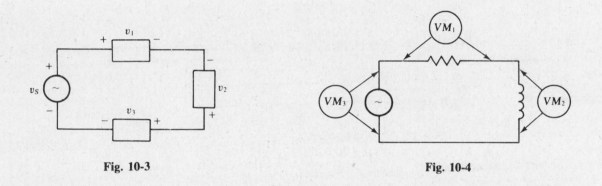

Fig. 10-3   Fig. 10-4

10.22   In the circuit shown in Fig. 10-4, voltmeters VM_1 and VM_2 have readings of 40 and 30 V, respectively.   Find the reading of voltmeter VM_3.

It is tempting to conclude that, by KVL, the reading of voltmeter VM_3 is the sum of the readings of voltmeters VM_1 and VM_2.   But this is *wrong* because KVL applies to phasor voltages and not to the rms voltages of the voltmeter readings.   The rms voltages, being positive real constants, do not have the angles that the phasor voltages have.

For the phasors required for KVL, angles must be associated with the given rms voltages.   One angle can be arbitrarily selected because only the magnitude of the sum is desired.   If 0° is selected for the resistor voltage phasor, this phasor is $40\underline{/0°}$ V and then that for the inductor voltage must be $30\underline{/90°}$ V.   The inductor voltage phasor has a 90° greater angle because this voltage leads the current by 90°, but the resistor voltage is in phase with the current.   By KVL, the phasor voltage for the source is $40 + 30\underline{/90°} = 40 + j30 = 50\underline{/36.9°}$ V, which has an rms value of 50 V.   So, the reading of voltmeter VM_3 is 50 V, and not the $30 + 40 = 70$ V   that might at first be supposed.

10.23   Find v_S for the circuit shown in Fig. 10-5.

The voltage v_S can be determined from $v_S = v_R + v_L + v_C$ after these component voltages are found.   By Ohm's law,

$$v_R = [0.234 \sin (3000t - 10°)](270) = 63.2 \sin (3000t - 10°) \text{ V}$$

The inductor voltage v_L leads the current by 90° and has a peak value of $\omega L = 3000(120 \times 10^{-3}) =$

360   times the peak value of the current:

$$v_L = 360(0.234) \sin (3000t - 10° + 90°) = 84.2 \sin (3000t + 80°) \text{ V}$$

The capacitor voltage v_C lags the current by 90° and has a peak value that is $1/\omega C = 1/(3000 \times 6 \times 10^{-6}) = 55.6$ times the peak value of the current:

$$v_C = 55.6(0.234) \sin (3000t - 10° - 90°) = 13 \sin (3000t - 100°) \text{ V}$$

Phasors, which are conveniently based on peak values, can be used to find the sum sinusoid:

$$\mathbf{V}_S = \mathbf{V}_R + \mathbf{V}_L + \mathbf{V}_C = 63.2\underline{/-10°} + 84.2\underline{/80°} + 13\underline{/-100°} = 95.2\underline{/38.4°} \text{ V}$$
$$\rightarrow \quad v_S = 95.2 \sin (3000t + 38.4°) \text{ V}$$

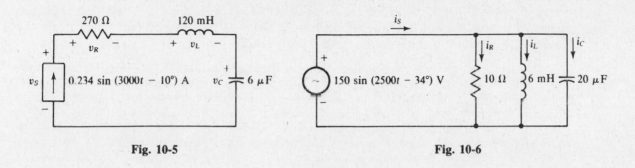

Fig. 10-5   Fig. 10-6

10.24 Find i_S for the circuit shown in Fig. 10-6.

The current i_S can be determined from $i_S = i_R + i_L + i_C$ after these component currents are found. By Ohm's law,

$$i_R = \frac{150 \sin (2500t - 34°)}{10} = 15 \sin (2500t - 34°) \text{ A}$$

The inductor current i_L lags the voltage by 90° and has a peak value that is $1/\omega L = 1/(2500 \times 6 \times 10^{-3}) = 1/15$ times the peak value of the voltage:

$$i_L = \frac{150 \sin (2500t - 34° - 90°)}{15} = 10 \sin (2500t - 124°) \text{ A}$$

The capacitor current i_C leads the voltage by 90° and has a peak value that is $\omega C = 2500(20 \times 10^{-6}) = 0.05$ times the peak value of the voltage:

$$i_C = 0.05(150) \sin (2500t - 34° + 90°) = 7.5 \sin (2500t + 56°) \text{ A}$$

Phasors, which are conveniently based on peak values, can be used to find the sum sinusoid:

$$\mathbf{I}_S = \mathbf{I}_R + \mathbf{I}_L + \mathbf{I}_C = 15\underline{/-34°} + 10\underline{/-124°} + 7.5\underline{/56°} = 15.2\underline{/-43.5°} \text{ A} \quad \rightarrow \quad i_S = 15.2 \sin (2500t - 43.5°) \text{ A}$$

10.25 If two currents have phasors of $10\underline{/0°}$ and $7\underline{/30°}$ mA, what is the angle and rms value of the current that is the sum of these currents? Solve by using a funicular diagram. Check the answer by using complex algebra.

Figure 10-7 shows the tail of the 7-mA phasor at the tip of the 10-mA phasor, as required for vector addition. The sum phasor, extending from the tail of the 10-mA phasor to the tip of the 7-mA phasor, has a length corresponding to approximately 16.5 mA and an angle of approximately 13°. In comparison, the result from complex algebra is

$$10\underline{/0°} + 7\underline{/30°} = 10 + 6.06 + j3.5 = 16.06 + j3.5 = 16.4\underline{/12.3°} \text{ mA}$$

which is, of course, considerably more accurate than the graphical result.

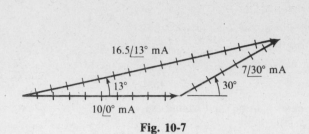

Fig. 10-7

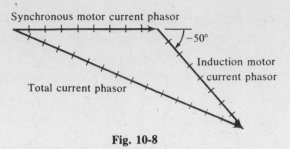

Fig. 10-8

10.26 A synchronous motor draws a 9-A current from a 240-V, 60-Hz source.   A parallel induction motor draws 8 A.   If the synchronous motor current leads the applied voltage by 20°, and if the induction motor current lags this voltage by 30°, what is the total current drawn from the source?   Find this current graphically and algebraically.

The choice of the reference phasor—the one arranged horizontally at 0°—is somewhat arbitrary.   The voltage phasor or either current phasor could be used.   In fact, no phasor has to be at 0°, but it is usually convenient to have one at this angle.   In Fig. 10-8 the synchronous motor current phasor is arbitrarily positioned horizontally, and the induction motor current phasor at its tip is positioned at an angle of 50° with it since there is a $20° - (-30°) = 50°$ phase angle difference between the two currents.   Also shown is the sum phasor, which has a measured length corresponding to 15.4 A.   In comparison, from complex algebra,

$$\mathbf{I} = 9\underline{/0°} + 8\underline{/-50°} = 9 + 5.14 - j6.13 = 14.14 - j6.13 = 15.4\underline{/-23.4°} \text{ A}$$

and
$$I = |\mathbf{I}| = |15.4\underline{/-23.4°}| = 15.4 \text{ A}$$

in agreement with the graphical result to three significant digits.   Usually, agreement to only two significant digits should be expected because of the comparative lack of accuracy with the graphical approach.

Supplementary Problems

10.27 Perform the following operations:

(a) $j6 - j7 + j4 - j8 + j9$ (b) $j2^2(-j3)(j7)(-j8)(j0.9)$ (c) $\dfrac{-j100}{5}$ (d) $\dfrac{8}{-j4}$

Ans.   (a) j4,   (b) −604.8,   (c) −j20,   (d) j2

10.28 Perform the following operations and express the results in rectangular form:

(a) $(4.59 + j6.28) + (5.21 - j4.63)$

(b) $(8.21 + j4.31) - (4.92 - j6.23) - (-5.16 + j7.21)$

(c) $3 + j4 - 5 + j6 - 7 + j8 - 9 + j10 - 11$

Ans.   (a) 9.8 + j1.65,   (b) 8.45 + j3.33,   (c) −29 + j28

10.29 Find the following products and express them in rectangular form:

(a) $(6 - j7)(4 + j2)$

(b) $(5 + j1)(-7 - j4)(-6 + j9)$

(c) $(-2 + j6)(-4 - j4)(-6 + j8)(7 + j3)$

Ans.   (a) 38 − j16,   (b) 429 − j117,   (c) −1504 + j2272

10.30 Find the following products and express them in rectangular form:

(a) $(4+j3)^2(4-j3)^2$ (b) $(0.6-j0.3)^2(-2+j4)^3$

Ans. (a) 625,   (b) $18-j36$

10.31 Evaluate $\begin{vmatrix} 6-j8 & 2-j3 \\ -4+j2 & -5+j9 \end{vmatrix}$.

Ans. $44+j78$

10.32 Evaluate $\begin{vmatrix} 6+j5 & -j2 & -4 \\ -j2 & 10-j8 & -6 \\ -4 & -6 & 5-j6 \end{vmatrix}$.

Ans. $156-j762$

10.33 Evaluate $\begin{vmatrix} 10-j2 & -2+j1 & -3-j4 \\ -2+j1 & 9-j8 & -6+j2 \\ -3-j4 & -6+j2 & 12-j4 \end{vmatrix}$.

Ans. $-65-j1400$

10.34 Find the following quotients in rectangular form:

(a) $\dfrac{1}{0.1-j0.4}$ (b) $\dfrac{1}{-0.4+j0.5}$ (c) $\dfrac{7-j2}{6-j3}$

Ans.   (a) $0.588+j2.35$,   (b) $-0.976-j1.22$,   (c) $1.07+j0.2$

10.35 Express each of the following as a ratio of two complex numbers in rectangular form:

(a) $\dfrac{5+j3}{6-j7}+\dfrac{8-j2}{-4+j8}$ (b) $\dfrac{11-j8}{-2+j4}-\dfrac{6+j5}{3-j2}$

Ans.   (a) $\dfrac{-10-j40}{32+j76}$,   (b) $\dfrac{49-j60}{2+j16}$

10.36 Convert each of the following to polar form:

(a) $8.1+j11$ (c) $-33.4+j14.7$ (e) $16.2+j16.2$

(b) $16.3-j12.2$ (d) $-12.7-j17.3$ (f) $-19.1+j19.1$

Ans.   (a) $13.7\underline{/53.6°}$, (b) $20.4\underline{/-36.8°}$, (c) $36.5\underline{/156°}$, (d) $21.5\underline{/-126°}$, (e) $22.9\underline{/45°}$, (f) $27\underline{/135°}$

10.37 Convert each of the following to rectangular form:

(a) $11.8\underline{/51°}$ (c) $15.8\underline{/215°}$ (e) $-16.9\underline{/-36°}$

(b) $13.7\underline{/142°}$ (d) $27.4\underline{/-73°}$ (f) $-24.1\underline{/-1200°}$

Ans.   (a) $7.43+j9.17$,   (b) $-10.8+j8.43$,   (c) $-12.9-j9.06$,   (d) $8.01-j26.2$,   (e) $-13.7+j9.93$,
(f) $12.1+j20.9$

10.38 Perform the following operations and express the results in polar form:

(a) $6.31-j8.23+7.14\underline{/23.1°}-8.92\underline{/-47.5°}$

(b) $45.7\underline{/-34.6°}-68.9\underline{/-76.3°}-48.9\underline{/121°}$

(c) $-56.1\underline{/-49.8°}+73.1\underline{/-74.2°}-8-j6$

Ans.   (a) $6.95\underline{/9.51°}$,   (b) $46.5\underline{/-1.14°}$,   (c) $41.4\underline{/-126°}$

10.39 Find the following products in polar form:

(a) $(5.21\underline{/-36.1°})(0.141\underline{/110°})(-6.31\underline{/-116°})(1.72\underline{/210°})$

(b) $(5 + j3)(-6 + j1)(0.23\underline{/-17.1°})$

(c) $(0.2 - j0.5)(1.4 - j0.72)(-2.3 + j1.3)(-1.62 - j1.13)$

Ans.   (a) $7.97\underline{/-12.1°}$,   (b) $-8.16\underline{/4.4°}$,   (c) $4.42\underline{/-90°}$

10.40 Find the following quotients in polar form:

(a) $\dfrac{173\underline{/62.1°}}{38.9\underline{/-14.1°}}$ (b) $\dfrac{4.13 - j3.21}{-7.12\underline{/23.1°}}$ (c) $\dfrac{26.1\underline{/37.8°}}{-4.91 - j5.32}$

Ans.   (a) $4.45\underline{/76.2°}$,   (b) $-0.735\underline{/-61°}$,   (c) $-3.61\underline{/-9.5°}$

10.41 Find the following quotients in polar form:

(a) $\dfrac{(6.21 - j9.23)(-7.21 + j3.62)(21.3\underline{/35.1°})}{(-14.1 + j6.82)(6.97\underline{/68°})(10.2\underline{/-41°})}$

(b) $\dfrac{(6\underline{/-45°})(3 - j8) - (-7 + j4)(8 - j4)(3.62\underline{/70°})}{(-4.1 + j2)(3.4 + j6.1)(11\underline{/-27°})}$

Ans.   (a) $-1.72\underline{/61°}$,   (b) $-0.665\underline{/-4.14°}$

10.42 Find the following quotient in polar form:

$$\frac{(-6.29\underline{/-70.1°})^4(8.4\underline{/17°})^3(8.1\underline{/44°})^{1/2}}{(13.4\underline{/-16°})^2(-62.9\underline{/-107°})(0.729\underline{/93°})^{1/3}}$$

Ans. $260\underline{/80.6°}$

10.43 Find the corresponding phasor voltages and currents of the following in polar form:

(a) $v = \sqrt{2}(42.1)\sin(400t - 30°)$ V   (d) $i = -38.1\cos(754t - 72°)$ A

(b) $i = \sqrt{2}(36.9)\sin(6000t + 72°)$ A   (e) $v = -86.4\cos(672t + 34°)$ V

(c) $v = -64.3\sin(377t - 34°)$ V

Ans.   (a) $\mathbf{V} = 42.1\underline{/-30°}$ V,   (b) $\mathbf{I} = 36.9\underline{/72°}$ A,   (c) $\mathbf{V} = -45.5\underline{/-34°}$ V,   (d) $\mathbf{I} = -26.9\underline{/18°}$ A,   (e) $\mathbf{V} = 61.1\underline{/-56°}$ V

10.44 Find the voltages and currents corresponding to the following phasor voltages and currents (each sinusoid has a radian frequency of 754 rad/s):

(a) $\mathbf{V} = 15.1\underline{/62°}$ V   (c) $\mathbf{V} = -14.3\underline{/-69.7°}$ V   (e) $\mathbf{V} = -7 - j8$ V

(b) $\mathbf{I} = 9.62\underline{/-31°}$ A   (d) $\mathbf{I} = 4 - j6$ A   (f) $\mathbf{I} = -8.96 + j7.61$ A

Ans.   (a) $v = 21.4\sin(754t + 62°)$ V   (d) $i = 10.2\sin(754t - 56.3°)$ A

     (b) $i = 13.6\sin(754t - 31°)$ A   (e) $v = -15\sin(754t + 48.8°)$ V

     (c) $v = -20.2\sin(754t - 69.7°)$ V   (f) $i = -16.6\sin(754t - 40.3°)$ A

10.45 Find a single sinusoid that is the equivalent of each of the following:

(a) $7.21\sin \omega t + 11.2\cos \omega t$

(b) $-8.63\sin 377t - 4.19\cos 377t$

(c) $4.12\sin(64t - 10°) - 6.23\sin(64t - 35°) + 7.26\cos(64t - 35°) - 8.92\cos(64t + 17°)$

Ans.   (a) $13.3\sin(\omega t + 57.2°)$,   (b) $-9.59\sin(377t + 25.9°)$,   (c) $5.73\sin(64t + 2.75°)$

10.46 In Fig. 10-9, find i_1 if $i_2 = 14.6 \sin(377t - 15°)$ mA, $i_3 = 21.3 \sin(377t + 30°)$ mA, and $i_4 = 13.7 \cos(377t + 15°)$ mA.   *Ans.* $i_1 = -27.7 \cos(377t + 88.3°)$ mA

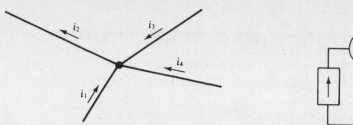

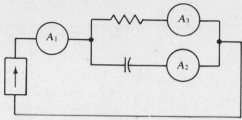

Fig. 10-9
Fig. 10-10

10.47 In a certain Y-connected three-phase circuit, the current in the neutral wire is $i_N = 6.21 \sin(377t + 10°) + 6.21 \sin(377t + 130°) + 6.21 \sin(377t - 110°)$ A. Reduce i_N to a single sinusoid. *Ans.* $i_N = 0$ A

10.48 In the circuit shown in Fig. 10-10, ammeters A_1 and A_2 have readings of 4 and 3 A, respectively. What is the reading of ammeter A_3?   *Ans.* 2.65 A

10.49 A current $i = 0.621 \sin(400t + 30°)$ mA flows through a 3.3-kΩ resistor in series with a 0.5-μF capacitor. Find the voltage across the series combination. Of course, as always, assume associated references when, as here, there is no statement to the contrary.   *Ans.* $v = 3.72 \sin(400t - 26.6°)$ V

10.50 A voltage $v = 240 \sin(400t + 10°)$ V is across a 680-Ω resistor in parallel with a 1-H inductor. Find the current flowing into this parallel combination.   *Ans.* $i = 0.696 \sin(400t - 49.5°)$ A

10.51 A current $i = 0.248 \cos(377t - 15°)$ A flows through the series combination of a 91-Ω resistor, a 120-mH inductor, and a 20-μF capacitor. Find the voltage across the series combination. *Ans.* $v = 31.3 \sin(377t + 31.2°)$ V

10.52 The voltage $v = 120 \sin(1000t + 20°)$ V is across the parallel combination of a 10-kΩ resistor, a 100-mH inductor, and a 10-μF capacitor. Find the total current i_T flowing into the parallel combination. Also, find the inductor current i_L and compare peak values of i_L and i_T.
Ans. $i_T = 0.012 \sin(1000t + 20°)$ A and $i_L = 1.2 \sin(1000t - 70°)$ A. The inductor current peak is 100 times the input current peak

Chapter 11

Basic AC Circuit Analysis, Impedance, and Admittance

INTRODUCTION

In the analysis of an ac circuit, voltage and current phasors are used with resistances and reactances in much the same way that voltages and currents are used with resistances in the analysis of a dc circuit. The original ac circuit, called a *time-domain circuit*, is transformed into a *frequency-domain circuit* that has phasors instead of sinusoidal voltages and currents, and that has reactances instead of inductances and capacitances. Resistances remain unchanged. The frequency-domain circuit is the circuit that is actually analyzed. It has the advantage that the resistances and reactances have the same ohm unit and so combine similarly to the way that resistors combine in a dc circuit analysis. Also, the analysis of the frequency-domain circuit requires no calculus, but only complex algebra. Finally, all the dc circuit analysis concepts apply to the analysis of a frequency-domain circuit, but, of course, complex numbers are used instead of real numbers.

FREQUENCY-DOMAIN CIRCUIT ELEMENTS

The transformation of a time-domain circuit into a frequency-domain circuit requires relations between the voltage and current phasors for resistors, inductors, and capacitors. First, consider getting this relation for a resistor of R ohms. For a current $i = I_m \sin(\omega t + \theta)$, the resistor voltage is, of course, $v = RI_m \sin(\omega t + \theta)$, with associated references assumed. The corresponding phasors are

$$\mathbf{I} = \frac{I_m}{\sqrt{2}}\underline{/\theta} \text{ A} \quad \text{and} \quad \mathbf{V} = \frac{RI_m}{\sqrt{2}}\underline{/\theta} \text{ V}$$

Dividing the voltage equation by the current equation eliminates I_m, θ, and $\sqrt{2}$ and produces a relation between the voltage and current phasors:

$$\frac{\mathbf{V}}{\mathbf{I}} = \frac{(I_m R/\sqrt{2})\underline{/\theta}}{(I_m/\sqrt{2})\underline{/\theta}} = R$$

This result shows that the resistance R of a resistor relates the resistor voltage and current phasors in the same way that it relates the resistor voltage and current ($R = v/i$). Because of this similarity, the relation $\mathbf{V}/\mathbf{I} = R$ can be represented in a frequency-domain circuit in the same way that $v/i = R$ is represented in the original time-domain circuit. Figure 11-1 shows this.

Next, consider an inductor of L henries. As shown in Chap. 9, for a current $i = I_m \sin(\omega t + \theta)$, the inductor voltage is $v = \omega L I_m \cos(\omega t + \theta) = \omega L I_m \sin(\omega t + \theta + 90°)$. The corresponding phasors are

$$\mathbf{I} = \frac{I_m}{\sqrt{2}}\underline{/\theta} \text{ A} \quad \text{and} \quad \mathbf{V} = \frac{\omega L I_m}{\sqrt{2}}\underline{/\theta + 90°} \text{ V}$$

Fig. 11-1

Dividing the voltage equation by the current equation results in a phasor relation of

$$\frac{\mathbf{V}}{\mathbf{I}} = \frac{(\omega L I_m/\sqrt{2})\,\underline{/\theta + 90^\circ}}{(I_m/\sqrt{2})\underline{/\theta}} = \omega L\underline{/90^\circ}$$

This result of $\omega L\underline{/90^\circ}$ in polar form is $j\omega L$ in rectangular form. Since ωL is the inductive reactance X_L, as defined in Chap. 9,

$$\frac{\mathbf{V}}{\mathbf{I}} = j\omega L = jX_L$$

Note that $j\omega L$ relates the inductor voltage and current phasors in the same way that R relates the resistor voltage and current phasors. Consequently, $j\omega L$ has a similar current-limiting action and the same ohm unit. In addition, because of its $j1$ multiplier, it produces a phase shift of 90° ($j1 = 1\underline{/90^\circ}$).

From the resistor discussion and the similarity of $\mathbf{V}/\mathbf{I} = R$ and $\mathbf{V}/\mathbf{I} = j\omega L$, the time-domain circuit to frequency-domain circuit transformation for an inductor, as shown in Fig. 11-2, should be apparent. The usual inductor circuit symbol is used in the frequency-domain circuit, but it is associated with $j\omega L$ ohms instead of with the L henries of the original time-domain circuit. Of course, the inductor voltage and current are transformed into corresponding phasors.

Fig. 11-2

The same approach can be used for a capacitor. For a voltage $v = V_m \sin(\omega t + \theta)$, a capacitor of C farads has a current of $i = \omega C V_m \sin(\omega t + \theta + 90^\circ)$. The corresponding phasors are

$$\mathbf{V} = \frac{V_m}{\sqrt{2}}\underline{/\theta}\ \text{V} \qquad \text{and} \qquad \mathbf{I} = \frac{\omega C V_m}{\sqrt{2}}\underline{/\theta + 90^\circ}\ \text{A}$$

and

$$\frac{\mathbf{V}}{\mathbf{I}} = \frac{(V_m/\sqrt{2})\,\underline{/\theta}}{(\omega C V_m/\sqrt{2})\,\underline{/\theta + 90^\circ}} = \frac{1}{\omega C\underline{/90^\circ}} = \frac{1}{j\omega C} = \frac{-j1}{\omega C}$$

As defined in Chap. 9, $-1/\omega C$ is the reactance X_C of a capacitor. Therefore,

$$\frac{\mathbf{V}}{\mathbf{I}} = \frac{-j1}{\omega C} = jX_C$$

(Remember that many circuits books have capacitive reactance defined as $X_C = 1/\omega C$, in which case $\mathbf{V}/\mathbf{I} = -jX_C$.) The $-j1/\omega C$ has a current-limiting action similar to that of a resistance. In addition, the $-j1$ multiplier produces a -90° phase shift.

Figure 11-3 shows the time-domain circuit to frequency-domain circuit transformation for a capacitor. In the frequency-domain circuit the conventional capacitor circuit symbol is used, but it is associated with $-j1/\omega C$ ohms instead of with the C farads of the original time-domain circuit.

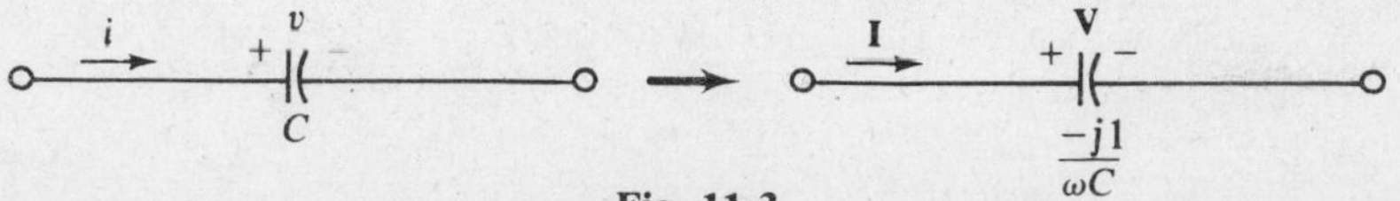

Fig. 11-3

AC SERIES CIRCUIT ANALYSIS

A method for analyzing a series ac circuit can be understood from a simple example. Suppose that the sinusoidal current i is to be found in the series circuit shown in Fig. 11-4a, in which the source has a radian frequency of $\omega = 4$ rad/s. The first step is to draw the corresponding frequency-domain circuit shown in Fig. 11-4b, in which the current and voltages are replaced by corresponding phasors, the inductance is replaced by

$$j\omega L = j4(2) = j8 \ \Omega$$

and the capacitance is replaced by

$$\frac{-j1}{\omega C} = \frac{-j1}{4(1/16)} = -j4 \ \Omega$$

The resistance, of course, is not changed.

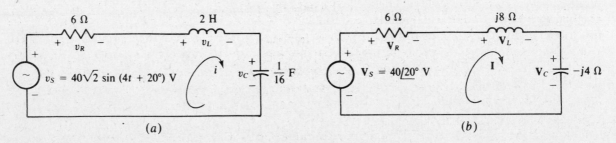

(a)   (b)

Fig. 11-4

The next step is to apply KVL to this frequency-domain circuit. Although it is not obvious, KVL applies to voltage phasors as well as to voltages because it applies to the sinusoidal voltages, and these sinusoids can be summed using phasors. (For similar reasons, KCL applies to the current phasors of frequency-domain circuits.) The result of applying KVL is

$$\mathbf{V}_S = \mathbf{V}_R + \mathbf{V}_L + \mathbf{V}_C$$

The third step is to substitute for the $\mathbf{V}$'s using $\mathbf{V}_S = 40\underline{/20°}$, $\mathbf{V}_R = 6\mathbf{I}$, $\mathbf{V}_L = j8\mathbf{I}$, and $\mathbf{V}_C = -j4\mathbf{I}$. With these substitutions the KVL equation becomes

$$40\underline{/20°} = 6\mathbf{I} + j8\mathbf{I} - j4\mathbf{I} = (6 + j4)\mathbf{I}$$

from which $$\mathbf{I} = \frac{40\underline{/20°}}{6 + j4} = \frac{40\underline{/20°}}{7.211\underline{/33.7°}} = 5.547\underline{/-13.7°} \ \text{A}$$

and $$i = 5.547\sqrt{2} \sin (4t - 13.7°) = 7.84 \sin (4t - 13.7°) \ \text{A}$$

IMPEDANCE

The KVL analysis method of the last section requires much more work than is necessary. Some of the initial steps can be eliminated by using *impedance*. Impedance has the quantity symbol $\mathbf{Z}$ and the unit ohm (Ω). For a two-terminal circuit with an input voltage phasor $\mathbf{V}$ and an input current phasor $\mathbf{I}$, as shown in Fig. 11-5, the impedance $\mathbf{Z}$ of the circuit is defined as

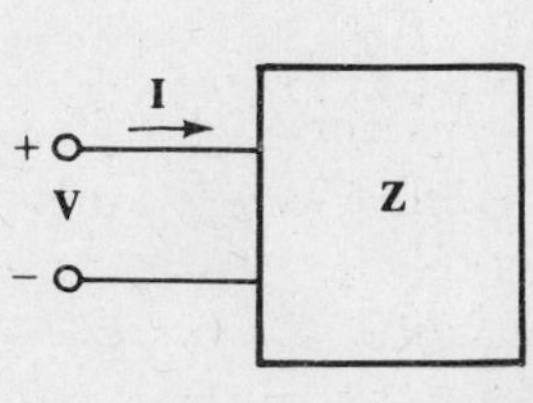

Fig. 11-5

$$\mathbf{Z} = \frac{\mathbf{V}}{\mathbf{I}}$$

For this impedance to exist, the circuit cannot have any independent sources, although it can have any number of dependent sources. This impedance is often called the *total* or *equivalent impedance*. It is also called the *input impedance*, especially for a circuit that has dependent sources or transformers. (Transformers will be discussed in Chap. 15.)

In general, and not just for series circuits,

$$\mathbf{Z} = R + jX$$

in which R, the real part, is the *resistance* and X, the imaginary part, is the *reactance* of the impedance. For the series frequency-domain circuit shown in Fig. 11-4b, $R = 6 \ \Omega$ and $X = 8 - 4 = 4 \ \Omega$. For this circuit, the resistance R depends only on the resistance of the resistor, and the reactance X depends only on the reactances of the inductor and capacitor. But for a more complex circuit, R and X are usually both dependent on the individual resistances and reactances.

Being a complex quantity, impedance can be expressed in polar form. From complex algebra,

$$\mathbf{Z} = R + jX = \sqrt{R^2 + X^2} \underline{/\tan^{-1}(X/R)}$$

in which $\sqrt{R^2 + X^2} = |\mathbf{Z}| = Z$ is the magnitude of impedance and $\tan^{-1}(X/R)$ is the angle of impedance.

As should be evident from $\mathbf{Z} = \mathbf{V}/\mathbf{I}$, the impedance angle is the angle by which the input voltage *leads* the input current, provided that this angle is positive. If it is negative, then the current leads the voltage. A circuit with a positive impedance angle is sometimes called an *inductive circuit* because the inductive reactances dominate the capacitive reactances to cause the input current to lag the input voltage. Similarly, a circuit that has a negative impedance angle is sometimes called a *capacitive circuit*.

Because impedances relate to voltage and current phasors in the same way that resistances relate to dc voltages and currents, it follows that impedances combine in the same way as resistances. Consequently, the total impedance $\mathbf{Z}_T$ of electrical components connected in series equals the sum of the impedances of the individual components:

$$\mathbf{Z}_T = \mathbf{Z}_1 + \mathbf{Z}_2 + \mathbf{Z}_3 + \cdots + \mathbf{Z}_N$$

And, for two parallel components with impedances $\mathbf{Z}_1$ and $\mathbf{Z}_2$,

$$\mathbf{Z}_T = \frac{\mathbf{Z}_1 \mathbf{Z}_2}{\mathbf{Z}_1 + \mathbf{Z}_2}$$

Often, the T subscript in $\mathbf{Z}_T$ is omitted.

The total impedance of an ac circuit is used in the same way as the total resistance of a dc circuit. For example, for the circuit shown in Fig. 11-4a, the first step after drawing the frequency-domain circuit illustrated in Fig. 11-4b, is to find the impedance of the circuit at the terminals of the source. This being a series circuit, the total impedance is equal to the sum of the individual impedances:

$$\mathbf{Z} = 6 + j(8 - 4) = 6 + j4 = 7.211 \underline{/33.7°} \ \Omega$$

Then, this is divided into the voltage phasor of the source to obtain the current phasor:

$$\mathbf{I} = \frac{\mathbf{V}}{\mathbf{Z}} = \frac{40 \underline{/20°}}{7.211 \underline{/33.7°}} = 5.547 \underline{/-13.7°} \ \text{A}$$

And, of course, the current i can be found from its phasor $\mathbf{I}$, as has been done.

An *impedance diagram* is an aid to understanding impedance. This diagram is constructed on an impedance plane which, as illustrated in Fig. 11-6, has a horizontal resistance axis designated by R and a vertical reactance axis designated by jX. Both axes have the same scale. Shown is a diagram of $\mathbf{Z}_1 = 6 + j5 = 7.81 \underline{/39.8°} \ \Omega$ for an inductive circuit and $\mathbf{Z}_2 = 8 - j6 = 10 \underline{/-36.9°} \ \Omega$ for a capacitive circuit. An inductive circuit has an impedance diagram in the first quadrant and a capacitive circuit has one in the fourth quadrant. For a diagram to be in either the second or third quadrant, a circuit must have a negative resistance produced by one or more dependent sources.

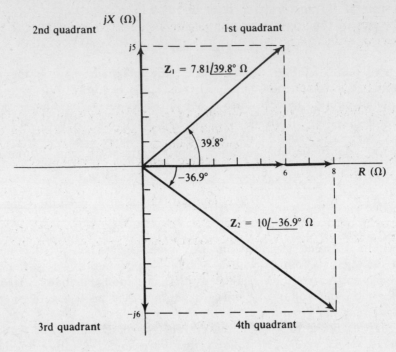

Fig. 11-6

An *impedance triangle* is often a more convenient graphical representation. The triangle contains vectors corresponding to R, jX, and Z, with the vector for jX drawn at the end of the R vector and the vector for Z drawn as the sum of these two vectors, as in Fig. 11-7a. Figure 11-7b shows an impedance triangle for $Z = 6 + j8 = 10\underline{/53.1°}$ Ω and Fig. 11-7c one for $Z = 6 - j8 = 10\underline{/-53.1°}$ Ω.

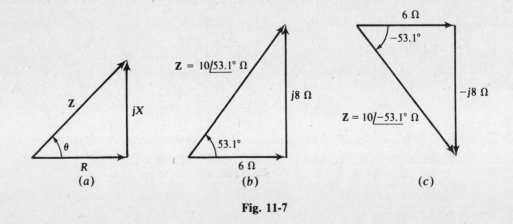

Fig. 11-7

VOLTAGE DIVISION

The voltage division or divider rule for ac circuits should be apparent from this rule for dc circuits. Of course, voltage and current phasors must be used instead of voltages and currents, and impedances instead of resistances. So, for a series circuit with an applied voltage that has the phasor $\mathbf{V}_S$, the voltage phasor $\mathbf{V}_X$ across a component with impedance $\mathbf{Z}_X$ is

$$\mathbf{V}_X = \frac{\mathbf{Z}_X}{\mathbf{Z}_T} \mathbf{V}_S$$

in which $\mathbf{Z}_T$ is the sum of the impedances.   A negative sign must be included if $\mathbf{V}_X$ and $\mathbf{V}_S$ do not have opposing polarity references.

AC PARALLEL CIRCUIT ANALYSIS

A method for analyzing a parallel ac circuit can be understood from a simple example.   Suppose that the sinusoidal voltage v is to be found in the parallel circuit shown in Fig. 11-8a.   With the techniques presented so far, the first step in finding v is to make the corresponding frequency-domain circuit shown in Fig. 11-8b, using the source frequency of 5000 rad/s.   The next step is to apply KCL to this circuit:

$$\mathbf{I}_S = \mathbf{I}_R + \mathbf{I}_L + \mathbf{I}_C$$

The third step is to substitute for the $\mathbf{I}$'s, using $\mathbf{I}_S = 10\underline{/0^\circ}$, $\mathbf{I}_R = \mathbf{V}/1000$, $\mathbf{I}_L = \mathbf{V}/j2500$,   and $\mathbf{I}_C = \mathbf{V}/(-j1000)$.   With these substitutions, the equation becomes

$$10\underline{/0^\circ} = \frac{\mathbf{V}}{1000} + \frac{\mathbf{V}}{j2500} + \frac{\mathbf{V}}{-j1000}$$

which simplifies to

$$10\underline{/0^\circ} = (0.001 + j0.0006)\mathbf{V}$$

from which

$$\mathbf{V} = \frac{10\underline{/0^\circ}}{0.001 + j0.0006} = \frac{10\underline{/0^\circ}}{0.001\ 166\underline{/31^\circ}}\ \mathbf{V} = 8.6\underline{/-31^\circ}\ \text{kV}$$

The corresponding voltage is

$$v = 8.6\sqrt{2}\ \sin(5000t - 31^\circ) = 12 \sin(5000t - 31^\circ)\ \text{kV}$$

Since this voltage lags the input current, the circuit is capacitive.   This is the result of the capacitive reactance being smaller than the inductive reactance—directly opposite the effect for a series circuit.

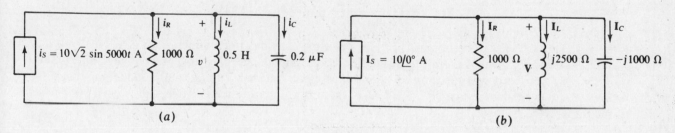

Fig. 11-8

ADMITTANCE

The analysis method of the last section can be improved upon by using *admittance*, which has the quantity symbol $\mathbf{Y}$ and the unit siemens (S).   By definition, admittance is the reciprocal of impedance:

$$\mathbf{Y} = \frac{1}{\mathbf{Z}}$$

From this it follows that

$$\mathbf{I} = \mathbf{Y}\mathbf{V}$$

Also, it follows that the admittance of a resistor is $\mathbf{Y} = 1/R = G$, that of an inductor is $\mathbf{Y} = 1/j\omega L = -j1/\omega L$, and that of a capacitor is $\mathbf{Y} = 1/(-j1/\omega C) = j\omega C$.

Being the reciprocal of impedance, the admittance of an ac circuit corresponds to the conductance of a dc resistive circuit. Consequently, admittances of *parallel* components add:

$$\mathbf{Y}_T = \mathbf{Y}_1 + \mathbf{Y}_2 + \mathbf{Y}_3 + \cdots + \mathbf{Y}_N$$

In general, and not just for parallel circuits,

$$\mathbf{Y} = G + jB$$

in which G, the real part, is the *conductance* and B, the imaginary part, is the *susceptance* of the admittance. For the parallel frequency-domain circuit shown in Fig. 11-8b,

$$\mathbf{Y} = \frac{1}{1000} + \frac{1}{j2500} + \frac{1}{-j1000} = 0.001 + j0.0006 \text{ S}$$

from which $G = 0.001$ S and $B = 0.0006$ S. For this simple parallel circuit, the conductance G depends only on the conductance of the resistor, and the susceptance B depends only on the susceptances of the inductor and capacitor. But for a more complex circuit, G and B usually both depend on the individual conductances and susceptances.

Being a complex quantity, admittance can be expressed in polar form. From complex algebra,

$$\mathbf{Y} = G + jB = \sqrt{G^2 + B^2}\underline{/\tan^{-1}(B/G)}$$

in which $\sqrt{G^2 + B^2} = |\mathbf{Y}| = Y$ is the magnitude and $\tan^{-1}(B/G)$ is the angle of admittance.

Since admittance is the reciprocal of impedance, the angle of an admittance is the negative of the angle for the corresponding impedance. Consequently, an admittance angle is positive for capacitive circuits and negative for inductive circuits. Also, B, the susceptance, has these same signs.

The total admittance of an ac circuit is used in the same way as the total conductance of a dc circuit. To illustrate, for the circuit shown in Fig. 11-8a, the first step after drawing the frequency-domain circuit illustrated in Fig. 11-8b is to find the admittance of the circuit at the terminals of the source. As has been found, $\mathbf{Y} = 0.001 + j0.0006 = 0.001\,166\underline{/31°}$ S. Then, this is divided into the current phasor to find the voltage phasor:

$$\mathbf{V} = \frac{\mathbf{I}}{\mathbf{Y}} = \frac{10\underline{/0°}}{0.001\,166\underline{/31°}} \text{ V} = 8.6\underline{/-31°} \text{ kV}$$

Finally, the voltage v can be obtained from its phasor $\mathbf{V}$, as has been done.

As should be expected from the discussion of an impedance diagram, there is an *admittance diagram* that can be constructed on an admittance plane that has a horizontal conductance axis G and a vertical susceptance axis jB. And, also, there is an *admittance triangle* that is used similarly to the impedance triangle.

CURRENT DIVISION

Current division applies to ac frequency-domain circuits in the same way as to dc resistive circuits. So, if a parallel frequency-domain circuit has a current phasor $\mathbf{I}_S$ directed into it, the current phasor $\mathbf{I}_X$ for a branch that has an admittance $\mathbf{Y}_X$ is

$$\mathbf{I}_X = \frac{\mathbf{Y}_X}{\mathbf{Y}_T}\mathbf{I}_S$$

in which $\mathbf{Y}_T$ is the sum of the admittances. A negative sign must be included if $\mathbf{I}_X$ and $\mathbf{I}_S$ do not have opposite reference directions into one of the nodes. For the special case of two parallel branches with impedances $\mathbf{Z}_1$ and $\mathbf{Z}_2$, this formula reduces to

$$\mathbf{I}_1 = \frac{\mathbf{Z}_2}{\mathbf{Z}_1 + \mathbf{Z}_2}\,\mathbf{I}_S$$

in which $\mathbf{I}_1$ is the current phasor for the $\mathbf{Z}_1$ branch.

For convenience, from this point on the word "phasor" in voltage phasor and current phasor will often be omitted. That is, the $\mathbf{V}$'s and $\mathbf{I}$'s will often be referred to as voltages and currents, respectively, as is common practice.

Solved Problems

11.1 Find the total impedance in polar form of a 0.5-H inductor and a series 20-Ω resistor at   (a) 0 Hz,   (b) 10 Hz,   and   (c) 10 kHz.

The total impedance is $\mathbf{Z} = R + j\omega L = R + j2\pi fL$.

(a) For $f = 0$ Hz,

$$\mathbf{Z} = 20 + 2\pi(0)(0.5) = 20 = 20\underline{/0°}\ \Omega$$

The impedance is purely resistive because 0 Hz corresponds to dc, and an inductor is a short circuit to dc.

(b) For $f = 10$ Hz,

$$\mathbf{Z} = 20 + j2\pi(10)(0.5) = 20 + j31.4 = 37.2\underline{/57.5°}\ \Omega$$

(c) For $f = 10$ kHz,

$$\mathbf{Z} = 20 + j2\pi(10^4)(0.5) = 20 + j3.14 \times 10^4\ \Omega = 31.4\underline{/89.96°}\ k\Omega$$

At 10 kHz the reactance is so much larger than the resistance that the resistance is negligible for most purposes.

11.2 A 200-Ω resistor, a 150-mH inductor, and a 2-μF capacitor are in series. Find the total impedance in polar form at 400 Hz. Also, draw the impedance diagram and the impedance triangle.

The total impedance is

$$\mathbf{Z} = R + j2\pi fL + \frac{-j1}{2\pi fC} = 200 + j2\pi(400)(150 \times 10^{-3}) + \frac{-j1}{2\pi(400)(2 \times 10^{-6})}$$
$$= 200 + j377 - j199 = 200 + j178 = 268\underline{/41.7°}\ \Omega$$

The impedance diagram is shown in Fig. 11-9a and the impedance triangle is shown in Fig. 11-9b. In the impedance diagram, the end point for the $\mathbf{Z}$ arrow is found by starting at the origin and moving up the vertical axis to $j377\ \Omega$ (jX_L), then moving horizontally to the right to over $200\ \Omega$ (R), and finally moving vertically down by $199\ \Omega$, the magnitude of the capacitive reactance $(|X_C|)$. The impedance triangle construction is obvious from the calculated $R = 200\ \Omega$ and $X = 178\ \Omega$.

11.3 A 2000-Ω resistor, a 1-H inductor, and a 0.01-μF capacitor are in series. Find the total impedance in polar form at   (a) 5 krad/s,   (b) 10 krad/s,   and   (c) 20 krad/s.

The formula for the total impedance is $\mathbf{Z} = R + j\omega L - j1/\omega C$.

(a) $\mathbf{Z} = 2000 + j5000(1) - \dfrac{j1}{5000(10^{-8})} = 2000 - j15\,000\ \Omega = 15.1\underline{/-82.4°}\ k\Omega$

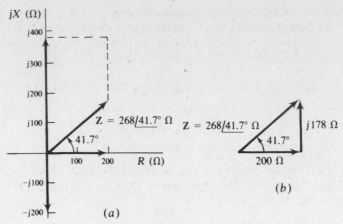

Fig. 11-9

(b) $\mathbf{Z} = 2000 + j10\,000(1) - \dfrac{j1}{10\,000(10^{-8})} = 2000\ \Omega = 2\underline{/0^\circ}$ kΩ

(c) $\mathbf{Z} = 2000 + j20\,000(1) - \dfrac{j1}{20\,000(10^{-8})} = 2000 + j15\,000\ \Omega = 15.1\underline{/82.4^\circ}$ kΩ

Notice that for $\omega = 10$ krad/s in part (b), the impedance is purely resistive because the inductive and capacitive terms cancel. This is the *resonant radian frequency* of the circuit. For lower frequencies, the circuit is capacitive, as is verified in part (a). And for higher frequencies, the circuit is inductive, as is verified in part (c).

11.4   A coil energized by 120 V at 60 Hz draws a 2-A current that lags the applied voltage by 40°. What are the coil resistance and inductance?

The magnitude of the impedance can be found by dividing the rms voltage by the rms current: $Z = 120/2 = 60\ \Omega$. The angle of the impedance is the 40° angle by which the voltage leads the current. Consequently, $\mathbf{Z} = 60\underline{/40^\circ} = 46 + j38.6\ \Omega$. From the real part, the resistance of the coil is 46 Ω, and from the imaginary part, the reactance is 38.6 Ω. Since ωL is the reactance, and $\omega = 2\pi(60) = 377$ rad/s, the inductance is $L = 38.6/377 = 0.102$ H.

11.5   A load has a voltage of $\mathbf{V} = 120\underline{/30^\circ}$ V and a current of $\mathbf{I} = 30\underline{/50^\circ}$ A at a frequency of 400 Hz. Find the two-element series circuit that the load could be. Of course, assume associated references.

The impedance is

$$\mathbf{Z} = \frac{\mathbf{V}}{\mathbf{I}} = \frac{120\underline{/30^\circ}}{30\underline{/50^\circ}} = 4\underline{/-20^\circ} = 3.76 - j1.37\ \Omega$$

Because the imaginary part is negative, the circuit is capacitive, which means that the two series elements are a resistor and a capacitor. The real part is the resistance of the resistor: $R = 3.76\ \Omega$. And, the imaginary part is the reactance of the capacitor: $-1/\omega C = -1.37$, from which

$$C = \frac{1}{1.37\omega} = \frac{1}{1.37(2\pi)(400)}\ \text{F} = 291\ \mu\text{F}$$

11.6   A 20-Ω resistor is in series with a 0.1-μF capacitor. At what radian frequency are the circuit voltage and current out of phase by 40°?

A good approach is to find the reactance from the impedance angle and the resistance, and then find the radian frequency from the reactance and the capacitance. The impedance angle has a magnitude of 40° because this is the phase angle difference between the voltage and the current. Also, the angle is negative because this is a capacitive circuit.   So, $\theta = -40°$.   As should be apparent from the impedance triangle shown in Fig. 11-7a, and also from the complex algebra presentation, reactance and resistance are related by the tangent of the impedance angle: $X = R \tan \theta$.   Here, $X_C = 20 \tan (-40°) = -16.8 \ \Omega$.   Finally, from $X_C = -1/\omega C$,

$$\omega = \frac{-1}{CX_C} = \frac{-1}{10^{-7}(-16.8)} \ \text{rad/s} = 0.596 \ \text{Mrad/s}$$

11.7   A 200-mH inductor and a resistor in series draw 0.6 A when 120 V at 100 Hz is applied.   Find the impedance in polar form.

The magnitude of the impedance can be found by dividing the voltage by the current: $Z = 120/0.6 = 200 \ \Omega$.   The angle of the impedance is $\theta = \sin^{-1} (X_L/Z)$, as is evident from the impedance triangle shown in Fig. 11-7a.   Here,

$$\frac{X_L}{Z} = \frac{2\pi(100)(0.2)}{200} = 0.2\pi \quad \text{and so} \quad \theta = \sin^{-1} 0.2\pi = 38.9°$$

The impedance is $Z = 200/\underline{38.9°} \ \Omega$.

11.8   What capacitor in series with a 750-Ω resistor limits the current to 0.2 A when 240 V at 400 Hz is applied?

When the capacitor is in the circuit, the impedance has a magnitude of $Z = V/I = 240/0.2 = 1200 \ \Omega$.   This is related to the resistance and reactance by $Z = \sqrt{R^2 + X^2}$.   If both sides are squared and X solved for, the result is

$$X^2 = Z^2 - R^2 \rightarrow X = \pm\sqrt{Z^2 - R^2}$$

The negative sign must be selected because the circuit is capacitive and therefore has a negative reactance.   Substituting for Z and R gives

$$X = -\sqrt{Z^2 - R^2} = -\sqrt{1200^2 - 750^2} = -937 \ \Omega$$

Finally, since $X = -1/\omega C$,

$$C = \frac{-1}{\omega X} = \frac{-1}{2\pi(400)(-937)} \ \text{F} = 0.425 \ \mu\text{F}$$

Incidentally, another way of finding X is from the impedance magnitude times the sine of the impedance angle:

$$X = Z \sin [-\cos^{-1} (R/Z)] = 1200 \sin [-\cos^{-1} (750/1200)] = -937 \ \Omega$$

11.9   A capacitor is in series with a coil that has 1.5 H of inductance and 5 Ω of resistance.   Find the capacitance that makes the combination purely resistive at 60 Hz.

For the circuit to be purely resistive, the reactances must add to zero.   And, since the reactance of the inductor is $2\pi(60)(1.5) = 565 \ \Omega$,   the reactance of the capacitor must be $-565 \ \Omega$.   From $X_C = -1/\omega C$,

$$C = -\frac{1}{\omega X_C} = \frac{-1}{2\pi(60)(-565)} \ \text{F} = 4.69 \ \mu\text{F}$$

11.10   Three circuit elements in series draw a current of 10 sin (400t + 70°) A in response to an applied voltage of 50 sin (400t + 15°) V.   If one element is a 16-mH inductor, what are the two other elements?

The unknown elements can be found from the impedance.   It has a magnitude that is equal to the voltage peak divided by the current peak: $Z = 50/10 = 5 \ \Omega$,   and an angle that is the voltage phase angle minus the current phase angle: $\theta = 15° - 70° = -55°$.   Therefore, the impedance is $Z = 5/\underline{-55°} = 2.87 - j4.1 \ \Omega$.   The real part must be produced by a 2.87-Ω resistor.   The third element must be a capacitor because the imaginary part, the reactance, is negative. Of course, the capacitive reactance plus the inductive reactance equals the impedance reactance:

$$\frac{-1}{400C} + 400(16 \times 10^{-3}) = -4.1$$

which solves to $C = 238 \ \mu F$.

11.11 Find the input impedance at 5 krad/s of the circuit shown in Fig. 11-10a.

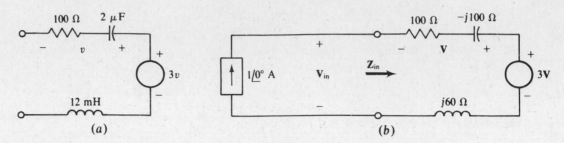

Fig. 11-10

The first step is to use $j\omega L$, $-j1/\omega C$, and phasors to construct the corresponding frequency-domain circuit that is shown in Fig. 11-10b along with a $1\underline{/0°}$-A source. The presence of the dependent source makes it necessary to apply a source to find $\mathbf{Z}_{in}$, and the best source is a $1\underline{/0°}$-A current source because with it, $\mathbf{Z}_{in} = \mathbf{V}_{in}/1\underline{/0°} = \mathbf{V}_{in}$. Note that the controlling voltage for the dependent source is the drop across the resistor and capacitor:

$$\mathbf{V} = -(1\underline{/0°})(100 - j100) = -100 + j100 \text{ V}$$

The initial negative sign is required because the voltage and current references are not associated. By KVL,

$$\mathbf{V}_{in} = (1\underline{/0°})(100) + (1\underline{/0°})(-j100) + 3(-100 + j100) + (1\underline{/0°})(j60)$$
$$= 100 - j100 - 300 + j300 + j60 = -200 + j260 = 328\underline{/128°} \text{ V}$$

Finally, $\mathbf{Z}_{in} = \mathbf{V}_{in} = 328\underline{/128°} \ \Omega$.

11.12 A 240-V source is connected in series with two components, one of which has an impedance of $80\underline{/60°} \ \Omega$. What is the impedance of the other component if the current that flows is 2 A and if it leads the source voltage by 40°?

Since the total impedance is the sum of the known and unknown impedances, the unknown impedance is the total impedance minus the known impedance. The total impedance has a magnitude of

$$Z_T = \frac{V}{I} = \frac{240}{2} = 120 \ \Omega$$

and an angle of $-40°$, the angle by which the voltage leads the current. (This angle is negative because the voltage lags, instead of leads, the current.) Therefore, the total impedance is $\mathbf{Z}_T = 120\underline{/-40°} \ \Omega$. Subtracting the known impedance of $80\underline{/60°} \ \Omega$ results in the desired impedance:

$$\mathbf{Z} = 120\underline{/-40°} - 80\underline{/60°} = 91.9 - j77.1 - (40 + j69.3) = 51.9 - j146.3 = 155\underline{/-70.5°} \ \Omega$$

11.13 Find the total impedance of two parallel components that have impedances of $\mathbf{Z}_1 = 300\underline{/30°} \ \Omega$ and $\mathbf{Z}_2 = 400\underline{/-50°} \ \Omega$.

The total impedance is the product of the individual impedances divided by the sum:

$$\mathbf{Z}_T = \frac{\mathbf{Z}_1\mathbf{Z}_2}{\mathbf{Z}_1 + \mathbf{Z}_2} = \frac{(300\underline{/30°})(400\underline{/-50°})}{300\underline{/30°} + 400\underline{/-50°}} = \frac{120\ 000\underline{/-20°}}{540\underline{/-16.8°}} = 222\underline{/-3.2°} \ \Omega$$

11.14 Find the total impedances at 1 krad/s of a 1-H inductor and a 1-μF capacitor connected in series and also in parallel.

The inductor and capacitor impedances are

$$j\omega L = j1000(1) = j1000 \ \Omega \quad \text{and} \quad \frac{-j1}{\omega C} = \frac{-j1}{1000(10^{-6})} = -j1000 \ \Omega$$

The total impedance of the elements in series is the sum of the individual impedances: $\mathbf{Z} = j1000 - j1000 = 0 \ \Omega$, which is a short circuit. For the two in parallel, the total impedance is

$$\mathbf{Z} = \frac{j1000(-j1000)}{j1000 - j1000} = \frac{10^6}{0} \rightarrow \infty \ \Omega$$

which is an open circuit.

11.15 What capacitor and resistor connected in series have the same total impedance at 400 rad/s as a 10-μF capacitor and a 500-Ω resistor connected in parallel?

At 400 rad/s, the impedance of the capacitor is

$$\frac{-j1}{\omega C} = \frac{-j1}{400(10 \times 10^{-6})} = -j250 \ \Omega$$

Of course, the total impedance of the parallel combination is the product of the individual impedances divided by the sum:

$$\frac{500(-j250)}{500 - j250} = \frac{125\,000/\!-90°}{559/\!-26.6°} = 224/\!-63.4° = 100 - j200 \ \Omega$$

For the series resistor and capacitor to have this impedance, the resistor resistance must be 100 Ω, the real part, and the capacitor reactance must be -200 Ω, the imaginary part. So, $R = 100 \ \Omega$, and by the capacitor reactance formula,

$$\frac{-1}{\omega C} = \frac{-1}{400C} = -200 \ \Omega \quad \text{from which} \quad C = \frac{1}{200(400)} \text{F} = 12.5 \ \mu\text{F}$$

11.16 Find the two circuit elements that when connected in series have the same total impedance at 4 krad/s as the parallel combination of a 50-μF capacitor and a 2-mH coil with a 10-Ω winding resistance.

The impedance of the coil is

$$10 + j4000(2 \times 10^{-3}) = 10 + j8 = 12.8/\!38.7° \ \Omega$$

and that of the capacitor is

$$\frac{-j1}{4000(50 \times 10^{-6})} = -j5 = 5/\!-90° \ \Omega$$

The impedance of the parallel combination is the product of these impedances divided by the sum:

$$\frac{(12.8/\!38.7°)(5/\!-90°)}{10 + j8 - j5} = \frac{64/\!-51.3°}{10.44/\!16.7°} = 6.13/\!-68° = 2.29 - j5.69 \ \Omega$$

To produce an impedance of $2.29 - j5.69 \ \Omega$, the two series components must be a resistor that has a resistance of 2.29 Ω and a capacitor that has a reactance of -5.69 Ω. Since $X_C = -1/\omega C$,

$$C = \frac{-1}{\omega X_C} = \frac{-1}{4000(-5.69)} \text{F} = 44 \ \mu\text{F}$$

11.17 For the circuit shown in Fig. 11-11, find the indicated unknown phasors and the corresponding sinusoids. The frequency is 60 Hz. Also, find the average power delivered by the source.

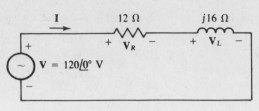

Fig. 11-11

Since this is a series circuit, the current should be found first and then used to find the voltages:

$$I = \frac{V}{Z} = \frac{120\underline{/0°}}{12 + j16} = \frac{120\underline{/0°}}{20\underline{/53.1°}} = 6\underline{/-53.1°} \text{ A}$$

The resistor and inductor voltage drops are the products of this current and the individual impedances:

$$V_R = (6\underline{/-53.1°})(12) = 72\underline{/-53.1°} \text{ V}$$
$$V_L = (6\underline{/-53.1°})(j16) = (6\underline{/-53.1°})(16\underline{/90°}) = 96\underline{/36.9°} \text{ V}$$

The radian frequency needed for the corresponding sinusoids is $\omega = 2\pi(60) = 377$ rad/s. Of course, the peak values of the sinusoids are the magnitudes of the corresponding phasors times $\sqrt{2}$. Thus, have

$$i = 6\sqrt{2} \sin(377t - 53.1°) = 8.49 \sin(377t - 53.1°) \text{ A}$$
$$v_R = 72\sqrt{2} \sin(377t - 53.1°) = 102 \sin(377t - 53.1°) \text{ V}$$
$$v_L = 96\sqrt{2} \sin(377t + 36.9°) = 136 \sin(377t + 36.9°) \text{ V}$$

Since the average power absorbed by the inductor is zero, the average power delivered by the source is the same as that absorbed by the resistor, which is $I^2R = 6^2 \times 12 = 432$ W.

11.18 Find the current and unknown voltages in the circuit shown in Fig. 11-12a.

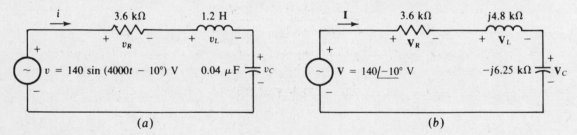

Fig. 11-12

The first step is to draw the corresponding frequency-domain circuit shown in Fig. 11-12b using the $\omega = 4000$ rad/s of the source. Since sinusoidal results are desired, it is best to use phasors based on peak rather than on rms values. That is why the source in Fig. 11-12b has a voltage of $140\underline{/-10°}$ V instead of $99\underline{/-10°}$ V ($99 = 140/\sqrt{2}$). The current is

$$I = \frac{V}{Z} = \frac{140\underline{/-10°}}{3600 + j4800 - j6250} = \frac{140\underline{/-10°}}{3881\underline{/-21.9°}} \text{ A} = 36.1\underline{/11.9°} \text{ mA}$$

This current can be used to obtain the voltage phasors:

$$V_R = (0.0361\underline{/11.9°})(3600) = 130\underline{/11.9°} \text{ V}$$
$$V_L = (0.0361\underline{/11.9°})(4800\underline{/90°}) = 173\underline{/102°} \text{ V}$$
$$V_C = (0.0361\underline{/11.9°})(6250\underline{/-90°}) = 225\underline{/-78.1°} \text{ V}$$

The corresponding sinusoidal quantities are

$$i = 36.1 \sin(4000t + 11.9°) \text{ mA}$$
$$v_R = 130 \sin(4000t + 11.9°) \text{ V}$$
$$v_L = 173 \sin(4000t + 102°) = 173 \cos(4000t + 12°) \text{ V}$$
$$v_C = 225 \sin(4000t - 78.1°) \text{ V}$$

11.19 A voltage $100\underline{/30°}$ V is applied across a resistor and inductor that are in series. If the resistor rms voltage drop is 40 V, what is the inductor voltage phasor?

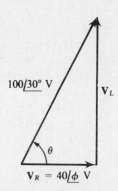

A funicular diagram is useful here.   Since the resistor voltage is in phase with the current, and the inductor voltage leads the current by 90°, the phasor funicular diagram is a right triangle, as shown in Fig. 11-13.   This particular diagram is useful only for finding the phasor magnitudes and the *relative* phasor angular relations, the latter because the phasors are not at the correct angles.   By Pythagoras' theorem, $V_L = \sqrt{100^2 - 40^2} = 91.7$ V.   The shown angle θ is $\theta = \tan^{-1}(91.7/40) = 66.4°$.   The angle of the resistor voltage is less than the source voltage angle by this 66.4°: $\phi = 30° - 66.4° = -36.4°$.   Of course, the angle of the inductor voltage is 90° greater than the resistor voltage angle: $90° + (-36.4°) = 53.6°$.   So, the inductor voltage phasor is $V_L = 91.7\underline{/53.6°}$ V.

Fig. 11-13

11.20 In a frequency-domain circuit, $220\underline{/30°}$ V is applied across two series components, one of which is a 20-Ω resistor and the other of which is a coil with an impedance of $40\underline{/20°}$ Ω. Use current to find the individual component voltage drops.

The current is

$$\mathbf{I} = \frac{\mathbf{V}}{\mathbf{Z}} = \frac{220\underline{/30°}}{20 + 40\underline{/20°}} = \frac{220\underline{/30°}}{59.2\underline{/13.4°}} = 3.72\underline{/16.6°} \text{ A}$$

Each component voltage drop is the product of the current and the component impedance:

$$\mathbf{V}_R = (3.72\underline{/16.6°})(20) = 74\underline{/16.6°} \text{ V}$$
$$\mathbf{V}_Z = (3.72\underline{/16.6°})(40\underline{/20°}) = 149\underline{/36.6°} \text{ V}$$

11.21 Repeat Prob. 11.20 using voltage division.

Voltage division eliminates the step of finding the current.   Instead, the voltages are found directly from the applied voltage and the impedances:

$$\mathbf{V}_R = \frac{R}{\mathbf{Z}_T}\mathbf{V}_S = \frac{20}{59.2\underline{/13.4°}} \times 220\underline{/30°} = 74\underline{/16.6°} \text{ V}$$

$$\mathbf{V}_Z = \frac{\mathbf{Z}_Z}{\mathbf{Z}_T}\mathbf{V}_S = \frac{40\underline{/20°}}{59.2\underline{/13.4°}} \times 220\underline{/30°} = 149\underline{/36.6°} \text{ V}$$

11.22 A frequency-domain circuit has $200\underline{/15°}$ V applied across three series-connected components having impedances of $20\underline{/15°}$, $30\underline{/-40°}$, and $40\underline{/50°}$ Ω. Use voltage division to find the voltage drop $\mathbf{V}$ across the component with the $40\underline{/50°}$-Ω impedance.

$$\mathbf{V} = \frac{40\underline{/50°}}{20\underline{/15°} + 30\underline{/-40°} + 40\underline{/50°}} \times 200\underline{/15°} = \frac{8000\underline{/65°}}{70\underline{/13.7°}} = 114\underline{/51.3°} \text{ V}$$

11.23 Use voltage division to find $\mathbf{V}_R$, $\mathbf{V}_L$, and $\mathbf{V}_C$ in the circuit shown in Fig. 11-14.

For voltage division, the total impedance $\mathbf{Z}$ is needed: $\mathbf{Z} = 20 + j1000 - j1000 = 20 \ \Omega$. Incidentally, since this impedance is purely resistive, the circuit is in *resonance*. By the voltage division formula,

$$\mathbf{V}_R = \frac{20}{20} \times 100\underline{/30°} = 100\underline{/30°} \ \text{V}$$

$$\mathbf{V}_L = \frac{j1000}{20} \times 100\underline{/30°} = (50\underline{/90°})(100\underline{/30°}) = 5000\underline{/120°} \ \text{V}$$

$$\mathbf{V}_C = \frac{-j1000}{20} \times 100\underline{/30°} = (50\underline{/-90°})(100\underline{/30°}) = 5000\underline{/-60°} \ \text{V}$$

Notice that the rms inductor and capacitor voltages are 50 times greater than the rms source voltage. This voltage rise, although impossible in a dc resistive circuit, is common in a series resonant ac circuit.

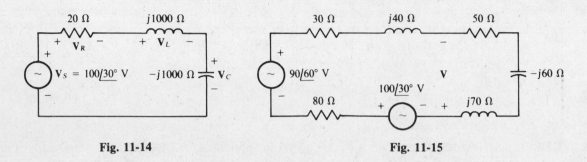

Fig. 11-14   Fig. 11-15

11.24 Use voltage division to find the voltage $\mathbf{V}$ in the circuit shown in Fig. 11-15.

Because the two voltage sources are in series, they produce a net applied voltage that is the sum of the individual source voltages: $\mathbf{V}_S = 90\underline{/60°} + 100\underline{/30°} = 184\underline{/44.2°} \ \text{V}$, which is the voltage needed for the voltage division formula. The series components that $\mathbf{V}$ is across have a combined impedance of $\mathbf{Z} = 50 - j60 + j70 = 50 + j10 = 51\underline{/11.3°} \ \Omega$. And, the total circuit impedance is

$$\mathbf{Z}_T = 30 + j40 + 50 - j60 + j70 + 80 = 160 + j50 = 168\underline{/17.4°} \ \Omega$$

Now, all of the quantities have been calculated that are needed for the voltage division formula, which is

$$\mathbf{V} = -\frac{\mathbf{Z}}{\mathbf{Z}_T} \mathbf{V}_S = -\frac{51\underline{/11.3°}}{168\underline{/17.4°}} \times 184\underline{/44.2°} = -55.8\underline{/38.1°} \ \text{V}$$

The negative sign is required in the formula because the reference polarity of $\mathbf{V}$ does not oppose the reference polarities of the sources.

11.25 Find the current $\mathbf{I}$ in the circuit shown in Fig. 11-16.

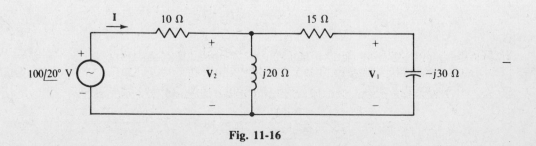

Fig. 11-16

The current can be found by dividing the voltage by the total impedance. This impedance can be found by combining impedances starting at the end of the circuit opposite the source. There, the series resistor and capacitor have a combined impedance of $15 - j30 = 33.5\underline{/-63.4°}$ Ω. This combines in parallel fashion with the $j20$ Ω of the parallel inductor:

$$\frac{j20(33.5\underline{/-63.4°})}{j20 + 15 - j30} = \frac{671\underline{/26.6°}}{18\underline{/-33.7°}} = 37.2\underline{/60.3°} = 18.5 + j32.3 \; Ω$$

This plus the 10 Ω of the series resistor is the total impedance:

$$\mathbf{Z} = 10 + 18.5 + j32.3 = 43.1\underline{/48.6°} \; Ω$$

Finally, the current $\mathbf{I}$ is

$$\mathbf{I} = \frac{\mathbf{V}}{\mathbf{Z}} = \frac{100\underline{/20°}}{43.1\underline{/48.6°}} = 2.32\underline{/-28.6°} \; A$$

11.26 Use voltage division twice to find $\mathbf{V}_1$ in the circuit shown in Fig. 11-16.

Voltage division can be used to find $\mathbf{V}_2$ from the source voltage, and used again to find $\mathbf{V}_1$ from $\mathbf{V}_2$. For the calculation of $\mathbf{V}_2$, the equivalent impedance to the right of the 10-Ω resistor is needed. It is $37.2\underline{/60.3°} = 18.5 + j32.3$ Ω, as was found in the solution to Prob. 11.25. By voltage division,

$$\mathbf{V}_2 = \frac{37.2\underline{/60.3°}}{10 + 18.5 + j32.3} \times 100\underline{/20°} = \frac{3720\underline{/80.3°}}{43.1\underline{/48.6°}} = 86.4\underline{/32°} \; V$$

And, by voltage division again,

$$\mathbf{V}_1 = \frac{-j30}{15 - j30} \times 86.4\underline{/32°} = \frac{2590\underline{/-58°}}{33.5\underline{/-63°}} = 77.3\underline{/5°} \; V$$

11.27 Derive expressions for the conductance and the susceptance of an admittance in terms of the resistance and reactance of the corresponding impedance.

In general,

$$\mathbf{Y} = \frac{1}{\mathbf{Z}} = \frac{1}{R + jX}$$

Rationalizing,

$$\mathbf{Y} = \frac{1}{R + jX} \times \frac{R - jX}{R - jX} = \frac{R}{R^2 + X^2} + j\frac{-X}{R^2 + X^2}$$

Since $\mathbf{Y} = G + jB$,

$$G = \frac{R}{R^2 + X^2} \quad \text{and} \quad B = \frac{-X}{R^2 + X^2}$$

Notice from $G = R/(R^2 + X^2)$ and $B = -X/(R^2 + X^2)$ that the conductance and the susceptance are both functions of the resistance and reactance. Also, $G \neq 1/R$ except if $X = 0$. And, $B \neq 1/X$. However, $B = -1/X$ if $R = 0$.

11.28 The impedance of a circuit has 2 Ω of resistance and 4 Ω of reactance. What are the conductance and susceptance of the admittance?

The formulas developed in the solution to Prob. 11.27 can be used:

$$G = \frac{2}{2^2 + 4^2} = \frac{2}{20} = 0.1 \; S \quad \text{and} \quad B = \frac{-4}{2^2 + 4^2} = \frac{-4}{20} = -0.2 \; S$$

But, in general, it is easier to use the inverse of impedance:

$$\mathbf{Y} = \frac{1}{\mathbf{Z}} = \frac{1}{2 + j4} = \frac{1}{4.47\underline{/63.4°}} = 0.224\underline{/-63.4°} = 0.1 - j0.2 \text{ S}$$

The real part is the conductance: $G = 0.1$ S;   the imaginary part is the susceptance: $B = -0.2$ S.

11.29 Find the total admittances in polar form of a 0.2-μF capacitor and a parallel 5.1-Ω resistor at frequencies of   (*a*) 0 Hz,   (*b*) 100 kHz,   and   (*c*) 40 MHz.

The total admittance is $\mathbf{Y} = G + j\omega C = 1/R + j2\pi f C$.

(*a*)   For $f = 0$ Hz,

$$\mathbf{Y} = 1/5.1 + j2\pi(0)(0.2 \times 10^{-6}) = 0.196 = 0.196\underline{/0°} \text{ S}$$

(*b*)   For $f = 100$ kHz,

$$\mathbf{Y} = 1/5.1 + j2\pi(100 \times 10^{3})(0.2 \times 10^{-6}) = 0.196 + j0.126 = 0.233\underline{/32.7°} \text{ S}$$

(*c*)   For $f = 40$ MHz,

$$\mathbf{Y} = 1/5.1 + j2\pi(40 \times 10^{6})(0.2 \times 10^{-6}) = 0.196 + j50.3 = 50.3\underline{/89.8°} \text{ S}$$

At 40 MHz, the susceptance is so much larger than the conductance that the conductance is negligible for most purposes.

11.30 A 200-Ω resistor, a 1-μF capacitor, and a 75-mH inductor are in parallel.   Find the total admittance in polar form at 400 Hz.   Also, draw the admittance diagram and the admittance triangle.

The total admittance is

$$\mathbf{Y} = \frac{1}{R} + j2\pi f C + \frac{-j1}{2\pi f L} = \frac{1}{200} + j2\pi(400)(1 \times 10^{-6}) + \frac{-j1}{2\pi(400)(75 \times 10^{-3})}$$

$$= 5 \times 10^{-3} + j2.51 \times 10^{-3} - j5.31 \times 10^{-3} = (5 - j2.8)(10^{-3}) \text{ S} = 5.73\underline{/-29.2°} \text{ mS}$$

The admittance diagram is shown in Fig. 11-17*a* and the admittance triangle in Fig. 11-17*b*.   In the admittance diagram, the end point for the **Y** arrow is found by starting at the origin and moving down the vertical axis to $-j5.31$ mS (jB_L), and then by moving horizontally to the right to over 5 mS (G) and vertically up by 2.51 mS (B_C).

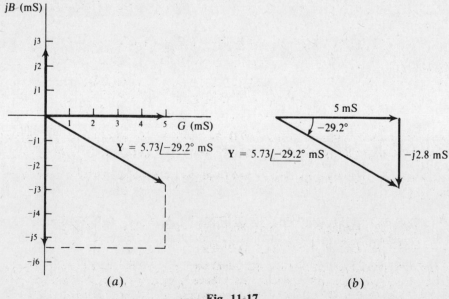

Fig. 11-17

11.31 A 100-Ω resistor, a 1-mH inductor, and a 0.1-μF capacitor are in parallel. Find the total admittances in polar form at radian frequencies of (a) 50 krad/s, (b) 100 krad/s, and (c) 200 krad/s.

The formula for the total admittance is $\mathbf{Y} = 1/R + j\omega C - j1/\omega L$.

(a) $\mathbf{Y} = \dfrac{1}{100} + j(50 \times 10^3)(0.1 \times 10^{-6}) - \dfrac{j1}{(50 \times 10^3)(10^{-3})}$

$= 0.01 + j0.005 - j0.02 = 0.01 - j0.015 = 0.018\underline{/-56.3^\circ}$ S

(b) $\mathbf{Y} = \dfrac{1}{100} + j(10^5)(0.1 \times 10^{-6}) - \dfrac{j1}{10^5(10^{-3})} = 0.01 + j0.01 - j0.01 = 0.01\underline{/0^\circ}$ S

(c) $\mathbf{Y} = \dfrac{1}{100} + j(2 \times 10^5)(0.1 \times 10^{-6}) - \dfrac{j1}{(2 \times 10^5)(10^{-3})} = 0.01 + j0.02 - j0.005$

$= 0.01 + j0.015 = 0.018\underline{/56.3^\circ}$ S

Notice that, for $\omega = 100$ krad/s in part (b), the admittance is real because the inductive and capacitive susceptance terms cancel. This is the resonant radian frequency of the circuit. For lower frequencies, the circuit is inductive, as is verified in part (a). And for greater frequencies, the circuit is capacitive, as is verified in part (c). This response is opposite that for a series RLC circuit.

11.32 Three components in parallel have a total admittance of $\mathbf{Y}_T = 6\underline{/30^\circ}$ S. If the admittances of two of the components are $\mathbf{Y}_1 = 4\underline{/45^\circ}$ S and $\mathbf{Y}_2 = 7\underline{/60^\circ}$ S, what is the admittance $\mathbf{Y}_3$ of the third component?

Since $\mathbf{Y}_T = \mathbf{Y}_1 + \mathbf{Y}_2 + \mathbf{Y}_3$,

$$\mathbf{Y}_3 = \mathbf{Y}_T - \mathbf{Y}_1 - \mathbf{Y}_2 = 6\underline{/30^\circ} - 4\underline{/45^\circ} - 7\underline{/60^\circ} = 6\underline{/-101^\circ} \text{ S}$$

11.33 What is the total impedance of three parallel components that have impedances of $\mathbf{Z}_1 = 2.5\underline{/75^\circ}$ Ω, $\mathbf{Z}_2 = 4\underline{/-50^\circ}$ Ω, and $\mathbf{Z}_3 = 5\underline{/45^\circ}$ Ω?

Perhaps the best approach is to invert each impedance to find the corresponding admittance, add the individual admittances to obtain the total admittance, and then invert the total admittance to find the total impedance.

Inverting,

$$\mathbf{Y}_1 = \dfrac{1}{\mathbf{Z}_1} = \dfrac{1}{2.5\underline{/75^\circ}} = 0.4\underline{/-75^\circ} \text{ S} \qquad \mathbf{Y}_2 = \dfrac{1}{\mathbf{Z}_2} = \dfrac{1}{4\underline{/-50^\circ}} = 0.25\underline{/50^\circ} \text{ S} \qquad \mathbf{Y}_3 = \dfrac{1}{\mathbf{Z}_3} = \dfrac{1}{5\underline{/45^\circ}} = 0.2\underline{/-45^\circ} \text{ S}$$

Adding,

$$\mathbf{Y}_T = \mathbf{Y}_1 + \mathbf{Y}_2 + \mathbf{Y}_3 = 0.4\underline{/-75^\circ} + 0.25\underline{/50^\circ} + 0.2\underline{/-45^\circ} = 0.527\underline{/-39.7^\circ} \text{ S}$$

Inverting,

$$\mathbf{Z}_T = \dfrac{1}{\mathbf{Y}_T} = \dfrac{1}{0.527\underline{/-39.7^\circ}} = 1.9\underline{/39.7^\circ} \ \Omega$$

11.34 Find the simplest parallel circuit that has the same impedance at 400 Hz as the series combination of a 300-Ω resistor, a 0.25-H inductor, and a 1-μF capacitor.

The parallel circuit can be determined from the admittance, which can be found by inverting the impedance:

$$\mathbf{Y} = \dfrac{1}{300 + j2\pi(400)(0.25) - j1/[2\pi(400)(10^{-6})]} = \dfrac{1}{300 + j230} = \dfrac{1}{378\underline{/37.5^\circ}}$$

$$= 2.64 \times 10^{-3}\underline{/-37.5^\circ} \text{ S} = 2.096 - j1.61 \text{ mS}$$

The simplest parallel circuit that has this admittance is a parallel resistor and inductor. From the real part of the admittance, this resistor must have a conductance of 2.096 mS and so a resistance of $1/(2.096 \times 10^{-3}) = 477$ Ω. And from the imaginary part, the inductor must have a susceptance of

-1.61 mS.   The corresponding inductance is, from $B_L = -1/\omega L$,

$$L = \frac{-1}{\omega B_L} = \frac{-1}{2\pi(400)(-1.61 \times 10^{-3})} \text{ H} = 247 \text{ mH}$$

11.35 A load has a voltage of $\mathbf{V} = 120\underline{/20°}$ V and a current of $\mathbf{I} = 48\underline{/60°}$ A, both at 2 kHz. Find the two-element parallel circuit that the load can be. As always, assume associated references because there is no statement to the contrary.

Because the two elements are in parallel, the load admittance should be used to find them:

$$\mathbf{Y} = \frac{\mathbf{I}}{\mathbf{V}} = \frac{48\underline{/60°}}{120\underline{/20°}} = 0.4\underline{/40°} = 0.3064 + j0.2571 \text{ S}$$

The real part 0.3064 is, of course, the conductance of a resistor. The corresponding resistance is $R = 1/0.3064 = 3.26\ \Omega$. The imaginary part 0.2571, being positive, is the susceptance of a capacitor. From $B_C = \omega C$,

$$C = \frac{B_C}{\omega} = \frac{0.2571}{2\pi(2000)} \text{ F} = 20.5 \ \mu\text{F}$$

11.36 A 0.5-Ω resistor is in parallel with a 10-mH inductor. At what radian frequency do the circuit voltage and current have a phase angle difference of 40°?

A good approach is to find the susceptance from the admittance angle and the conductance, and then find the radian frequency from the susceptance and the inductance. The admittance angle has a magnitude of 40° because this is the phase angle difference between the voltage and current, and it is negative because this is an inductive circuit. So, $\theta = -40°$. From $\theta = \tan^{-1}(B_L/G)$,

$$B_L = G \tan \theta = (1/0.5) \tan(-40°) = -1.678 \text{ S}$$

And, from the formula for inductive susceptance, $B_L = -1/\omega L$,

$$\omega = \frac{-1}{LB_L} = \frac{-1}{0.01(-1.678)} = 59.6 \text{ rad/s}$$

11.37 A resistor and a parallel 1-μF capacitor draw 0.48 A when 120 V at 400 Hz is applied. Find the admittance in polar form.

The magnitude of the admittance is $Y = I/V = 0.48/120$ S $= 4$ mS. From admittance triangle considerations, the angle of the admittance is $\theta = \sin^{-1}(B/Y)$. Since $B = \omega C$,

$$\frac{B}{Y} = \frac{\omega C}{Y} = \frac{2\pi(400)(10^{-6})}{0.004} = 0.2\pi$$

and $\theta = \sin^{-1} 0.2\pi = 38.9°$. Therefore, the admittance is $\mathbf{Y} = 4\underline{/38.9°}$ mS.

11.38 Capacitors are sometimes connected in parallel with inductive industrial loads to decrease the current drawn from the source without affecting the load current. To verify this concept, consider connecting a capacitor across a coil that has 10 mH of inductance and 2 Ω of resistance and that is energized by a 60-Hz, 120-V source. What is the capacitance required to make the source current a minimum, and what is the decrease in this current?

Since $\mathbf{I} = \mathbf{YV}$, the current magnitude will be a minimum when the admittance magnitude Y is a minimum. The total admittance $\mathbf{Y}$ is the sum of the admittances of the coil and capacitor:

$$\mathbf{Y} = \frac{1}{R + j\omega L} + j\omega C = \frac{1}{2 + j2\pi(60)(10 \times 10^{-3})} + j2\pi(60)C = \frac{1}{2 + j3.77} + j377C$$
$$= 0.110 - j0.207 + j377C$$

Since the capacitance can affect only the susceptance, the admittance magnitude is a minimum for zero susceptance.   For this,

$$377C = 0.207 \quad \text{from which} \quad C = \frac{0.207}{377} \text{ F} = 549 \ \mu\text{F}$$

With zero susceptance, $\mathbf{Y} = 0.110$ S   and $|\mathbf{I}| = |\mathbf{Y}| \, |\mathbf{V}| = 0.110(120) = 13.2$ A.   In comparison, before the capacitor was added, the magnitude of the current was equal to the product of the magnitudes of the coil admittance and voltage: $|0.110 - j0.207|(120) = 0.234(120) = 28.1$ A.   So, the parallel capacitor causes a decrease in source current of $28.1 - 13.2 = 14.9$ A   even though the coil current remains the same 28.1 A.   What happens is that some of the coil current flows through the capacitor instead of through the source.   Incidentally, since the susceptance is zero, the circuit is in resonance.

11.39   Find the total impedance $\mathbf{Z}_T$ of the circuit shown in Fig. 11-18.

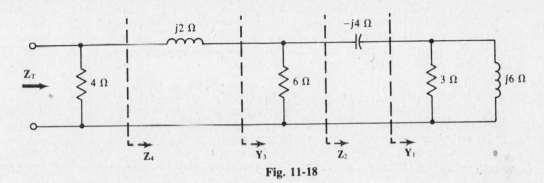

Fig. 11-18

This is, of course, a ladder circuit.   Although for such a circuit it is possible to find $\mathbf{Z}_T$ by using only impedance (or admittance), it is usually best to alternate admittance and impedance, using admittance for parallel branches and impedance for series branches.   This will be done starting at the end opposite the input.   There, the 3- and $j6$-Ω elements have a combined admittance of

$$\mathbf{Y}_1 = \tfrac{1}{3} - j\tfrac{1}{6} = 0.373\underline{/-26.6^\circ} \text{ S}$$

which corresponds to an impedance of

$$\frac{1}{0.373\underline{/-26.6^\circ}} = 2.68\underline{/26.6^\circ} = 2.4 + j1.2 \ \Omega$$

This adds to the $-j4$ Ω of the series capacitor for an impedance of

$$\mathbf{Z}_2 = 2.4 + j1.2 - j4 = 2.4 - j2.8 = 3.69\underline{/-49.4^\circ} \ \Omega$$

The inverse of this added to the conductance of the parallel 6-Ω resistor is

$$\mathbf{Y}_3 = \frac{1}{3.69\underline{/-49.4^\circ}} + \frac{1}{6} = 0.176 + j0.206 + 0.167 = 0.4\underline{/31^\circ} \text{ S}$$

The corresponding impedance adds to the $j2$ Ω of the series inductor:

$$\mathbf{Z}_4 = \frac{1}{0.4\underline{/31^\circ}} + j2 = 2.14 - j1.29 + j2 = 2.26\underline{/18.4^\circ} \ \Omega$$

The corresponding admittance plus the conductance of the 4-Ω resistor is $\mathbf{Y}_T$:

$$\mathbf{Y}_T = \frac{1}{2.26\underline{/18.4^\circ}} + \frac{1}{4} = 0.42 - j0.14 + 0.25 = 0.684\underline{/-11.8^\circ} \text{ S}$$

Finally,

$$\mathbf{Z}_T = \frac{1}{\mathbf{Y}_T} = \frac{1}{0.684\underline{/-11.8^\circ}} = 1.46\underline{/11.8^\circ} \ \Omega$$

11.40 Find the input admittance at 50 krad/s of the circuit shown in Fig. 11-19a.

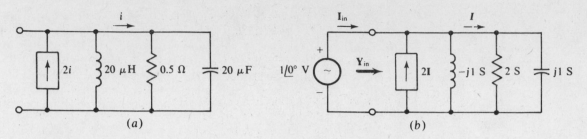

Fig. 11-19

The first step is to use $-j1/\omega L$, G, $j\omega C$, and phasors to construct the corresponding frequency-domain circuit shown in Fig. 11-19b along with a $1\underline{/0°}$-V source. With this source, the circuit has an input admittance of $Y_{in} = I_{in}/1\underline{/0°} = I_{in}$. Note that the controlling current I is the sum of the currents in the two right-hand branches:

$$I = (1\underline{/0°})(2) + (1\underline{/0°})(j1) = 2 + j1 \text{ A}$$

And so the dependent source current flowing down is $-2I = -2(2 + j1)$. This can be used in a KCL equation at the top node to obtain I_{in}:

$$I_{in} = -2(2 + j1) + (1\underline{/0°})(-j1) + 2 + j1 = -2 - j2 = 2.83\underline{/-135°} \text{ A}$$

Finally, $Y_{in} = I_{in} = 2.83\underline{/-135°}$ S.

11.41 Find I_{in} and I_L for the circuit shown in Fig. 11-20.

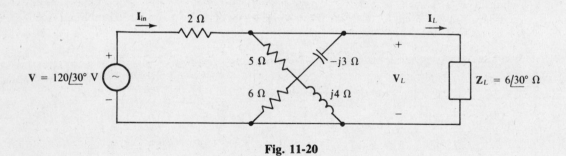

Fig. 11-20

The current I_{in} can be found from the source voltage divided by the input impedance Z_{in}, which equals the 2 Ω of the series resistor plus the total impedance of the three branches to the right of this resistor. Since these branches extend between the same two nodes, they are in parallel and have a total admittance Y that is the sum of the individual admittances:

$$Y = \frac{1}{5 + j4} + \frac{1}{6 - j3} + \frac{1}{6\underline{/30°}} = 0.156\underline{/-38.7°} + 0.149\underline{/26.6°} + 0.167\underline{/-30°} = 0.416\underline{/-16°} \text{ S}$$

Adding the 2 Ω to the inverse of this admittance results in

$$Z_{in} = 2 + \frac{1}{Y} = 2 + \frac{1}{0.416\underline{/-16°}} = 2 + 2.41\underline{/16°} = 4.36\underline{/8.72°} \text{ Ω}$$

from which $$I_{in} = \frac{V}{Z_{in}} = \frac{120\underline{/30°}}{4.36\underline{/8.72°}} = 27.5\underline{/21.3°} \text{ A}$$

The current $\mathbf{I}_L$ can be found from the load voltage and impedance. The load voltage $\mathbf{V}_L$ is equal to the current $\mathbf{I}_{in}$ divided by the total admittance of the three parallel branches:

$$\mathbf{V}_L = \frac{\mathbf{I}_{in}}{\mathbf{Y}} = \frac{27.5\underline{/21.3^\circ}}{0.416\underline{/-16^\circ}} = 66.2\underline{/37^\circ}\ \text{V}$$

and

$$\mathbf{I}_L = \frac{\mathbf{V}_L}{\mathbf{Z}_L} = \frac{66.2\underline{/37^\circ}}{6\underline{/30^\circ}} = 11\underline{/7^\circ}\ \text{A}$$

Alternatively, $\mathbf{I}_L$ can be found directly from $\mathbf{I}_{in}$ by current division. $\mathbf{I}_L$ is equal to the product of $\mathbf{I}_{in}$ and the admittance of the load divided by the total admittance of the three parallel branches:

$$\mathbf{I}_L = 27.5\underline{/21.3^\circ} \times \frac{0.167\underline{/-30^\circ}}{0.416\underline{/-16^\circ}} = 11\underline{/7^\circ}\ \text{A}$$

11.42 A current $4\underline{/30^\circ}$ A flows into four parallel branches that have admittances of $6\underline{/-70^\circ}$, $5\underline{/30^\circ}$, $7\underline{/60^\circ}$, and $9\underline{/45^\circ}$ S. Use current division to find the current $\mathbf{I}$ in the $5\underline{/30^\circ}$-S branch. Of course, since there is no statement to the contrary, assume that the current references are such that the current division formula does not have an initial negative sign.

The current $\mathbf{I}$ in the branch with the $5\underline{/30^\circ}$-S admittance is equal to this admittance divided by the sum of the admittances, all times the input current:

$$\mathbf{I} = \frac{5\underline{/30^\circ}}{6\underline{/-70^\circ} + 5\underline{/30^\circ} + 7\underline{/60^\circ} + 9\underline{/45^\circ}} \times 4\underline{/30^\circ} = \frac{20\underline{/60^\circ}}{18.7\underline{/29.8^\circ}} = 1.07\underline{/30.2^\circ}\ \text{A}$$

11.43 Use current division to find $\mathbf{I}_L$ for the circuit shown in Fig. 11-21.

Since there are just two branches and the branch impedances are specified, the impedance form of the current division formula is preferable: The current in one branch is equal to the impedance of the other branch divided by the sum of the impedances, all times the input current. For this circuit, though, a negative sign is required because the input current and $\mathbf{I}_L$ have reference directions into the same node—the bottom node:

$$\mathbf{I}_L = \frac{6}{6 + j9} \times 4\underline{/20^\circ} = \frac{-24\underline{/20^\circ}}{10.8\underline{/56.3^\circ}} = -2.22\underline{/-36.3^\circ}\ \text{A}$$

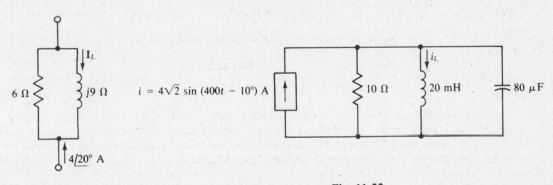

Fig. 11-21   **Fig. 11-22**

11.44 Use current division to find i_L for the circuit shown in Fig. 11-22.

The individual admittances are

$$G = \tfrac{1}{10} = 0.1\ \text{S} \qquad jB_L = -\frac{j1}{\omega L} = \frac{-j1}{400(20 \times 10^{-3})} = -j0.125\ \text{S} \qquad jB_C = j\omega C = j400(80 \times 10^{-6})$$
$$= j0.032\ \text{S}$$

These substituted into the current division formula give

$$I_L = \frac{jB_L}{G + jB_L + jB_C} \times I = \frac{-j0.125}{0.1 - j0.125 + j0.032} \times 4\underline{/-10°} = \frac{(0.125\underline{/-90°})(4\underline{/-10°})}{0.1366\underline{/-42.9°}} = \frac{0.5\underline{/-100°}}{0.1366\underline{/-42.9°}}$$
$$= 3.66\underline{/-57.1°} \text{ A}$$

from which $i_L = 3.66\sqrt{2} \sin(400t - 57.1°) = 5.18 \sin(400t - 57.1°)$ A

11.45 Use current division twice to find the current I_L for the circuit shown in Fig. 11-23.

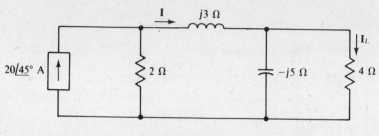

Fig. 11-23

The approach is to find I from the source current by current division, and then find I_L from I by current division. For the I current division formula, the impedance to the right of the 2-Ω resistor is needed. It is

$$j3 + \frac{4(-j5)}{4 - j5} = j3 + 3.12\underline{/-38.7°} = 2.65\underline{/23.3°} \ \Omega$$

By current division,

$$I = \frac{2}{2 + 2.65\underline{/23.3°}} \times 20\underline{/45°} = \frac{40\underline{/45°}}{4.56\underline{/13.3°}} = 8.77\underline{/31.7°} \text{ A}$$

By current division again,

$$I_L = \frac{-j5}{4 - j5} \times 8.77\underline{/31.7°} = \frac{43.8\underline{/-58.3°}}{6.4\underline{/-51.3°}} = 6.85\underline{/-7°} \text{ A}$$

Supplementary Problems

11.46 A 0.5-μF capacitor and a 2-kΩ resistor are in series. Find the total impedance in polar form at   (a) 0 Hz,   (b) 60 Hz,   and   (c) 10 kHz.   Ans.   (a) $\infty\underline{/-90°}$ Ω,   (b) $5.67\underline{/-69.3°}$ kΩ,   (c) $2\underline{/-0.912°}$ kΩ

11.47 A 300-Ω resistor, a 1-H inductor, and a 1-μF capacitor are in series. Find the total impedance in polar form and whether the circuit is inductive or capacitive at   (a) 833 rad/s,   (b) 1000 rad/s,   and   (c) 1200 rad/s.
Ans.   (a) $474\underline{/-50.8°}$ Ω, capacitive;   (b) $300\underline{/0°}$ Ω, neither capacitive nor inductive;   (c) $474\underline{/50.7°}$ Ω, inductive

11.48 A capacitor and resistor in series have an impedance of $1.34\underline{/-45°}$ kΩ at 400 Hz. Find the capacitance and resistance.   Ans. 0.42 μF, 948 Ω

11.49   A load has a voltage of $240\underline{/75°}$ V and a current of $20\underline{/60°}$ A at a frequency of 60 Hz.   Find the two-element series circuit that the load can be.   *Ans.* An 11.6-Ω resistor and an 8.24-mH inductor

11.50   Two circuit elements in series draw a current of 16 sin (200t + 35°) A in response to an applied voltage of 80 cos 200t V.   Find the two elements.   *Ans.* A 2.87-Ω resistor and a 20.5-mH inductor

11.51   A 100-Ω resistor is in series with a 120-mH inductor.   At what frequency do the circuit voltage and current have a phase angle difference of 35°?   *Ans.* 92.9 Hz

11.52   A 750-Ω resistor is in series with a 0.1-μF capacitor.   At what frequency does the total impedance have a magnitude of 1000 Ω?   *Ans.* 2.41 kHz

11.53   Find the total impedance in polar form of three series-connected components that have impedances of $10\underline{/-40°}$, $12\underline{/65°}$, and $15\underline{/-30°}$ Ω.   *Ans.* $25.9\underline{/-6.77°}$ Ω

11.54   What resistor in series with a 2-H inductor limits the current to 120 mA when 120 V at 60 Hz is applied?   *Ans.* 657 Ω

11.55   Two circuit elements in series draw a current of 24 sin (5000t − 10°) mA in response to an applied voltage of $120\sqrt{2}$ sin (5000t + 30°) V.   Find the two elements.
Ans. A 5.42-kΩ resistor and a 0.909-H inductor

11.56   Find the input impedance at 20 krad/s for the circuit shown in Fig. 11-24.   *Ans.* $228\underline{/28.8°}$ Ω

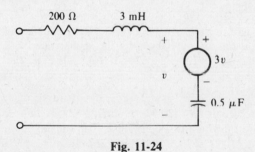

Fig. 11-24

11.57   A 300-V source is connected in series with three components, two of which have impedances of $40\underline{/30°}$ and $30\underline{/-60°}$ Ω.   Find the impedance of the third component if the current that flows is 5 A and if it lags the source voltage by 20°.   *Ans.* $27.3\underline{/75.7°}$ Ω

11.58   Find the total impedance of two parallel components that have identical impedances of $100\underline{/60°}$ Ω.   *Ans.* $50\underline{/60°}$ Ω

11.59   What is the total impedance of two parallel components that have impedances of $80\underline{/-30°}$ and $60\underline{/40°}$ Ω?   *Ans.* $41.6\underline{/10.7°}$ Ω

11.60   A 120-mH coil with a 30-Ω winding resistance is in parallel with a 20-Ω resistor.   What series resistor and inductor produce the same impedance at 60 Hz as this parallel combination?
Ans. 15.6 Ω, 10.6 mH

11.61   A 2-mH coil with a 10-Ω winding resistance is in parallel with a 10-μF capacitor.   What two series circuit elements have the same impedance at 8 krad/s?
   Ans. A 13.9-Ω resistor and a 7.2-μF capacitor

11.62   For the circuit shown in Fig. 11-25, find **I**, $\mathbf{V}_R$, and $\mathbf{V}_C$, and the corresponding sinusoidal quantities if the frequency is 50 Hz.   Also, find the average power delivered by the source.
   Ans. $\mathbf{I} = 7.5\underline{/81.3°}$ A $\mathbf{V}_R = 150\underline{/81.3°}$ V
      $\mathbf{V}_C = 187\underline{/-8.66°}$ V $i = 10.6 \sin(314t + 81.3°)$ A
      $v_R = 212 \sin(314t + 81.3°)$ V $v_C = 265 \sin(314t - 8.66°)$ V
      Average power delivered = 1.12 kW

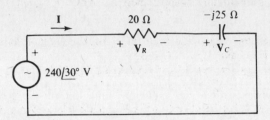

Fig. 11-25

11.63   A voltage source of $340 \sin(1000t + 25°)$ V, a 2-Ω resistor, a 1-H inductor, and a 1-μF capacitor are in series.   Find the current out of the positive terminal of the source.   Also, find the resistor, inductor, and capacitor voltage drops.
   Ans. $i = 170 \sin(1000t + 25°)$ A $v_R = 340 \sin(1000t + 25°)$ V
      $v_L = 170 \cos(1000t + 25°)$ kV $v_C = 170 \sin(1000t - 65°)$ kV

11.64   A voltage that has a phasor of $200\underline{/-40°}$ V is applied across a resistor and capacitor that are in series.   If the capacitor rms voltage is 120 V, what is the resistor voltage phasor?
   Ans. $160\underline{/-3.13°}$ V

11.65   A frequency-domain circuit has $220\underline{/30°}$ V applied across two components, one of which is a 30-Ω resistor and the other of which is a coil that has an impedance of $30\underline{/40°}$ Ω.   Find the voltage drops across the resistor and the coil.   *Ans.* Resistor voltage = $117\underline{/10°}$ V, coil voltage = $117\underline{/50°}$ V

11.66   A voltage source of $170 \sin(377t - 30°)$ V, a 200-Ω resistor, and a 10-μF capacitor are in series.   Find the resistor and capacitor voltage drops.
   Ans. $v_R = 102 \sin(377t + 23°)$ V, $v_C = 136 \sin(377t - 67°)$ V

11.67   Repeat Prob. 11.66 with an added series 1-H inductor.   Also, find the inductor voltage.
   Ans. $v_R = 148 \sin(377t - 59°)$ V, $v_C = 197 \sin(377t - 149°)$ V, $v_L = 280 \sin(377t + 31°)$ V

11.68   A frequency-domain circuit has $500\underline{/40°}$ V applied across three series-connected components that have impedances of $20\underline{/40°}$, $30\underline{/-60°}$, and $40\underline{/70°}$ Ω.   Find the component voltage drops.
   Ans. $\mathbf{V}_{20} = 199\underline{/50.9°}$ V, $\mathbf{V}_{30} = 298\underline{/-49.1°}$ V, $\mathbf{V}_{40} = 397\underline{/80.9°}$ V

11.69   What is the current **I** for the circuit shown in Fig. 11-26?   *Ans.* $\mathbf{I} = 7.93\underline{/45.8°}$ A

11.70   Use voltage division twice to find **V** in the circuit shown in Fig. 11-26.   *Ans.* $\mathbf{V} = 81.2\underline{/6.04°}$ V

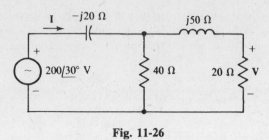

Fig. 11-26

11.71 Derive expressions for the resistance and reactance of an impedance in terms of the conductance and susceptance of the corresponding admittance.   *Ans.* $R = G/(G^2 + B^2)$, $X = -B/(G^2 + B^2)$

11.72 Find the total admittance in polar form of a 1-μF capacitor and a parallel 3.6-kΩ resistor at (*a*) 5 Hz, (*b*) 44.2 Hz, and (*c*) 450 Hz.
Ans. (*a*) 0.28$\underline{/6.45°}$ mS, (*b*) 0.393$\underline{/45°}$ mS, (*c*) 2.84$\underline{/84.4°}$ mS

11.73 A 1-kΩ resistor, a 1-H inductor, and a 1-μF capacitor are in parallel.   Find the total admittance in polar form at (*a*) 500 rad/s, (*b*) 1000 rad/s, and (*c*) 5000 rad/s.
Ans. (*a*) 1.8$\underline{/-56.3°}$ mS, (*b*) 1$\underline{/0°}$ mS, (*c*) 4.9$\underline{/78.2°}$ mS

11.74 An inductor and a parallel resistor have an admittance of 100$\underline{/-30°}$ mS at 400 Hz.   What are the inductance and resistance?   *Ans.* 7.96 mH, 11.5 Ω

11.75 Find the simplest series circuit that has the same total impedance at 400 Hz as the parallel arrangement of a 620-Ω resistor, a 0.5-H inductor, and a 0.5-μF capacitor.
Ans. A 573-Ω resistor and a 2.43-μF capacitor

11.76 A load has a voltage of 240$\underline{/60°}$ V and a current of 120$\underline{/20°}$ mA.   What two-element parallel circuit can this load be at 400 Hz?   *Ans.* A 2.61-kΩ resistor and a 1.24-H inductor

11.77 A resistor and a parallel 0.5-μF capacitor draw 50 mA when 120 V at 60 Hz is applied.   What is the total admittance in polar form and what is the resistance of the resistor?
Ans. 0.417$\underline{/26.9°}$ mS, 2.69 kΩ

11.78 What two circuit elements in parallel have an admittance of 0.4$\underline{/-50°}$ S at 60 Hz?
Ans. A 3.89-Ω resistor and an 8.66-mH inductor

11.79 What two circuit elements in parallel have an admittance of 2.5$\underline{/30°}$ mS at 400 Hz?
Ans. A 462-Ω resistor and a 0.497-μF capacitor

11.80 Three circuit elements in parallel have an admittance of 6.3$\underline{/-40°}$ mS at a frequency of 2 kHz.   If one is a 60-mH inductor, what are the two other elements?
Ans. A 207-Ω resistor and a 29.2-mH inductor

11.81 A 2-kΩ resistor is in parallel with a 0.1-μF capacitor.   At what frequency does the total admittance have an angle of 40°?   *Ans.* 668 Hz

11.82 A resistor and a parallel 120-mH inductor draw 3 A when 100 V at 60 Hz is applied.   What is the total admittance?   *Ans.* 30$\underline{/-47.5°}$ mS

11.83 A certain industrial load has an impedance of $0.6\underline{/30°}\ \Omega$ at a frequency of 60 Hz. What capacitor connected in parallel with this load causes the angle of the total impedance to decrease to 15°? Also, if the load voltage is 120 V, what is the decrease in line current produced by adding the capacitor? *Ans.* 1.18 mF, 20.7 A

11.84 Find the admittance **Y** of the circuit shown in Fig. 11-27. *Ans.* **Y** $= 2.29\underline{/-42.2°}$ S

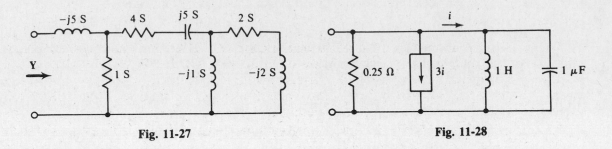

Fig. 11-27   Fig. 11-28

11.85 Find the input admittance at 1 krad/s of the circuit shown in Fig. 11-28. *Ans.* 4 S

11.86 Repeat Prob. 11.85 for a radian frequency of 1 Mrad/s. *Ans.* $5.66\underline{/45°}$ S

11.87 A $20\underline{/30°}$-A current flows into three parallel branches that have impedances of 200, $j10$, and $-j10\ \Omega$. Find the current in the $j10$-Ω branch. *Ans.* $400\underline{/-60°}$ A

11.88 A $20\sin(200t - 30°)$-A current flows into the parallel combination of a 100-Ω resistor and a 25-μF capacitor. Find the capacitor current. *Ans.* $8.94\sin(200t + 33.4°)$ A

11.89 A $20\underline{/-45°}$-A current flows into three parallel branches that have impedances of $16\underline{/30°}$, $20\underline{/-45°}$, and $25\underline{/-60°}\ \Omega$. What is the current in the $25\underline{/-60°}$-Ω branch? *Ans.* $6.89\underline{/-4.49°}$ A

11.90 Use current division twice to find **I** for the circuit shown in Fig. 11-29. *Ans.* **I** $= 1.41\underline{/-19.5°}$ A

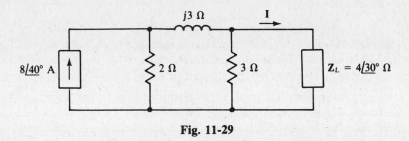

Fig. 11-29

Chapter 12

Mesh, Loop, and Nodal Analyses
of AC Circuits

INTRODUCTION

The material in this chapter is similar to that in Chap. 5. Of course, the analysis techniques apply to ac frequency-domain circuits instead of to dc resistive circuits and so to voltage and current phasors instead of to voltages and currents and to impedances and admittances instead of just to resistances and to conductances. Also, an analysis is often considered completed after the unknown voltage or current phasors are determined. The final step of finding the actual time-function voltages and currents is seldom done because they are not usually important. Besides, it is a simple matter to get them from the phasors.

One other introductory note: From this point on, the terms "impedances" and "admittances" will often be used to mean *components with impedances* and *components with admittances*, as is common practice.

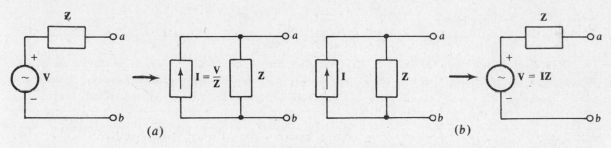

Fig. 12-1

SOURCE CONVERSIONS

As has been explained, mesh and loop analyses are usually easier to do with all current sources converted to voltage sources. And nodal analysis is usually easier to do with all voltage sources converted to current sources. Figure 12-1a shows the rather obvious conversion from a voltage source to a current source, and Fig. 12-1b shows the conversion from a current source to a voltage source. In each circuit the rectangle next to **Z** indicates components that have a total impedance of **Z**. These components can be in any configuration and can, of course, include dependent sources—but not independent sources.

MESH AND LOOP ANALYSES

Mesh analysis for frequency-domain circuits should be apparent from the presentation of mesh analysis for dc circuits in Chap. 5. Preferably all current sources are converted to voltage sources, then clockwise-referenced mesh currents are assigned, and finally KVL is applied to each mesh.

As an illustration, consider the frequency-domain circuit shown in Fig. 12-2. The KVL equation for mesh 1 is

$$\mathbf{I_1Z_1} + (\mathbf{I_1} - \mathbf{I_3})\mathbf{Z_2} + (\mathbf{I_1} - \mathbf{I_2})\mathbf{Z_3} = \mathbf{V_1} + \mathbf{V_2} - \mathbf{V_3}$$

where I_1Z_1, $(I_1 - I_3)Z_2$, and $(I_1 - I_2)Z_3$ are the voltage drops across the impedances Z_1, Z_2, and Z_3. Of course, $V_1 + V_2 - V_3$ is the sum of the voltage rises from voltage sources in mesh 1. As a memory aid, a source voltage is added if it "aids" current flow—that is, if the principal current has a direction out of the positive terminal of the source. Otherwise, the source voltage is subtracted.

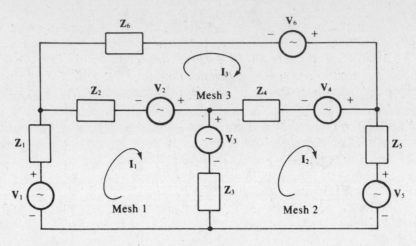

Fig. 12-2

This equation simplifies to

$$(Z_1 + Z_2 + Z_3)I_1 - Z_3I_2 - Z_2I_3 = V_1 + V_2 - V_3$$

The $Z_1 + Z_2 + Z_3$ coefficient of I_1 is the *self-impedance* of mesh 1, which is the sum of the impedances of mesh 1. The $-Z_3$ coefficient of I_2 is the negative of the impedance in the branch common to meshes 1 and 2. This impedance Z_3 is a *mutual impedance*—it is mutual to meshes 1 and 2. Likewise, the $-Z_2$ coefficient of I_3 is the negative of the impedance in the branch mutual to meshes 1 and 3, and so Z_2 is also a mutual impedance. It is important to remember in mesh analysis that the mutual terms have initial negative signs.

It is, of course, easier to write mesh equations using self-impedances and mutual impedances than it is to directly apply KVL. Doing this for meshes 2 and 3 results in

$$-Z_3I_1 + (Z_3 + Z_4 + Z_5)I_2 - Z_4I_3 = V_3 + V_4 - V_5$$

and

$$-Z_2I_1 - Z_4I_2 + (Z_2 + Z_4 + Z_6)I_3 = -V_2 - V_4 + V_6$$

Placing the equations together shows the symmetry of the I coefficients about the principal diagonal:

$$(Z_1 + Z_2 + Z_3)I_1 - \qquad\qquad Z_3I_2 - \qquad\qquad Z_2I_3 = \quad V_1 + V_2 - V_3$$
$$-Z_3I_1 + (Z_3 + Z_4 + Z_5)I_2 - \qquad\qquad Z_4I_3 = \quad V_3 + V_4 - V_5$$
$$-Z_2I_1 - \qquad\qquad Z_4I_2 + (Z_2 + Z_4 + Z_6)I_3 = -V_2 - V_4 + V_6$$

Usually, there is no such symmetry if the corresponding circuit has dependent sources. Also, some of the off-diagonal coefficients may not have initial negative signs.

Loop analysis is similar except that the paths around which KVL is applied are not necessarily meshes, and the loop currents may not all be referenced clockwise. So, even if a circuit has no dependent sources, some of the mutual impedance coefficients may not have initial negative signs. Preferably, the loop current paths are selected such that each current source has just one loop current through it. Then, these loop currents become known quantities with the result that it is unnecessary to write KVL equations for the loops or to convert any current sources to voltage sources. Finally, the required number of loop currents is $B - N + 1$ where B is the number of branches and N is the number of nodes. For a planar circuit, which is a circuit that can be drawn on a flat surface with no wires crossing, this number of loop currents is the same as the number of meshes.

NODAL ANALYSIS

Nodal analysis for frequency-domain circuits is similar to nodal analysis for dc circuits. Preferably, all voltage sources are converted to current sources. Then, a reference node is selected and all other nodes are referenced positive in potential with respect to this reference node. Finally, KCL is applied to each nonreference node. Usually the polarity signs for the node voltages are not shown because it is conventional to reference these voltages positive with respect to the reference node.

For an illustration of nodal analysis applied to a frequency-domain circuit, consider the circuit shown in Fig. 12-3. The KCL equation for node 1 is

$$\mathbf{V}_1\mathbf{Y}_1 + (\mathbf{V}_1 - \mathbf{V}_2)\mathbf{Y}_2 + (\mathbf{V}_1 - \mathbf{V}_3)\mathbf{Y}_6 = \mathbf{I}_1 + \mathbf{I}_2 - \mathbf{I}_6$$

where $\mathbf{V}_1\mathbf{Y}_1$, $(\mathbf{V}_1 - \mathbf{V}_2)\mathbf{Y}_2$, and $(\mathbf{V}_1 - \mathbf{V}_3)\mathbf{Y}_6$ are the currents flowing away from node 1 through the admittances $\mathbf{Y}_1$, $\mathbf{Y}_2$, and $\mathbf{Y}_6$. Of course, $\mathbf{I}_1 + \mathbf{I}_2 - \mathbf{I}_6$ is the sum of the currents flowing into node 1 from current sources,

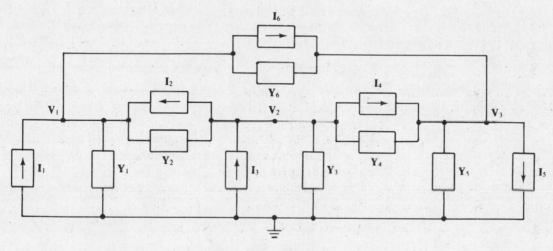

Fig. 12-3

This equation simplifies to

$$(\mathbf{Y}_1 + \mathbf{Y}_2 + \mathbf{Y}_6)\mathbf{V}_1 - \mathbf{Y}_2\mathbf{V}_2 - \mathbf{Y}_6\mathbf{V}_3 = \mathbf{I}_1 + \mathbf{I}_2 - \mathbf{I}_6$$

The coefficient $\mathbf{Y}_1 + \mathbf{Y}_2 + \mathbf{Y}_6$ of $\mathbf{V}_1$ is the *self-admittance* of node 1, which is the sum of the admittances connected to node 1. The coefficient $-\mathbf{Y}_2$ of $\mathbf{V}_2$ is the negative of the admittance connected between nodes 1 and 2. So, $\mathbf{Y}_2$ is a *mutual admittance*. Similarly, the coefficient $-\mathbf{Y}_6$ of $\mathbf{V}_3$ is the negative of the admittance connected between nodes 1 and 3, and so $\mathbf{Y}_6$ is also a mutual admittance.

It is, of course, easier to write nodal equations using self-admittances and mutual admittances than it is to directly apply KCL. Doing this for nodes 2 and 3 produces

$$-\mathbf{Y}_2\mathbf{V}_1 + (\mathbf{Y}_2 + \mathbf{Y}_3 + \mathbf{Y}_4)\mathbf{V}_2 - \mathbf{Y}_4\mathbf{V}_3 = -\mathbf{I}_2 + \mathbf{I}_3 - \mathbf{I}_4$$

and

$$-\mathbf{Y}_6\mathbf{V}_1 - \mathbf{Y}_4\mathbf{V}_2 + (\mathbf{Y}_4 + \mathbf{Y}_5 + \mathbf{Y}_6)\mathbf{V}_3 = \mathbf{I}_4 - \mathbf{I}_5 + \mathbf{I}_6$$

Placing the equations together shows the symmetry of the $\mathbf{V}$ coefficients about the principal diagonal:

$$(\mathbf{Y}_1 + \mathbf{Y}_2 + \mathbf{Y}_6)\mathbf{V}_1 - \qquad\qquad \mathbf{Y}_2\mathbf{V}_2 - \qquad\qquad \mathbf{Y}_6\mathbf{V}_3 = \quad \mathbf{I}_1 + \mathbf{I}_2 - \mathbf{I}_6$$
$$-\mathbf{Y}_2\mathbf{V}_1 + (\mathbf{Y}_2 + \mathbf{Y}_3 + \mathbf{Y}_4)\mathbf{V}_2 - \qquad\qquad \mathbf{Y}_4\mathbf{V}_3 = -\mathbf{I}_2 + \mathbf{I}_3 - \mathbf{I}_4$$
$$-\mathbf{Y}_6\mathbf{V}_1 - \qquad\qquad \mathbf{Y}_4\mathbf{V}_2 + (\mathbf{Y}_4 + \mathbf{Y}_5 + \mathbf{Y}_6)\mathbf{V}_3 = \quad \mathbf{I}_4 - \mathbf{I}_5 + \mathbf{I}_6$$

Usually, there is no such symmetry if the corresponding circuit has dependent sources. Also, some of the off-diagonal coefficients may not have initial negative signs.

Solved Problems

12.1 Perform a source conversion on the circuit shown in Fig. 12-4.

The series impedance is $3 + j4 + 6\|(-j5) = 5.56\underline{/10.9^\circ}\ \Omega$, which divided into the voltage of the original source gives the source current of the equivalent circuit:

$$\frac{20\underline{/30^\circ}}{5.56\underline{/10.9^\circ}} = 3.6\underline{/19.1^\circ}\ \text{A}$$

As shown in Fig. 12-5, the current direction is toward node a, as it must be because the positive terminal of the voltage source is toward that node also. The parallel impedance is, of course, the series impedance of the original circuit.

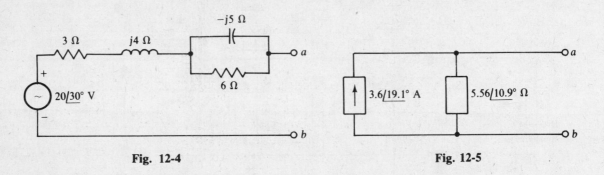

Fig. 12-4   Fig. 12-5

12.2 Perform a source conversion on the circuit shown in Fig. 12-6.

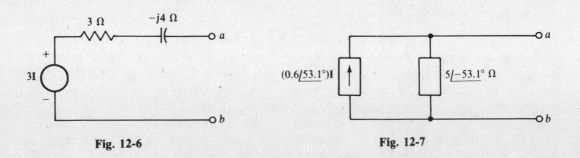

Fig. 12-6   Fig. 12-7

This circuit has a dependent voltage source that produces a voltage in volts that is three times the current **I** flowing *elsewhere* (not shown) in the complete circuit. When, as here, the controlling quantity is not in the circuit being converted, the conversion is the same as for a circuit with an independent source. Therefore, the parallel impedance is $3 - j4 = 5\underline{/-53.1^\circ}\ \Omega$, and the source current directed toward node a is

$$\frac{3\mathbf{I}}{5\underline{/-53.1°}} = (0.6\underline{/53.1°})\mathbf{I}$$

as shown in Fig. 12-7.

When the controlling quantity is in the portion of the circuit being converted, then a different method must be used, as is explained in Chap. 13 in the section on Thevenin's and Norton's theorems.

12.3   Perform a source conversion on the circuit shown in Fig. 12-8.

The parallel impedance is $6\|(5 + j3) = 3.07\underline{/15.7°}\ \Omega$,   which when multiplied by the current gives the voltage of the equivalent voltage source:

$$(4\underline{/-35°})(3.07\underline{/15.7°}) = 12.3\underline{/-19.3°}\ \text{V}$$

As shown in Fig. 12-9, the positive terminal of the voltage source is toward node a, as it must be since the current of the original circuit is toward that node also. Of course, the source impedance is the same $3.07\underline{/15.7°}\ \Omega$, but is in series with the source instead of in parallel with it.

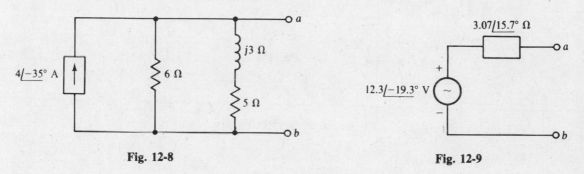

Fig. 12-8   Fig. 12-9

12.4   Perform a source conversion on the circuit shown in Fig. 12-10.

This circuit has a dependent current source that produces a current flow in amperes that is six times the voltage $\mathbf{V}$ across a component *elsewhere* (not shown) in the complete circuit. Since the controlling quantity is not in the circuit being converted, the conversion is the same as for a circuit with an independent source. Consequently, the series impedance is $5\|(4 - j6) = 3.33\underline{/-22.6°}\ \Omega$,   and the source voltage is

$$6\mathbf{V} \times 3.33\underline{/-22.6°} = (20\underline{/-22.6°})\mathbf{V}$$

with, as shown in Fig. 12-11, the positive polarity toward node a because the current of the current source is also toward that node. Of course, the same source impedance is in the circuit, but is in series with the source instead of in parallel with it.

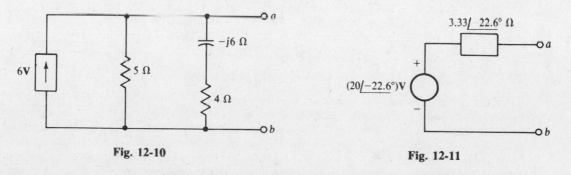

Fig. 12-10   Fig. 12-11

12.5   Find the quantities that go in the blanks.   Assume that these are mesh equations for a circuit that does not have any dependent sources.

$$(16 - j5)\mathbf{I}_1 \underline{\qquad} \mathbf{I}_2 - (3 + j2)\mathbf{I}_3 = 4 - j2$$
$$-(4 + j3)\mathbf{I}_1 + (18 + j9)\mathbf{I}_2 - (6 - j8)\mathbf{I}_3 = 10\underline{/20°}$$
$$\underline{\qquad}\mathbf{I}_1 \underline{\qquad}\mathbf{I}_2 + (20 + j10)\mathbf{I}_3 = 14 + j11$$

The key is the required symmetry of the **I** coefficients about the principal diagonal. Because of this symmetry, the coefficient of $\mathbf{I}_2$ in the first equation must be $-(4 + j3)$, the same as the coefficient of $\mathbf{I}_1$ in the second equation. Also, the coefficient of $\mathbf{I}_1$ in the third equation must be $-(3 + j2)$, the same as the coefficient of $\mathbf{I}_3$ in the first equation. And the coefficient of $\mathbf{I}_2$ in the third equation must be $-(6 - j8)$, the same as the coefficient of $\mathbf{I}_3$ in the second equation.

12.6 Find the voltages across the impedances in the circuit shown in Fig. 12-12a. Then convert the voltage source and $10\underline{/30°}$-Ω component to an equivalent current source and again find the voltages. Compare results.

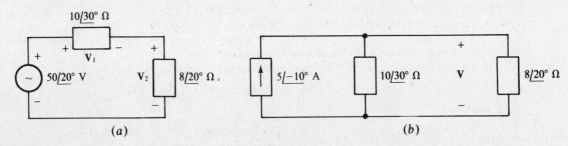

Fig. 12-12

By voltage division,

$$\mathbf{V}_1 = \frac{10\underline{/30°}}{10\underline{/30°} + 8\underline{/20°}} \times 50\underline{/20°} = \frac{500\underline{/50°}}{17.9\underline{/25.6°}} = 27.9\underline{/24.4°} \text{ V}$$

By KVL,

$$\mathbf{V}_2 = 50\underline{/20°} - 27.9\underline{/24.4°} = 22.3\underline{/14.4°} \text{ V}$$

Conversion of the voltage source results in a current source of $(50\underline{/20°})/(10\underline{/30°}) = 5\underline{/-10°}$ A in parallel with a $10\underline{/30°}$-Ω component, both in parallel with the $8\underline{/20°}$-Ω component, as shown in Fig. 12-12b. In this parallel circuit, the same voltage **V** is across all three components. That voltage can be found from the product of the total impedance and the current:

$$\mathbf{V} = \frac{(10\underline{/30°})(8\underline{/20°})}{10\underline{/30°} + 8\underline{/20°}} \times 5\underline{/-10°} = \frac{400\underline{/40°}}{17.9\underline{/25.6°}} = 22.3\underline{/14.4°} \text{ V}$$

Notice that the $8\underline{/20°}$-Ω component voltage is the same as for the original circuit, but that the $10\underline{/30°}$-Ω component voltage is different. This result illustrates the fact that a converted source produces the same voltages and currents outside the source, but not inside it.

12.7 Find the mesh currents for the circuit shown in Fig. 12-13.

The self-impedance and mutual-impedance approach is almost always best for getting mesh equations. The self-impedance of mesh 1 is $4 + j15 + 6 - j7 = 10 + j8$ Ω, and the impedance mutual with mesh 2 is $6 - j7$ Ω. The sum of the source voltage rises in the direction of $\mathbf{I}_1$ is $15\underline{/-30°} - 10\underline{/20°} = 11.5\underline{/-71.8°}$ V. In this sum the $10\underline{/20°}$-V voltage is subtracted because it is a voltage drop instead of a rise. Of course, the mesh 1 equation has a left-hand side that is the product of the self-impedance and $\mathbf{I}_1$ minus the product of the mutual impedance and $\mathbf{I}_2$. The right-hand side is the sum of the source voltage rises. Thus,

$$(10 + j8)\mathbf{I}_1 - (6 - j7)\mathbf{I}_2 = 11.5\underline{/-71.8°}$$

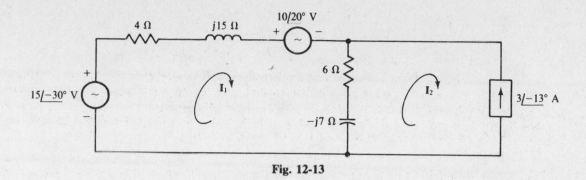

Fig. 12-13

No KVL equation is needed for mesh 2 because I_2 is the only loop current through the $3\underline{/-13°}$-A current source. As a result, $I_2 = -3\underline{/-13°}$ A.   The initial negative sign is required because I_2 has a positive direction down through the source, but the specified $3\underline{/-13°}$-A current is up. Remember that, if for some reason a KVL equation for mesh 2 is wanted, a variable must be included for the voltage across the current source since this voltage is not known.

The substitution of $I_2 = -3\underline{/-13°}$ A   into the mesh 1 equation produces

$$(10 + j8)I_1 - (6 - j7)(-3\underline{/-13°}) = 11.5\underline{/-71.8°}$$

from which

$$I_1 = \frac{11.5\underline{/-71.8°} + (6 - j7)(-3\underline{/-13°})}{10 + j8} = \frac{11.5\underline{/-71.8°} - 27.7\underline{/-62.4°}}{12.8\underline{/38.7°}} = \frac{16.4\underline{/124.2°}}{12.8\underline{/38.7°}} = 1.28\underline{/85.5°} \text{ A}$$

Another good analysis approach is to first convert the current source and parallel impedance to an equivalent voltage source and series impedance, and then find I_1 from the resulting single mesh circuit. If this is done, the equation for I_1 will be identical to the one above.

12.8   Solve for the mesh currents I_1 and I_2 in the circuit shown in Fig. 12-14.

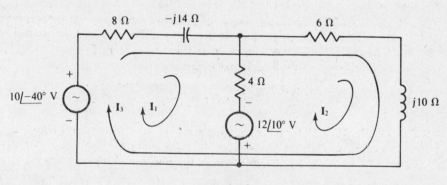

Fig. 12-14

The self-impedance and mutual-impedance approach is the best for mesh analysis. The self-impedance of mesh 1 is $8 - j14 + 4 = 12 - j14$ Ω,   the mutual impedance with mesh 2 is 4 Ω, and the sum of the source voltage rises in the direction of I_1 is $10\underline{/-40°} + 12\underline{/10°} = 20\underline{/-12.6°}$ V.   so, the mesh 1 KVL equation is

$$(12 - j14)I_1 - 4I_2 = 20\underline{/-12.6°}$$

For mesh 2 the self-impedance is $6 + j10 + 4 = 10 + j10$ Ω,   the mutual impedance is 4 Ω, and the sum of the voltage rises from voltage sources is $-12\underline{/10°}$ V.   These give a mesh 2 KVL equation of

$$-4\mathbf{I}_1 + (10 + j10)\mathbf{I}_2 = -12\underline{/10°}$$

Placing the two mesh equations together shows the symmetry of coefficients (here -4) about the principal diagonal as a result of the common mutual impedance:

$$(12 - j14)\mathbf{I}_1 - \qquad 4\mathbf{I}_2 = 20\underline{/-12.6°}$$
$$-4\mathbf{I}_1 + (10 + j10)\mathbf{I}_2 = -12\underline{/10°}$$

By Cramer's rule,

$$\mathbf{I}_1 = \frac{\begin{vmatrix} 20\underline{/-12.6°} & -4 \\ -12\underline{/10°} & 10 + j10 \end{vmatrix}}{\begin{vmatrix} 12 - j14 & -4 \\ -4 & 10 + j10 \end{vmatrix}} = \frac{(20\underline{/-12.6°})(10 + j10) - (-12\underline{/10°})(-4)}{(12 - j14)(10 + j10) - (-4)(-4)}$$

$$= \frac{282\underline{/32.4°} - 48\underline{/10°}}{244 - j20} = \frac{239\underline{/36.8°}}{245\underline{/-4.7°}} = 0.974\underline{/41.5°} \text{ A}$$

and since $\mathbf{I}_2$ has the same denominator as $\mathbf{I}_1$,

$$\mathbf{I}_2 = \frac{\begin{vmatrix} 12 - j14 & 20\underline{/-12.6°} \\ -4 & -12\underline{/10°} \end{vmatrix}}{245\underline{/-4.7°}} = \frac{(12 - j14)(-12\underline{/10°}) - (-4)(20\underline{/-12.6°})}{245\underline{/-4.7°}} = \frac{-221\underline{/-39.4°} + 80\underline{/-12.6°}}{245\underline{/-4.7°}}$$

$$= \frac{154\underline{/127.1°}}{245\underline{/-4.7°}} = 0.63\underline{/131.8°} = -0.63\underline{/-48.2°} \text{ A}$$

12.9   Use loop analysis to find the current down through the 4-Ω resistor in the circuit shown in Fig. 12-14.

The preferable selection of loop currents is $\mathbf{I}_1$ and $\mathbf{I}_3$ because then $\mathbf{I}_1$ is the desired current since it is the only current in the 4-Ω resistor and has a downward direction.   Of course, the self-impedance and mutual-impedance approach should be used.

The self-impedance of the $\mathbf{I}_1$ loop is $8 - j14 + 4 = 12 - j14 \ \Omega$,   the mutual impedance with the $\mathbf{I}_3$ loop is $8 - j14 \ \Omega$, and the sum of the source voltage rises in the direction of $\mathbf{I}_1$ is $10\underline{/-40°} + 12\underline{/10°} = 20\underline{/-12.6°}$ V.   The self-impedance of the $\mathbf{I}_3$ loop is $8 - j14 + 6 + j10 = 14 - j4 \ \Omega$,   of which $8 - j14 \ \Omega$ is mutual with the $\mathbf{I}_1$ loop.   The source voltage rise in the direction of $\mathbf{I}_3$ is $10\underline{/-40°}$ V.   Therefore, the loop equations are

$$(12 - j14)\mathbf{I}_1 + (8 - j14)\mathbf{I}_3 = 20\underline{/-12.6°}$$
$$(8 - j14)\mathbf{I}_1 + (14 - j4)\mathbf{I}_3 = 10\underline{/-40°}$$

The mutual terms are positive because the $\mathbf{I}_1$ and $\mathbf{I}_3$ loop currents have the same direction through the mutual impedance.

By Cramer's rule,

$$\mathbf{I}_1 = \frac{\begin{vmatrix} 20\underline{/-12.6°} & 8 - j14 \\ 10\underline{/-40°} & 14 - j4 \end{vmatrix}}{\begin{vmatrix} 12 - j14 & 8 - j14 \\ 8 - j14 & 14 - j4 \end{vmatrix}} = \frac{(20\underline{/-12.6°})(14 - j4) - (10\underline{/-40°})(8 - j14)}{(12 - j14)(14 - j4) - (8 - j14)(8 - j14)} = \frac{291\underline{/-28.5°} - 161\underline{/-100°}}{244 - j20}$$

$$= \frac{285\underline{/4°}}{245\underline{/-4.7°}} = 1.16\underline{/8.7°} \text{ A}$$

As a check, notice that this loop current should be equal to the difference in the mesh currents $\mathbf{I}_1$ and $\mathbf{I}_2$ found in the solution to Prob. 12.8.   It is, since $\mathbf{I}_1 - \mathbf{I}_2 = 0.974\underline{/41.5°} - (-0.63\underline{/-48.2°}) = 1.16\underline{/8.7°}$ A.

12.10   Find the mesh currents for the circuit shown in Fig. 12-15a.

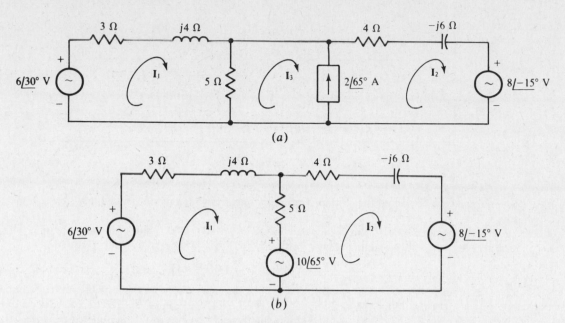

Fig. 12-15

The first step is the conversion of the $2\underline{/65°}$-A current source and parallel 5-Ω resistor into a voltage source and series resistor, as shown in the circuit of Fig. 12-15b. Note that this conversion eliminates mesh 3. The self-impedance of mesh 1 is $3 + j4 + 5 = 8 + j4$ Ω, and that of mesh 2 is $4 - j6 + 5 = 9 - j6$ Ω. The mutual impedance is 5 Ω. The sum of the voltage rises from sources is $6\underline{/30°} - 10\underline{/65°} = 6.14\underline{/-80.9°}$ V for mesh 1 and $10\underline{/65°} - 8\underline{/15°} = 11.7\underline{/107°}$ V for mesh 2. The corresponding mesh equations are

$$(8 + j4)\mathbf{I}_1 - 5\mathbf{I}_2 = 6.14\underline{/-80.9°}$$
$$-5\mathbf{I}_1 + (9 - j6)\mathbf{I}_2 = 11.7\underline{/107°}$$

By Cramer's rule,

$$\mathbf{I}_1 = \frac{\begin{vmatrix} 6.14\underline{/-80.9°} & -5 \\ 11.7\underline{/107°} & 9 - j6 \end{vmatrix}}{\begin{vmatrix} 8 + j4 & -5 \\ -5 & 9 - j6 \end{vmatrix}} = \frac{(6.14\underline{/-80.9°})(9 - j6) - (11.7\underline{/107°})(-5)}{(8 + j4)(9 - j6) - (-5)(-5)} = \frac{45.4\underline{/-174°}}{72\underline{/-9.59°}}$$

$$= 0.631\underline{/-164.4°} = -0.631\underline{/15.6°} \text{ A}$$

and since $\mathbf{I}_2$ has the same denominator as $\mathbf{I}_1$,

$$\mathbf{I}_2 = \frac{\begin{vmatrix} 8 + j4 & 6.14\underline{/-80.9°} \\ -5 & 11.7\underline{/107°} \end{vmatrix}}{72\underline{/-9.59°}} = \frac{(8 + j4)(11.7\underline{/107°}) - (-5)(6.14\underline{/-80.9°})}{72\underline{/-9.59°}} = \frac{81.1\underline{/146.5°}}{72\underline{/-9.59°}}$$

$$= 1.13\underline{/156.1°} = -1.13\underline{/-23.9°} \text{ A}$$

From the original circuit shown in Fig. 12-15a, the current in the current source is $\mathbf{I}_2 - \mathbf{I}_3 = 2\underline{/65°}$ A. Consequently,

$$\mathbf{I}_3 = \mathbf{I}_2 - 2\underline{/65°} = -1.13\underline{/-23.9°} - 2\underline{/65°} = 2.31\underline{/-144.1°} = -2.31\underline{/35.9°} \text{ A}$$

12.11 Use loop analysis to solve for the current flowing down through the 5-Ω resistor in the circuit shown in Fig. 12-15a.

Because this circuit has three meshes, the analysis requires three loop currents. The loops can be selected as in Fig. 12-16 with only one current $\mathbf{I}_1$ flowing through the 5-Ω resistor so that only one

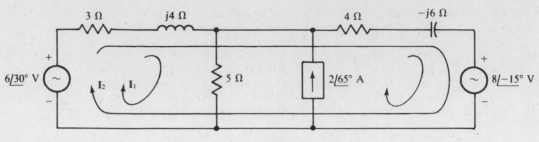

Fig. 12-16

current needs to be solved for. Also, preferably only one loop current should flow through the current source.

The self-impedance of the I_1 loop is $3 + j4 + 5 = 8 + j4 \ \Omega$, the impedance mutual with the I_2 loop is $3 + j4 \ \Omega$, and the aiding source voltage is $6\underline{/30°}$ V. These give a loop 1 equation of

$$(8 + j4)I_1 + (3 + j4)I_2 = 6\underline{/30°}$$

The I_2 coefficient is positive because I_2 and I_1 have the same direction through the mutual components.

For the second loop, the self-impedance is $3 + j4 + 4 - j6 = 7 - j2 \ \Omega$, of which $3 + j4 \ \Omega$ is mutual with loop 1. The $2\underline{/65°}$-A current flowing through the $4 - j6$-Ω components produces a voltage drop of $(4 - j6)(2\underline{/65°}) = 14.4\underline{/8.69°}$ V that has the same effect as the voltage from an opposing voltage source. In addition, the voltage sources have a net aiding voltage of $6\underline{/30°} - 8\underline{/-15°} = 5.67\underline{/117°}$ V. The resulting loop 2 equation is

$$(3 + j4)I_1 + (7 - j2)I_2 = 5.67\underline{/117°} - 14.4\underline{/8.69°} = 17\underline{/170°}$$

For convenience, the two loop equations should be placed together before applying Cramer's rule:

$$(8 + j4)I_1 + (3 + j4)I_2 = 6\underline{/30°}$$
$$(3 + j4)I_1 + (7 - j2)I_2 = 17\underline{/170°}$$

By Cramer's rule,

$$I_1 = \frac{\begin{vmatrix} 6\underline{/30°} & 3 + j4 \\ 17\underline{/170°} & 7 - j2 \end{vmatrix}}{\begin{vmatrix} 8 + j4 & 3 + j4 \\ 3 + j4 & 7 - j2 \end{vmatrix}} = \frac{(6\underline{/30°})(7 - j2) - (17\underline{/170°})(3 + j4)}{(8 + j4)(7 - j2) - (3 + j4)(3 + j4)} = \frac{125\underline{/33.5°}}{72\underline{/-9.59°}} = 1.74\underline{/43.1°} \text{ A}$$

As a check, this loop current I_1 should be equal to the difference in the mesh currents I_1 and I_3 found in the solution to Prob. 12.10. It is, since $I_1 - I_3 = -0.631\underline{/15.6°} - (-2.31\underline{/35.9°}) = 1.74\underline{/43.1°}$ A.

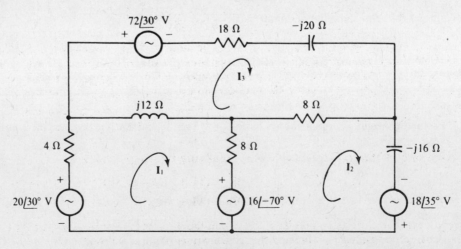

Fig. 12-17

Use mesh analysis to solve for I_1 in the circuit shown in Fig. 12-17. Also, show the determinants for solving for I_2 and I_3.

The self-impedances are $4 + j12 + 8 = 12 + j12 \ \Omega$ for mesh 1, $8 + 8 - j16 = 16 - j16 \ \Omega$ for mesh 2, and $18 - j20 + 8 + j12 = 26 - j8 \ \Omega$ for mesh 3. The mutual impedances are $8 \ \Omega$ for meshes 1 and 2, $8 \ \Omega$ for meshes 2 and 3, and $j12 \ \Omega$ for meshes 1 and 3. The sum of the aiding source voltages is $20\underline{/30°} - 16\underline{/-70°} = 27.7\underline{/64.7°}$ V for mesh 1, $16\underline{/-70°} + 18\underline{/35°} = 20.8\underline{/-13.1°}$ V for mesh 2, and $-72\underline{/30°}$ V for mesh 3. So, the mesh equations are

$$(12 + j12)I_1 - \quad\quad 8I_2 - \quad\quad j12I_3 = 27.7\underline{/64.7°}$$
$$-8I_1 + (16 - j16)I_2 - \quad\quad 8I_3 = 20.8\underline{/-13.1°}$$
$$-j12I_1 - \quad\quad 8I_2 + (26 - j8)I_3 = -72\underline{/30°}$$

By Cramer's rule,

$$I_1 = \frac{\begin{vmatrix} 27.7\underline{/64.7°} & -8 & -j12 \\ 20.8\underline{/-13.1°} & 16 - j16 & -8 \\ -72\underline{/30°} & -8 & 26 - j8 \end{vmatrix}}{\begin{vmatrix} 12 + j12 & -8 & -j12 \\ -8 & 16 - j16 & -8 \\ -j12 & -8 & 26 - j8 \end{vmatrix}}$$

Both determinants can be evaluated using the diagonal method:

$$I_1 = \frac{(17\underline{/2.57°} - 4.61\underline{/30°} + 1.99\underline{/76.9°} - 19.6\underline{/75°} - 1.77\underline{/64.7°} + 4.52\underline{/-30.2°})(10^3)}{9984 - j3072 - 768 - j768 + 2304 - j2304 - 768 - j768 - 1664 + j512}$$

$$= \frac{25.2 \times 10^3\underline{/-62.6°}}{12.2 \times 10^3\underline{/-36°}} = 2.07\underline{/-26.6°} \ \text{A}$$

Since I_2 and I_3 have the same denominator as I_1,

$$I_2 = \frac{\begin{vmatrix} 12 + j12 & 27.7\underline{/64.7°} & -j12 \\ -8 & 20.8\underline{/-13.1°} & -8 \\ -j12 & -72\underline{/30°} & 26 - j8 \end{vmatrix}}{12.2 \times 10^3\underline{/-36°}}$$

and

$$I_3 = \frac{\begin{vmatrix} 12 + j12 & -8 & 27.7\underline{/64.7°} \\ -8 & 16 - j16 & 20.8\underline{/-13.1°} \\ -j12 & -8 & -72\underline{/30°} \end{vmatrix}}{12.2 \times 10^3\underline{/-36°}}$$

This analysis is about as complex as should be attempted by hand since it is difficult to evaluate a third-order determinant of complex elements without making errors, unless, of course, one has a good programmable calculator. Also, there are easy-to-use computer programs for evaluating third- and much higher-order determinants. There are even easy-to-use computer programs, such as ECAP and PCAP, for analyzing ac circuits of a wide range of complexity.

12.13 Show a circuit that corresponds to the following mesh equations:

$$(17 - j4)I_1 - (11 + j5)I_2 = 6\underline{/30°}$$
$$-(11 + j5)I_1 + (18 + j7)I_2 = -8\underline{/30°}$$

Because there are two equations, the circuit has two meshes: mesh 1 for which I_1 is the principal mesh current, and mesh 2 for which I_2 is the principal mesh current. The $-(11 + j5)$ coefficients indicate that meshes 1 and 2 have a $11 + j5$-Ω mutual impedance that could be from an 11-Ω resistor in series with an inductor that has a reactance of 5 Ω. In the first equation the I_1 coefficient indicates that the resistors in mesh 1 have a total resistance of 17 Ω. Since 11 Ω of this is in the mutual impedance, there is $17 - 11 = 6$ Ω of resistance in mesh 1 that is not mutual. The $-j4$ of the I_1 coefficient indicates that mesh 1 has a total reactance of -4 Ω. Since the mutual branch has a reactance of 5 Ω, the remainder of mesh 1 must have a reactance of $-4 - 5 = -9$ Ω, which can be from a single capacitor. The $6\underline{/30°}$ on the right-hand side of the mesh 1 equation is the result of a total of $6\underline{/30°}$ V of voltage source rises (aiding source voltages). One way to obtain this is with a single $6\underline{/30°}$-V source that is not in the mutual branch and that has a polarity such that I_1 flows out of its positive terminal.

Similarly, from the second equation, mesh 2 has a nonmutual resistance of $18 - 11 = 7$ Ω that can be from a resistor that is not in the mutual branch. And from the $j7$ part of the I_2 coefficient, mesh 2 has a total reactance of 7 Ω. Since 5 Ω of this is in the mutual branch, there is $7 - 5 = 2$ Ω remaining that could be from a single inductor that is not in the mutual branch. The $-8\underline{/30°}$ on the right-hand side is the result of a total of $8\underline{/30°}$ V of voltage source drops—opposing source voltages. One way to obtain this is with a single $8\underline{/30°}$-V source that is not in the mutual branch and that has a polarity such that I_2 flows *into* its positive terminal.

Figure 12-18 shows the corresponding circuit. This is just one of an infinite number of circuits from which the equations could have been written.

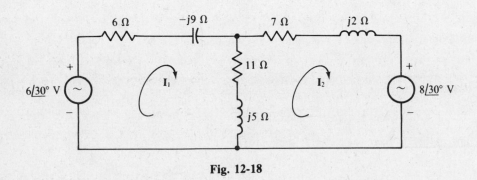

Fig. 12-18

12.14 Show a circuit corresponding to

$$I_2 = \frac{\begin{vmatrix} 16 + j10 & 5 - j3 & -(5 + j6) \\ -(3 - j4) & -4 + j1 & -(10 + j4) \\ -(5 + j6) & 0 & 20 - j8 \end{vmatrix}}{\begin{vmatrix} 16 + j10 & -(3 - j4) & -(5 + j6) \\ -(3 - j4) & 13 - j8 & -(10 + j4) \\ -(5 + j6) & -(10 + j4) & 20 - j8 \end{vmatrix}}$$

Because current is being solved for, and because the bottom determinant has negative elements that are symmetrical about the principal diagonal, I_2 can be a mesh current. And because the determinants are third-order, the circuit has three meshes—1, 2, and 3.

The mutual impedances should be considered first. From the first row of the bottom determinant, there is $3 - j4$ Ω mutual to meshes 1 and 2 and $5 + j6$ Ω mutual to meshes 1 and 3, leaving $16 + j10 - (3 - j4) - (5 + j6) = 8 + j8$ Ω in mesh 1 that is not mutual with other meshes. Similarly, from the second row there is $10 + j4$ Ω common to meshes 2 and 3 and $13 - j8 - (10 + j4) - (3 - j4) = -j8$ Ω in mesh 2 that is not mutual. From row 3 there is $20 - j8 - (5 + j6) - (10 + j4) = 5 - j18$ Ω in mesh 3 that is not mutual.

The second column in the numerator determinant, being the only column that differs from corresponding columns in the denominator determinant, shows that the voltage sources are $5 - j3 = 5.83\underline{/-31°}$ V aiding in mesh 1; $-4 + j1 = -(4 - j1) = -4.12\underline{/-14°}$ V, which is $4.12\underline{/-14°}$ V opposing, in mesh 2; and 0 V in mesh 3.

Figure 12-19 shows one circuit that satisfies these conditions.

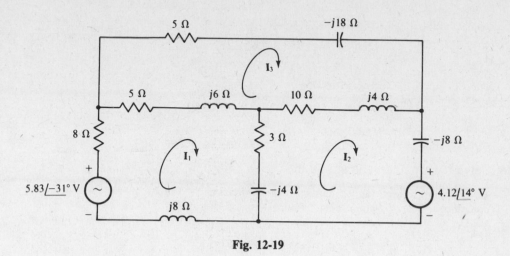

Fig. 12-19

12.15 Use loop analysis to solve for the current flowing to the right through the 6-Ω resistor in the circuit shown in Fig. 12-20.

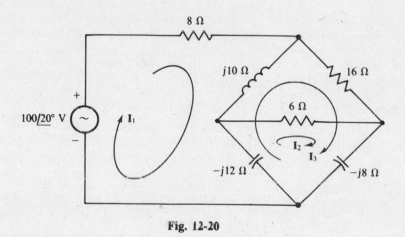

Fig. 12-20

Three loop currents are required because the circuit has three meshes. Only one of the loop currents should flow through the 6-Ω resistor so that only one current has to be solved for. This current is I_2, as shown. The paths for the two other loop currents can be selected as shown, but there are other suitable paths.

It is rather easy to fill in the determinants for solving for I_2 without writing the loop equations. The loop self-impedances and mutual impedances can be used to fill in the denominator determinant. The numerator determinant differs from the denominator determinant in only one column, which is the second column since I_2 is being solved for. This column has the sum of the aiding source voltages, which is $100\underline{/20°}$ V for loop 1 and 0 V for the two other loops. From these considerations, the determinant equation for I_2 is

$$I_2 = \frac{\begin{vmatrix} 8 - j2 & 100\underline{/20°} & j2 \\ j12 & 0 & -j20 \\ j2 & 0 & 16 - j10 \end{vmatrix}}{\begin{vmatrix} 8 - j2 & j12 & j2 \\ j12 & 6 - j20 & -j20 \\ j2 & -j20 & 16 - j10 \end{vmatrix}}$$

which, by the diagonal approach, evaluates to

$$I_2 = \frac{4000\underline{/20°} - 22.6 \times 10^3\underline{/78°}}{3936 - j4192}$$

$$= \frac{20.8 \times 10^3\underline{/-92.6°}}{5.75 \times 10^3\underline{/-46.8°}} = 3.62\underline{/-45.8°} \text{ A}$$

12.16 Solve for the node voltages in the circuit shown in Fig. 12-21.

Using self-admittances and mutual admittances is almost always best for obtaining the nodal equations. The self-admittance of node 1 is

$$\frac{1}{0.25} + \frac{1}{j0.5} = 4 - j2 \text{ S}$$

of which 4 S is mutual conductance. The sum of the currents from current sources into node 1 is $20\underline{/10°} + 15\underline{/-30°} = 32.9\underline{/-7.02°}$ A. So, the node 1 KCL equation is

$$(4 + j2)V_1 - 4V_2 = 32.9\underline{/-7.02°}$$

No KCL equation is needed for node 2 because a grounded voltage source is connected to it, making $V_2 = -12\underline{/-15°}$ V. If, however, for some reason a KCL equation is wanted for node 2, a variable has to be introduced for the current through the voltage source because this current is unknown. Note that, because the voltage source does not have a series impedance, it cannot be converted to a current source with the source conversion techniques presented in this chapter.

The substitution of $V_2 = -12\underline{/-15°}$ into the node 1 equation results in

$$(4 - j2)V_1 - 4(-12\underline{/-15°}) = 32.9\underline{/-7.02°}$$

from which

$$V_1 = \frac{32.9\underline{/-7.02°} - 48\underline{/-15°}}{4 - j2} = \frac{16.05\underline{/148°}}{4.47\underline{/-27°}} = 3.59\underline{/175°} = -3.59\underline{/-5°} \text{ A}$$

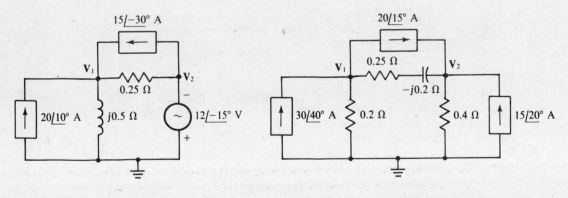

Fig. 12-21   Fig. 12-22

12.17 Solve for the node voltages in the circuit shown in Fig. 12-22.

The self-admittance of node 1 is

$$\frac{1}{0.2} + \frac{1}{0.25 - j0.2} = 5 + 2.44 + j1.95 = 7.69\underline{/14.7°} \text{ S}$$

of which $2.44 + j1.95 = 3.12\underline{/38.7°}$ S is mutual admittance. The sum of the currents into node 1 from current sources is $30\underline{/40°} - 20\underline{/15°} = 14.6\underline{/75.4°}$ A.   Therefore, the node 1 KCL equation is

$$(7.69\underline{/14.7°})V_1 - (3.12\underline{/38.7°})V_2 = 14.6\underline{/75.4°}$$

The self-admittance of node 2 is

$$\frac{1}{0.4} + \frac{1}{0.25 - j0.2} = 2.5 + 2.44 + j1.95 = 5.31\underline{/21.6°}\text{ S}$$

of which $3.12\underline{/38.7°}$ S is mutual admittance.   The sum of the currents into node 2 from current sources is $20\underline{/15°} + 15\underline{/20°} = 35\underline{/17.1°}$ A.   The result is a node 2 KCL equation of

$$-(3.12\underline{/38.7°})V_1 + (5.31\underline{/21.6°})V_2 = 35\underline{/17.1°}$$

Placing the two nodal equations together shows the symmetry of the coefficients $-3.12\underline{/38.7°}$ about the principal diagonal as a result of the same mutual admittance being in both equations:

$$(7.69\underline{/14.7°})V_1 - (3.12\underline{/38.7°})V_2 = 14.6\underline{/75.4°}$$
$$-(3.12\underline{/38.7°})V_1 + (5.31\underline{/21.6°})V_2 = 35\underline{/17.1°}$$

These can be solved by Cramer's rule:

$$V_1 = \frac{\begin{vmatrix} 14.6\underline{/75.4°} & -3.12\underline{/38.7°} \\ 35\underline{/17.1°} & 5.31\underline{/21.6°} \end{vmatrix}}{\begin{vmatrix} 7.69\underline{/14.7°} & -3.12\underline{/38.7°} \\ -3.12\underline{/38.7°} & 5.31\underline{/21.6°} \end{vmatrix}} = \frac{175\underline{/72.7°}}{34.1\underline{/25.4°}} = 5.13\underline{/47.3°}\text{ V}$$

$$V_2 = \frac{\begin{vmatrix} 7.69\underline{/14.7°} & 14.6\underline{/75.4°} \\ -3.12\underline{/38.7°} & 35\underline{/17.1°} \end{vmatrix}}{34.1\underline{/25.4°}} = \frac{279\underline{/41.1°}}{34.1\underline{/25.4°}} = 8.18\underline{/15.7°}\text{ V}$$

12.18 Use nodal analysis to find **V** for the circuit shown in Fig. 12-23.

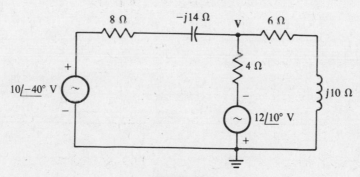

Fig. 12-23

Although a good approach is to convert both voltage sources to current sources, this conversion is not essential because both voltage sources are grounded.   (Actually, source conversions are never absolutely necessary.)   Leaving the circuit as it stands and summing currents away from the V node in the form of voltages divided by impedances gives the equation of

$$\frac{V - 10\underline{/-40°}}{8 - j14} + \frac{V - (-12\underline{/10°})}{4} + \frac{V}{6 + j10} = 0$$

The first term is the current flowing to the left through the $8 - j14$-Ω components, the second is the current flowing down through the 4-Ω resistor, and the third is the current flowing to the right through the $6 + j10$-Ω components.
    This equation simplifies to

$$(0.062\underline{/60.3°} + 0.25 + 0.0857\underline{/-59°})V = 0.62\underline{/20.3°} - 3\underline{/10°}$$

This is the same equation that would have been found if source conversions had been made. The coefficient of **V** is the sum of the admittances connected to the **V** node (i.e., they make up the self-admittance), the first term on the right-hand side is the source current from a conversion of the $10/{-40}°$-V source, and the second term on the right-hand side is the source current from a conversion of the $12/10°$-V source.

Further simplification reduces the equation to

$$(0.325/{-3.47°})\mathbf{V} = 2.392/{-173°}$$

from which
$$\mathbf{V} = \frac{2.392/{-173°}}{0.325/{-3.47°}} = 7.35/{-169.2°} = -7.35/10.8° \text{ V}$$

Incidentally, this result can be checked since the circuit shown in Fig. 12-23 is the same as that shown in Fig. 12-14 for which, in the solution to Prob. 12.9, the current down through the 4-Ω resistor was found to be $1.16/8.7°$ A. the voltage **V** across the center branch can be calculated from this current: $\mathbf{V} = 4(1.16/8.7°) - 12/10° = -7.35/10.8°$ V, which checks.

12.19 Find the node voltages in the circuit shown in Fig. 12-24a.

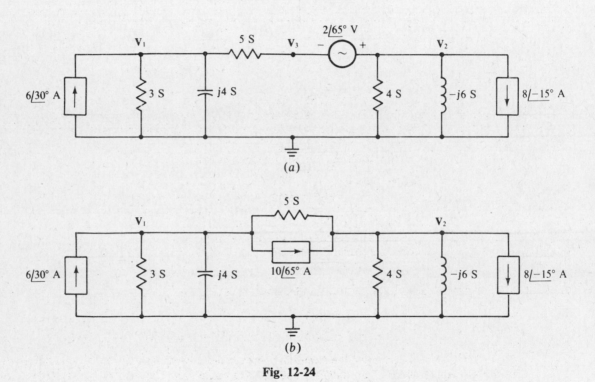

Fig. 12-24

Since the voltage source does not have a grounded terminal, a good first step for nodal analysis is the conversion of this source and the series resistor to a current source and parallel resistor, as shown in Fig. 12-24b. Note that this conversion eliminates node 3. In the circuit shown in Fig. 12-24b the self-admittance of node 1 is $3 + j4 + 5 = 8 + j4$ S, and that of node 2 is $5 + 4 - j6 = 9 - j6$ S. The mutual admittance is 5 S. The sum of the currents into node 1 from current sources is $6/30° - 10/65° = 6.14/{-80.9°}$ A, and that into node 2 is $10/65° - 8/{-15°} = 11.7/107°$ A. Thus, the corresponding nodal equations are

$$(8 + j4)\mathbf{V}_1 - \qquad 5\mathbf{V}_2 = 6.14/{-80.9°}$$
$$-5\mathbf{V}_1 + (9 - j6)\mathbf{V}_2 = 11.7/107°$$

Except for having **V**'s instead of **I**'s, these are the same equations as for Prob. 12.10. Consequently, the answers are the same: $\mathbf{V}_1 = -0.631/15.6°$ V, and $\mathbf{V}_2 = -1.13/{-23.9°}$ V.

From the original circuit shown in Fig. 12-24a, the voltage at node 3 is $2\underline{/65°}$ V more negative than the voltage at node 2.   So,

$$\mathbf{V}_3 = \mathbf{V}_2 - 2\underline{/65°} = -1.13\underline{/-23.9°} - 2\underline{/65°} = 2.31\underline{/-144.1°} = -2.31\underline{/35.9°} \text{ V}$$

12.20 Obtain the nodal equations for the circuit shown in Fig. 12-25 and find $\mathbf{V}_1$.

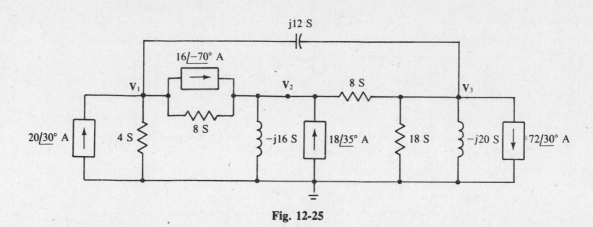

Fig. 12-25

The self-admittances are $4 + 8 + j12 = 12 + j12$ S   for node 1, $8 - j16 + 8 = 16 - j16$ S   for node 2, and $8 + 18 - j20 + j12 = 26 - j8$ S   for node 3.   The mutual admittances are 8 S for nodes 1 and 2, $j12$ S for nodes 1 and 3, and 8 S for nodes 2 and 3.   The currents flowing into the nodes from current sources are $20\underline{/30°} - 16\underline{/-70°} = 27.7\underline{/64.7°}$ A   for node 1, $16\underline{/-70°} + 18\underline{/35°} = 20.8\underline{/-13.1°}$ A for node 2, and $-72\underline{/30°}$ A for node 3.   So, the nodal equations are

$$
\begin{aligned}
(12 + j12)\mathbf{V}_1 - \quad\quad\quad 8\mathbf{V}_2 - \quad\quad j12\mathbf{V}_3 &= 27.7\underline{/64.7°} \\
-8\mathbf{V}_1 + (16 - j16)\mathbf{V}_2 - \quad\quad 8\mathbf{V}_3 &= 20.8\underline{/-13.1°} \\
-j12\mathbf{V}_1 - \quad\quad\quad 8\mathbf{V}_2 + (26 - j8)\mathbf{V}_3 &= -72\underline{/30°}
\end{aligned}
$$

Except for having $\mathbf{V}$'s instead of $\mathbf{I}$'s, this set of equations is the same as that for Prob. 12.12.   So, the answer for $\mathbf{V}_1$ here is the same as that for $\mathbf{I}_1$ of Prob. 12.12: $\mathbf{V}_1 = 2.07\underline{/26.6°}$ V.

12.21 Show a circuit corresponding to the nodal equations

$$
\begin{aligned}
(8 + j6)\mathbf{V}_1 - (3 - j4)\mathbf{V}_2 &= 4 + j2 \\
-(3 - j4)\mathbf{V}_1 + (11 - j6)\mathbf{V}_2 &= -6\underline{/-50°}
\end{aligned}
$$

Since there are two equations, the circuit has three nodes, one of which is the ground or reference node, and the others of which are nodes 1 and 2.   The circuit admittances can be found by starting with the mutual admittance.   From the $-(3 - j4)$ coefficients, nodes 1 and 2 have a mutual admittance of $3 - j4$ S, which can be from a resistor and inductor connected in parallel between nodes 1 and 2.   The $8 + j6$ coefficient of $\mathbf{V}_1$ in the first equation is the self-admittance of node 1.   Since $3 - j4$ S of this is in mutual admittance, there must be components connected between node 1 and ground that have a total of $8 + j6 - (3 - j4) = 5 + j10$ S   of admittance.   This can be from a resistor and parallel capacitor.   Similarly, from the second equation, components connected between node 2 and ground have a total admittance of $11 - j6 - (3 - j4) = 8 - j2$ S.   This can be from a resistor and parallel inductor.

The $4 + j2$ on the right-hand side of the first equation can be from a total current of $4 + j2 = 4.47\underline{/26.6°}$ A entering node 1 from current sources.   The easiest way to get this is with a single current source connected between node 1 and ground with the source arrow directed into node 1.   Similarly, from the second equation, the $-6\underline{/-50°}$ can be from a single $6\underline{/-50°}$-A current source connected between node 2 and ground with the source arrow directed *away* from node 2 because of the initial negative sign in $-6\underline{/-50°}$.

The resulting circuit is shown in Fig. 12-26.

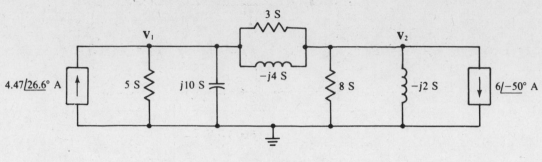

Fig. 12-26

12.22 For the circuit shown in Fig. 12-27, which contains a transistor model, first find **V** as a function of **I**.   Then, find **V** as a numerical value.

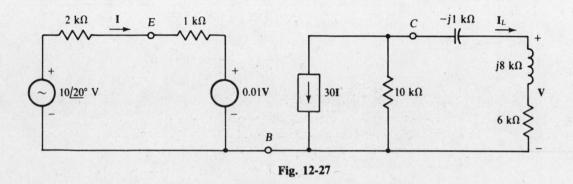

Fig. 12-27

In the right-hand section of the circuit, the current I_L is, by current division,

$$I_L = - \frac{10^4}{10\,000 + 6000 + j8000 - j1000} \times 30I = \frac{-3 \times 10^5 I}{17.46 \times 10^3 \underline{/23.6^\circ}} = -(17.2\underline{/-23.6^\circ})I$$

And, by Ohm's law,

$$\mathbf{V} = (6000 + j8000)\mathbf{I}_L = (10^4\underline{/53.1^\circ})(-17.2\underline{/-23.6^\circ})\mathbf{I} = (-17.2 \times 10^4\underline{/29.5^\circ})\mathbf{I}$$

which shows that the magnitude of **V** is 17.2×10^4 times that of **I**, and the angle of **V** is $29.5^\circ - 180^\circ = -150.5^\circ$ plus that of **I**.   (The -180° is from the negative sign.)

If this value of **V** is used in the 0.01**V** of the dependent source in the left-hand section of the circuit, and then KVL applied, the result is

$$2000\mathbf{I} + 1000\mathbf{I} + 0.01(-17.2 \times 10^4\underline{/29.5^\circ})\mathbf{I} = 10\underline{/20^\circ}$$

from which

$$\mathbf{I} = \frac{10\underline{/20^\circ}}{2000 + 1000 - 17.2 \times 10^2\underline{/29.5^\circ}} = \frac{10\underline{/20^\circ}}{1.73 \times 10^3\underline{/-29.3^\circ}} = 5.79 \times 10^{-3}\underline{/49.3^\circ} \text{ A}$$

This substituted into the equation for **V** gives

$$\mathbf{V} = (-17.2 \times 10^4\underline{/29.5^\circ})(5.79 \times 10^{-3}\underline{/49.3^\circ}) = -995\underline{/78.8^\circ} \text{ V}$$

12.23 Solve for **I** in the circuit shown in Fig. 12-28.

What analysis method is best for this circuit?   A brief consideration of the circuit shows that two equations are necessary whether mesh, loop, or nodal analysis is used.   Arbitrarily, nodal analysis will

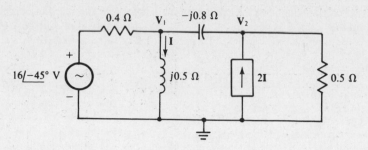

Fig. 12-28

be used to find V_1, and then I will be found from V_1.   For nodal analysis, the voltage source and series resistor are preferably converted to a current source with parallel resistor.   The current source has a current of $(16/-45°)/0.4 = 40/-45°$ A   directed into node 1, and the parallel resistor has a resistance of 0.4 Ω.

The self-admittances are

$$\frac{1}{0.4} + \frac{1}{j0.5} + \frac{1}{-j0.8} = 2.5 - j0.75 \text{ S}$$

for node 1, and

$$\frac{1}{0.5} + \frac{1}{-j0.8} = 2 + j1.25 \text{ S}$$

for node 2.   The mutual admittance is $1/(-j0.8) = j1.25$ S.

The controlling current I in terms of V_1 is $I = V_1/j0.5 = -j2V_1$,   which means that $2I = -j4V_1$ is the current into node 2 from the dependent current source.

From the admittances and the source currents, the nodal equations are

$$(2.5 - j0.75)V_1 - \qquad j1.25V_2 = 40/-45°$$
$$-j1.25V_1 + (2 + j1.25)V_2 = -j4V_1$$

which, with $j4V_1$ added to both sides of the second equation, simplify to

$$(2.5 - j0.75)V_1 - \qquad j1.25V_2 = 40/-45°$$
$$j2.75V_1 + (2 + j1.25)V_2 = 0$$

The lack of symmetry of the coefficients about the principal diagonal and the lack of an initial negative sign for the V_1 term in the second equation are caused by the dependent source.

By Cramer's rule,

$$V_1 = \frac{\begin{vmatrix} 40/-45° & -j1.25 \\ 0 & 2 + j1.25 \end{vmatrix}}{\begin{vmatrix} 2.5 - j0.75 & -j1.25 \\ j2.75 & 2 + j1.25 \end{vmatrix}} = \frac{94.3/-13°}{2.98/33°} = 31.64/-46° \text{ V}$$

Finally,

$$I = \frac{V_1}{j0.5} = \frac{31.64/-46°}{0.5/90°} = 63.3/-136° = -63.3/44° \text{ A}$$

Supplementary Problems

12.24   A 30-Ω resistor and a 0.1-H inductor are in series with a voltage source that produces a voltage of $120 \sin (377t + 10°)$ V.   Make a source conversion and find the components for the corresponding frequency-domain current-source circuit.

Ans. A $1.76/-41.5°$-A current source in parallel with a $48.2/51.5°$-Ω impedance

12.25 A $40/45°$-V voltage source is in series with a 6-Ω resistor and the parallel combination of a 10-Ω resistor and an inductor with a reactance of 8 Ω. Find the equivalent current-source circuit.
Ans. A $3.62/18.8°$-A current source and a parallel $11/26.2°$-Ω impedance

12.26 A $2/30°$-MV voltage source is in series with the parallel arrangement of an inductor that has a reactance of 100 Ω and a capacitor that has a reactance of -100 Ω. Find the current-source equivalent circuit.   *Ans.* An open circuit

12.27 Find the voltage-source circuit equivalent of the parallel arrangement of a $30.4/-24°$-mA current source, a 60-Ω resistor, and an inductor with an 80-Ω reactance.
Ans. A $1.46/12.9°$-V voltage source in series with a $48/36.9°$-Ω impedance

12.28 A $20.1/45°$-MA current source is in parallel with the series arrangement of an inductor that has a reactance of 100 Ω and a capacitor that has a reactance of -100Ω. Find the equivalent voltage-source circuit.   *Ans.* A short circuit

12.29 In the circuit shown in Fig. 12-29, find the currents I_1 and I_2. Then do a source conversion on the current source and parallel $4/30°$-Ω impedance and find the currents in the impedances. Compare.
Ans. $I_1 = 4.06/14.4°$ A, $I_2 = 3.25/84.4°$ A. After the conversion both are $3.25/84.4°$ A. So, the current does not remain the same in the $4/30°$-Ω impedance involved in the source conversion.

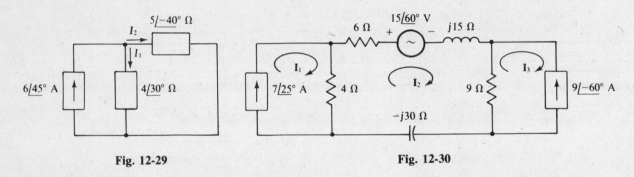

Fig. 12-29   **Fig. 12-30**

12.30 Find the mesh currents in the circuit shown in Fig. 12-30.
Ans. $I_1 = 7/25°$ A, $I_2 = -3/-33.6°$ A, $I_3 = -9/-60°$ A

12.31 Find **I** in the circuit shown in Fig. 12-31.   *Ans.* $3.86/-34.5°$ A

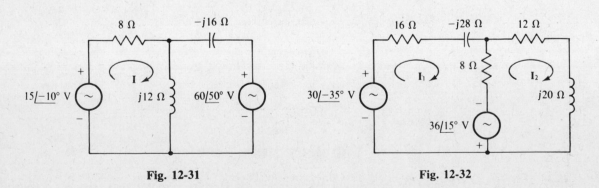

Fig. 12-31   **Fig. 12-32**

12.32 Find the mesh currents in the circuit shown in Fig. 12-32.
Ans. $I_1 = 1.46/46.5°$ A, $I_2 = -0.945/-43.2°$ A

12.33   Find the mesh currents for the circuit shown in Fig. 12-33.
   Ans. $I_1 = 1.26\underline{/10.6°}$ A, $I_2 = 4.63\underline{/30.9°}$ A, $I_3 = 2.25\underline{/-28.9°}$ A

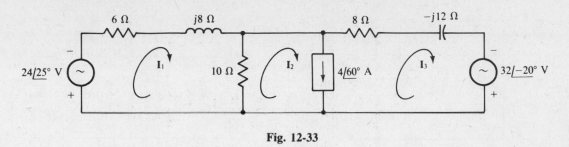

Fig. 12-33

12.34   Use loop analysis to solve for the current that flows down in the 10-Ω resistor in the circuit shown in Fig. 12-33.   *Ans.* $-3.47\underline{/38.1°}$ A

12.35   Use mesh analysis to find the current **I** for the circuit shown in Fig. 12-34.   *Ans.* $40.6\underline{/12.9°}$ A

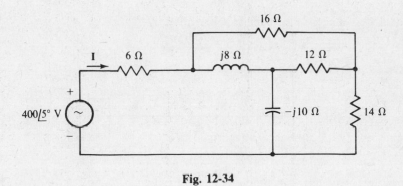

Fig. 12-34

12.36   Use loop analysis to find the current flowing down through the capacitor in the circuit shown in Fig. 12-34.   *Ans.* $36.1\underline{/29.9°}$ A

12.37   Find the current **I** for the circuit shown in Fig. 12-35.   *Ans.* $-13.1\underline{/-53.7°}$ A

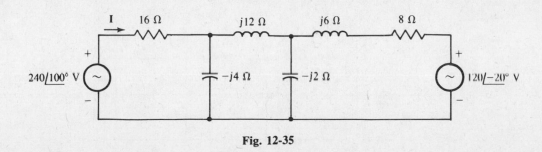

Fig. 12-35

12.38   For the circuit shown in Fig. 12-35, use loop analysis to find the current flowing down through the capacitor that has the reactance of $-j2$ Ω.   *Ans.* $28.5\underline{/-41.5°}$ A

12.39   Use loop analysis to find **I** in the circuit shown in Fig. 12-36.   *Ans.* $2.71\underline{/-55.8°}$ A

12.40 Rework Prob. 12.39 with all impedances doubled.   *Ans.* 1.36/−55.8° A

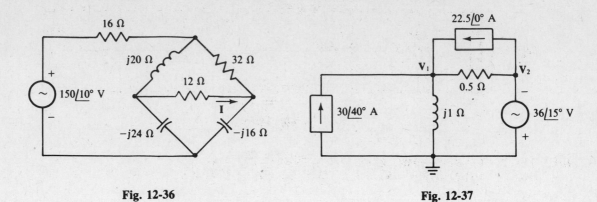

Fig. 12-36   **Fig. 12-37**

12.41 Find the node voltages in the circuit shown in Fig. 12-37.
Ans. $V_1 = -10.8/25°$ V, $V_2 = -36/15°$ V

12.42 Find the node voltages in the circuit shown in Fig. 12-38.
Ans. $V_1 = 1.17/−22.1°$ V, $V_2 = 0.675/−7.33°$ V

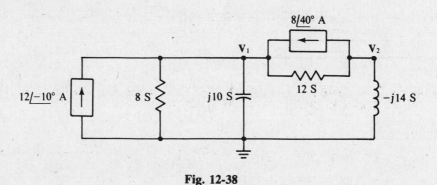

Fig. 12-38

12.43 Solve for the node voltages in the circuit shown in Fig. 12-39.
Ans. $V_1 = -51.9/−19.1°$ V, $V_2 = 58.7/73.9°$ V

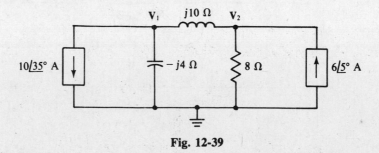

Fig. 12-39

12.44 Find the node voltages in the circuit shown in Fig. 12-40.
Ans. $V_1 = -1.26/20.6°$ V, $V_2 = -2.25/−18.9°$ V, $V_3 = -4.63/40.9°$ V

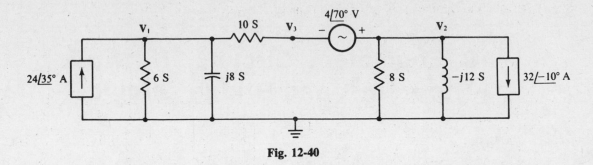

Fig. 12-40

12.45 Solve for the node voltages of the circuit shown in Fig. 12-41.
Ans. $V_1 = 1.75\underline{/50.9°}$ V, $V_2 = 2.47\underline{/-24.6°}$ V, $V_3 = 1.53\underline{/2.36°}$ V

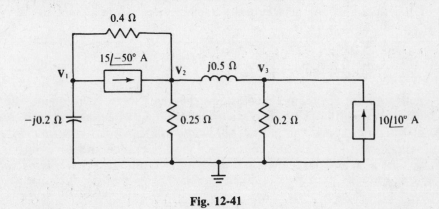

Fig. 12-41

12.46 For the circuit shown in Fig. 12-42, find **V** as a function of **I**, and then find **V** as a numerical value.
Ans. $V = (-6.87 \times 10^3\underline{/29.5°})I$, $V = -99.5\underline{/68.8°}$ V

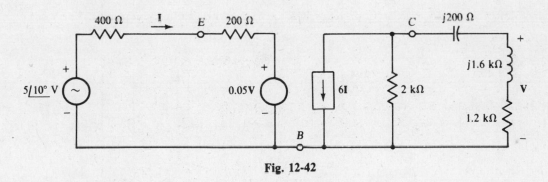

Fig. 12-42

12.47 Solve for **I** in the circuit shown in Fig. 12-43.   *Ans.* $-253\underline{/34°}$ A

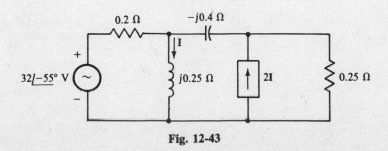

Fig. 12-43

Chapter 13

AC Equivalent Circuits, Network Theorems, and Bridge Circuits

INTRODUCTION

With two minor modifications, the dc network theorems discussed in Chap. 6 apply as well to ac frequency-domain circuits: The maximum power transfer theorem has to be modified slightly for circuits containing inductors or capacitors, and the same is true of the superposition theorem if the time-domain circuits have sources of different frequencies. Otherwise, though, the applications of the theorems for ac frequency-domain circuits are essentially the same as for dc circuits.

THEVENIN'S AND NORTON'S THEOREMS

In the application of Thevenin's or Norton's theorems to an ac frequency-domain circuit, the circuit is divided into two parts, A and B, with two joining wires, as shown in Fig. 13-1a. Then, for Thevenin's theorem applied to part A, the wires are separated at terminals a and b, and the open-circuit voltage $\mathbf{V}_{Th}$, the *Thevenin voltage*, is found referenced positive at terminal a, as shown in Fig. 13-1b. The next step, as shown in Fig. 13-1c, is to find *Thevenin's impedance* $\mathbf{Z}_{Th}$ of part A at terminals a and b.

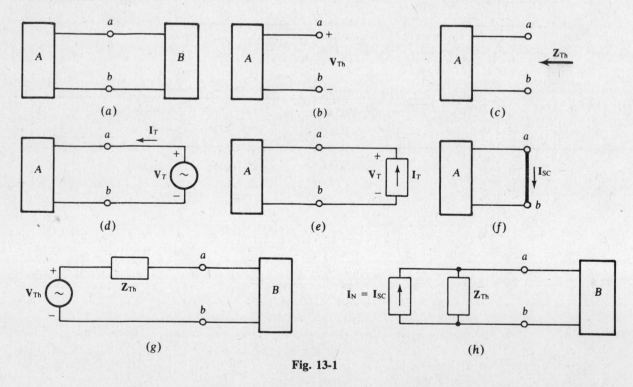

Fig. 13-1

There are three ways to find $\mathbf{Z}_{Th}$. In the most popular, the independent sources in part A are killed, but the dependent sources are not killed unless the controlling quantities are in part B, a situation which rarely occurs. Then, if the impedances in part A are in series and parallel, $\mathbf{Z}_{Th}$ is found by combining impedances and admittances—that is, by circuit reduction.

If the impedances of part A are not arranged series-parallel, it may not be convenient to use circuit reduction, or it may be impossible, especially if part A has dependent sources.   In this case, Z_{Th} can be found in a second way by applying a voltage source as shown in Fig. 13-1d or a current source as shown in Fig. 13-1e, and finding $Z_{Th} = V_T/I_T$.   Often, the most convenient source voltage is $V_T = 1\underline{/0°}$ V   and the most convenient source current is $I_T = 1\underline{/0°}$ A.

The third way to find Z_{Th} is to apply a short circuit across terminals a and b, as shown in Fig. 13-1f, then find the short-circuit current I_{SC}, and use it in $Z_{Th} = V_{Th}/I_{SC}$.   Of course, V_{Th} must also be known.   For this approach, part A must have independent sources, and they must not be killed.

In the circuit shown in Fig. 13-1g, the Thevenin equivalent produces the same voltages and currents in part B that the original part A does.   But only the part B voltages and currents remain the same; those in part A change, except at the a and b terminals.

For the Norton equivalent circuit shown in Fig. 13-1h, the Thevenin impedance is in parallel with a current source that produces a current *up* that is equal to the short-circuit current *down* in the circuit shown in Fig. 13-1f.   Of course, the Norton equivalent circuit also produces the same part B voltages and currents that the original part A does.

Because of the relation $V_{Th} = I_{SC}Z_{Th}$,   any two of the three quantities V_{Th}, I_{SC}, and Z_{Th} can be found from part A and then this equation used to find the third quantity if it is needed for the application of either Thevenin's or Norton's theorem.

MAXIMUM POWER TRANSFER THEOREM

The load that absorbs maximum average power from a circuit can be found from the Thevenin equivalent of this circuit at the load terminals.   The load should have a reactance that cancels the reactance of this Thevenin impedance because reactance does not absorb any average power but *does* limit the current.   Obviously, for maximum power transfer, there should be no reactance limiting the current flow to the resistance part of the load. This, in turn, means that the load and Thevenin reactances must be equal in magnitude but opposite in sign.

With the reactance cancellation, the overall circuit becomes essentially purely resistive.   As a result, the rule for maximum power transfer for the resistances is the same as that for a dc circuit: The load resistance must be equal to the resistance part of the Thevenin impedance.   Having the same resistance but a reactance that differs only in sign, *the load impedance for maximum power transfer is the conjugate of the Thevenin impedance of the circuit connected to the load*: $Z_L = Z_{Th}^*$.   Also, because the overall circuit is purely resistive, the maximum power absorbed by the load is the same as for a dc circuit: $V_{Th}^2/4R_{Th}$, in which V_{Th} is the rms value of the Thevenin voltage V_{Th} and R_{Th} is the resistance part of Z_{Th}.

SUPERPOSITION THEOREM

If, in an ac time-domain circuit, the independent sources operate at the *same* frequency, the superposition theorem for the corresponding frequency-domain circuit is the same as for a dc circuit.   That is, the desired voltage or current phasor contribution is found from each individual source or combination of sources, and then the various contributions are algebraically added to obtain the desired voltage or current phasor.   Independent sources not involved in a particular solution are killed, but dependent sources are left in the circuit.

For a circuit in which all sources have the same frequency, an analysis with the superposition theorem is usually more work than a standard mesh, loop, or nodal analysis with all sources present.   But the superposition theorem is essential if a time-domain circuit has inductors or capacitors and has sources operating at *different* frequencies.   Since the reactances depend on the radian frequency, the same frequency-domain circuit cannot be used for all sources if they do not have the same frequency.   There must be a different frequency-domain circuit for each different radian frequency, with the differences being in the reactances and in the killing of the various independent sources.   Preferably, all independent sources having the same radian frequency are considered at a time, while the other independent sources are killed.   This radian frequency is used to find the inductive and capacitive reactances for the corresponding frequency-domain circuit, and this circuit is analyzed to find the desired phasor.   Then, the phasor is transformed to a sinusoid.   This process is repeated for each different radian frequency of the sources.   Finally, the individual sinusoidal responses are added to get the total response.   Note that the adding is of the sinusoids and not of the phasors.   This is because phasors of different frequencies cannot be added.

AC Y-Δ and Δ-Y CONVERSIONS

Chapter 6 gives the Y-Δ and Δ-Y conversion formulas for resistances.   The only difference for impedances is in the use of Z's instead of R's.   Specifically, for the Δ-Y arrangement shown in Fig. 13-2, the Y-to-Δ conversion formulas are

$$Z_1 = \frac{Z_A Z_B + Z_A Z_C + Z_B Z_C}{Z_B} \qquad Z_2 = \frac{Z_A Z_B + Z_A Z_C + Z_B Z_C}{Z_C} \qquad Z_3 = \frac{Z_A Z_B + Z_A Z_C + Z_B Z_C}{Z_A}$$

and the Δ-to-Y conversion formulas are

$$Z_A = \frac{Z_1 Z_2}{Z_1 + Z_2 + Z_3} \qquad Z_B = \frac{Z_2 Z_3}{Z_1 + Z_2 + Z_3} \qquad Z_C = \frac{Z_1 Z_3}{Z_1 + Z_2 + Z_3}$$

The Y-to-Δ conversion formulas all have the same numerator, which is the sum of the different products of the pairs of the Y impedances.   Each denominator is the Y impedance shown in Fig. 13-2 that is opposite the impedance being found.   The Δ-to-Y conversion formulas, on the other hand, have the same denominator, which is the sum of the Δ impedances.   Each numerator is the product of the two Δ impedances shown in Fig. 13-2 that are adjacent to the Y impedance being found.

If all three Y impedances are the same Z_Y, the Y-to-Δ conversion formulas are the same: $Z_\Delta = 3Z_Y$.   And if all three Δ impedances are the same Z_Δ, the Δ-to-Y conversion formulas are the same: $Z_Y = Z_\Delta/3$.

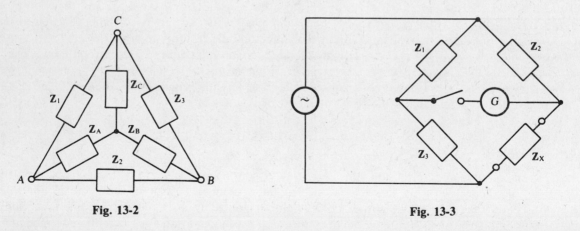

Fig. 13-2   Fig. 13-3

AC BRIDGE CIRCUITS

An ac bridge circuit, as shown in Fig. 13-3, can be used to measure inductance or capacitance in the same way that a Wheatstone bridge can be used to measure resistance, as explained in Chap. 6.   The bridge components, except for the unknown impedance Z_X, are typically just resistors and a capacitance standard—a capacitor the capacitance of which is known to great precision.   For a measurement, two of the resistors are varied until the galvanometer in the center arm reads zero when the switch is closed.   Then the bridge is balanced, and the unknown impedance Z_X can be found from the bridge balance equation $Z_X = Z_2 Z_3/Z_1$, which is the same as that for a Wheatstone bridge except for having Z's instead of R's.

Solved Problems

In those Thevenin and Norton equivalent circuit problems in which the equivalent circuits are not shown, the equivalent circuits are as shown in Fig. 13-1g and h with V_{Th} referenced positive at terminal a and $I_N = I_{SC}$ referenced toward the same terminal. Of course, the Thevenin impedance is in series with the Thevenin voltage source in the Thevenin equivalent circuit, and is in parallel with the Norton current source in the Norton equivalent circuit.

13.1 Find Z_{Th}, V_{Th}, and I_N for the Thevenin and Norton equivalents of the circuit external to the load impedance Z_L in the circuit shown in Fig. 13-4.

The Thevenin impedance Z_{Th} is the impedance at terminals a and b with the load impedance removed and the voltage source replaced by a short circuit. From combining impedances,

$$Z_{Th} = -j4 + \frac{6(j8)}{6 + j8} = -j4 + \frac{48\underline{/90°}}{10\underline{/53.13°}} = -j4 + 4.8\underline{/36.87°} = 4\underline{/-16.26°} \ \Omega$$

Although either V_{Th} or I_N can be found next, V_{Th} should be found because the $-j4$-Ω series branch makes I_N more difficult to find. With an open circuit at terminals a and b, this branch has zero current and so zero voltage. Consequently, V_{Th} is equal to the voltage drop across the $j8$-Ω impedance. By voltage division,

$$V_{Th} = \frac{j8}{6 + j8} \times 1\underline{/30°} = \frac{8\underline{/120°}}{10\underline{/53.13°}} = 0.8\underline{/66.87°} \ V$$

Finally,

$$I_N = \frac{V_{Th}}{Z_{Th}} = \frac{0.8\underline{/66.87°}}{4\underline{/-16.26°}} = 0.2\underline{/83.1°} \ A$$

13.2 If in the circuit shown in Fig. 13-4 the load is a resistor with resistance R, what value of R causes a 0.1-A rms current to flow through the load?

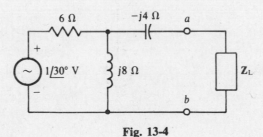

Fig. 13-4

As is evident from Fig. 13-1g, the load current is equal to the Thevenin voltage divided by the sum of the Thevenin and load impedances:

$$I_L = \frac{V_{Th}}{Z_{Th} + Z_L} \quad \text{from which} \quad Z_{Th} + Z_L = \frac{V_{Th}}{I_L}$$

Since only the rms load current is specified, angles are not known, which means that magnitudes must be used. Substituting $V_{Th} = 0.8$ V from the solution to Prob. 13.1,

$$|Z_{Th} + Z_L| = \frac{V_{Th}}{I_L} = \frac{0.8}{0.1} = 8 \ \Omega$$

Also from this solution, $Z_{Th} = 4\underline{/-16.26°} \ \Omega$. So,

$$|4\underline{/-16.26°} + R| = 8 \quad \text{or} \quad |3.84 - j1.12 + R| = 8$$

Because the magnitude of a complex number is equal to the square root of the sum of the squares of the real and imaginary parts,

$$\sqrt{(3.84 + R)^2 + (-1.12)^2} = 8$$

Squaring and simplifying,

$$R^2 + 7.68R + 16 = 64 \qquad \text{or} \qquad R^2 + 7.68R - 48 = 0$$

Applying the quadratic formula,

$$R = \frac{-7.68 \pm \sqrt{7.68^2 - 4(-48)}}{2} = \frac{-7.68 \pm 15.84}{2}$$

The positive sign must be used to get a physically significant positive resistance.   So,

$$R = \frac{-7.68 + 15.84}{2} = 4.08 \ \Omega$$

Note in the solution that the Thevenin and load impedances must be added before and not after the magnitudes are taken.   This is because $|\mathbf{Z}_{Th}| + |\mathbf{Z}_L| \neq |\mathbf{Z}_L + \mathbf{Z}_{Th}|$.

13.3   Find $\mathbf{Z}_{Th}$, $\mathbf{V}_{Th}$, and $\mathbf{I}_N$ for the Thevenin and Norton equivalents of the circuit shown in Fig. 13-5.

The Thevenin impedance $\mathbf{Z}_{Th}$ is the impedance at terminals a and b with the current source replaced by an open circuit.   By circuit reduction,

$$\mathbf{Z}_{Th} = 4\|[j2 + 3\|(-j4)] = \frac{4[j2 + 3(-j4)/(3 - j4)]}{4 + j2 + 3(-j4)/(3 - j4)}$$

Multiplying the numerator and denominator by $3 - j4$ gives

$$\mathbf{Z}_{Th} = \frac{4[j2(3 - j4) - j12]}{(4 + j2)(3 - j4) - j12} = \frac{40/\!-36.87°}{29.7/\!-47.73°} = 1.35/\!10.9° \ \Omega$$

The short-circuit current is easy to find because, if a short circuit is placed across terminals a and b, all the source current flows through this short circuit: $\mathbf{I}_{SC} = \mathbf{I}_N = 3/\!60°$ A.   None of the source current can flow through the impedances because the short circuit places zero voltage across them.   Finally,

$$\mathbf{V}_{Th} = \mathbf{I}_N\mathbf{Z}_{Th} = (3/\!60°)(1.35/\!10.9°) = 4.04/\!70.9° \ \text{V}$$

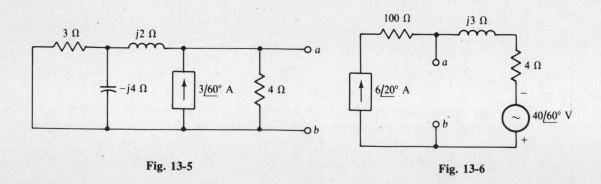

Fig. 13-5   Fig. 13-6

13.4   Find $\mathbf{Z}_{Th}$, $\mathbf{V}_{Th}$, and $\mathbf{I}_N$ for the Thevenin and Norton equivalents of the circuit shown in Fig. 13-6.

The Thevenin impedance $\mathbf{Z}_{Th}$ is the impedance at terminals a and b, with the current source replaced by an open circuit and the voltage source replaced by a short circuit.   The 100-Ω resistor is then in series with the open circuit that replaced the current source.   Consequently, this resistor has no effect on Z_{Th}.   The $j3$- and 4-Ω impedances are placed across terminals a and b by the short circuit that replaces the voltage source.   As a result, $Z_{Th} = 4 + j3 = 5\underline{/36.9°}\ \Omega$.

The short-circuit current $I_{SC} = I_N$ will be found and used to obtain V_{Th}.   If a short circuit is placed across terminals a and b, the current to the right through the $j3$-Ω impedance is

$$\frac{40\underline{/60°}}{4 + j3} = \frac{40\underline{/60°}}{5\underline{/36.9°}} = 8\underline{/23.1°}\ A$$

because the short circuit places all the $40\underline{/60°}$ V of the voltage source across the 4- and $j3$-Ω impedances.   Of course, the current to the right through the 100-Ω resistor is the $6\underline{/20°}$-A source current.   By KCL applied at terminal a, the short-circuit current is the difference between these currents:

$$I_{SC} = I_N = 6\underline{/20°} - 8\underline{/23.1°} = 2.04\underline{/-147.6°} = -2.04\underline{/32.4°}\ A$$

Finally,

$$V_{Th} = I_N Z_{Th} = (-2.04\underline{/32.4°})(5\underline{/36.9°}) = -10.2\underline{/69.3°}\ V$$

The negative signs for I_N and V_{Th} can, of course, be eliminated by reversing the references—that is, by having the Thevenin voltage source positive toward terminal b and the Norton current directed toward terminal b.

As a check, V_{Th} can be found from the open-circuit voltage across terminals a and b.   Because of this open circuit, all the $6\underline{/20°}$-A source current must flow through the 4- and $j3$-Ω impedances.   Consequently, from the right-hand half of the circuit, the voltage drop from terminal a to b is

$$V_{Th} = (6\underline{/20°})(4 + j3) - 40\underline{/60°} = 30\underline{/56.9°} - 40\underline{/60°} = 10.2\underline{/-110.7°} = -10.2\underline{/69.3°}\ V$$

which checks.

13.5   Find Z_{Th} and V_{Th} for the Thevenin equivalent of the circuit shown in Fig. 13-7.

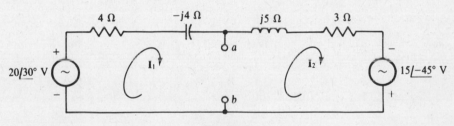

Fig. 13-7

The Thevenin impedance Z_{Th} can be easily found by replacing the voltage sources with short circuits and finding the impedances at terminals a and b.   Since the short circuit places the right- and left-hand halves of the circuit in parallel,

$$Z_{Th} = \frac{(4 - j4)(3 + j5)}{4 - j4 + 3 + j5} = \frac{32 + j8}{7 + j1} = \frac{32.98\underline{/14.04°}}{7.07\underline{/8.13°}} = 4.66\underline{/5.91°}\ \Omega$$

A brief inspection of the circuit shows that the short-circuit current is easier to find than the open-circuit voltage.   This current from terminal a to b is

$$I_{SC} = I_1 - I_2 = \frac{20\underline{/30°}}{4 - j4} - \frac{15\underline{/-45°}}{3 + j5} = 3.54\underline{/75°} - 2.57\underline{/-104°} = 6.11\underline{/75.4°}\ A$$

Last,

$$V_{Th} = I_{SC} Z_{Th} = (6.11\underline{/75.4°})(4.66\underline{/5.91°}) = 28.5\underline{/81.3°}\ V$$

13.6   Find Z_{Th} and V_{Th} for the Thevenin equivalent of the circuit shown in Fig. 13-8.

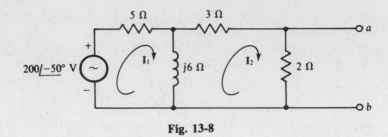

Fig. 13-8

If the voltage source is replaced by a short circuit, the impedance $\mathbf{Z}_{Th}$ at terminals a and b is, by circuit reduction,

$$\mathbf{Z}_{Th} = 2\|(3 + j6\|5) = \frac{2[3 + 5(j6)/(5 + j6)]}{2 + 3 + 5(j6)/(5 + j6)}$$

Multiplying the numerator and denominator by $5 + j6$, and simplifying, gives

$$\mathbf{Z}_{Th} = \frac{2[3(5 + j6) + j30]}{5(5 + j6) + j30} = \frac{30 + j96}{25 + j60} = \frac{100.6\underline{/72.65°}}{65\underline{/67.38°}} = 1.55\underline{/5.27°} \ \Omega$$

The Thevenin voltage can be found from $\mathbf{I}_2$, and $\mathbf{I}_2$ can be found from mesh analysis. The mesh equations are, from the self-impedance and mutual-impedance approaches,

$$(5 + j6)\mathbf{I}_1 - \qquad j6\mathbf{I}_2 = 200\underline{/-50°}$$
$$-j6\mathbf{I}_1 + (5 + j6)\mathbf{I}_2 = 0$$

which solve to

$$\mathbf{I}_2 = \frac{\begin{vmatrix} 5 + j6 & 200\underline{/-50°} \\ -j6 & 0 \end{vmatrix}}{\begin{vmatrix} 5 + j6 & -j6 \\ -j6 & 5 + j6 \end{vmatrix}} = \frac{-(-j6)(200\underline{/-50°})}{(5 + j6)^2 - (-j6)^2} = \frac{1200\underline{/40°}}{65\underline{/67.4°}} = 18.46\underline{/-27.4°} \ \text{A}$$

And
$$\mathbf{V}_{Th} = 2\mathbf{I}_2 = 2(18.46\underline{/-27.4°}) = 36.9\underline{/-27.4°} \ \text{V}$$

13.7   Find $\mathbf{Z}_{Th}$ and $\mathbf{I}_N$ for the Norton equivalent of the circuit shown in Fig. 13-9.

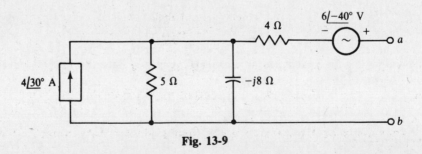

Fig. 13-9

When the current source is replaced by an open circuit and the voltage source is replaced by a short circuit, the impedance at terminals a and b is

$$\mathbf{Z}_{Th} = 4 + \frac{5(-j8)}{5 - j8} = \frac{20 - j72}{5 - j8} = \frac{74.7\underline{/-74.5°}}{9.43\underline{/-58°}} = 7.92\underline{/-16.5°} \ \Omega$$

Because of the series arm connected to terminal a and the voltage source in it, the Norton current is best found from the Thevenin voltage and impedance. The Thevenin voltage is equal to the voltage drop across the parallel components plus the voltage of the voltage source:

$$\mathbf{V}_{Th} = \frac{5(-j8)}{5 - j8} \times 4\underline{/30°} + 6\underline{/-40°} = \frac{160\underline{/-60°}}{9.43\underline{/-58°}} + 6\underline{/-40°} = 16.96\underline{/-2°} + 6\underline{/-40°} = 22\underline{/-11.67°} \text{ V}$$

And
$$\mathbf{I}_N = \frac{\mathbf{V}_{Th}}{\mathbf{Z}_{Th}} = \frac{22\underline{/-11.67°}}{7.92\underline{/-16.48°}} = 2.78\underline{/4.81°} \text{ A}$$

13.8 Find $\mathbf{Z}_{Th}$ and $\mathbf{V}_{Th}$ for the Thevenin equivalent of the circuit shown in Fig. 13-10.

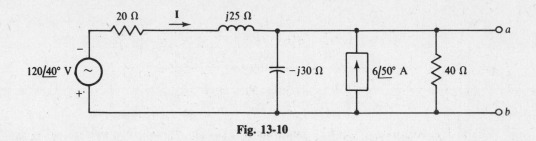

Fig. 13-10

When the voltage source is replaced by a short circuit and the current source by an open circuit, the admittance at terminals a and b is

$$\frac{1}{40} + \frac{1}{-j30} + \frac{1}{20 + j25} = 0.025 + j0.0333 + 0.0195 - j0.0244 = 0.0454\underline{/11.36°} \text{ S}$$

The inverse of this is $\mathbf{Z}_{Th}$:

$$\mathbf{Z}_{Th} = \frac{1}{0.0454\underline{/11.36°}} = 22\underline{/-11.36°} \text{ } \Omega$$

Because of the generally parallel configurations of the circuit, it may be better not to find $\mathbf{V}_{Th}$ directly, but rather to get $\mathbf{I}_N$ first and then find $\mathbf{V}_{Th}$ from $\mathbf{V}_{Th} = \mathbf{I}_N\mathbf{Z}_{Th}$. If a short circuit is placed across terminals a and b, the short-circuit current is $\mathbf{I} + 6\underline{/50°}$ since the short circuit prevents any current flow through the two parallel impedances. The current $\mathbf{I}$ can be found from the source voltage divided by the sum of the series impedances since the short circuit places this voltage across these impedances:

$$\mathbf{I} = -\frac{120\underline{/40°}}{20 + j25} = -\frac{120\underline{/40°}}{32\underline{/51.3°}} = -3.75\underline{/-11.3°} \text{ A}$$

And so
$$\mathbf{I}_N = \mathbf{I} + 6\underline{/50°} = -3.75\underline{/-11.3°} + 6\underline{/50°} = 5.34\underline{/88.05°} \text{ A}$$

Finally,
$$\mathbf{V}_{Th} = \mathbf{I}_N\mathbf{Z}_{Th} = (5.34\underline{/88.05°})(22\underline{/-11.36°}) = 118\underline{/76.7°} \text{ V}$$

13.9 Using Thevenin's or Norton's theorem, find $\mathbf{I}$ in the bridge circuit shown in Fig. 13-11 if $\mathbf{I}_S = 0$ A.

Since the current source produces 0 A, it is equivalent to an open circuit and can be removed from the circuit. Also, the 2- and $j3$-Ω impedances need to be removed in finding an equivalent circuit because these are the load impedances. With this done, $\mathbf{Z}_{Th}$ can be found after replacing the voltage source with a short circuit. This short circuit places the 3- and $j5$-Ω impedances in parallel and also the $-j4$- and 4-Ω impedances in parallel. Since these two parallel arrangements are in series between terminals a and b,

$$\mathbf{Z}_{Th} = 3\|j5 + 4\|(-j4) = \frac{3(j5)}{3 + j5} + \frac{4(-j4)}{4 - j4} = 2.572\underline{/30.96°} + 2.828\underline{/-45°} = 4.26\underline{/-9.14°} \text{ } \Omega$$

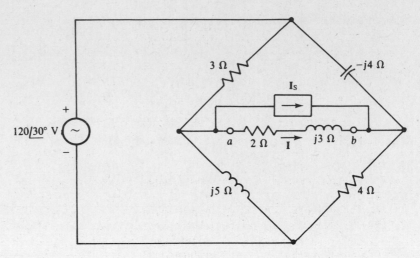

Fig. 13-11

The open-circuit voltage is easier to find than the short-circuit current. By KVL applied at the bottom half of the bridge, V_{Th} is equal to the difference in voltage drops across the $j5$- and 4-Ω impedances, which drops can be found by voltage division. Thus,

$$V_{Th} = \frac{j5}{3 + j5} \times 120\underline{/30°} - \frac{4}{4 - j4} \times 120\underline{/30°} = \frac{600\underline{/120°}}{5.83\underline{/59°}} - \frac{480\underline{/30°}}{5.66\underline{/-45°}} =$$

$$= 103\underline{/61°} - 84.9\underline{/75°} = 29.1\underline{/16°} \text{ V}$$

As should be evident from the Thevenin discussion and also from Fig. 13-1g, I is equal to the Thevenin voltage divided by the sum of the Thevenin and load impedances:

$$I = \frac{29.1\underline{/16°}}{4.26\underline{/-9.14°} + 2 + j3} = \frac{29.1\underline{/16°}}{6.63\underline{/20.5°}} = 4.39\underline{/-4.5°} \text{ A}$$

13.10 Find I for the circuit shown in Fig. 13-11 if $I_S = 10\underline{/-50°}$ A.

The current source does not, of course, affect Z_{Th}, which has the same value as found in the solution to Prob. 13.9: $Z_{Th} = 4.26\underline{/-9.14}$ Ω. The current source does, however, contribute to the Thevenin voltage. By superposition, it contributes a voltage equal to the source current times the impedance at terminals a and b with the load replaced by an open circuit. Since this impedance is Z_{Th}, the voltage contribution of the current source is $(10\underline{/-50°})(4.26\underline{/-9.14°}) = 42.6\underline{/-59.1°}$ V, which is a voltage drop from terminal b to a because the direction of the source current is into terminal b. Consequently, the Thevenin voltage is, by superposition, the Thevenin voltage obtained in the solution to Prob. 13.9 minus this voltage:

$$V_{Th} = 29.1\underline{/16°} - 42.6\underline{/-59.1°} = 45\underline{/82.1°} \text{ V}$$

and
$$I = \frac{V_{Th}}{Z_{Th} + Z_L} = \frac{45\underline{/82.1°}}{6.63\underline{/20.5°}} = = 6.79\underline{/61.6°} \text{ A}$$

13.11 Find the output impedance of the circuit to the left of terminals a and b for the circuit shown in Fig. 13-12.

The output impedance is the same as the Thevenin impedance. The only way of finding Z_{Th} is by applying a source and finding the ratio of the voltage and current at the source terminals. This impedance cannot be found from $Z_{Th} = V_{Th}/I_N$ because V_{Th} and I_N are both zero since there are no independent sources to the left of terminals a and b. And, of course, circuit reduction cannot be used because of the presence of the dependent source. The most convenient source to apply is a $1\underline{/0°}$-A

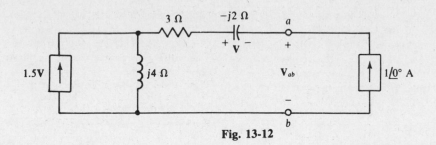

Fig. 13-12

current source with a current direction into terminal a, as shown in Fig. 13-12. Then, $\mathbf{Z}_{Th} = \mathbf{V}_{ab}/1\underline{/0^\circ} = \mathbf{V}_{ab}$.

The first step in calculating $\mathbf{Z}_{Th}$ is to find the control voltage $\mathbf{V}$. It is $\mathbf{V} = -(-j2)(1\underline{/0^\circ}) = j2$ V, with the initial negative sign occurring because the capacitor voltage and current references are not associated (the $1\underline{/0^\circ}$-A current is directed into the negative terminal of $\mathbf{V}$). The next step is to find the current flowing down through the $j4$-Ω impedance. This is the $1\underline{/0^\circ}$-A current from the independent current source plus the $1.5\mathbf{V} = 1.5(j2) = j3$-A current from the dependent current source, a total of $1 + j3$ A. With this current known, the voltage $\mathbf{V}_{ab}$ can be found from the sum of the voltage drops across the three impedances:

$$\mathbf{V}_{ab} = (1\underline{/0^\circ})(3 - j2) + (1 + j3)(j4) = 3 - j2 + j4 - 12 = -9 + j2 \text{ V}$$

which, as mentioned, means that $\mathbf{Z}_{Th} = -9 + j2$ Ω. The negative resistance of -9 Ω is caused by the dependent source. In polar form this impedance is

$$\mathbf{Z}_{Th} = -9 + j2 = 9.22\underline{/167.5^\circ} = -9.22\underline{/-12.5^\circ} \ \Omega$$

13.12 Find $\mathbf{Z}_{Th}$ and $\mathbf{I}_N$ for the Norton equivalent of the circuit shown in Fig. 13-13.

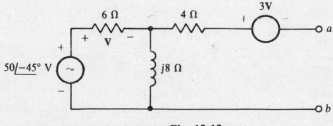

Fig. 13-13

Because of the series arm with dependent source connected to terminal a, $\mathbf{V}_{Th}$ is easier to find than $\mathbf{I}_N$. This voltage is equal to the sum of the voltage drops across the $j8$-Ω impedance and the $3\mathbf{V}$ dependent voltage source. (Of course, the 4-Ω resistor has a 0-V drop.) It is usually best to first solve for the controlling quantity, which here is the voltage $\mathbf{V}$ across the 6-Ω resistor. By voltage division,

$$\mathbf{V} = \frac{6}{6 + j8} \times 50\underline{/-45^\circ} = \frac{300\underline{/-45^\circ}}{10\underline{/53.1^\circ}} = 30\underline{/-98.1^\circ} \text{ V}$$

Also by voltage division, the voltage drop top to bottom across the $j8$-Ω impedance is

$$\frac{j8}{6 + j8} \times 50\underline{/-45^\circ} = \frac{400\underline{/45^\circ}}{10\underline{/53.1^\circ}} = 40\underline{/-8.1^\circ} \text{ V}$$

So $\qquad \mathbf{V}_{Th} = 40\underline{/-8.1^\circ} - 3(30\underline{/-98.1^\circ}) = 98.49\underline{/57.91^\circ} \text{ V}$

In this equation the dependent source voltage has an initial negative sign because it is a voltage rise from terminal a to b instead of a voltage drop.

The Thevenin impedance can be found by applying a $1/\underline{0°}$-A current source at terminals a and b, as shown in the circuit in Fig. 13-14, and finding the voltage $\mathbf{V}_{ab}$. Then, $\mathbf{Z}_{Th} = \mathbf{V}_{ab}/1/\underline{0°} = \mathbf{V}_{ab}$. The control voltage $\mathbf{V}$ must be found first, as to be expected. It has a different value than in the $\mathbf{V}_{Th}$ calculation because the circuit is different. The voltage $\mathbf{V}$ can be found from the current $\mathbf{I}$ flowing through the 6-Ω resistor across which $\mathbf{V}$ is taken. Since the 6- and j8-Ω impedances are in parallel, and since $1/\underline{0°}$ A from the current source flows into this parallel arrangement, $\mathbf{I}$ is, by current division,

$$\mathbf{I} = \frac{j8}{6 + j8} \times 1/\underline{0°} = \frac{8/\underline{90°}}{10/\underline{53.1°}} = 0.8/\underline{36.9°} \text{ A}$$

And, by Ohm's law,

$$\mathbf{V} = -6\mathbf{I} = -6(0.8/\underline{36.9°}) = -4.8/\underline{36.9°} \text{ V}$$

The negative sign occurs because the $\mathbf{V}$ and $\mathbf{I}$ references are not associated.

With $\mathbf{V}$ known, $\mathbf{V}_{ab}$ can be found by summing the voltage drops from terminal a to terminal b:

$$\mathbf{V}_{ab} = -3(-4.8/\underline{36.9°}) + (1/\underline{0°})(4) - (-4.8/\underline{36.9°}) = 22.53/\underline{30.75°} \text{ V}$$

from which $\mathbf{Z}_{Th} = 22.53/\underline{30.75°} \ \Omega$.

Finally,
$$\mathbf{I}_N = \frac{\mathbf{V}_{Th}}{\mathbf{Z}_{Th}} = \frac{98.49/\underline{57.91°}}{22.53/\underline{30.75°}} = 4.37/\underline{27.2°} \text{ A}$$

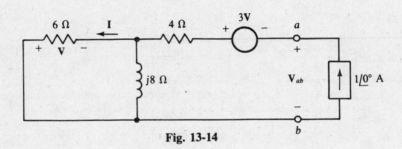

Fig. 13-14

13.13 Find $\mathbf{Z}_{Th}$ and $\mathbf{I}_N$ for the Norton equivalent of the transistor circuit shown in Fig. 13-15.

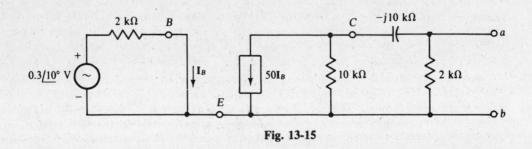

Fig. 13-15

The Thevenin impedance $\mathbf{Z}_{Th}$ can be found directly by replacing the independent voltage source by a short circuit. Since with this replacement there is no source of voltage in the base circuit, $\mathbf{I}_B = 0$ A and so the $50\mathbf{I}_B$ of the dependent current source is also 0 A. And this means that this dependent source is equivalent to an open circuit. Notice that the dependent source was not killed, as an independent source would be. Instead, it is equivalent to an open circuit because its control current is 0 A. With this current source replaced by an open circuit, $\mathbf{Z}_{Th}$ can be found by combining impedances:

$$\mathbf{Z}_{Th} = \frac{2000(10\,000 - j10\,000)}{2000 + 10\,000 - j10\,000} = \frac{28.3 \times 10^6/\underline{-45°}}{15.6 \times 10^3/\underline{-39.81°}} \ \Omega = 1.81/\underline{-5.19°} \text{ k}\Omega$$

The current $\mathbf{I}_N$ can be found from the current flowing through a short circuit placed across terminals a and b. Because this short circuit places the 10- and $-j10$-kΩ impedances in parallel, and since $\mathbf{I}_N$ is the current through the $-j10$-Ω impedance, by current division $\mathbf{I}_N$ is

$$\mathbf{I}_N = -\frac{10\,000}{10\,000 - j10\,000} \times 50\mathbf{I}_B = \frac{-50\mathbf{I}_B}{\sqrt{2}\underline{/-45°}}$$

The initial negative sign is necessary because both $50\mathbf{I}_B$ and $\mathbf{I}_N$ have directions into terminal b. The 2-kΩ resistance across terminals a and b does not appear because it is in parallel with the short circuit.

From the base circuit,

$$\mathbf{I}_B = \frac{0.3\underline{/10°}}{2000} \text{ A} = 0.15\underline{/10°} \text{ mA}$$

Finally, $$\mathbf{I}_N = \frac{-50(0.15\underline{/10°})}{\sqrt{2}\underline{/-45°}} = -5.3\underline{/55°} \text{ mA}$$

13.14 In the circuit shown in Fig. 13-16, v_{in} is the sum of an infinite number of sinusoidal voltages each of which has the same peak value but a different frequency. Find the frequency at which the peak of v_{out} is $1/\sqrt{2}$ of its maximum possible value at any frequency.

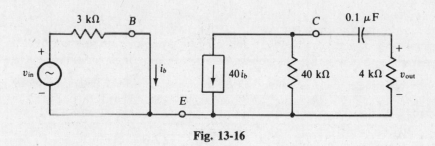

Fig. 13-16

Of course, the presence of the capacitor makes it necessary to use a frequency-domain circuit that has a phasor $\mathbf{V}_{out}$ instead of the voltage v_{out}. Fortunately, $\mathbf{V}_{out}$ can be used for the analysis because its magnitude is proportional to the peak of v_{out}.

A good approach is to consider the 0.1-μF capacitor and the 4-kΩ resistor as the load, find the Thevenin equivalent of the remainder of the circuit, and then use voltage division to obtain $\mathbf{V}_{out}$. For the Thevenin equivalent, the open-circuit voltage across the 40-kΩ resistor could be found at any frequency, but the value of this voltage is not important because it is the same for each frequency as a result of the identical peak values and also the lack of reactive elements in this part of the circuit. So, this voltage will just be called $\mathbf{V}_{Th}$. The Thevenin impedance can be found by replacing the independent voltage source by a short circuit. This makes $i_b = 0$ A and also $40i_b = 0$ A, which means that the dependent current source acts like an open circuit. Consequently, $\mathbf{Z}_{Th} = 40$ kΩ.

Figure 13-17 shows the Thevenin equivalent circuit with the load, in which $-j1/\omega C = -j10^7/\omega$. The radian frequency ω is a variable since it can take on any real positive value. By

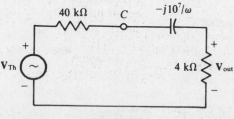

Fig. 13-17

voltage division,

$$\mathbf{V}_{out} = \frac{4000}{40\,000 + 4000 - j10^7/\omega} \times \mathbf{V}_{Th} = \frac{4000}{44\,000 - j10^7/\omega} \times \mathbf{V}_{Th}$$

From this equation it should be apparent that $|\mathbf{V}_{out}| = V_{out}$ has a maximum value for the smallest possible denominator, which occurs for ω approaching infinity because $-j10^7/\omega \to 0$ as $\omega \to \infty$. This maximum value is

$$V_{out(max)} = \frac{4000}{44\,000} \times V_{Th} = \frac{V_{Th}}{11}$$

At the frequency at which V_{out} is $1/\sqrt{2}$ of this maximum value, the denominator of $\mathbf{V}_{out}$ must have a magnitude that is $\sqrt{2}$ times the minimum denominator. Specifically,

$$|44\,000 - j10^7/\omega| = \sqrt{2}(44\,000)$$

which, from complex algebra considerations, means that $10^7/\omega = 44\,000$. So,

$$\omega = \frac{10^7}{44\,000} = 227 \text{ rad/s} \quad \text{and} \quad f = \frac{227}{2\pi} = 36 \text{ Hz}$$

For lower frequencies, $\mathbf{V}_{out}$ has less magnitude, and in fact the lower the frequency the less the magnitude. And for 0 Hz (dc), $|\mathbf{V}_{out}|$ is zero.

13.15 What is the maximum average power that can be drawn from an ac generator that has an internal impedance of $150\underline{/60°}\ \Omega$ and an rms open-circuit voltage of 12.5 kV? Do not be concerned about whether the generator power rating may be exceeded.

The maximum average power will be absorbed by a load that is the conjugate of the internal impedance, which is also the Thevenin impedance. The formula for this power is $P_{max} = V_{Th}^2/4R_{Th}$. Here, $V_{Th} = 12.5$ kV and $R_{Th} = 150 \cos 60° = 75\ \Omega$. So,

$$P_{max} = \frac{(12.5 \times 10^3)^2}{4(75)} \text{ W} = 521 \text{ kW}$$

13.16 A signal generator operating at 2 MHz has an rms open-circuit voltage of 0.5 V and an internal impedance of $50\underline{/30°}\ \Omega$. If it energizes a capacitor and parallel resistor, find the capacitance and resistance of these components for maximum average power absorption by this resistor. Also, find this power.

The load that absorbs maximum average power has an impedance $\mathbf{Z}_L$ that is the conjugate of the internal impedance of the generator. So, $\mathbf{Z}_L = 50\underline{/-30°}\ \Omega$ since the conjugate has the same magnitude and an angle that differs only in sign. Being in parallel, the load resistor and capacitor can best be determined from the load admittance, which is

$$\mathbf{Y}_L = \frac{1}{\mathbf{Z}_L} = \frac{1}{50\underline{/-30°}} = 0.02\underline{/30°} \text{ S} = 17.3 + j10 \text{ mS}$$

But $\mathbf{Y}_L = G + j\omega C$ in which $\omega = 2\pi f = 2\pi(2 \times 10^6)$ rad/s $= 12.6$ Mrad/s

So $G = \dfrac{1}{R} = 17.3$ mS   from which $R = \dfrac{1}{17.3 \times 10^{-3}} = 57.7\ \Omega$

and $j\omega C = j(12.6 \times 10^6)C = j10 \times 10^{-3}$ S   from which $C = \dfrac{10 \times 10^{-3}}{12.6 \times 10^6}$ F $= 796$ pF

The maximum average power absorbed by the 57.7-Ω resistor can be found from $P_{max} = V_{Th}^2/4R_{Th}$ in which R_{Th} is the resistance of $50\underline{/30°} = 43.3 + j25\ \Omega$:

$$P_{max} = \frac{0.5^2}{4(43.3)} \text{ W} = 1.44 \text{ mW}$$

Of course, 43.3 Ω is used instead of the 57.7 Ω of the load resistor because 43.3 Ω is the Thevenin resistance of the source as well as the resistance of the impedance of the parallel resistor-capacitor load.

13.17 For the circuit shown in Fig. 13-18, what load impedance $\mathbf{Z}_L$ absorbs maximum average power, and what is this power?

The Thevenin equivalent of the source circuit at the load terminals is needed. By voltage division,

$$\mathbf{V}_{\text{Th}} = \frac{4 + j2 - j8}{4 + j2 - j8 + 3 + j8} \times 240\underline{/30°} = \frac{1731\underline{/-26.31°}}{7.28\underline{/15.95°}} = 237.7\underline{/-42.3°} \text{ V}$$

The Thevenin impedance is

$$\mathbf{Z}_{\text{Th}} = \frac{(3 + j8)(4 + j2 - j8)}{3 + j8 + 4 + j2 - j8} = \frac{60 + j14}{7 + j2} = \frac{61.6\underline{/13.134°}}{7.28\underline{/15.945°}} = 8.46\underline{/-2.81°} \text{ Ω}$$

For maximum average power absorption, $\mathbf{Z}_L = \mathbf{Z}_{\text{Th}}^* = 8.46\underline{/2.81°}$ Ω, the resistive part of which is $R_{\text{Th}} = 8.46 \cos 2.81° = 8.45$ Ω.   Finally, the maximum average power absorbed is

$$P_{\text{max}} = \frac{V_{\text{Th}}^2}{4R_{\text{Th}}} = \frac{237.7^2}{4(8.45)} \text{ W} = 1.67 \text{ kW}$$

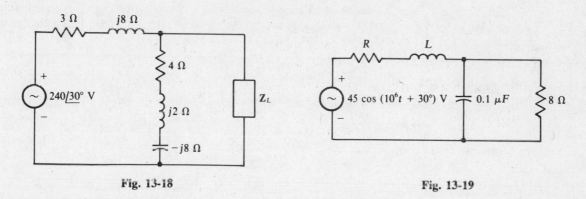

Fig. 13-18   Fig. 13-19

13.18 In the circuit shown in Fig. 13-19, find R and L for maximum average power absorption by the parallel resistor and capacitor load, and also find this power.

A good first step is to find the load impedance.   Since the impedance of the capacitor is

$$jX_C = \frac{-j1}{\omega C} = \frac{-j1}{10^6(0.1 \times 10^{-6})} = -j10 \text{ Ω}$$

the impedance of the load is

$$\mathbf{Z}_L = \frac{8(-j10)}{8 - j10} = \frac{80\underline{/-90°}}{12.8\underline{/-51.3°}} = 6.25\underline{/-38.7°} = 4.88 - j3.9 \text{ Ω}$$

Since for maximum average power absorption there should be no reactance limiting the current to the resistive part of the load, the inductance L should be selected such that its inductive reactance cancels the capacitive reactance of the load. So, $\omega L = 3.9$ Ω, from which $L = 3.9/10^6$ H $= 3.9$ μH. With the cancellation of the reactances, the circuit is essentially the voltage source, the resistance R, and the 4.88 Ω of the load, all in series. As should be apparent, for maximum average power absorption by the 4.88 Ω of the load, the source resistance should be zero: $R = 0$ Ω. Then, all the source voltage is across the 4.88 Ω and the power absorbed is

$$P_{\text{max}} = \frac{(45/\sqrt{2})^2}{4.88} = 208 \text{ W}$$

Notice that the source impedance is not the conjugate of the load impedance.   The reason is that here the load resistance is fixed while the source resistance is a variable.   The conjugate result occurs in the much more common situation in which the load impedance can be varied but the source impedance is fixed.

13.19 Use superposition to find **V** in the circuit shown in Fig. 13-20.

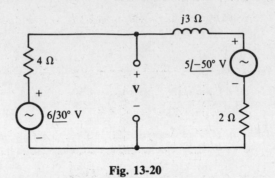

Fig. 13-20

The voltage **V** can be considered to have a component **V′** from the $6\underline{/30°}$-V source and another component **V″** from the $5\underline{/-50°}$-V source such that   **V = V′ + V″**.   The component **V′** can be found by replacing the $5\underline{/-50°}$-V source with a short circuit and using voltage division:

$$\mathbf{V'} = \frac{2 + j3}{2 + j3 + 4} \times 6\underline{/30°} = \frac{21.6\underline{/86.3°}}{6.71\underline{/26.6°}} = 3.22\underline{/59.7°} \text{ V}$$

Similarly, **V″** can be found by replacing the $6\underline{/30°}$-V source with a short circuit and using voltage division:

$$\mathbf{V''} = \frac{4}{2 + j3 + 4} \times 5\underline{/-50°} = \frac{20\underline{/-50°}}{6.71\underline{/26.6°}} = 2.98\underline{/-76.6°} \text{ V}$$

Adding

$$\mathbf{V} = \mathbf{V'} + \mathbf{V''} = 3.22\underline{/59.7°} + 2.98\underline{/-76.6°} = 2.32\underline{/-2.82°} \text{ V}$$

13.20 Use superposition to find i in the circuit shown in Fig. 13-21.

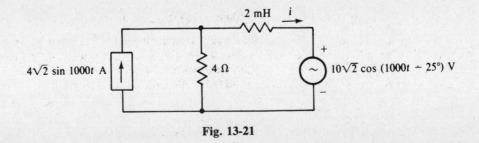

Fig. 13-21

It is necessary to construct the corresponding frequency-domain circuit, as shown in Fig. 13-22.   The current **I** can be considered to have a component **I′** from the current source and a component **I″** from the voltage source such that   **I = I′ + I″**.   The component **I′** can be found by replacing the voltage source with a short circuit and using current division:

$$\mathbf{I'} = \frac{4}{4 + j2} \times 4\underline{/0°} = \frac{16\underline{/0°}}{4.47\underline{/26.6°}} = 3.58\underline{/-26.6°} \text{ A}$$

And $\mathbf{I}''$ can be found by replacing the current source with an open circuit and using Ohm's law:

$$\mathbf{I}'' = -\frac{10\underline{/65°}}{4 + j2} = \frac{-10\underline{/65°}}{4.47\underline{/26.6°}} = -2.24\underline{/38.4°}\ \text{A}$$

The negative sign is necessary because the voltage and current references are not associated.   Adding,

$$\mathbf{I} = \mathbf{I}' + \mathbf{I}'' = 3.58\underline{/-26.6°} - 2.24\underline{/38.4°} = 3.32\underline{/-64.2°}\ \text{A}$$

Finally, the corresponding sinusoidal current is

$$i = \sqrt{2}(3.32)\sin(1000t - 64.2°) = 4.7\sin(1000t - 64.2°)\ \text{A}$$

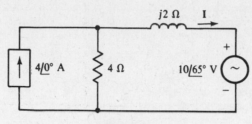

Fig. 13-22

13.21 Use superposition to find i for the circuit shown in Fig. 13-21 if the voltage of the voltage source is changed to $10\sqrt{2}\cos(2000t - 25°)$ V.

The current i can be considered to have a component i' from the current source and a component i'' from the voltage source.   Because these two sources have different frequencies, two different frequency-domain circuits are necessary.   The frequency-domain circuit for the current source is the same as that shown in Fig. 13-22, but with the voltage source replaced by a short circuit.   As a result, the current phasor $\mathbf{I}'$ is the same as that found in the solution to Prob. 13.21: $\mathbf{I}' = 3.58\underline{/26.6°}$ A.   The corresponding current is

$$i = \sqrt{2}(3.58)\sin(1000t - 26.6°) = 5.06\sin(1000t - 26.6°)\ \text{A}$$

The frequency-domain circuit for the voltage source and $\omega = 2000$ rad/s is shown in Fig. 13-23. By Ohm's law,

$$\mathbf{I}'' = -\frac{10\underline{/65°}}{4 + j4} = -1.77\underline{/20°}\ \text{A}$$

from which $i'' = \sqrt{2}(-1.77)\sin(2000t + 20°) = -2.5\sin(2000t + 20°)\ \text{A}$

Finally,

$$i = i' + i'' = 5.06\sin(1000t - 26.6°) - 2.5\sin(2000t + 20°)\ \text{A}$$

Notice in this solution that the phasors $\mathbf{I}'$ and $\mathbf{I}''$ cannot be added, as they could be in the solution to Prob. 13.20.   The reason is that here the phasors are for different frequencies while in the solution to Prob. 13.20 they are for the same frequency.   When the phasors are for different frequencies, the corresponding sinusoids must be found first, and then these added.   Also, the sinuoids cannot be combined into a single term.

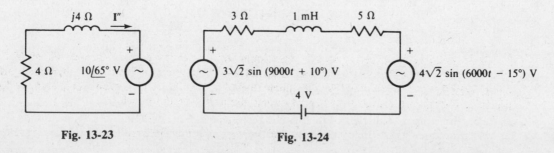

Fig. 13-23   **Fig. 13-24**

13.22 Although superposition does not apply to power calculations in general, it does apply to the calculation of *average* power absorbed when all sources are sinusoids of *different* frequencies.   Use this fact to find the average power absorbed by the 5-Ω resistor in the circuit shown in Fig. 13-24.

Consider first the dc component of average power absorbed by the 5-Ω resistor.   Of course, for this calculation the ac voltage sources are replaced by short circuits.   Also, the inductor is replaced by a short circuit because an inductor is a short circuit to dc.   So,

$$I_{dc} = \frac{4}{3 + 5} = 0.5 \text{ A}$$

This 0.5-A current produces a power dissipation in the 5-Ω resistor of $P_{dc} = 0.5^2(5) = 1.25 \text{ W}$.
The rms current from the 6000-rad/s voltage source is, by superposition,

$$I_{6000} = \frac{|4\underline{/-15°}|}{|3 + j6 + 5|} = \frac{4}{10} = 0.4 \text{ A}$$

It produces a power dissipation of $P_{6000} = 0.4^2(5) = 0.8 \text{ W}$ in the 5-Ω resistor.   And the rms current from the 9000-rad/s voltage source is

$$I_{9000} = \frac{|3\underline{/10°}|}{|3 + j9 + 5|} = \frac{3}{12.04} = 0.249 \text{ A}$$

It produces a power dissipation of $P_{9000} = 0.249^2(5) = 0.31 \text{ W}$ in the 5-Ω resistor.
The total average power P_{av} absorbed is the sum of these powers:

$$P_{av} = P_{dc} + P_{6000} + P_{9000} = 1.25 + 0.8 + 0.31 = 2.36 \text{ W}$$

13.23 Use superposition to find **V** in the circuit shown in Fig. 13-25.

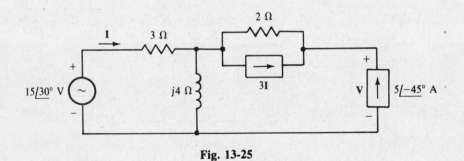

Fig. 13-25

If the independent current source is replaced by an open circuit, the circuit is as shown in Fig. 13-26, in which **V'** is the component of **V** from the voltage source.   Because of the open-circuited terminals, no part of **I** can flow through the 2-Ω resistor and the 3**I** dependent current source.   Instead, all of **I** flows through the j4-Ω impedance as well as through the 3-Ω resistance.   Thus,

$$\mathbf{I} = \frac{15\underline{/30°}}{3 + j4} = \frac{15\underline{/30°}}{5\underline{/53.1°}} = 3\underline{/-23.1°} \text{ A}$$

With this **I** known, **V'** can be found from the voltage drops across the 2- and j4-Ω impedances:

$$\mathbf{V'} = \mathbf{V_1} + \mathbf{V_2} = 2(3\mathbf{I}) + j4\mathbf{I} = (6 + j4)(3\underline{/-23.1°}) = (7.21\underline{/33.7°})(3\underline{/-23.1°}) = 21.6\underline{/10.6°} \text{ V}$$

If the voltage source in the circuit of Fig. 13-25 is replaced by a short circuit, the circuit is as shown in Fig. 13-27, where **V''** is the component of **V** from the independent current source.   As a reminder, the current to the left of the parallel resistor and dependent-source combination is shown as $5\underline{/-45°}$ A, the same as the independent source current, as it must be.   Because this current flows into

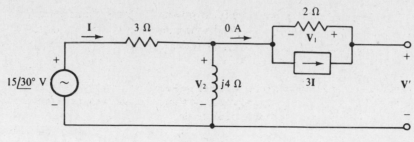

Fig. 13-26

the parallel 3- and $j4$-Ω combination, $\mathbf{I}$ in the 3-Ω resistance can be found by current division:

$$\mathbf{I} = -\frac{j4}{3 + j4} \times 5\underline{/-45°} = \frac{20\underline{/-135°}}{5\underline{/53.1°}} = 4\underline{/-188.1°} \text{ A}$$

With $\mathbf{I}$ known, $\mathbf{V}''$ can be found from the voltage drops $\mathbf{V}_1$ and $\mathbf{V}_2$ across the 2-Ω resistance and the parallel 3- and $j4$-Ω impedances.   Since the 2-Ω resistance current is $3\mathbf{I} + 5\underline{/-45°}$,

$$\mathbf{V}_1 = [3(4\underline{/-188.1°}) + 5\underline{/-45°}](2) = -17.1\underline{/12.4°} \text{ V}$$

Also, since the current in the 3- and $j4$-Ω parallel combination is $5\underline{/-45°}$ A,

$$\mathbf{V}_2 = \frac{3(j4)}{3 + j4} \times 5\underline{/-45°} = \frac{60\underline{/45°}}{5\underline{/53.1°}} = 12\underline{/-8.1°} \text{ V}$$

So
$$\mathbf{V}'' = \mathbf{V}_1 + \mathbf{V}_2 = -17.1\underline{/12.4°} + 12\underline{/-8.1°} = 7.21\underline{/-132°} \text{ V}$$

Finally, by superposition,

$$\mathbf{V} = \mathbf{V}' + \mathbf{V}'' = 21.6\underline{/10.6°} + 7.21\underline{/-132°} = 16.5\underline{/-4.89°} \text{ V}$$

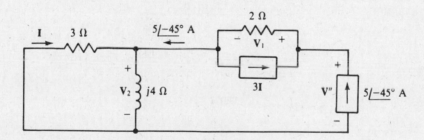

Fig. 13-27

The main purpose of this problem is to illustrate the fact that dependent sources are not killed in using superposition.   Actually, using superposition on the circuit shown in Fig. 13-25 requires much more work than using loop or nodal analysis.

13.24 Convert the Δ shown in Fig. 13-28a to the Y in Fig. 13-28b for (a) $\mathbf{Z}_1 = \mathbf{Z}_2 = \mathbf{Z}_3 = 12\underline{/36°}$ Ω,   and for (b) $\mathbf{Z}_1 = 3 + j5$ Ω, $\mathbf{Z}_2 = 6\underline{/20°}$ Ω,   and $\mathbf{Z}_3 = 4\underline{/-30°}$ Ω.

(a) Because all three Δ impedances are the same, all three Y impedances are the same and each is equal to one-third of the common Δ impedance:

$$\mathbf{Z}_A = \mathbf{Z}_B = \mathbf{Z}_C = \frac{12\underline{/36°}}{3} = 4\underline{/36°} \ \Omega$$

(b) All three Δ-to-Y conversion formulas have the same denominator, which is

$$\mathbf{Z}_1 + \mathbf{Z}_2 + \mathbf{Z}_3 = (3 + j5) + 6\underline{/20°} + 4\underline{/-30°} = 13.1\underline{/22.66°} \ \Omega$$

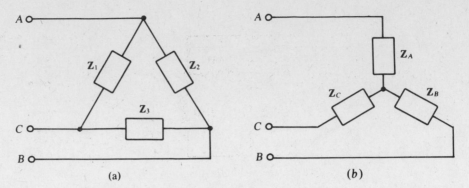

Fig. 13-28

By these formulas,

$$\mathbf{Z}_A = \frac{\mathbf{Z}_1\mathbf{Z}_2}{\mathbf{Z}_1 + \mathbf{Z}_2 + \mathbf{Z}_3} = \frac{(3 + j5)(6\underline{/20°})}{13.1\underline{/22.66°}} = \frac{35\underline{/79.04°}}{13.1\underline{/22.66°}} = 2.67\underline{/56.4°}\ \Omega$$

$$\mathbf{Z}_B = \frac{\mathbf{Z}_2\mathbf{Z}_3}{\mathbf{Z}_1 + \mathbf{Z}_2 + \mathbf{Z}_3} = \frac{(6\underline{/20°})(4\underline{/-30°})}{13.1\underline{/22.66°}} = \frac{24\underline{/-10°}}{13.1\underline{/22.66°}} = 1.83\underline{/-32.7°}\ \Omega$$

$$\mathbf{Z}_C = \frac{\mathbf{Z}_1\mathbf{Z}_3}{\mathbf{Z}_1 + \mathbf{Z}_2 + \mathbf{Z}_3} = \frac{(3 + j5)(4\underline{/-30°})}{13.1\underline{/22.66°}} = \frac{23.3\underline{/29.04°}}{13.1\underline{/22.66°}} = 1.78\underline{/6.38°}\ \Omega$$

13.25 Convert the Y shown in Fig. 13-28b to the Δ in Fig. 13-28a for (a) $\mathbf{Z}_A = \mathbf{Z}_B = \mathbf{Z}_C = 4 - j7\ \Omega$, and for (b) $\mathbf{Z}_A = 10\ \Omega$, $\mathbf{Z}_B = 6 - j8\ \Omega$, and $\mathbf{Z}_C = 9\underline{/30°}\ \Omega$.

(a)   Because all three Y impedances are the same, all three Δ impedances are the same and each is equal to three times the common Y impedance. So,

$$\mathbf{Z}_1 = \mathbf{Z}_2 = \mathbf{Z}_3 = 3(4 - j7) = 12 - j21 = 24.2\underline{/-60.3°}\ \Omega$$

(b)   All three Y-to-Δ conversion formulas have the same numerator, which here is

$$\mathbf{Z}_A\mathbf{Z}_B + \mathbf{Z}_A\mathbf{Z}_C + \mathbf{Z}_B\mathbf{Z}_C = 10(6 - j8) + 10(9\underline{/30°}) + (6 - j8)(9\underline{/30°}) = 231.6\underline{/-17.7°}$$

By these formulas,

$$\mathbf{Z}_1 = \frac{\mathbf{Z}_A\mathbf{Z}_B + \mathbf{Z}_A\mathbf{Z}_C + \mathbf{Z}_B\mathbf{Z}_C}{\mathbf{Z}_B} = \frac{231.6\underline{/-17.7°}}{6 - j8} = \frac{231.6\underline{/-17.7°}}{10\underline{/-53.1°}} = 23.2\underline{/35.4°}\ \Omega$$

$$\mathbf{Z}_2 = \frac{\mathbf{Z}_A\mathbf{Z}_B + \mathbf{Z}_A\mathbf{Z}_C + \mathbf{Z}_B\mathbf{Z}_C}{\mathbf{Z}_C} = \frac{231.6\underline{/-17.7°}}{9\underline{/30°}} = 25.7\underline{/-47.7°}\ \Omega$$

$$\mathbf{Z}_3 = \frac{\mathbf{Z}_A\mathbf{Z}_B + \mathbf{Z}_A\mathbf{Z}_C + \mathbf{Z}_B\mathbf{Z}_C}{\mathbf{Z}_A} = \frac{231.6\underline{/-17.7°}}{10} = 23.2\underline{/-17.7°}\ \Omega$$

13.26 Using a Δ-to-Y conversion, find **I** for the circuit shown in Fig. 13-29.

Extending between nodes A, B, and C there is a Δ, as shown in Fig. 13-30, that can be converted to the shown Y, with the result that the entire circuit becomes series-parallel and so can be reduced by combining impedances. The denominator of each Δ-to-Y conversion formula is $3 + 4 - j4 = 7 - j4 = 8.062\underline{/-29.7°}\ \Omega$. And by these formulas,

$$\mathbf{Z}_A = \frac{3(-j4)}{8.062\underline{/-29.7°}} = \frac{12\underline{/-90°}}{8.062\underline{/-29.7°}} = 1.49\underline{/-60.3°}\ \Omega \qquad \mathbf{Z}_B = \frac{3(4)}{8.062\underline{/-29.7°}} = 1.49\underline{/29.7°}\ \Omega$$

$$\mathbf{Z}_C = \frac{4(-j4)}{8.062\underline{/-29.7°}} = \frac{16\underline{/-90°}}{8.062\underline{/-29.7°}} = 1.98\underline{/-60.3°}\ \Omega$$

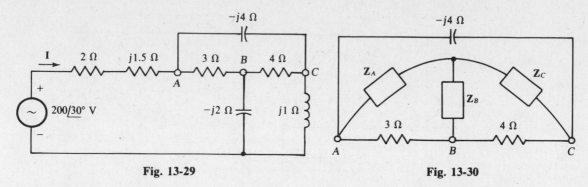

Fig. 13-29   **Fig. 13-30**

With this Δ-to-Y conversion, the circuit is as shown in Fig. 13-31. Since this circuit is in series-parallel form, the input impedance $\mathbf{Z}_{in}$ can be found by circuit reduction. And then $\mathbf{Z}_{in}$ can be divided into the applied voltage to get the current $\mathbf{I}$:

$$\mathbf{Z}_{in} = 2 + j1.5 + 1.49\underline{/-60.3°} + \frac{(1.49\underline{/29.7°} - j2)(1.98\underline{/-60.3°} + j1)}{1.49\underline{/29.7°} - j2 + 1.98\underline{/-60.3°} + j1}$$

$$= 2.74 + j0.208 + \frac{2.21\underline{/-80.6°}}{3.02\underline{/-41.1°}} = 3.31\underline{/-4.5°} \ \Omega$$

Finally,
$$\mathbf{I} = \frac{\mathbf{V}}{\mathbf{Z}_{in}} = \frac{200\underline{/30°}}{3.31\underline{/-4.5°}} = 60.4\underline{/34.5°} \text{ A}$$

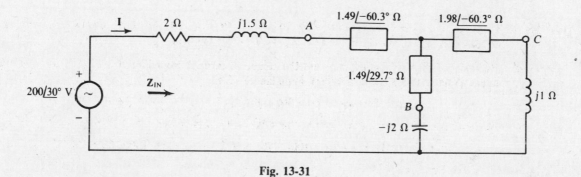

Fig. 13-31

Incidentally, the circuit shown in Fig. 13-29 can also be reduced to series-parallel form by converting the Δ of the $-j2$-, 4-, and $j1$-Ω impedances to a Y, or by converting to a Δ either the Y of the 3-, $-j2$-, and 4-Ω impedances or that of the $-j4$-, 4-, and $j1$-Ω impedances.

13.27 Find the current $\mathbf{I}$ for the circuit shown in Fig. 13-32.

As the circuit stands, a considerable number of mesh or nodal equations are required to find $\mathbf{I}$. But the circuit, which has a Δ and a Y, can easily be reduced to just two meshes by using Δ-Y conversions. Although these conversions do not always lessen the work required, they do here because they are so simple as a result of the common impedances of the Y branches and also of the Δ branches.

One way to reduce the Δ-Y configuration is shown in Fig. 13-33. If the Y of $9 + j12$-Ω impedances is converted to a Δ, the result is a Δ of $3(9 + j12) = 27 + j36$-Ω impedances in parallel with the $-j36$-Ω impedances of the original Δ, as shown in Fig. 13-33a. Combining the parallel impedances produces a Δ with impedances of

$$\frac{(27 + j36)(-j36)}{27 + j36 - j36} = 48 - j36 \ \Omega$$

as shown in Fig. 13-33b. Then, if this is converted to a Y, the Y has impedances of $(48 - j36)/3 = 16 - j12 \ \Omega$, as shown in Fig. 13-33c.

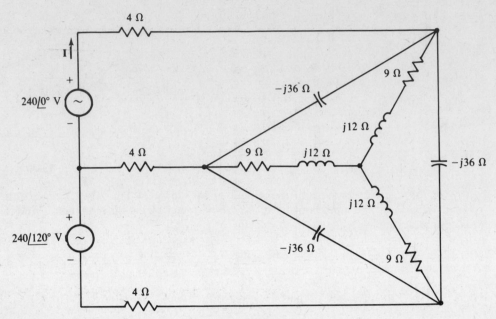

Fig. 13-32

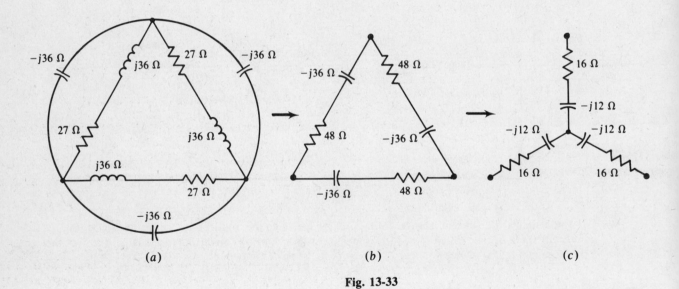

(a)   (b)   (c)

Fig. 13-33

Figure 13-34 shows the circuit with this Y replacing the Δ-Y combination.   The self-impedances of both meshes are the same: $4 + 16 - j12 - j12 + 16 + 4 = 40 - j24\ \Omega$,   and the mutual impedance is $20 - j12\ \Omega$.   So, the mesh equations are

$$(40 - j24)\mathbf{I} - (20 - j12)\mathbf{I}' = 240\underline{/0°}$$

$$-(20 - j12)\mathbf{I} + (40 - j24)\mathbf{I}' = 240\underline{/120°}$$

By Cramer's rule,

$$\mathbf{I} = \frac{\begin{vmatrix} 240\underline{/0°} & -(20 - j12) \\ 240\underline{/120°} & 40 - j24 \end{vmatrix}}{\begin{vmatrix} 40 - j24 & -(20 - j12) \\ -(20 - j12) & 40 - j24 \end{vmatrix}} = \frac{9696\underline{/-0.96°}}{1632\underline{/-61.93°}} = 5.94\underline{/61°}\ \text{A}$$

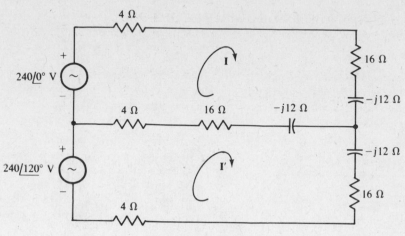

Fig. 13-34

In reducing the Δ-Y circuit, it would have been easier to convert the Δ of $j36$-Ω impedances to a Y of $-j36/3 = -j12$-Ω impedances.   Then, although not obvious, the impedances of this Y would be in parallel with corresponding impedances of the other Y as a result of the two center nodes being at the same potential, which occurs because of equal impedance arms in each Y.   If the parallel impedances are combined, the result is a Y of equal impedances of

$$\frac{-j12(9 + j12)}{-j12 + 9 + j12} = 16 - j12 \ \Omega$$

the same as shown in Fig. 13-33c.

13.28 Assume that the bridge circuit of Fig. 13-3 is balanced for $Z_1 = 5 \ \Omega$, $Z_2 = 4\underline{/30°} \ \Omega$, and $Z_3 = 8.2 \ \Omega$,   and for a source frequency of 2 kHz.   If branch Z_X consists of two components in series, what are they?

The two components can be determined from the real and imaginary parts of Z_X.   From the bridge balance equation,

$$Z_X = \frac{Z_2 Z_3}{Z_1} = \frac{(4\underline{/30°})(8.2)}{5} = 6.56\underline{/30°} = 5.68 + j3.28 \ \Omega$$

which corresponds to a 5.68-Ω resistor and a series inductor that has a reactance of 3.28 Ω.   The corresponding inductance is

$$L = \frac{X_L}{\omega} = \frac{3.28}{2\pi(2000)} \ \text{H} = 261 \ \mu\text{H}$$

13.29 The bridge circuit shown in Fig. 13-35 is a *capacitance comparison bridge* that is used for measuring the capacitance C_X of a capacitor and any resistance R_X inherent to the capacitor or in series with it.   The bridge has a standard capacitor, the capacitance C_S of which is known.   Find R_X and C_X if the bridge is in balance for $R_1 = 500 \ \Omega$, $R_2 = 2 \ \text{k}\Omega$, $R_3 = 1 \ \text{k}\Omega$, $C_S = 0.02 \ \mu\text{F}$,   and a source radian frequency of 1 krad/s.

The bridge balance equation can be used to determine R_X and C_X.   From a comparison of Figs. 13-3 and 13-35, $Z_1 = 500 \ \Omega$, $Z_2 = 2000 \ \Omega$,

$$Z_3 = 1000 - \frac{j1}{1000(0.02 \times 10^{-6})} = 1000 - j50\,000 \ \Omega$$

and

$$Z_X = R_X - \frac{j1}{1000C_X}$$

From the bridge balance equation $\mathbf{Z}_X = \mathbf{Z}_2\mathbf{Z}_3/\mathbf{Z}_1$,

$$R_X - \frac{j1}{1000C_X} = \frac{2000(1000 - j50\,000)}{500} = 4000 - j200\,000 \ \Omega$$

For two complex quantities in rectangular form to be equal, as here, both the real parts must be equal and the imaginary parts must be equal. This means that $R_X = 4000 \ \Omega$ and

$$-\frac{1}{1000C_X} = -200\,000 \qquad \text{from which} \qquad C_X = \frac{1}{1000(200\,000)} \ \text{F} = 5 \ \text{nF}$$

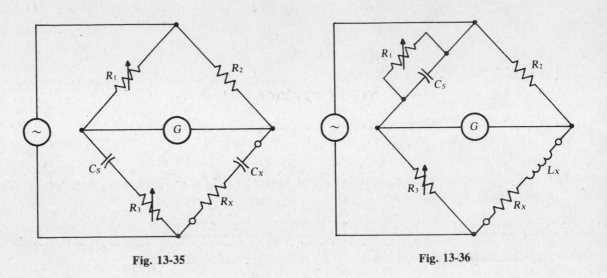

Fig. 13-35   **Fig. 13-36**

13.30 For the capacitance comparison bridge shown in Fig. 13-35, derive general formulas for R_X and C_X in terms of the other bridge components.

For bridge balance, $\mathbf{Z}_1\mathbf{Z}_X = \mathbf{Z}_2\mathbf{Z}_3$,   which in terms of the bridge components is

$$R_1\left(R_X - \frac{j1}{\omega C_X}\right) = R_2\left(R_3 - \frac{j1}{\omega C_S}\right) \qquad \text{or} \qquad R_1 R_X - j\frac{R_1}{\omega C_X} = R_2 R_3 - j\frac{R_2}{\omega C_S}$$

From equating real parts, $R_1 R_X = R_2 R_3$, or $R_X = R_2 R_3/R_1$. And from equating imaginary parts, $-R_1/(\omega C_X) = -R_2/(\omega C_S)$, or $C_X = R_1 C_S/R_2$.

13.31 The bridge circuit shown in Fig. 13-36, called a *Maxwell bridge*, is used for measuring the inductance and resistance of a coil in terms of a capacitance standard. Find L_X and R_X if the bridge is in balance for $R_1 = 500$ kΩ, $R_2 = 6.2$ kΩ, $R_3 = 5$ kΩ, $C_S = 0.1 \ \mu$F.

First, general formulas will be derived for R_X and L_X in terms of the other bridge components. Then, values will be substituted into these formulas to find R_X and L_X for the specified bridge. From a comparison of Figs. 13-3 and 13-36, $\mathbf{Z}_2 = R_2$, $\mathbf{Z}_3 = R_3$, $\mathbf{Z}_X = R_X + j\omega L_X$, and

$$\mathbf{Z}_1 = \frac{R_1(-j1/\omega C_S)}{R_1 - j1/\omega C_S} = \frac{-jR_1}{R_1\omega C_S - j1}$$

Substituting these into the bridge balance equation $\mathbf{Z}_1\mathbf{Z}_X = \mathbf{Z}_2\mathbf{Z}_3$ gives

$$\left(\frac{-jR_1}{R_1\omega C_S - j1}\right)(R_X + j\omega L_X) = R_2 R_3$$

which upon being multiplied by $R\omega C_S - j1$ and simplified becomes

$$R_1\omega L_X - jR_1 R_X = R_2 R_3 R_1 \omega C_S - jR_2 R_3$$

From equating real parts,

$$R_1 \omega L_x = R_2 R_3 R_1 \omega C_S \qquad \text{from which} \qquad L_x = R_2 R_3 C_S$$

and from equating imaginary parts,

$$-R_1 R_x = -R_2 R_3 \qquad \text{from which} \qquad R_x = \frac{R_2 R_3}{R_1}$$

which are the general formulas for L_x and R_x. For the values of the specified bridge, these formulas give

$$R_x = \frac{(6.2 \times 10^3)(5 \times 10^3)}{500 \times 10^3} = 62 \ \Omega \quad \text{and} \quad L_x = (6.2 \times 10^3)(5 \times 10^3)(0.1 \times 10^{-6}) = 3.1 \text{ H}$$

Supplementary Problems

13.32 Find V_{Th} and Z_{Th} for the Thevenin equivalent of the circuit shown in Fig. 13-37.
     Ans. $133\underline{/-88.4°}$ V, $8.36\underline{/17.6°}$ Ω

13.33 What resistor will draw an 8-A rms current when connected across terminals a and b of the circuit shown in Fig. 13-37?   Ans. 8.44 Ω

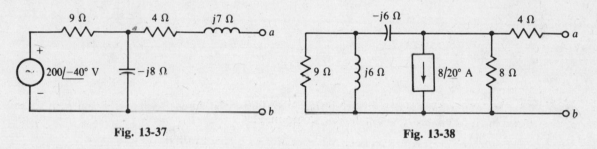

Fig. 13-37   Fig. 13-38

13.34 Find I_N and Z_{Th} for the Norton equivalent of the circuit shown in Fig. 13-38.
     Ans. $-3.09\underline{/5.07°}$ A, $6.3\underline{/-9.03°}$ Ω

13.35 Find V_{Th} and Z_{Th} for the Thevenin equivalent of the circuit shown in Fig. 13-39 for $R = 0 \ \Omega$.
     Ans. $3.47\underline{/123°}$ V, $3.05\underline{/29.2°}$ Ω

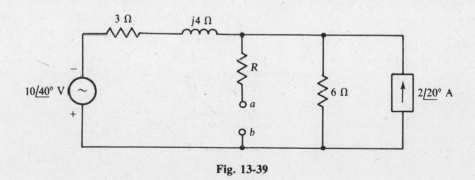

Fig. 13-39

13.36 Find I_N and Z_{Th} for the Norton equivalent of the circuit shown in Fig. 13-39 for $R = 2 \ \Omega$.
     Ans. $0.71\underline{/105°}$ A, $4.89\underline{/17.7°}$ Ω

13.37 Find V_{Th} and Z_{Th} for the Thevenin equivalent of the circuit shown in Fig. 13-40 for $R_1 = R_2 = 0\ \Omega$ and $V_S = 0\ V$.   *Ans.* $-40.4\underline{/-41.4°}$ V, $1.92\underline{/19.4°}\ \Omega$

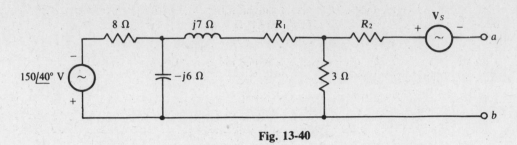

Fig. 13-40

13.38 Find V_{Th} and Z_{Th} for the Thevenin equivalent of the circuit shown in Fig. 13-40 for $R_1 = 5\ \Omega$, $R_2 = 4\ \Omega$, and $V_S = 50\underline{/-60°}$ V. *Ans.* $-71.5\underline{/-50.2°}$ V, $6.24\underline{/2.03°}\ \Omega$

13.39 Find V_{Th} and Z_{Th} for the Thevenin equivalent of the circuit shown in Fig. 13-41.
Ans. $11.8\underline{/25.3°}$ V, $4.67\underline{/5.25°}\ \Omega$

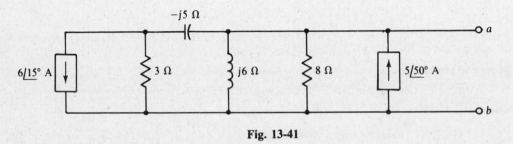

Fig. 13-41

13.40 What resistor will draw a 2-A rms current when connected across terminals a and b of the circuit shown in Fig. 13-41?   *Ans.* $1.21\ \Omega$

13.41 Using Thevenin's or Norton's theorem, find I for the bridge circuit shown in Fig. 13-42 if $I_S = 0$ A and $Z_L = 60\underline{/30°}\ \Omega$.   *Ans.* $10.4\underline{/-43.5°}$ A

13.42 Find I for the bridge circuit shown in Fig. 13-42 if $I_S = 10\underline{/30°}$ A and $Z_L = 40\underline{/-40°}\ \Omega$.
Ans. $15\underline{/6.3°}$ A

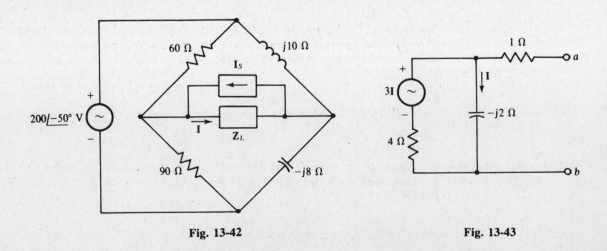

Fig. 13-42   **Fig. 13-43**

13.43   Find the output impedance of the circuit shown in Fig. 13-43.   *Ans.* 4.49$\underline{/-20.9°}$ Ω

13.44   Find the output impedance of the circuit shown in Fig. 13-43 with the **I** reference direction reversed—being up instead of down.   *Ans.* 1.68$\underline{/-39.1°}$ Ω

13.45   Find V_{Th} and Z_{Th} for the Thevenin equivalent of the circuit shown in Fig. 13-44.
Ans. −1.75$\underline{/23°}$ V, 0.361$\underline{/19.4}$ Ω

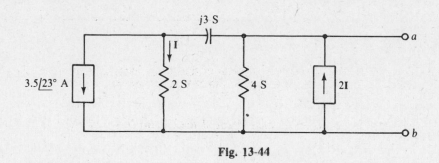

Fig. 13-44

13.46   In the circuit shown in Fig. 13-44, reverse the **I** reference direction—have it up instead of down—and find I_N and Z_{Th} for the Norton equivalent circuit.   *Ans.* 4.85$\underline{/-70.2°}$ A, 0.116$\underline{/-18.8°}$ Ω

13.47   Find the output impedance at 10^4 rad/s of the circuit shown in Fig. 13-45.   *Ans.* 11.9$\underline{/-4.7°}$ kΩ

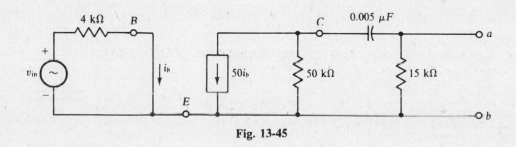

Fig. 13-45

13.48   What is the maximum average power that can be drawn from an ac generator that has an internal impedance of 100$\underline{/20°}$ Ω and an open-circuit voltage of 25 kV rms?   *Ans.* 1.66 MW

13.49   A signal generator operating at 5 MHz has an rms short-circuit current of 100 mA and an internal impedance of 80$\underline{/20°}$ Ω. If it energizes a capacitor and a parallel resistor, find the capacitance and resistance for maximum average power absorption by the resistor. Also, find this power.
Ans. 136 pF, 85.1 Ω, 0.213 W

13.50   For the circuit shown in Fig. 13-46, what Z_L draws maximum average power and what is this power?   *Ans.* 12.8$\underline{/-51.3°}$ Ω, 48.5 W

13.51   In the circuit shown in Fig. 13-46, move the −j8-Ω impedance from in series with the current source to in parallel with it. Then, find both the Z_L that absorbs maximum average power and this power.   *Ans.* 14$\underline{/-1.69°}$ Ω, 61 W

13.52   Use superposition to find **I** for the circuit shown in Fig. 13-47.   *Ans.* 2.27$\underline{/65.2°}$ A

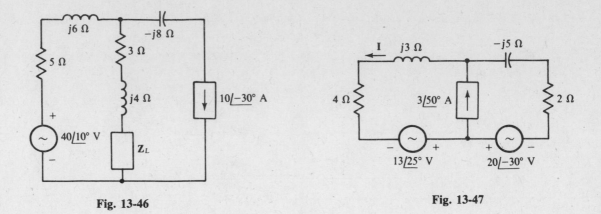

Fig. 13-46   **Fig. 13-47**

13.53 For the circuit shown in Fig. 13-48, find the average power dissipated in the 3-Ω resistor using superposition and also without using superposition. Then, repeat this with the 10° phase angle changed to 40° for the one voltage source. This problem illustrates the fact that superposition can be used to find the average power absorbed by a resistor from two sources of the *same* frequency only if these sources produce resistor currents that have a 90° difference in phase angle.

Ans. 34.7 W with and without using superposition; an incorrect 34.7 W with superposition and a correct 20.3 W without using superposition

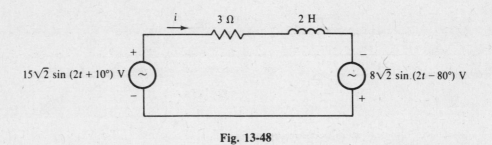

Fig. 13-48

13.54 Find v for the circuit shown in Fig. 13-49.   *Ans.* 5.24 sin (5000t − 61.6°) − 4.39 sin (8000t − 34.6°) V

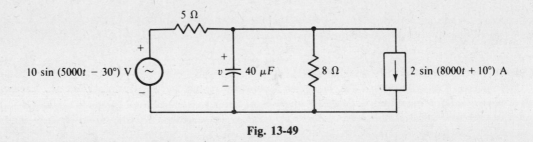

Fig. 13-49

13.55 Find the average power dissipated in the 5-Ω resistor of the circuit shown in Fig. 13-49.
Ans. 5.74 W

13.56 Find i for the circuit shown in Fig. 13-50.   *Ans.* −2 sin (5000t + 23.1°) − 4.96 sin (10^4t − 2.87°) A

13.57 Find the average power absorbed by the 200-Ω resistor in the circuit shown in Fig. 13-50.
Ans. 523 W

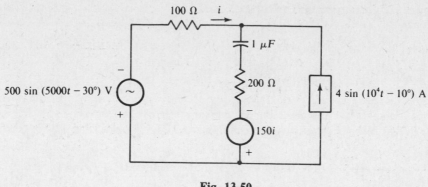

Fig. 13-50

13.58   Convert the T shown in Fig. 13-51a to the Π in Fig. 13-51b for   (a) $\mathbf{Z}_A = \mathbf{Z}_B = \mathbf{Z}_C = 10\underline{/-50°}\ \Omega$,   and for   ($b$) $\mathbf{Z}_A = 5\underline{/-30°}\ \Omega$, $\mathbf{Z}_B = 6\underline{/40°}\ \Omega$, $\mathbf{Z}_C = 6 - j7\ \Omega$.
   Ans.   (a) $\mathbf{Z}_1 = \mathbf{Z}_2 = \mathbf{Z}_3 = 30\underline{/-50°}\ \Omega$;   ($b$) $\mathbf{Z}_1 = 17.5\underline{/-68°}\ \Omega$, $\mathbf{Z}_2 = 11.4\underline{/21.4°}\ \Omega$, $\mathbf{Z}_3 = 21\underline{/2.05°}\ \Omega$

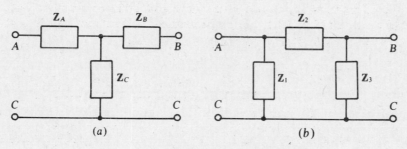

Fig. 13-51

13.59   Convert the Π shown in Fig. 13-51b to the T in Fig. 13-51a for   (a) $\mathbf{Z}_1 = \mathbf{Z}_2 = \mathbf{Z}_3 = 36\underline{/-24°}\ \Omega$,   and for   ($b$) $\mathbf{Z}_1 = 15\underline{/-24°}\ \Omega$, $\mathbf{Z}_2 = 14 - j20\ \Omega$, $\mathbf{Z}_3 = 10 + j16\ \Omega$.
   Ans.   (a) $\mathbf{Z}_A = \mathbf{Z}_B = \mathbf{Z}_C = 12\underline{/-24°}\ \Omega$;   ($b$) $\mathbf{Z}_A = 9.38\underline{/-64°}\ \Omega$, $\mathbf{Z}_B = 11.8\underline{/18°}\ \Omega$, $\mathbf{Z}_C = 7.25\underline{/49°}\ \Omega$

13.60   Using a Δ-Y conversion, find **I** for the circuit shown in Fig. 13-52.   *Ans.* $26.9\underline{/22°}$ A

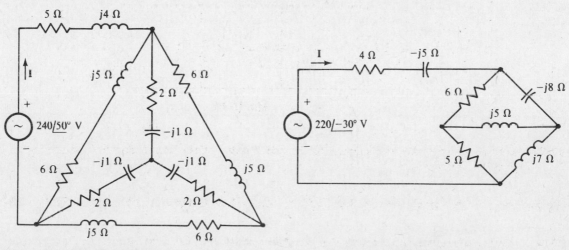

Fig. 13-52   **Fig. 13-53**

13.61 Using a Δ-Y conversion, find $\mathbf{I}$ for the circuit shown in Fig. 13-53.   *Ans.* 17.6$\underline{/13.1°}$ A

13.62 Assume that the bridge circuit shown in Fig. 13-3 is balanced for $\mathbf{Z}_1 = 10\underline{/-30°}\ \Omega$, $\mathbf{Z}_2 = 15\underline{/40°}\ \Omega$, and $\mathbf{Z}_3 = 9.1\ \Omega$, and for a source frequency of 5 kHz. If branch $\mathbf{Z}_X$ consists of two components in parallel, what are they?   *Ans.* A 39.9-Ω resistor and a 462-μH inductor

13.63 Find C_X and R_X for the capacitor comparison bridge shown in Fig. 13-35 if this bridge is balanced for $R_1 = 1\ k\Omega$, $R_2 = 4\ k\Omega$, $R_3 = 2\ k\Omega$, and $C_S = 0.1\ \mu$F.   *Ans.* 25 nF, 8 kΩ

13.64 Find L_X and R_X for the Maxwell bridge shown in Fig. 13-36 if this bridge is balanced for $R_1 = 50\ k\Omega$, $R_2 = 8.2\ k\Omega$, $R_3 = 4\ k\Omega$, and $C_S = 0.05\ \mu$F.   *Ans.* 1.64 H, 656 Ω

Chapter 14

Power in AC Circuits

INTRODUCTION

The major topic of this chapter is the average power absorbed by ac components and circuits. Consequently, it will not be necessary to always use the adjective *average* with *power* to avoid misunderstanding. Also, it is not necessary to use the subscript notation "av" with the symbol P. Similarly, since the popular power formulas have only effective or rms values of voltage and current, the subscript notation "eff" can be deleted from V_{eff} and I_{eff} (or "rms" from V_{rms} and I_{rms}) and just the lightface V and I used to designate effective or rms values.

As a final introductory point, in the following text material and problems the specified voltages and currents always have associated references unless there are statements or designations to the contrary.

CIRCUIT POWER ABSORPTION

The average power absorbed by a two-terminal ac circuit can be derived from the instantaneous power absorbed. If the circuit has an applied voltage $v = V_m \sin(\omega t + \theta)$ and an input current $i = I_m \sin \omega t$, the instantaneous power absorbed by the circuit is

$$p = vi = V_m \sin(\omega t + \theta) \times I_m \sin \omega t = V_m I_m \sin(\omega t + \theta) \sin \omega t$$

This can be simplified by using the trigonometric identity

$$\sin A \sin B = \tfrac{1}{2}[\cos(A - B) - \cos(A + B)]$$

and the substitutions $A = \omega t + \theta$ and $B = \omega t$:

$$p = \frac{V_m I_m}{2}[\cos \theta - \cos(2\omega t + \theta)]$$

Since

$$\frac{V_m I_m}{2} = \frac{V_m}{\sqrt{2}} \times \frac{I_m}{\sqrt{2}} = VI$$

the instantaneous power can be expressed as

$$p = VI \cos \theta - VI \cos(2\omega t + \theta)$$

The average value of this power is the sum of the average values of the two terms. The second term, being sinusoidal, has a zero average value. The first term, though, is a constant, and so must be the average power absorbed by the circuit:

$$P = VI \cos \theta$$

It is important to remember that in this formula the angle θ is the angle by which the input voltage leads the input current. For a circuit that does not contain any independent sources, this is the impedance angle.

For a purely resistive circuit, $\theta = 0°$ and $\cos 0° = 1$ and so $P = VI \cos \theta = VI$. For a purely inductive circuit, $\theta = 90°$ and $\cos \theta = \cos 90° = 0$, and so $P = 0$ W, which means that a purely inductive circuit absorbs zero average power. The same is true for a purely capacitive circuit since, for it, $\theta = -90°$ and $\cos(-90°) = 0$.

The term "$\cos \theta$" is called the *power factor*. It is often symbolized by PF, as in $P = VI \times PF$. The angle θ is called the *power factor angle*. Of course, it is often also the impedance angle.

The power factor angle has different signs for inductive and capacitive circuits, but since $\cos \theta = \cos (-\theta)$, the sign of the power factor angle has no effect on the power factor. Because the power factors of inductive and capacitive circuits cannot be distinguished mathematically, they are distinguished by name. The power factor of an inductive circuit is called a *lagging power factor* and that of a capacitive circuit is called a *leading power factor*. These names can be remembered from the fact that for an inductive circuit the current lags the voltage, and for a capacitive circuit the current leads the voltage.

Another power formula can be obtained by substituting $V = IZ$ into $P = VI \cos \theta$:

$$P = VI \cos \theta = (IZ)I \cos \theta = I^2(Z \cos \theta) = I^2R$$

Of course, $R = Z \cos \theta$ is the input resistance, the same as the real part of the input impedance. The formula $P = I^2R$ may seem obvious from dc considerations, but remember that R is usually not the resistance of a physical resistor. Rather, it is the real part of the input impedance and is usually dependent on inductive and capacitive reactances as well as on resistances.

Similarly, with the substitution of $I = YV$,

$$P = VI \cos \theta = V(VY) \cos \theta = V^2(Y \cos \theta) = V^2G$$

in which $G = Y \cos \theta$ is the input conductance. In using this formula $P = V^2G$, remember that, except for a purely resistive circuit, the input conductance G is not the inverse of the input resistance R. If, however, V is the voltage across a resistor of R ohms, then $P = V^2G = V^2/R$.

WATTMETERS

Average power can be measured by an instrument called a *wattmeter*, as shown in Fig. 14-1. It has two pairs of terminals: a pair of voltage terminals on the left-hand side and a pair of current terminals on the right-hand side. The bottom terminal of each pair has a ± designation for aiding in connecting up the wattmeter, as will be explained.

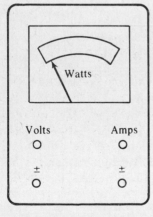

Fig. 14-1

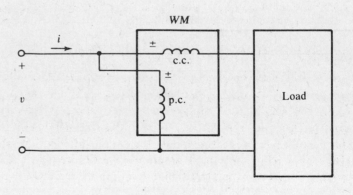

Fig. 14-2

For a measurement of power absorbed by a load, the voltage terminals are connected in parallel with the load and the current terminals are connected in series with the load. Because the voltage circuit inside the wattmeter has a very high resistance and the current circuit has a very low resistance, the voltage circuit can be considered an open circuit and the current circuit a short circuit for the power measurements of almost all loads. As a result, inserting a wattmeter in a circuit seldom has a significant effect on the power absorbed. For convenience, in circuit diagrams the voltage circuit will be shown as a coil labeled "p.c." for potential coil, and the current circuit will be shown as a coil labeled "c.c." for current coil, as shown in Fig. 14-2.

The ± designations help in making wattmeter connections so that the wattmeter reads upscale, to the right in Fig. 14-1, for positive absorbed power. A wattmeter will read upscale with the connection in Fig. 14-2 if the load absorbs average power. Notice that, for the associated voltage and current references, the reference current enters the ± current terminal and the positive reference of the voltage is at the ± voltage terminal. The effect is the same, though, if both coils are reversed. If a load is active—a source of average power—then one coil connection, but not both, should be reversed for an upscale reading. The wattmeter reading is considered to be negative for this connection. Incidentally, in the circuit shown in Fig. 14-2, the wattmeter reads the same with the potential coil connected on the load side of the current coil instead of on the source side.

REACTIVE POWER

For industrial power considerations, a quantity called *reactive power* is often useful. It has the quantity symbol Q and the unit of *voltampere reactive*, the symbol for which is the var. (The var is often considered to be also the unit.) The reactive power is defined as

$$Q = VI \sin \theta$$

for a two-terminal circuit with an input rms voltage V and an input rms current I. θ is the angle by which the input voltage leads the input current—the power factor angle. The quantity "$\sin \theta$" is called the *reactive factor* of the load and has the symbol RF. Notice that it is negative for capacitive loads and is positive for inductive loads. A load that absorbs negative vars is considered to be producing vars—that is, it is a source of reactive power.

As was done for real power P, other formulas for Q can be found by substituting from $V = IZ$ and $I = YV$ into $Q = VI \sin \theta$. These formulas are

$$Q = I^2 X \qquad \text{and} \qquad Q = -V^2 B$$

where X is the reactance or imaginary part of the input impedance and B is the susceptance or imaginary part of the input admittance. Remember that B is not the inverse of X. If, however, V is the voltage across an inductor or capacitor with reactance X, then $Q = V^2/X$.

COMPLEX POWER AND APPARENT POWER

There is a relation among the real power of a load, the reactive power, and another power called the *complex power*. For the derivation of this relation, consider the load impedance triangle shown in Fig. 14-3a. If each side is multiplied by the square of the rms current I to the load, the result is the triangle shown in Fig. 14-3b. Notice that this multiplication does not affect the impedance angle θ since each side is multiplied by the same quantity. The horizontal side is the real power $P = I^2 R$, the vertical side is $j1$ times the reactive power, $jI^2 X = jQ$, and the hypotenuse $I^2 Z$ is the complex power of the load. Complex power has the quantity symbol S and the unit of *voltampere*

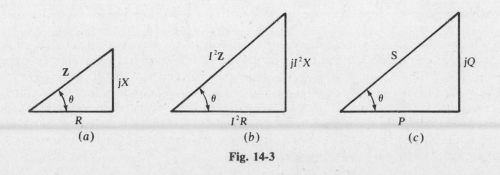

Fig. 14-3

with symbol VA. These power quantities are shown in Fig. 14-3c, which is known as the *power triangle.* From this triangle, clearly $\mathbf{S} = P + jQ$.

The length of the hypotenuse, $|\mathbf{S}| = S$, is called the *apparent power.* Its name comes from the fact that it is equal to the product of the input rms voltage and current:

$$S = |I^2\mathbf{Z}| = |I\mathbf{Z}| \times I = VI$$

and from the fact that in dc circuits this product VI is the power absorbed. The substitution of $V = IZ$ and $I = V/Z$ into $S = VI$ produces two other formulas: $S = I^2Z$ and $S = V^2/Z$.

The VI formula for apparent power leads to another popular formula for complex power. Since $\mathbf{S} = S\underline{/\theta}$, and $S = VI$, then $\mathbf{S} = VI\underline{/\theta}$.

A third formula for complex power is $\mathbf{S} = \mathbf{V}\mathbf{I}^*$, where $\mathbf{I}^*$ is the conjugate of the input current $\mathbf{I}$. This is a valid formula since the magnitude of $\mathbf{V}\mathbf{I}^*$ is the product of the applied rms voltage and current, and, consequently, is the apparent power. Also, the angle of this product is the angle of the voltage phasor *minus* the angle of the current phasor, with the subtraction occurring because of the use of the conjugate of the current phasor. Of course, this difference in angles is the complex power angle θ—the angle by which the input voltage leads the input current—and also the power factor angle.

One use of complex power is for getting the total complex power of several loads energized by the same source, usually in parallel. It can be shown that the total complex power is the sum of the individual complex powers. It follows that the total real power is the sum of the individual real powers, and that the total reactive power is the sum of the individual reactive powers. To repeat for emphasis: Complex powers, real powers, and reactive powers can be added to obtain the total complex power, real power, and reactive power, respectively. The same is *not* true for apparent powers. In general, apparent powers cannot be added to find a total apparent power any more than rms voltages or currents can be added to find a total rms voltage or current.

The total complex power can be used to find the total input current, as should be apparent from the fact that the magnitude of the total complex power, the apparent power, is the product of the input voltage and current.   Another use for complex power is in power factor correction, which is the subject of the next section.

POWER FACTOR CORRECTION

In the consumption of a large amount of power, a large power factor is desirable—the larger the better. The reason is that the current required to deliver a given amount of power to a load is inversely proportional to the load power factor, as is evident from rearranging $P = VI \cos \theta$ to

$$I = \frac{P}{V \cos \theta} = \frac{P}{V \times PF}$$

So, for a given power P absorbed and applied voltage V, the smaller the power factor the greater the current I to the load. Larger than necessary currents are undesirable because of the accompanying larger voltage losses and I^2R power losses in power lines and other power distribution equipment.

As a practical matter, low power factors are always the result of inductive loads because almost all loads are inductive. From a power triangle viewpoint, the vars that such loads consume make the power triangle have a large vertical side and so a large angle θ. The result is a small $\cos \theta$, which is the power factor. Improving the power factor of a load requires adding capacitors across the power line at the load to produce the vars consumed by the inductive load. From another point of view, these capacitors supply current to the load inductors, which current, without the capacitors, would have to come over the power line. More accurately, there is a current interchange between these capacitors and the load inductors.

Although adding sufficient capacitance to increase the power factor to unity is possible, it may not be economical. For finding the minimum capacitance required to improve the power factor to the extent desired, the general procedure is to first calculate the initial number of vars Q_i being consumed by the load. This can be calculated from $Q_i = P \tan \theta_i$, which formula should be apparent from the power triangle shown in Fig. 14-3c. Of course, θ_i is the load impedance angle. The next step is to determine the final impedance angle θ_f from the final desired power factor: $\theta_f = \cos^{-1} PF_f$. This angle is used in $Q_f = P \tan \theta_f$ to find the total number of vars Q_f for the combined load. This formula is valid since adding the parallel capacitor or capacitors does not change P. The next step is to find the vars that must be provided by the capacitors: $\Delta Q = Q_f - Q_i$. Finally, ΔQ is used to find the required amount of capacitance:

$$\Delta Q = \frac{V^2}{X} = \frac{V^2}{-1/\omega C} = -\omega C V^2 \quad \text{from which} \quad C = -\frac{\Delta Q}{\omega V^2}$$

If ΔQ is defined as $Q_i - Q_f$, the negative sign can be eliminated in the formula for C; then, $C = \Delta Q / \omega V^2$.

Although calculating the capacitance required for power factor correction may be a good academic exercise, it is not necessary on the job. Manufacturers specify their power factor correction capacitors by operating voltages and the kilovars the capacitors produce. So, for power factor correction, it is only necessary to know the voltages of the lines across which the capacitors will be placed and the kilovars required.

Solved Problems

14.1 The instantaneous power absorbed by a circuit is $p = 10 + 8 \sin (377t + 40°)$ W. Find the maximum, minimum, and average absorbed powers.

   The maximum value occurs at those times when the sinusoidal term is a maximum. Since this term has a maximum value of 8, $p_{max} = 10 + 8 = 18$ W. The minimum value occurs when the sinusoidal term is at its minimum value of -8: $p_{min} = 10 - 8 = 2$ W. Because the sinusoidal term has a zero average value, the average power absorbed is $P = 10 + 0 = 10$ W.

14.2 With $v = 300 \cos (20t + 30°)$ V applied, a circuit draws $i = 15 \cos (20t - 25°)$ A. Find the power factor and also the average, maximum, and minimum absorbed powers.

   The power factor of the circuit is the cosine of the power factor angle, which is the angle by which the voltage leads the current:

$$PF = \cos [30° - (-25°)] = \cos 55° = 0.574$$

It is lagging because the current lags the voltage.

   The average power absorbed is the product of the rms voltage and current and the power factor:

$$P = \frac{300}{\sqrt{2}} \times \frac{15}{\sqrt{2}} \times 0.574 = 1.29 \times 10^3 \text{ W} = 1.29 \text{ kW}$$

The maximum and minimum absorbed powers can be found from the instantaneous power, which is

$$p = vi = 300 \cos (20t + 30°) \times 15 \cos (20t - 25°) = 4500 \cos (20t + 30°) \cos (20t - 25°) \text{ W}$$

This can be simplified by using the trigonometric identity

$$\cos A \cos B = 0.5[\cos (A + B) + \cos (A - B)]$$

and the substitutions $A = 20t + 30°$ and $B = 20t - 25°$. The result is

$$p = 4500 \times 0.5[\cos (40t + 5°) + \cos 55°] = 2250[\cos (40t + 5°) + \cos 55°] \text{ W}$$

Clearly, the maximum value occurs when the first cosine term is 1 and the minimum value when this term is -1:

$$p_{max} = 2250(1 + \cos 55°) \text{ W} = 3.54 \text{ kW}$$

$$p_{min} = 2250(-1 + \cos 55°) = -959 \text{ W}$$

The negative minimum absorbed power indicates that the circuit is delivering power instead of absorbing it.

14.3 For each following load voltage and current pair find the corresponding power factor and average power absorbed:

(a) $v = 277\sqrt{2} \sin (377t + 30°) \text{ V}, \quad i = 5.1\sqrt{2} \sin (377t - 10°) \text{ A}$

(b) $v = 679 \sin (377t + 50°) \text{ V}, \quad i = 13 \cos (377t + 10°) \text{ A}$

(c) $v = -170 \sin (377t - 30°) \text{ V}, \quad i = 8.1 \cos (377t + 30°) \text{ A}$

(a)   Since the angle by which the voltage leads the current is $\theta = 30° - (-10°) = 40°$, the power factor is $PF = \cos 40° = 0.766$. It is lagging because the current lags the voltage, or, in other words, because the power factor angle θ is positive. The average power absorbed is the product of the rms voltage and current and the power factor:

$$P = VI \times PF = 277(5.1)(0.766) = 1.08 \times 10^3 \text{ W} = 1.08 \text{ kW}$$

(b)   The power factor angle θ can be found by phase angle subtraction only if v and i have the same sinusoidal form, which they do not have here. The cosine term of i can be converted to the sine form of v by using the identity $\cos x = \sin (x + 90°)$:

$$i = 13 \cos (377t + 10°) = 13 \sin (377t + 10° + 90°) = 13 \sin (377t + 100°) \text{ A}$$

So, the power factor angle is $\theta = 50° - 100° = -50°$, and the power factor is $PF = \cos (-50°) = 0.643$. It is a leading power factor because the current leads the voltage, and also because θ is negative, which is equivalent. The average power absorbed is

$$P = VI \times PF = \frac{679}{\sqrt{2}} \times \frac{13}{\sqrt{2}} \times 0.643 = 2.84 \times 10^3 \text{ W} = 2.84 \text{ kW}$$

(c)   The voltage sinusoid will be put in the same sinusoidal form as the current sinusoid as an aid in finding θ. The negative sign can be eliminated by using $-\sin x = \sin (x \pm 180°)$:

$$v = -170 \sin (377t - 30°) = 170 \sin (377t - 30° \pm 180°)$$

Then the identity $\sin x = \cos (x - 90°)$ can be used:

$$\begin{aligned} v &= 170 \sin (377t - 30° \pm 180°) = 170 \cos (377t - 30° \pm 180° - 90°) \\ &= 170 \cos (377t - 120° \pm 180°) \end{aligned}$$

The positive sign of $\pm 180°$ should be selected in order to have the voltage and current phase angles as close together as possible:

$$v = 170 \cos (377t - 120° + 180°) = 170 \cos (377t + 60°) \text{ V}$$

So, $\theta = 60° - 30° = 30°$, and the power factor is $PF = \cos 30° = 0.866$. Of course, it is lagging because θ is positive. Finally, the average power absorbed is

$$P = VI \times PF = \frac{170}{\sqrt{2}} \times \frac{8.1}{\sqrt{2}} \times 0.866 = 596 \text{ W}$$

14.4 Find the power factor of a circuit that absorbs 1.5 kW for an input voltage and current of 120 V and 16 A.

From $P = VI \times PF$, the power factor is

$$PF = \frac{P}{VI} = \frac{\text{average power}}{\text{apparent power}} = \frac{1500}{120(16)} = 0.781$$

There is not enough information given to determine whether this power factor is leading or lagging.

Note that the power factor is equal to the average power divided by the apparent power. Some authors of circuit analysis books use this for the definition of power factor because it is more general than $PF = \cos\theta$.

14.5 What is the power factor of a fully loaded 10-hp induction motor that operates at 80 percent efficiency while drawing 28 A from a 480-V line?

The motor power factor is equal to the power input divided by the apparent power input. And, the power input is the power output divided by the efficiency of operation:

$$P_{in} = \frac{P_{out}}{\eta} = \frac{10 \times 745.7}{0.8} \text{ W} = 9.321 \text{ kW}$$

in which 1 hp = 745.7 W is used. So,

$$PF = \frac{P_{in}}{VI} = \frac{9.321 \times 10^3}{480(28)} = 0.694$$

This power factor is lagging because induction motors are inductive loads.

14.6 Find the power absorbed by a $6\underline{/30°}$-Ω load when 42 V is applied.

The rms current needed for the power formulas is equal to the rms voltage divided by the magnitude of the impedance: $I = 42/6 = 7$ A. Of course, the power factor is the cosine of the impedance angle: $PF = \cos 30° = 0.866$. Thus,

$$P = VI \times PF = 42(7)(0.866) = 255 \text{ W}$$

The absorbed power can also be obtained from $P = I^2R$, in which $R = Z\cos\theta = 6\cos 30° = 5.2\ \Omega$:

$$P = 7^2 \times 5.2 = 255 \text{ W}$$

The power cannot be found from $P = V^2/R$, as is evident from the fact that $V^2/R = 42^2/5.2 = 339$ W and not the correct 255 W. The reason for the incorrect result is that the 42 V is across the entire impedance and not just the resistance part. For $P = V^2/R$ to be valid, the V used must be that across R.

14.7 What power is absorbed by a circuit that has an $0.4 + j0.5$-S input admittance and a 30-A input current?

The formula $P = V^2G$ can be used after the input voltage V is found. It is equal to the current divided by the magnitude of the admittance:

$$V = \frac{I}{|\mathbf{Y}|} = \frac{30}{|0.4 + j0.5|} = \frac{30}{0.64} = 46.85 \text{ V}$$

So $P = V^2G = (46.85)^2 0.4 = 878 \text{ W}$

Alternatively, the power formula $P = VI\cos\theta$ can be used. The power factor angle θ is the negative of the admittance angle: $\theta = -\tan^{-1}(0.5/0.4) = -51.34°$. So,

$$P = VI\cos\theta = 46.85(30)\cos(-51.34°) = 878 \text{ W}$$

14.8 A resistor in parallel with a capacitor absorbs 20 W when the combination is connected to a 240-V, 60-Hz source. If the power factor is 0.7 leading, what are the resistance and capacitance?

The resistance can be found by solving for R in $P = V^2/R$:

$$R = \frac{V^2}{P} = \frac{240^2}{20}\ \Omega = 2.88 \text{ k}\Omega$$

One way to find the capacitance is from the susceptance B, which can be found from $B =$

$G \tan \phi$ after the conductance G and admittance angle ϕ are known.   The conductance is

$$G = \frac{1}{R} = \frac{1}{2.88 \times 10^3} = 0.347 \times 10^{-3} \text{ S}$$

For this capacitive circuit, the admittance angle is the negative of the power factor angle: $\phi = -(-\cos^{-1} 0.7) = 45.6°$.   So,

$$B = G \tan \phi = 0.347 \times 10^{-3} \tan 45.6° = 0.354 \times 10^{-3} \text{ S}$$

Finally, since $B = \omega C$,

$$C = \frac{B}{\omega} = \frac{0.354 \times 10^{-3}}{2\pi(60)} \text{ F} = 0.94 \ \mu\text{F}$$

14.9   A resistor in series with a capacitor absorbs 10 W when the combination is connected to a 120-V, 400-Hz source.   If the power factor is 0.6 leading, what are the resistance and capacitance?

Because this is a series circuit, impedance should be used to find the resistance and capacitance.   The impedance can be found by using the input current, which from $P = VI \times PF$ is

$$I = \frac{P}{V \times PF} = \frac{10}{120(0.6)} = 0.1389 \text{ A}$$

The magnitude of the impedance is equal to the voltage divided by the current, and the impedance angle is, for this capacitive circuit, the negative of the arccosine of the power factor:

$$\mathbf{Z} = \frac{V}{I} \underline{/-\cos^{-1} PF} = \frac{120}{0.1389} \underline{/-\cos^{-1} 0.6} = 864\underline{/-53.13°} = 518 - j691 \ \Omega$$

From the real part the resistance is $R = 518 \ \Omega$,   and from the imaginary part and $X = -1/\omega C$,   the capacitance is

$$C = -\frac{1}{\omega X} = \frac{-1}{2\pi(400)(-691)} \text{ F} = 0.576 \ \mu\text{F}$$

14.10   If a coil draws 0.5 A from a 120-V, 60-Hz source at a 0.7 lagging power factor, what are the coil resistance and inductance?

The resistance and inductance can be obtained from the impedance.   The impedance magnitude is $Z = V/I = 120/0.5 = 240 \ \Omega$,   and the impedance angle is the power factor angle: $\theta = \cos^{-1} 0.7 = 45.57°$.   So, the coil impedance is $\mathbf{Z} = 240\underline{/45.57°} = 168 + j171.4 \ \Omega$.   From the real part, the coil resistance is $R = 168 \ \Omega$,   and from the imaginary part the coil reactance is 171.4 Ω.   The inductance can be found from $X = \omega L$.   It is $L = X/\omega = 171.4/2\pi(60) = 0.455$ H.

14.11   A resistor and parallel capacitor draw 0.2 A from a 24-V, 400-Hz source at a 0.8 leading power factor.   Find the resistance and capacitance.

Since the components are in parallel, admittance should be used to find the resistance and capacitance.   The admittance magnitude is $Y = I/V = 0.2/24$ S $= 8.33$ mS,   and the admittance angle is, for this capacitive circuit, the arccosine of the power factor: $\cos^{-1} 0.8 = 36.9°$.   Thus, the admittance is

$$\mathbf{Y} = 8.33\underline{/36.9°} = 6.67 + j5 \text{ mS}$$

From the real part, the conductance of the resistor is 6.67 mS, and so the resistance is $R = 1/(6.67 \times 10^{-3}) = 150 \ \Omega$.   From the imaginary part the capacitive susceptance is 5 mS, and so the capacitance is

$$C = \frac{B}{\omega} = \frac{5 \times 10^{-3}}{2\pi(400)} \text{ F} = 1.99 \ \mu\text{F}$$

14.12 Operating at maximum capacity, a 12 470-V alternator supplies 35 MW at a 0.7 lagging power factor. What is the maximum real power that the alternator can deliver?

The limitation on the alternator capacity is the maximum voltamperes—the apparent power, which is the real power divided by the power factor. For this alternator, the maximum apparent power is $P/PF = 35/0.7 = 50$ MVA. At unity power factor all of this would be real power, which means that the maximum real power that this alternator can supply is 50 MW.

14.13 An induction motor delivers 50 hp while operating at 80 percent efficiency from 480-V lines. If the power factor is 0.6, what current does the motor draw? If the power factor is 0.9, instead, what current does this motor draw?

The current can be found from $P = VI \times PF$, where P is the motor input power of $50 \times 745.7/0.8$ W $= 46.6$ kW. For a power factor of 0.6, the current to the motor is

$$I = \frac{P}{V \times PF} = \frac{46.6 \times 10^3}{480 \times 0.6} = 162 \text{ A}$$

And, for a power factor of 0.9, it is

$$I = \frac{P}{V \times PF} = \frac{46.6 \times 10^3}{480 \times 0.9} = 108 \text{ A}$$

The 54-A decrease in current for the same output power shows that a large power factor is desirable.

14.14 For the circuit shown in Fig. 14-4, find the wattmeter reading when the $\pm$ terminal of the potential coil is connected to node a, and also when it is connected to node b.

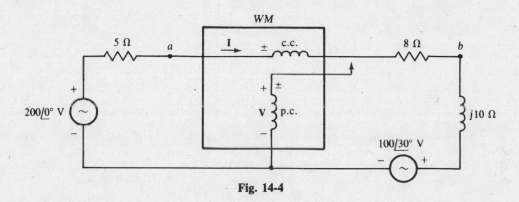

Fig. 14-4

The wattmeter reading is equal to $VI \cos\theta$, where V is the rms voltage across the potential coil, I is the rms current flowing through the current coil, and θ is the phase angle difference of the corresponding voltage and current phasors when they are referenced as shown with respect to the $\pm$ markings of the wattmeter coils. These three quantities must be found to determine the wattmeter reading.

The phasor current $\mathbf{I}$ is

$$\mathbf{I} = \frac{200\underline{/0°} - 100\underline{/30°}}{5 + 8 + j10} = \frac{124\underline{/-23.8°}}{16.4\underline{/37.6°}} = 7.56\underline{/-61.4°} \text{ A}$$

With the $\pm$ terminal of the potential coil at node a, the phasor voltage drop $\mathbf{V}$ across this coil is the $200\underline{/0°}$-V source voltage minus the drop across the 5-Ω resistor:

$$\mathbf{V} = 200\underline{/0°} - 5\mathbf{I} = 200\underline{/0°} - 5(7.56\underline{/-61.4°}) = 185\underline{/10.3°} \text{ V}$$

And the wattmeter reading is

$$P = VI \cos\theta = 185(7.56) \cos[10.3° - (-61.4°)] = 439 \text{ W}$$

With the $\pm$ terminal of the potential coil at node b, $\mathbf{V}$ is equal to the voltage drop across the $j10$-Ω

impedance and the $100\underline{/-30°}$-V source:

$$\mathbf{V} = j10(7.56\underline{/-61.4°}) + 100\underline{/30°} = 176\underline{/29.4°} \text{ V}$$

And so the wattmeter reading is

$$P = VI \cos\theta = 176(7.56) \cos[29.4° - (-61.4°)] = -18 \text{ W}$$

Probably the wattmeter cannot directly give a negative reading. If not, one wattmeter coil should be reversed so that the wattmeter reads upscale. Then the reading should be interpreted as being negative.

14.15 In the circuit shown in Fig. 14-5, find the total power absorbed by the three resistors. Then find the sum of the readings of the two wattmeters. Compare results.

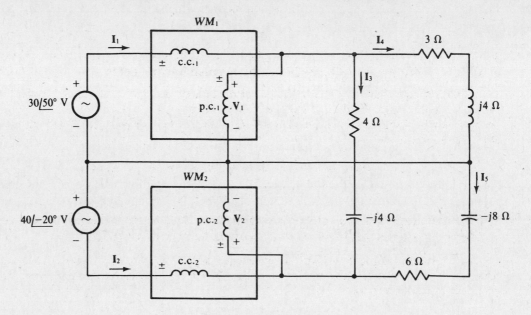

Fig. 14-5

The powers absorbed by the resistors can be found by using $P = I^2 R$. The currents through the resistors are

$$\mathbf{I}_3 = \frac{30\underline{/50°} + 40\underline{/-20°}}{4 - j4} = \frac{57.6\underline{/9.29°}}{5.66\underline{/-45°}} = 10.19\underline{/54.3°} \text{ A}$$

$$\mathbf{I}_4 = \frac{30\underline{/50°}}{3 + j4} = 6\underline{/-3.13°} \text{ A} \quad \text{and} \quad \mathbf{I}_5 = \frac{40\underline{/-20°}}{6 - j8} = 4\underline{/33.1°} \text{ A}$$

Of course, only the rms values of these currents are used in $P = I^2 R$:

$$P_T = I_3^2(4) + I_4^2(3) + I_5^2(6) = 10.19^2(4) + 6^2(3) + 4^2(6) = 619 \text{ W}$$

The currents $\mathbf{I}_1$ and $\mathbf{I}_2$ are needed in finding the wattmeter readings since these are the currents that flow through the current coils:

$$\mathbf{I}_1 = \mathbf{I}_3 + \mathbf{I}_4 = 10.19\underline{/54.3°} + 6\underline{/-3.13°} = 14.34\underline{/33.6°} \text{ A}$$

$$\mathbf{I}_2 = -\mathbf{I}_3 - \mathbf{I}_5 = -10.19\underline{/54.3°} - 4\underline{/33.1°} = 14\underline{/-131.6°} \text{ A}$$

Of course, the potential coil voltages are $\mathbf{V}_1 = 30\underline{/50°}$ V and $\mathbf{V}_2 = -40\underline{/-20°} = 40\underline{/160°}$ V. These potential coil voltages and current coil currents give wattmeter readings that have a sum of

$$P_T = 30(14.34) \cos(50° - 33.6°) + 40(14) \cos[160° - (-131.6°)] = 413 + 206 = 619 \text{ W}$$

Notice that this sum of the two wattmeter readings is equal to the total power absorbed. This should not be expected since each wattmeter reading cannot be associated with powers absorbed by certain resistors.   It can be shown, though, that this result is completely general for loads with three wires and for the connections shown.   This use of wattmeters is the famous *two-wattmeter method* that is popular for measuring power to three-phase loads, as will be considered in Chap. 16.

14.16 What is the reactive factor of an inductive load that has an apparent power input of 50 kVA while absorbing 30 kW?

The reactive factor is the sine of the power factor angle θ, which is

$$\theta = \cos^{-1}\frac{P}{S} = \cos^{-1}\frac{30\,000}{50\,000} = 53.1°$$

So $RF = \sin 53.1° = 0.8$

14.17 With $v = 200 \sin (377t + 30°)$ V   applied,   a   circuit   draws $i = 25 \sin (377t - 20°)$ A. What is the reactive factor and what is the reactive power absorbed?

The reactive factor is the sine of the power factor angle θ, which is the phase angle of the voltage minus the phase angle of the current: $\theta = 30° - (-20°) = 50°$.   So, $RF = \sin 50° = 0.766$.   The reactive power absorbed can be found from $Q = VI \times RF$,   where V and I are the rms values of the voltage and current:

$$Q = \frac{200}{\sqrt{2}} \times \frac{25}{\sqrt{2}} \times 0.766 = 1.92 \times 10^3 \text{ var} = 1.92 \text{ kvar}$$

14.18 What is the reactive factor of a circuit that has a $40\underline{/50°}$-Ω input impedance?   Also, what reactive power does the circuit absorb when the input current is 5 A?

The reactive factor is the sine of the impedance angle: $RF = \sin 50° = 0.766$.   One easy way to find the reactive power is with the formula $Q = I^2X$,   where X, the reactance, is equal to $40 \sin 50° = 30.64 \ \Omega$:

$$Q = I^2X = 5^2(30.64) = 766 \text{ var}$$

14.19 What is the reactive factor of a circuit that has a $20\underline{/-40°}$-Ω input impedance?   What is the reactive power absorbed with 240 V applied?

The reactive factor is the sine of the impedance angle: $RF = \sin (-40°) = -0.643$.   Perhaps the easiest way to find the reactive power absorbed is from $Q = VI \times RF$.   The only unknown in this formula is the rms current, which is equal to the rms voltage divided by the magnitude of impedance: $I = V/Z = 240/20 = 12$ A.   Then,

$$Q = VI \times RF = 240(12)(-0.643) \text{ var} = -1.85 \text{ kvar}$$

The negative sign indicates that the circuit delivers vars, as should be expected from this capacitive circuit.

As a check, the formula $Q = I^2X$ can be used, in which X, the imaginary part of the impedance, is $X = 20 \sin (-40°) = -12.86 \ \Omega$: $Q = 12^2(-12.86) \text{ var} = -1.85 \text{ kvar}$,   the same.

14.20 When 3 A flows through a circuit with a $0.4 + j0.5$-S input admittance, what reactive power does the circuit consume?

The reactive power consumed can be found from $Q = I^2X$ after X is found from the admittance.   Of course X is the imaginary part of the input impedance $\mathbf{Z}$.   We have

$$\mathbf{Z} = \frac{1}{\mathbf{Y}} = \frac{1}{0.4 + j0.5} = \frac{1}{0.64\underline{/51.3°}} = 1.56\underline{/-51.3°} = 0.976 - j1.22 \ \Omega$$

So, $X = -1.22 \, \Omega$, and

$$Q = I^2 X = 3^2(-1.22) = -11 \text{ var}$$

The negative sign indicates that the circuit delivers reactive power.

A check can be made by using $Q = -V^2 B$, where $V = IZ = 3(1.56) = 4.68$ V (of course, $B = 0.5$ S from the input admittance):

$$Q = -V^2 B = -(4.68)^2(0.5) = -11 \text{ var}$$

14.21 Two circuit elements in series consume 60 var when connected to a 120-V, 60-Hz source. If the reactive factor is 0.6, what are the two components and what are their values?

The two components can be found from the input impedance. The angle of this impedance is the arcsine of the reactive factor: $\theta = \sin^{-1} 0.6 = 36.9°$. The magnitude of the impedance can be found by substituting $I = V/Z$ into $Q = VI \times RF$:

$$Q = V\left(\frac{V}{Z}\right)(RF) \qquad \text{from which} \qquad Z = \frac{V^2(RF)}{Q} = \frac{120^2(0.6)}{60} = 144 \, \Omega$$

So
$$Z = 144\underline{/36.9°} = 115 + j86.4 \, \Omega$$

From this impedance, the two elements are a resistor with a resistance of $R = 115 \, \Omega$ and an inductor with a reactance of 86.4 Ω. The inductance is

$$L = \frac{X}{\omega} = \frac{86.4}{2\pi(60)} = 0.229 \text{ H}$$

14.22 What resistor and capacitor in parallel present the same load to a 480-V, 60-Hz source as a fully loaded 20-hp synchronous motor that operates at a 75 percent efficiency and a 0.8 leading power factor?

The resistance can be found from the motor input power, which is

$$P_{\text{in}} = \frac{P_{\text{out}}}{\eta} = \frac{20 \times 745.7}{0.75} \text{ W} = 19.9 \text{ kW}$$

From $P_{\text{in}} = V^2/R$,
$$R = \frac{V^2}{P_{\text{in}}} = \frac{480^2}{19.9 \times 10^3} = 11.6 \, \Omega$$

The corresponding conductance and the admittance angle, which is the negative of the power factor angle, can be used to find the capacitive susceptance. And then the capacitance can be found from this susceptance. The conductance is $G = 1/11.6 = 0.0863$ S, and the admittance angle is $\phi = \cos^{-1} 0.8 = 36.9°$. So, the susceptance is

$$B = G \tan \phi = 0.0863 \tan 36.9° = 0.0647 \text{ S}$$

Finally, the capacitance is this susceptance divided by the radian frequency:

$$C = \frac{B}{\omega} = \frac{0.0647}{2\pi(60)} \text{ F} = 172 \, \mu\text{F}$$

14.23 A 120-mH inductor is energized by 120 V at 60 Hz. Find the average, peak, and reactive powers absorbed.

Since the power factor is zero ($PF = \cos 90° = 0$), the inductor absorbs zero average power: $P = 0$ W. The peak power can be obtained from the instantaneous power. As derived in this chapter, the general expression for instantaneous power is

$$p = VI \cos \theta - VI \cos (2\omega t + \theta)$$

For an inductor, $\theta = 90°$, which means that the first term is zero. Consequently, the peak power is the peak value of the second term, which is VI: $p_{\text{max}} = VI$. The voltage V is given: $V = 120$ V. The current I can be found from this voltage divided by the inductive reactance:

$$I = \frac{V}{X} = \frac{120}{2\pi(60)(120 \times 10^{-3})} = \frac{120}{45.24} = 2.65 \text{ A}$$

So $p_{\max} = VI = 120(2.65) = 318 \text{ W}$

The reactive power absorbed is

$$Q = I^2 X = 2.65^2(45.24) = 318 \text{ var}$$

which has the same numerical value as the peak power absorbed by the inductor.   This is generally true because $Q = I^2 X = (IX)I = VI$,   which is the peak power absorbed.

14.24   What are the power components resulting from a 4-A current flowing into a $30\underline{/40°}$-Ω load?   In other words, what are the complex, real, reactive, and apparent powers of the load?

From Fig. 14-3b, the complex power **S** is

$$\mathbf{S} = I^2\mathbf{Z} = 4^2(30\underline{/40°}) = 480\underline{/40°} = 368 + j309 \text{ VA}$$

The real power is the real part, $P = 368 \text{ W}$,   the reactive power is the imaginary part, $Q = 309 \text{ var}$,   and the apparent power is the magnitude, $S = 480 \text{ VA}$.

14.25   Find the power components of an induction motor that delivers 5 hp while operating at an 85 percent efficiency and a 0.8 lagging power factor.

The input power is

$$P_{\text{in}} = \frac{P_{\text{out}}}{\eta} = \frac{5 \times 745.7}{0.85} \text{ W} = 4.386 \text{ kW}$$

The apparent power, which is the magnitude of the complex power, is the real power divided by the power factor: $S = 4.386/0.8 = 5.48 \text{ kVA}$.   The angle of the complex power is the power factor angle: $\theta = \cos^{-1} 0.8 = 36.9°$.   So, the complex power is

$$\mathbf{S} = 5.48\underline{/36.9°} = 4.386 + j3.29 \text{ kVA}$$

And, the reactive power is, of course, the imaginary part: $Q = 3.29 \text{ kvar}$.

14.26   Find the power components of a load that draws $20\underline{/-30°}$ A with $240\underline{/20°}$ V applied.

The complex power can be found from $\mathbf{S} = \mathbf{VI}^*$.   Since $\mathbf{I} = 20\underline{/-30°}$ A,   its conjugate is $\mathbf{I}^* = 20\underline{/30°}$ A,   and the complex power is

$$\mathbf{S} = (240\underline{/20°})(20\underline{/30°}) = 4800\underline{/50°} \text{ VA} = 3.09 + j3.68 \text{ kVA}$$

From the magnitude and real and imaginary parts, the apparent, real, and reactive powers are $S = 4.8 \text{ kVA}$, $P = 3.09 \text{ kW}$,   and $Q = 3.68 \text{ kvar}$.

14.27   A load, connected across a 12 470-V line, draws 20 A at a 0.75 lagging power factor.   Find the load impedance and the power components.

Since the impedance magnitude is equal to the voltage divided by the current, and the impedance angle is the power factor angle, the load impedance is

$$\mathbf{Z} = \frac{12\,470}{20}\underline{/\cos^{-1} 0.75} = 623.5\underline{/41.4°} \ \Omega$$

From $\mathbf{S} = I^2\mathbf{Z}$,   the complex power is

$$\mathbf{S} = 20^2(623.5\underline{/41.4°}) = 249.4 \times 10^3\underline{/41.4°} \text{ VA} = 187 + j165 \text{ kVA}$$

From the magnitude and the real and imaginary parts, $S = 249.4 \text{ kVA}$, $P = 187 \text{ kW}$,   and $Q = 165 \text{ kvar}$.

14.28 A 20-μF capacitor and a parallel 200-Ω resistor draw 4 A at 60 Hz. Find the power components.

Once the impedance is found, the complex power can be obtained from $S = I^2 Z$. The capacitive reactance is

$$X = -\frac{1}{\omega C} = \frac{-1}{2\pi(60)(20 \times 10^{-6})} = -132.6 \ \Omega$$

and the impedance of the parallel combination is

$$Z = \frac{200(-j132.6)}{200 - j132.6} = 110.5\underline{/-56.4°} \ \Omega$$

Substitution into $S = I^2 Z$ results in a complex power of

$$S = 4^2(110.5\underline{/-56.4°}) = 1.77 \times 10^3\underline{/-56.4°} \ VA = 0.98 - j1.47 \ kVA$$

So, $S = 1.77$ kVA, $P = 0.98$ kW, and $Q = -1.47$ kvar.

14.29 A fully loaded 10-hp induction motor operates from a 480-V, 60-Hz line at an efficiency of 85 percent and a 0.8 lagging power factor. Find the overall power factor when a 33.3-μF capacitor is placed in parallel with the motor.

The power factor can be determined from the power factor angle, which is $\theta = \tan^{-1}(Q_T/P_{in})$. For this, the input power P_{in} and the total reactive power Q_T are needed. The capacitor does not change the real power absorbed, which is

$$P_{in} = \frac{P_{out}}{\eta} = \frac{10 \times 745.7}{0.85} \ W = 8.77 \ kW$$

The total reactive power is the sum of the motor and capacitive reactive powers. As is evident from power triangle considerations, the reactive power Q_M of the motor is equal to the power times the tangent of the motor power factor angle, which is the arccosine of the motor power factor:

$$Q_M = P_{in} \tan \theta_M = 8.77 \tan (\cos^{-1} 0.8) = 6.58 \ kvar$$

The reactive power absorbed by the capacitor is

$$Q_C = -\omega C V^2 = -2\pi(60)(33.3 \times 10^{-6})(480)^2 \ var = -2.89 \ kvar$$

And the total reactive power is

$$Q_T = Q_M + Q_C = 6.58 - 2.89 = 3.69 \ kvar$$

With Q_T and P_{in} known, the power factor angle θ can be determined:

$$\theta = \tan^{-1} \frac{Q_T}{P_{in}} = \tan^{-1} \frac{3.69 \times 10^3}{8.77 \times 10^3} = 22.8°$$

And the overall power factor is $PF = \cos 22.8° = 0.922$. It is lagging because the power factor angle is positive.

14.30 A 240-V source energizes the parallel combination of a purely resistive 6-kW heater and an induction motor that draws 7 kVA at a 0.8 lagging power factor. Find the overall load power factor and also the current from the source.

The power factor and current can be determined from the total complex power S_T, which is the sum of the complex powers of the heater and motor:

$$S_T = S_H + S_M = 6000\underline{/0°} + 7000\underline{/\cos^{-1} 0.8} = 6000\underline{/0°} + 7000\underline{/36.9°}$$
$$= 6000 + (5600 + j4200) \ VA = 12.34\underline{/19.9°} \ kVA$$

The overall power factor is the cosine of the angle of the total complex power: $PF = \cos 19.9° = 0.94$. Of course, it is lagging because the power factor angle is positive. The source current is equal to the total apparent power divided by the voltage:

$$I = \frac{12.34 \times 10^3}{240} = 51.4 \text{ A}$$

Notice that the total apparent power of 12.34 kVA is not the sum of the load apparent powers of 6 and 7 kVA. This is generally true except in the unusual situation in which all complex powers have the same angle.

14.31 A 480-V source energizes two loads in parallel, supplying 2 kVA at a 0.5 lagging power factor to one load and 4 kVA at a 0.6 leading power factor to the other load. Find the source current and also the total impedance of the combination.

The current can be found from the total apparent power, which is the magnitude of the total complex power:

$$S = 2000\underline{/\cos^{-1} 0.5} + 4000\underline{/-\cos^{-1} 0.6} = 2000\underline{/60°} + 4000\underline{/-53.13°} \text{ VA} = 3.703\underline{/-23.4°} \text{ kVA}$$

Of course, the power factor angle for the 4-kVA load is negative because the power factor is leading, which means that the current leads the voltage.

The current is equal to the apparent power divided by the voltage:

$$I = \frac{S}{V} = \frac{3.703 \times 10^3}{480} = 7.715 \text{ A}$$

From $S = I^2 Z$, the impedance is equal to the complex power divided by the square of the current:

$$Z = \frac{S}{I^2} = \frac{3.703 \times 10^3\underline{/-23.4°}}{7.715^2} = 62.2\underline{/-23.4°} \ \Omega$$

14.32 Three loads are connected across a 277-V line. One is a fully loaded 5-hp induction motor operating at a 75 percent efficiency and a 0.7 lagging power factor. Another is a fully loaded 7-hp synchronous motor operating at an 80 percent efficiency and a 0.4 leading power factor. And the third is a 5-kW resistive heater. Find the total line current and the overall power factor.

The line current and power factor can be determined from the total complex power, which is the sum of the individual complex powers. The complex power of the induction motor has a magnitude that is equal to the input power divided by the power factor, and an angle that is the power factor angle. The same is true for the synchronous motor. Of course, the complex power for the heater is the same as the real power. So,

$$S = \frac{5 \times 745.7}{0.75 \times 0.7}\underline{/\cos^{-1} 0.7} + \frac{7 \times 745.7}{0.8 \times 0.4}\underline{/-\cos^{-1} 0.4} + 5000\underline{/0°}$$
$$= 7.1 \times 10^3\underline{/45.6°} + 16.3 \times 10^3\underline{/-66.4°} + 5000\underline{/0°} \text{ VA} = 19.23\underline{/-30.9°} \text{ kVA}$$

The total line current is equal to the apparent power divided by the line voltage: $I = (19.23 \times 10^3)/277 = 69.4$ A. And, the overall power factor is the cosine of the angle of the total complex power: $PF = \cos(-30.9°) = 0.858$. It is leading because the power factor angle is negative.

14.33 In the circuit shown in Fig. 14-6, load 1 absorbs 2.4 kW and 1.8 kvar, load 2 absorbs 1.3 kW and 2.6 kvar, and load 3 absorbs 1 kW and generates 1.2 kvar. Find the total power components, the source current I_1, and the impedance of each load.

The total complex power is the sum of the individual complex powers:

$$S_T = S_1 + S_2 + S_3 = (2400 + j1800) + (1300 + j2600) + (1000 - j1200)$$
$$= 4700 + j3200 \text{ VA} = 5.69\underline{/34.2°} \text{ kVA}$$

From the total complex power, the total apparent power is $S_T = 5.69$ kVA, the total real power is $P_T = 4.7$ kW, and the total reactive power is $Q_T = 3.2$ kvar. Of course, the source current

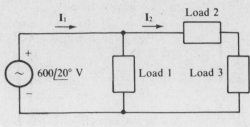

Fig. 14-6

magnitude I_1 is equal to the apparent power divided by the source voltage: $I_1 = (5.69 \times 10^3)/600 = 9.48$ A. And the angle of I_1 is the angle of the voltage minus the power factor angle: $20° - 34.2° = -14.2°$. So, $I_1 = 9.48\underline{/-14.2°}$ A.

The angle of the load 1 impedance Z_1 is the load power factor angle, which is also the angle of the complex power S_1. Since $S_1 = 2400 + j1800 = 3000\underline{/36.9°}$ VA, this impedance angle is $\theta = 36.9°$. Because the load 1 voltage is known, the magnitude Z_1 can be found from $S_1 = V^2/Z_1$:

$$Z_1 = \frac{V^2}{S_1} = \frac{600^2}{3000} = 120 \ \Omega$$

So, $Z_1 = Z_1\underline{/\theta} = 120\underline{/36.9°}$ Ω. The impedances Z_2 and Z_3 of loads 2 and 3 cannot be found in a similar manner because the load voltages are not known. But the rms current I_2 can be found from the sum of the complex powers of loads 2 and 3, and used in $S = I^2Z$ to find the impedances. This sum is

$$S_{23} = (1300 + j2600) + (1000 - j1200) = 2300 + j1400 \text{ VA} = 2.693\underline{/31.3°} \text{ kVA}$$

The apparent power S_{23} can be used to obtain I_2 from $S_{23} = VI_2$:

$$I_2 = \frac{S_{23}}{V} = \frac{2.693 \times 10^3}{600} = 4.49 \text{ A}$$

Since, $S_2 = 1300 + j2600 \text{ VA} = 2.91\underline{/63.4°} \text{ kVA}$, the impedance of load 2 is

$$Z_2 = \frac{S_2}{I_2^2} = \frac{2.91 \times 10^3\underline{/63.4°}}{4.49^2} = 144\underline{/63.4°} \ \Omega$$

Similarly, $S_3 = 1000 - j1200 \text{ VA} = 1.562\underline{/-50.2°} \text{ kVA}$, and

$$Z_3 = \frac{S_3}{I_2^2} = \frac{1.562 \times 10^3\underline{/-50.2°}}{4.49^2} = 77.6\underline{/-50.2°} \ \Omega$$

14.34 A load that absorbs 100-kW at a 0.7 lagging power factor has capacitors placed across it to produce an overall power factor of 0.9 lagging. The line voltage is 480 V. How much reactive power must the capacitors produce, and what is the resulting decrease in line current?

The initial reactive power is $Q_i = P \tan \theta_i$, where θ_i is the initial power factor angle: $\theta_i = \cos^{-1} 0.7 = 45.6°$. Therefore

$$Q_i = 100 \times 10^3 \tan 45.6° \text{ var} = 102 \text{ kvar}$$

The final reactive power is

$$Q_f = P \tan \theta_f = 100 \times 10^3 \tan (\cos^{-1} 0.9) \text{ var} = 48.4 \text{ kvar}$$

Consequently, the capacitors must supply $102 - 48.4 = 53.6$ kvar.

The initial and final currents can be obtained from $P = VI \times PF$:

$$I_i = \frac{P}{V \times PF_i} = \frac{100 \times 10^3}{480 \times 0.7} = 297.6 \text{ A} \quad \text{and} \quad I_f = \frac{P}{V \times PF_f} = \frac{100 \times 10^3}{480 \times 0.9} = 231.5 \text{ A}$$

The resulting decrease in line current is $297.6 - 231.5 = 66.1$ A.

14.35 A synchronous motor that draws 20 kW is in parallel with an induction motor that draws 50 kW at a lagging power factor of 0.7. If the synchronous motor is operated at a leading power factor, how much reactive power must it provide to cause the overall power factor to be 0.9 lagging, and what is its power factor?

Since the total power input is $P_T = 20 + 50 = 70$ kW, the total reactive power is

$$Q_T = P_T \tan(\cos^{-1} PF_T) = 70 \tan(\cos^{-1} 0.9) = 33.9 \text{ kvar}$$

Because the reactive power absorbed by the induction motor is

$$Q_{IM} = P_{IM} \tan\theta_{IM} = 50 \tan(\cos^{-1} 0.7) = 51 \text{ kvar}$$

the synchronous motor must supply $Q_{IM} - Q_T = 51 - 33.9 = 17.1$ kvar. Thus, $Q_{SM} = -17.1$ kvar. The resulting power factor of the synchronous motor is $\cos\theta_{SM}$ in which θ_{SM}, the synchronous motor power factor angle, is

$$\theta_{SM} = \tan^{-1}\frac{Q_{SM}}{P_{SM}} = \tan^{-1}\frac{-17.1 \times 10^3}{20 \times 10^3} = -40.5°$$

So, $PF_{SM} = \cos(-40.5°) = 0.76$ leading.

14.36 A factory draws 100 A at a 0.7 lagging power factor from a 12 470-V, 60-Hz line. What capacitor placed across the line at the input to the factory increases the overall power factor to unity? Also, what are the final currents for the factory, capacitor, and line?

The capacitance can be determined from the reactive power that the capacitor must provide to cause the power factor to be unity. The reactive power absorbed by the factory is the apparent power times the reactive factor, which is the sine of the arccosine of the power factor: $RF = \sin(\cos^{-1} 0.7) = 0.714$. Thus

$$Q = VI \times RF = 12\,470 \times 100 \times 0.714 \text{ var} = 890.5 \text{ kvar}$$

For a unity power factor, the capacitor must supply all this reactive power. Since the formula for capacitor reactive power is $Q = \omega C V^2$, the required capacitance is

$$C = \frac{Q}{\omega V^2} = \frac{890.5 \times 10^3}{2\pi(60)(12\,470)^2} \text{ F} = 15.2 \ \mu\text{F}$$

Adding the capacitor in parallel does not change the current input to the factory since there is no change in the factory load. This current remains at 100 A. The current to the capacitor can be found from $Q = VI_C \times RF$ with $RF = -1$ since the power factor angle is $-90°$ for the capacitor. The result is

$$I_C = \frac{Q}{V \times RF} = \frac{-890.5 \times 10^3}{(12\,470)(-1)} = 71.4 \text{ A}$$

The total final line current I_{fL} can be found from the power input which is

$$P = VI_{iL} \times PF_i = (12\,470)(100)(0.7) \text{ W} = 873 \text{ kW}$$

Adding the capacitor does not change this power, but it does change the power factor to 1. So, from $P = VI_{fL} \times PF_f$,

$$873 \times 10^3 = 12\,470(I_{fL})(1) \quad \text{from which} \quad I_{fL} = \frac{873 \times 10^3}{12\,470} = 70 \text{ A}$$

Notice that the 70-A rms final line current is not equal to the sum of the capacitor 71.4-A rms current and the factory 100-A rms current. This should not be surprising because, in general, rms quantities cannot be validly added since the phasor angles are not included.

14.37 A 240-V, 60-Hz source energizes a $30\underline{/50°}$-Ω load. What capacitor in parallel with this load produces an overall power factor of 0.95 lagging?

Although powers could be used in the solution, it is often easier to use admittance when a circuit or its impedance is specified.   The initial admittance is

$$\mathbf{Y} = \frac{1}{30\underline{/50°}} = 33.3 \times 10^{-3}\underline{/-50°} \text{ S} = 21.4 - j25.5 \text{ mS}$$

Adding the capacitor changes only the susceptance, which becomes

$$B = G \tan(-\theta) = 21.4 \tan(-\cos^{-1} 0.95) = -7.04 \text{ mS}$$

This formula $B = G \tan(-\theta)$ should be evident from admittance triangle considerations and the fact that the admittance angle is the negative of the power factor angle.   From $\Delta B = \omega C$,

$$C = \frac{\Delta B}{\omega} = \frac{25.5 \times 10^{-3} - 7.04 \times 10^{-3}}{2\pi(60)} = 49.1 \times 10^{-6} \text{ F} = 49.1 \text{ } \mu\text{F}$$

14.38 At 60 Hz, what is the power factor of the circuit shown in Fig. 14-7?   What capacitor connected across the input terminals causes the overall power factor to be 1 (unity)?   What capacitor causes the overall power factor to be 0.85 lagging?

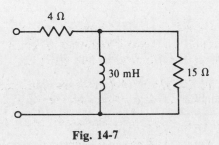

Fig. 14-7

Because a circuit is specified, the power factor and capacitor are probably easier to find using impedance and admittance instead of powers.   The power factor is the cosine of the impedance angle.   Since the reactance of the inductor is $2\pi(60)(0.03) = 11.3 \text{ } \Omega$,   the impedance of the circuit is

$$\mathbf{Z} = 4 + \frac{15(j11.3)}{15 + j11.3} = 11.9\underline{/37.38°} \text{ } \Omega$$

And the power factor is $PF = \cos 37.38° = 0.795$ lagging.

Because the capacitor is to be connected in parallel, the circuit admittance should be used to determine the capacitance.   Before the capacitor is added, this admittance is

$$\mathbf{Y} = \frac{1}{\mathbf{Z}} = \frac{1}{11.9\underline{/37.38°}} = 0.0842\underline{/-37.38°} \text{ S} = 66.9 - j51.1 \text{ mS}$$

For unity power factor, the imaginary part of the admittance must be zero, which means that the added capacitor must have a susceptance of 51.1 mS.   Consequently, its capacitance is

$$C = \frac{B}{\omega} = \frac{51.1 \times 10^{-3}}{2\pi(60)} = 136 \times 10^{-6} \text{ F} = 136 \text{ } \mu\text{F}$$

A different capacitor is required for a power factor of 0.85 lagging.   The new susceptance can be found from $B = G \tan(-\theta)$ (in this formula, G is the conductance, which does not change by adding a parallel capacitor, and θ is the new power factor angle):

$$B = 66.9 \tan(-\cos^{-1} 0.85) = -41.5 \text{ mS}$$

Because the added capacitor provides the change in susceptance, its capacitance is

$$C = \frac{\Delta B}{\omega} = \frac{51.1 \times 10^{-3} - 41.5 \times 10^{-3}}{2\pi(60)} = 25.6 \times 10^{-6} \text{ F} = 25.6 \text{ } \mu\text{F}$$

Of course, less capacitance is required to improve the power factor to 0.85 lagging than to 1.

14.39 An induction motor draws 50 kW at a 0.6 lagging power factor from a 480-V, 60-Hz source. What parallel capacitor will increase the overall power factor to 0.9 lagging? What is the resulting decrease in input current?

Of course, adding the capacitor causes a decrease in consumed reactive power, but does not change the consumed real power. The decrease in consumed reactive power is

$$\Delta Q = Q_i - Q_f = P \tan \theta_i - P \tan \theta_f$$

where

$$\theta_i = \cos^{-1} 0.6 = 53.1°$$

$$\theta_f = \cos^{-1} 0.9 = 25.8°$$

So

$$\Delta Q = 50\,000 \tan 53.1° - 50\,000 \tan 25.8° = 42.5 \times 10^3 \text{ var}$$

and

$$C = \frac{\Delta Q}{\omega V^2} = \frac{42.5 \times 10^3}{2\pi(60)(480)^2} = 489 \times 10^{-6} \text{ F} = 489 \ \mu\text{F}$$

From $P = VI \times PF$, the decrease in input current is

$$\Delta I = I_i - I_f = \frac{P}{V \times PF_i} - \frac{P}{V \times PF_f} = \frac{50\,000}{480(0.6)} - \frac{50\,000}{480(0.9)} = 57.9 \text{ A}$$

14.40 A factory draws 30 MVA at a 0.7 lagging power factor from a 12 470-V, 60-Hz line. Find the capacitance of the parallel capacitors required to improve the power factor to 0.85 lagging. Also, find the resulting decrease in line current.

Of course, adding parallel capacitors decreases the consumed reactive power, and this decrease can be used to find the capacitance required. The initial reactive power consumed is

$$Q_i = S \sin \theta_i = 30 \sin (\cos^{-1} 0.7) = 21.4 \text{ Mvar}$$

The final reactive power cannot be found in a similar way because the apparent power changes with adding the capacitors. But the real power does not, and so it can be used. This power is $P = 30(0.7) = 21$ MW. So,

$$Q_f = P \tan \theta_f = 21 \tan (\cos^{-1} 0.85) = 13 \text{ Mvar}$$

and the decrease in reactive power is $\Delta Q = Q_i - Q_f = 21.4 - 13 = 8.4$ Mvar. From this and $\Delta Q = \omega C V^2$, the capacitance required is

$$C = \frac{\Delta Q}{\omega V^2} = \frac{8.4 \times 10^6}{2\pi(60)(12\,470)^2} \text{ F} = 143 \ \mu\text{F}$$

The decrease in line current is equal to the decrease in apparent power divided by the line voltage. The initial apparent power is the specified 30 MVA, and the final apparent power is $P/PF_f = 21 \times 10^6/0.85 = 24.7 \times 10^6$ VA. So,

$$\Delta I = \frac{30 \times 10^6 - 24.7 \times 10^6}{12\,470} = 425 \text{ A}$$

14.41 A 20-MW industrial load supplied from a 12 470-V, 60-Hz line has its power factor improved to 0.9 lagging by the addition of a 230-μF bank of capacitors. Find the power factor of the original load.

The initial reactive power is needed. It is equal to the final reactive power plus that added by the capacitors:

$$Q_i = P \tan \theta_f + \omega C V^2 = 20 \times 10^6 \tan (\cos^{-1} 0.9) + 2\pi(60)(230 \times 10^{-6})(12\,470)^2$$
$$= 9.69 \times 10^6 + 13.5 \times 10^6 \text{ var} = 23.2 \text{ Mvar}$$

The real power and the initial reactive power can be used to find the initial power factor angle:

$$\theta_i = \tan^{-1} \frac{Q_i}{P} = \tan^{-1} \frac{23.2 \times 10^6}{20 \times 10^6} = 49.2°$$

Finally, the initial power factor is $PF_i = \cos \theta_i = \cos 49.2° = 0.653$ lagging.

14.42 A 480-V, 60-Hz source energizes a load consisting of an induction motor and a synchronous motor. The induction motor draws 50 kW at a 0.65 lagging power factor, and the synchronous motor draws 10 kW at a 0.6 leading power factor. Find the capacitance of the parallel capacitor required to produce an overall power factor of 0.9 lagging.

The required change in reactive power is needed. The initial absorbed reactive power is the sum of that of the two motors, which from $Q = P \tan \theta$ is

$$Q_i = 50 \tan (\cos^{-1} 0.65) + 10 \tan (-\cos^{-1} 0.6) = 58.456 - 13.333 = 45.12 \text{ kvar}$$

The final reactive power is, from $Q_f = P_T \tan (\cos^{-1} PF_f)$,

$$Q_f = (50 + 10) \tan (\cos^{-1} 0.9) = 29.06 \text{ kvar}$$

So the change ΔQ in reactive power is $\Delta Q = 45.12 - 29.06 = 16.1 \text{ kvar}$ and

$$C = \frac{\Delta Q}{\omega V^2} = \frac{16.1 \times 10^3}{2\pi (60)(480)^2} \text{ F} = 185 \ \mu\text{F}$$

Supplementary Problems

14.43 The instantaneous power absorbed by a circuit is $p = 6 + 4 \cos^2 (2t + 30°)$ W. Find the maximum, minimum, and average powers absorbed. *Ans.* $p_{max} = 10$ W, $p_{min} = 6$ W, $P = 8$ W

14.44 With $170 \sin (377t + 10°)$ V applied, a circuit draws $8 \sin (377t + 35°)$ A. Find the power factor and the maximum, minimum, and average powers absorbed.
Ans. $PF = 0.906$ leading, $p_{max} = 1.3$ kW, $p_{min} = -63.7$ W, $P = 616$ W

14.45 For each following load voltage and current pair, find the corresponding power factor and average power absorbed:

(a) $v = 170 \sin (50t - 40°)$ V, $i = 4.3 \sin (50t + 10°)$ A

(b) $v = 340 \cos (377t - 50°)$ V, $i = 6.1 \sin (377t + 30°)$ A

(c) $v = 679 \sin (377t + 40°)$ V, $i = -7.2 \cos (377t + 50°)$ A

Ans. (a) 0.643 leading, 235 W; (b) 0.985 lagging, 1.02 kW; (c) 0.174 lagging, 424 W

14.46 Find the power factor of a fully loaded 5-hp induction motor that operates at 85 percent efficiency while drawing 15 A from a 480-V line. *Ans.* 0.609 lagging

14.47 What is the power factor of a circuit that has a $5\underline{/-25°}$-Ω input impedance? Also, what is the power absorbed when 50 V is applied? *Ans.* 0.906 leading, 453 W

14.48 If a circuit has a $40 + j20$-S input admittance and a 180-V applied voltage, what is the power factor and the power absorbed? *Ans.* 0.894 leading, 1.3 MW

14.49 A resistor in parallel with an inductor absorbs 25 W when the combination is connected to a 120-V, 60-Hz source. If the total current is 0.3 A, what are the resistance and inductance?
Ans. 576 Ω, 1.47 H

14.50 A coil absorbs 20 W when connected to a 240-V, 400-Hz source. If 0.2 A flows, find the resistance and inductance of the coil. *Ans.* 500 Ω, 0.434 H

14.51 A resistor and series capacitor draw 1 A from a 120-V, 60-Hz source at a 0.6 leading power factor. Find the resistance and capacitance.   *Ans.* 72 Ω, 27.6 μF

14.52 A resistor and parallel capacitor draw 0.6 A from a 120-V, 400-Hz source at a 0.7 leading power factor. Find the resistance and capacitance.   *Ans.* 286 Ω, 1.42 μF

14.53 A 100-kW load operates at a 0.6 lagging power factor from a 480-V, 60-Hz line. What current does the load draw? What current does the load draw if it operates at unity power factor instead?
Ans. 347 A, 208 A

14.54 A fully loaded 100-hp induction motor operates at 85 percent efficiency from a 480-V line. If the power factor is 0.65 lagging, what current does the motor draw? If the power factor is 0.9 lagging, instead, what current does this motor draw?   *Ans.* 281 A, 203 A

14.55 Find the wattmeter reading for the circuit shown in Fig. 14-8.   *Ans.* 16 W

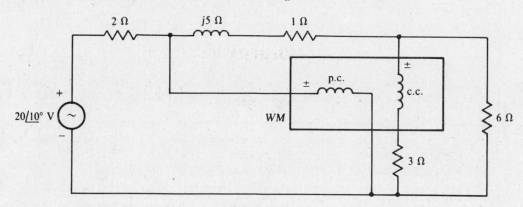

Fig. 14-8

14.56 Find each wattmeter reading for the circuit shown in Fig. 14-9.
    Ans. $WM_1 = 1.54$ kW, $WM_2 = 656$ W

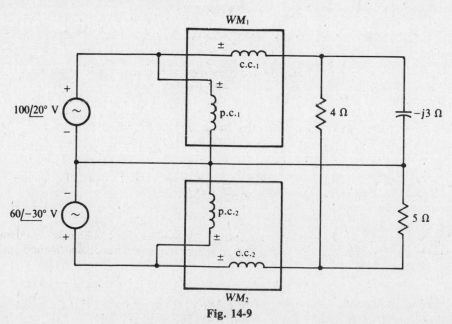

Fig. 14-9

14.57   With 200 sin (754t + 35°) V applied, a circuit draws 456 sin (754t + 15°) mA.   What is the reactive factor, and what is the reactive power absorbed?   *Ans.* 0.342, 15.6 var

14.58   With 300 cos (377t − 75°) V applied, a circuit draws 2.1 sin (377t + 70°) A.   What is the reactive factor, and what is the reactive power absorbed?   *Ans.* −0.819, −258 var

14.59   What is the reactive factor of a circuit that has a 50/35°-Ω input impedance?   What reactive power does the circuit absorb when the input current is 4 A?   *Ans.* 0.574, 459 var

14.60   What is the reactive factor of a circuit that has a 600/−30°-Ω input impedance?   What is the reactive power absorbed with 480 V applied?   *Ans.* −0.5, −192 var

14.61   When 120 V is applied across a circuit with a 1.23/40°-S input admittance, what reactive power does the circuit absorb?   *Ans.* −11.4 kvar

14.62   When 4.1 A flows into a circuit with a 0.7 − j1.1-S input admittance, what reactive power does the circuit absorb?   *Ans.* 10.9 var

14.63   A load consumes 500 var when energized from a 240-V source.   If the reactive factor is 0.35, what current does the load draw and what is the load impedance?   *Ans.* 5.95 A, 40.3/20.5° Ω

14.64   Two circuit elements in parallel consume 90 var when connected to a 120-V, 60-Hz source.   If the reactive factor is 0.8, what are the two components and what are their values?
Ans. A 213-Ω resistor and a 0.424-H inductor

14.65   Two circuit elements in series consume −80 var when connected to a 240-V, 60-Hz source.   If the reactive factor is −0.7, what are the two components and what are their values?
Ans. A 360-Ω resistor and a 7.52-μF capacitor

14.66   A 300-mA, 60-Hz current flows through a 10-μF capacitor.   Find the average, peak, and reactive powers absorbed.   *Ans.*   P = 0 W, p_{max} = 23.9 W,   Q = −23.9 var

14.67   What are the power components resulting from a 3.6-A current flowing through a 50/−30°-Ω load?   *Ans.*   **S** = 648/−30° VA,   S = 648 VA,   P = 561 W,   Q = −324 var

14.68   Find the power components of a fully loaded 10-hp synchronous motor operating at an 87 percent efficiency and a 0.7 leading power factor.   *Ans.*   **S** = 12.2/−45.6° kVA,   S = 12.2 kVA,   P = 8.57 kW,   Q = −8.74 kvar

14.69   A load draws 3 A with 75 V applied.   If the load power factor is 0.6 lagging, find the power components of the load.   *Ans.*   **S** = 225/53.1° VA,   S = 225 VA,   P = 135 W,   Q = 180 var

14.70   Find the power components of a load that draws 8.1/36° A with 480/10° V applied.
Ans.   **S** = 3.89/−26° kVA,   S = 3.89 kVA,   P = 3.49 kW,   Q = −1.7 kvar

14.71   A 120-mH inductor and a parallel 30-Ω resistor draw 6.1 A at 60-Hz.   Find the power components.
Ans.   **S** = 930/33.6° VA,   S = 930 VA,   P = 775 W,   Q = 514 var

14.72 A fully loaded 15-hp induction motor operates from a 480-V, 60-Hz line at an efficiency of 83 percent and a 0.7 lagging power factor. Find the overall power factor when a 75-μF capacitor is placed in parallel with the motor.   *Ans.* 0.881 lagging

14.73 Two loads are connected in parallel across a 277-V line. One is a fully loaded 5-hp induction motor that operates at an 80 percent efficiency and a 0.7 lagging power factor. The other is a 5-kW resistive heater. Find the overall power factor and line current.   *Ans.* 0.897 lagging, 38.9 A

14.74 Two loads are connected in parallel across a 12 470-V line. One load takes 23 kVA at a 0.75 lagging power factor and the other load takes 10 kVA at a 0.6 leading power factor. Find the total line current and also the impedance of the combination.   *Ans.* 1.95 A, 6.39$\underline{/17.2°}$ kΩ

14.75 Three loads are connected across a 480-V line. One is a fully loaded 10-hp induction motor operating at an 80 percent efficiency and a 0.6 lagging power factor. Another is a fully loaded 5-hp synchronous motor operating at a 75 percent efficiency and a 0.6 leading power factor. And the third is a 7-kW resistive heater. Find the total line current and the overall power factor.   *Ans.* 46 A, 0.965 lagging

14.76 In the circuit shown in Fig. 14-10, load 1 absorbs 6.3 kW and 9.27 kvar, and load 2 absorbs 5.26 kW and generates 2.17 kvar. Find the total power components, the source voltage **V**, and the impedance of each load.

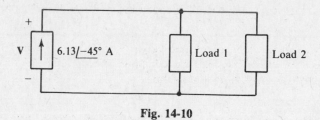

Fig. 14-10

Ans. $\mathbf{S}_T = 13.6\underline{/31.6°}$ kVA $S_T = 13.6$ kVA $P_T = 11.6$ kW $Q_T = 7.1$ kvar

   $\mathbf{V} = 2.21\underline{/-13.4°}$ kV $\mathbf{Z}_1 = 437\underline{/55.8°}\ \Omega$ $\mathbf{Z}_2 = 861\underline{/-22.4°}\ \Omega$

14.77 How much reactive power must be supplied by parallel capacitors to a 50-kVA load with a 0.65 lagging power factor in order to increase the overall power factor to 0.85 lagging?   *Ans.* 17.9 kvar

14.78 An electric motor delivers 50 hp while operating from a 480-V line at an 83 percent efficiency and a 0.65 lagging power factor. If it is paralleled with a capacitor that increases the overall power factor to 0.9 lagging, what is the decrease in line current?   *Ans.* 40 A

14.79 A load energized from a 480-V, 60-Hz line has a power factor of 0.6 lagging. If placing a 100-μF capacitor across the line raises the overall power factor to 0.85 lagging, find the real power of the load and the decrease in line current.   *Ans.* 12.2 kW, 12.4 A

14.80 A factory draws 90 A at a 0.75 lagging power factor from a 25 000-V, 60-Hz line. Find the capitance of a parallel capacitor that will increase the overall power factor to 0.9 lagging.   *Ans.* 2.85 μF

14.81 A fully loaded 75-hp induction motor operates from a 480-V, 60-Hz line at an 80 percent efficiency and a 0.65 lagging power factor. The power factor is to be raised to 0.9 lagging by placing a capacitor across the motor terminals. Find the capacitance required and the resulting decrease in line current.
Ans. 551 μF, 62.2 A

14.82 A $50\underline{/60°}$-Ω load is connected to a 480-V, 60-Hz source. What capacitor connected in parallel with the load will produce an overall power factor of 0.9 lagging? *Ans.* 33.1 μF

14.83 At 400 Hz, what is the power factor of the circuit shown in Fig. 14-11? What capacitor connected across the input terminals causes the overall power factor to be 0.9 lagging?
Ans. 0.77 lagging, 8.06 μF

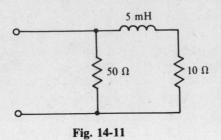

Fig. 14-11

14.84 For a load energized by a 277-V, 60-Hz source, an added parallel 5-μF capacitor improves the power factor from 0.65 lagging to 0.9 lagging. What is the source current both before and after the capacitor is added? *Ans.* 1.17 A, 0.847 A

Chapter 15

Transformers

INTRODUCTION

A *transformer* has two or more windings, also called coils, that are magnetically coupled. As shown in Fig. 15-1, a typical transformer has two windings wound on a core that may be made from iron. Here, winding 1 has $N_1 = 4$ number of turns, and winding 2 has $N_2 = 3$ number of turns. Of course, a turn is just one encirclement of the core. Circuit 1, connected to winding 1, is often a source, and circuit 2, connected to winding 2, is often a load. In this case, winding 1 is called the *primary winding* or just *primary*, and winding 2 is called the *secondary winding* or just *secondary*.

In the operation, current i_1 flowing in winding 1 produces a magnetic flux ϕ_{1m} that, for power transformers, is ideally confined to the core and so passes through or couples winding 2. The m in the subscript means "mutual"—the flux is *mutual* to both windings. Similarly, current i_2 flowing in winding 2 produces a flux ϕ_{2m} that, for power transformers, ideally couples winding 1. When these currents change in magnitude or direction, they produce corresponding changes in the fluxes. And these changing fluxes induce voltages in the windings. In this way, the transformer couples circuit 1 to circuit 2 so that electric energy can flow from one circuit to the other.

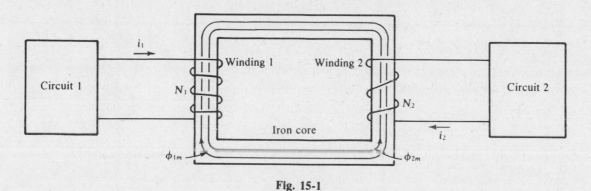

Fig. 15-1

Although flux is a convenient aid for understanding transformer operation, it is not used in the analyses of transformer circuits. Instead, either transformer turns ratios or inductances are used, as will be explained.

Transformers are very important electrical components. At high efficiencies, they change voltage and current levels, which is essential for electric power distribution. In electronic applications they match load impedances to source impedances for maximum power transfer. And, they couple amplifiers together without any direct metallic connections that would conduct dc currents. At the same time they may act with capacitors to filter signals.

RIGHT-HAND RULE

In Fig. 15-1 the flux ϕ_{1m} produced by i_1 has a clockwise direction, but ϕ_{2m} produced by i_2 has a counterclockwise direction. The direction of the flux produced by current flowing in a winding can be determined from a version of the *right-hand rule* that is different from that presented in Chap. 8 for a single wire. As shown in Fig. 15-2, if the fingers of a right hand encircle a winding in the direction of the current, the thumb points in the direction of the flux produced in the winding by the current.

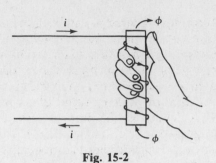

Fig. 15-2

DOT CONVENTION

Using dots at winding terminals in agreement with the *dot convention* is a convenient method for specifying winding direction relations. One terminal of each winding is dotted, or marked in some other way, with the dotted terminals selected such that *currents flowing into the dotted terminals produce adding fluxes.* Because these dots specify the transformer winding relations, they are used in circuit diagrams with inductor symbols in place of illustrated windings. A transformer circuit diagram symbol consists of two adjacent inductor symbols with dots. If the winding relations are not important, the dots may be omitted.

Figure 15-3 shows the use of dots. In a circuit diagram, the more convenient transformer representation with dots in Fig. 15-3*b* is used instead of the one with windings in Fig. 15-3*a*. But both are equivalent. An actual transformer may have some marking other than dots. In Fig. 15-3*b*, the two vertical lines between the inductor symbols mean that the transformer is either an iron-core transformer or an ideal transformer, which is considered next.

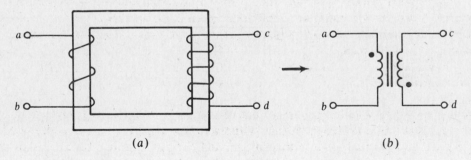

Fig. 15-3

THE IDEAL TRANSFORMER

In most respects, an *ideal transformer* is an excellent model for a transformer with an iron core—an *iron-core transformer.* Power transformers, the transformers used in electric power distribution systems, are iron-core transformers. Being a model, an ideal transformer is a convenient approximation of the real thing. The approximations are zero winding resistance, zero core loss, and infinite core permeability. Having windings of zero resistance, an ideal transformer has no winding ohmic power loss (I^2R loss) and no resistive voltage drops. The second property of zero core loss means that there is no power loss in the core—no hysteresis or eddy-current losses. And since there is no power loss in the windings either, then there is no power loss in the entire ideal transformer—the power out equals the power in. The third and last feature of infinite core permeability means that no current is required to establish the magnetic flux to produce the induced voltages. It also means that all the magnetic flux is confined to the core, coupling both windings. All flux is mutual, and there is no *leakage flux*, which is flux that couples only one winding.

In the analysis of a circuit containing an ideal transformer, the transformer *turns ratio*, also called *transformation ratio*, is used instead of flux. The turns ratio, with symbol a, is $a = N_1/N_2$. This is the ratio of the number of primary turns to secondary turns. In many electric circuits books, however, this ratio is defined as the number of secondary turns to primary turns, and sometimes the symbol n or N is used.

The turns ratio of an iron-core or ideal transformer is not usually directly specified on a circuit diagram. Instead, over the transformer symbol there is a designation such as $20:1$, which means that the winding on the left of the vertical bars has 20 times as many turns as the winding on the right. If the designation were $1:25$, instead, the winding on the right would have 25 times as many turns as the winding on the left.

The turns ratio is convenient because it relates the winding voltages. By Faraday's law, $v_1 = \pm N_1 \, d\phi/dt$ and $v_2 = \pm N_2 \, d\phi/dt$. (The same flux ϕ is in both equations because an ideal transformer has no leakage flux.) The ratio of these equations is

$$\frac{v_1}{v_2} = \pm \frac{N_1 (d\phi/dt)}{N_2 (d\phi/dt)} = \pm \frac{N_1}{N_2} = \pm a$$

The positive sign must be selected when both dotted terminals have the same reference voltage polarity. Otherwise the negative sign must be selected. The justification for this selection is that, as can be shown by Lenz's law, at any one time the dotted terminals of an ideal transformer always have the same actual polarities—either both positive or both negative with respect to the other terminals. A little thought will show that these sign selections conform to this physical reality. Incidentally, these actual polarities have nothing to do with the selection of voltage reference polarities, which is completely arbitrary.

It is obvious from $v_1/v_2 = \pm a$ that if a transformer has a turns ratio less than one ($a < 1$), the secondary rms voltage is greater than the primary rms voltage. Such a transformer is called a *step-up transformer*. But if the turns ratio is greater than one ($a > 1$), the secondary rms voltage is less than the primary rms voltage, and the transformer is called a *step-down transformer*.

As can be shown from the property of infinite permeability, and also from zero power loss, the primary and secondary currents have a relation that is the inverse of that for the primary and secondary voltages. Specifically,

$$\frac{i_2}{i_1} = \pm a$$

The positive sign must be selected if one current reference is into a dotted terminal and the other current reference is out of a dotted terminal. Otherwise the negative sign must be selected. The reason for this selection is that, at any one time, actual current flow is into the dotted terminal of one winding and out of the dotted terminal of the other. So, only the specified selection of signs will give the correct signs for the currents. But this selection of signs has nothing to do with the selection of current reference directions, which is completely arbitrary.

In the analysis of a circuit containing ideal transformers, the usual approach is to eliminate the transformers by reflecting impedances and also sources, if necessary. For an understanding of this reflecting, consider the circuit shown in Fig. 15-4a. The impedance $\mathbf{Z}_r$ "looking" into the primary winding is

$$\mathbf{Z}_r = \frac{\mathbf{V}_1}{\mathbf{I}_1} = \frac{-a\mathbf{V}_2}{(-1/a)\mathbf{I}_2} = a^2 \frac{\mathbf{V}_2}{\mathbf{I}_2} = a^2 \mathbf{Z}_2$$

which is the turns ratio squared times the secondary circuit impedance $\mathbf{Z}_2$. If $\mathbf{Z}_r$, called the *reflected impedance*, replaces the primary winding, as shown in Fig. 15-4b, the primary current $\mathbf{I}_1$ is unchanged. As can be proven by trying all different dot arrangements, the dot locations have no effect on this reflected impedance.

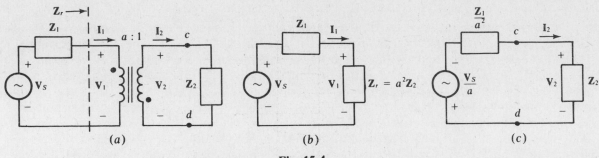

Fig. 15-4

If the secondary circuit is not a lumped impedance, but a circuit with individual resistive and reactive components, the total impedance can be found and reflected. Alternatively, the whole secondary circuit can be reflected into the primary circuit. In this reflection, the circuit configuration is kept the same and each individual impedance is multiplied by the square of the turns ratio. Of course, the transformer is eliminated.

Reflection can also be from the primary to the secondary. To see this, consider making cuts at terminals c and d in the circuit shown in Fig. 15-4a and finding the Thevenin equivalent of the circuit to the left. Because of the open circuit created by the cuts, the secondary current $I_2 = 0$ A, which in turn means that the primary current $I_1 = 0$ A. Consequently, there is 0 V across Z_1 and all the source voltage is at the primary winding. As a result, the Thevenin voltage referenced positive toward terminal c is $V_{Th} = V_2 = -V_1/a = -V_S/a$. From impedance reflection the Thevenin impedance is $Z_{Th} = Z_1/a^2$, with a^2 being in the denominator instead of the numerator because the winding being "looked" into is the secondary winding. The result is shown in the circuit of Fig. 15-4c. Note that the source voltage polarity reverses because the dots are at opposite ends of the windings. By use of Norton's theorem in a similar way, it can be shown that a source of current I_S would have reflected into the secondary as aI_S and would have been reversed in direction because the dots are not at the same ends of the windings. Whole circuits can be reflected in this way.

For ac voltages and currents, an ideal transformer gives results that are within a few percent—often less than 1 percent—of those of the corresponding actual power transformer. But for dc voltages and currents, an ideal transformer gives incorrect results. The reason is that an ideal transformer will transform dc voltages and currents while an actual transformer will not.

THE AIR-CORE TRANSFORMER

The ideal transformer approximation is not valid for a transformer with a core constructed of nonmagnetic material, as may be required for operation at radio and higher frequencies. A transformer with such a core is often called an *air-core transformer* or a *linear transformer*.

Figure 15-5 shows two circuits coupled by an air-core transformer. Current i_1 produces a mutual flux ϕ_{1m} and a leakage flux ϕ_{1l}, and current i_2 produces a mutual flux ϕ_{2m} and a leakage flux ϕ_{2l}. As mentioned, a mutual flux couples both windings, but a leakage flux couples only one winding.

The *coefficient of coupling*, with symbol k, indicates the closeness of coupling, which in turn means the fraction of total flux that is mutual. Specifically,

$$k = \sqrt{\frac{\phi_{1m}}{\phi_{1l} + \phi_{1m}} \times \frac{\phi_{2m}}{\phi_{2l} + \phi_{2m}}}$$

Clearly k cannot have a value greater than 1 nor less than 0. And the greater each fraction of mutual flux, the greater the coefficient of coupling. The coefficient of coupling of a good power transformer is very close to 1, but an air-core transformer typically has a coefficient of coupling less than 0.5.

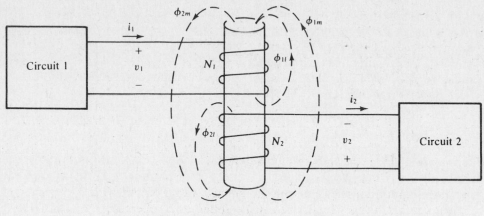

Fig. 15-5

The voltages induced by changing fluxes are given by Faraday's law:

$$v_1 = \pm N_1 \frac{d}{dt}(\psi_{1m} + \psi_{1l} \pm \psi_{2m}) \qquad v_2 = \pm N_2 \frac{d}{dt}(\psi_{2m} + \psi_{2l} \pm \psi_{1m})$$

The positive signs of $\pm\phi_{2m}$ and $\pm\phi_{1m}$ are selected if and only if these mutual fluxes have the same direction in each winding.

For circuit analysis, it is better to use inductances instead of fluxes. The *self-inductances* of the windings are

$$L_1 = \frac{N_1(\phi_{1m} + \phi_{1l})}{i_1} \qquad L_2 = \frac{N_2(\phi_{2m} + \phi_{2l})}{i_2}$$

These are just the ordinary winding inductances as defined in Chap. 8. There is, however, another inductance called the *mutual inductance* with symbol M. It accounts for the flux linkages of one winding caused by current flow in the other winding. Specifically,

$$M = \frac{N_1\phi_{2m}}{i_2} = \frac{N_2\phi_{1m}}{i_1}$$

With these substitutions, the voltage equations become

$$v_1 = L_1 \frac{di_1}{dt} \pm M \frac{di_2}{dt} \qquad \text{and} \qquad v_2 = L_2 \frac{di_2}{dt} \pm M \frac{di_1}{dt}$$

in which the initial $\pm$ signs have been eliminated because of the assumption of associated voltage and current references. For sinusoidal analysis the corresponding equations are

$$\mathbf{V}_1 = j\omega L_1 \mathbf{I}_1 \pm j\omega M \mathbf{I}_2 \qquad \text{and} \qquad \mathbf{V}_2 = j\omega L_2 \mathbf{I}_2 \pm j\omega M \mathbf{I}_1$$

In these equations, the negative signs of $\pm$ are used if one current has a reference into a dotted terminal and the other has a reference out of a dotted terminal. Otherwise the positive signs are used. Put another way, if positive i_1 and i_2 or $\mathbf{I}_1$ and $\mathbf{I}_2$ produce adding mutual fluxes, then the L and M terms add. As mentioned, these equations are based on associated voltage and current references. If a pair of these references are not associated, the v or $\mathbf{V}$ of the corresponding equation should have a negative sign. Everything else, though, remains the same.

On a time-domain circuit diagram the self-inductances are indicated adjacent to the corresponding windings in the usual manner. The mutual inductances are indicated between the coupled windings, with arrows to designate which pair of windings each mutual inductance is for. In a frequency-domain circuit, of course, $j\omega L_1$, $j\omega L_2$, and $j\omega M$ are used instead of L_1, L_2, and M.

If substitutions are made for the fluxes in the coefficient of coupling equation, the result is $k = M/\sqrt{L_1 L_2}$.

Mesh and loop analyses are best for analyzing circuits with air-core transformers. Writing the equations is the same as for other circuits except that it is necessary to include the $j\omega MI$ terms resulting from the magnetic coupling. Nodal analysis is difficult to use.

If the secondary circuit contains no independent sources, it is possible to reflect impedances in a manner similar to that used for ideal transformers. For an understanding of this reflection, consider the circuit shown in Fig. 15-6. The mesh equations are

$$V_S = (Z_1 + j\omega L_1)I_1 - j\omega MI_2$$
$$0 = -j\omega MI_1 + (j\omega L_2 + Z_L)I_2$$

The mutual terms are negative in both equations because one winding current is referenced into a dotted terminal while the other is referenced out of a dotted terminal. If the second equation is solved for I_2 and a substitution made for I_2 in the first equation, the result is

$$V_S = \left(Z_1 + j\omega L_1 + \frac{\omega^2 M^2}{j\omega L_2 + Z_L}\right)I_1$$

which indicates that the secondary circuit reflects into the primary circuit as an impedance $\omega^2 M^2/(j\omega L_2 + Z_L)$ in *series with the primary winding.* As can be found by trying different dot locations, this impedance does not depend on those locations. Some authors of circuits books call this impedance a "reflected impedance." Others, however, use the term "coupled impedance."

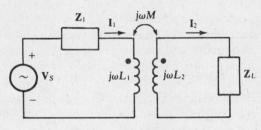

Fig. 15-6

THE AUTOTRANSFORMER

An *autotransformer* is a transformer with a single winding that has an intermediate terminal that divides the winding into two sections. For an understanding of autotransformer operation, it helps to consider the two sections of the winding to be the two windings of a power transformer, as is done next.

Consider a 50-kVA power transformer that has a voltage rating of 10 000/200 V. From the kilovoltampere and voltage ratings, the full-load current of the high voltage winding is 50 000/10 000 = 5 A and that of the low voltage winding is 50 000/200 = 250 A. Figure 15-7a shows such a transformer, fully loaded, with its windings connected in series such that the dotted end of one winding is connected to the undotted end of the other. As shown, the 10 000-V secondary circuit can be loaded to a maximum of 250 + 5 = 255 A with neither winding being current-overloaded. Since the source current is 250 A, the transformer can deliver 10 200 × 250 VA = 2550 kVA. This can also be determined from the secondary circuit: 10 000 × 255 VA = 2550 kVA. In effect, the autotransformer connection has increased the transformer kilovoltampere rating from 50 to 2550 kVA.

The explanation for this increase is that the original 50-kVA transformer had no metallic connections between the two windings, and so the 50 kVA of a full load had to pass through the transformer by magnetic coupling. However, with the windings connected to provide autotransformer operation, there is a metallic connection between the windings that passes 2550 − 50 = 2500 kVA without being magnetically transformed. So, it is the direct metallic connection that gives the kilovoltampere increase. Although advantageous in this respect, such a connection

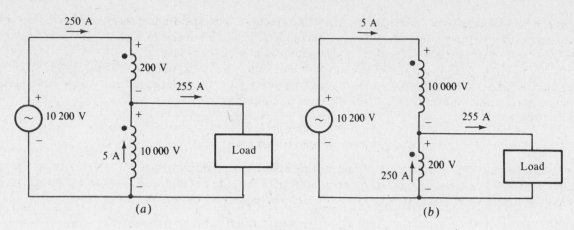

Fig. 15-7

destroys the isolation property that conventional transformers have, which in turn means that autotransformers cannot be used in every transformer application.

If the windings are connected as in Fig. 15-7b, the kilovoltampere rating is just $10\,200 \times 5 = 200 \times 255$ VA $= 51$ kVA. This slight increase of 2 percent in kilovoltampere rating is the result of the greatly different voltage levels of the two circuits connected to the autotransformer. In general, the closer the voltage levels are to being the same, the greater the increase in kilovoltampere rating. This is why autotransformers are so often used in power systems as links between systems operating at nearly the same voltage levels.

In the analysis of a circuit containing an autotransformer, an ideal transformer model can be assumed, and its turns ratio used in much the same way as for a conventional transformer connection. Along with this can be used the fact that the lower voltage lines carry the sum of the two winding currents. Also, part of the winding carries only the difference of the source and load currents. This is the part that is common to both the source and load circuits.

Contrary to what Fig. 15-7 suggests, autotransformers are preferably purchased as such and not constructed from conventional power transformers. An exception, however, is the "buck and boost" transformer. A typical one can be used to reduce 120 or 240 V to 12 or 24 V. The principal use, though, is as an autotransformer with the primary and secondary interconnected to give a slight adjustment in voltage, either greater or lesser.

Solved Problems

15.1 For the winding shown in Fig. 15-8a, what is the direction of flux produced in the core by current flowing into terminal a?

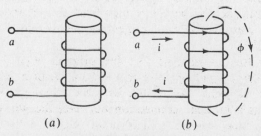

Fig. 15-8

Current that flows into terminal a flows over the core to the right, underneath to the left, then over the core to the right again, and so on, as is shown in Fig. 15-8b. For the application of the right-hand rule, fingers of a right hand should be imagined grasping the core with the fingers directed from left to right over the core. Then the thumb will point up, which means that the direction of the flux is up *inside* the core.

15.2   Supply the missing dots for the transformers shown in Fig. 15-9.

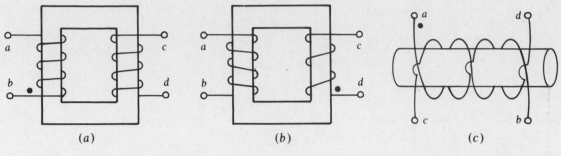

(a)   (b)   (c)

Fig. 15-9

(a) By the right-hand rule, current flowing into dotted terminal b produces clockwise flux. By trial and error it can be found that current flowing into terminal c also produces clockwise flux. So, terminal c should have a dot.

(b) Current flowing into dotted terminal d produces counterclockwise flux. Since current flowing into terminal b also produces counterclockwise flux, terminal b should have a dot.

(c) Current flowing into dotted terminal a produces flux to the right inside the core. Since current flowing into terminal d also produces flux to the right inside the core, terminal d should have a dot.

15.3   What is the turns ratio of a transformer that has a 684-turn primary winding and a 36-turn secondary winding?

The turns ratio a is the ratio of the number of primary turns to secondary turns: $a = 684/36 = 19$.

15.4   Find the turns ratio of a transformer that transforms the 12 470 V of a power line to the 240 V supplied to a house.

Since the high-voltage winding is connected to the power lines, it is the primary. The turns ratio is equal to the ratio of the primary to secondary voltages: $a = 12\,470/240 = 51.96$.

15.5   What are the full-load primary and secondary currents of a 25 000/240-V, 50-kVA transformer? Assume, of course, that the 25 000-V winding is the primary.

The current rating of a winding is the transformer kilovoltampere rating divided by the winding voltage rating. So, the full-load primary current is $50\,000/25\,000 = 2$ A, and the full-load secondary current is $50\,000/240 = 208$ A.

15.6   A power transformer with a 12 500/240-V voltage rating has a primary current rating of 50 A. Find the transformer kilovoltampere rating and the secondary current rating if the 240 V is the secondary voltage rating.

The transformer has a kilovoltampere rating of the primary voltage rating times the primary current rating: $12\,500(50) = 625\,000$ VA $= 625$ kVA. Since this is also the product of the secondary voltage and current ratings, the secondary current rating is $625\,000/240 = 2.6 \times 10^3$ A $= 2.6$ kA. As a check, the secondary current rating is equal to the primary current rating times the turns ratio, which is $a = 12\,500/240 = 52.1$. So, the secondary current rating is $52.1(50) = 2.6 \times 10^3$ A $= 2.6$ kA, which checks.

15.7 A transformer has a 500-turn winding linked by flux changing at the rate of 0.4 Wb/s. Find the induced voltage.

If the polarity of the voltage is temporarily ignored, then by Faraday's law, $v = N\,d\phi/dt$. The quantity $d\phi/dt$ is the time rate of change of flux, which is specified as 0.4 Wb/s. So, $v = 500(0.4) = 200$ V; the magnitude of the induced voltage is 200 V. The voltage polarity can be either positive or negative depending on the voltage reference polarity, the direction of the winding, and the direction in which the magnetic flux is either decreasing or increasing, none of which are specified. So the most that can be determined is that the magnitude of the induced voltage is 200 V at the time that the flux is changing at the rate of 0.4 Wb/s.

15.8 An iron-core transformer has 400 primary turns and 100 secondary turns. If the applied primary voltage is 240 V rms at 60 Hz, find the secondary rms voltage and the peak magnetic flux.

Since the transformer has an iron core, the turns ratio can be used to find the secondary rms voltage: $V_2 = (1/a)V_1 = (100/400)(240) = 60$ V rms. Because the voltages vary sinusoidally, they are induced by a sinusoidally varying flux that can be considered to be $\phi = \phi_m \sin \omega t$, where ϕ_m is the peak value of flux and ω is the radian frequency of $\omega = 2\pi(60) = 377$ rad/s. The time rate of change of flux is $d\phi/dt = d(\phi_m \sin \omega t)/dt = \omega\phi_m \cos \omega t$, which has a peak value of $\omega\phi_m$. Of course, the peak voltage is $\sqrt{2}V_{\text{rms}}$. It follows from $v = N\,d\phi/dt$ that the peak voltage and flux values are related by $\sqrt{2}V_{\text{rms}} = N\omega\phi_m$. If this equation is solved for ϕ_m and primary quantities used, the result is

$$\phi_m = \frac{\sqrt{2}V_{\text{rms}}}{N\omega} = \frac{\sqrt{2}(240)}{400(377)} = 2.25 \times 10^{-3} \text{ Wb} = 2.25 \text{ mWb}$$

Alternatively, the secondary voltage and turns could have been used since the same flux is assumed to couple both windings.

Incidentally, the equation $\sqrt{2}V_{\text{rms}} = N\omega\phi_m$ solved for V_{rms} is

$$V_{\text{rms}} = \frac{N(2\pi f)\phi_m}{\sqrt{2}} = 4.44fN\phi_m$$

This is called the *general transformer equation*.

15.9 If a 50-turn transformer winding has a 120-V rms applied voltage, and if the peak coupling flux is 20 mWb, find the frequency of the applied voltage.

From rearranging the general transformer equation,

$$f = \frac{V_{\text{rms}}}{4.44N\phi_m} = \frac{120}{4.44(50)(20 \times 10^{-3})} = 27 \text{ Hz}$$

15.10 An iron-core transformer has 1500 primary turns and 500 secondary turns. A 12-Ω resistor is connected across the secondary winding. Find the resistor voltage when the primary current is 3 A.

Since no voltage or current references are specified, only rms values are of interest and are to be assumed without specific mention of them. The secondary current is equal to the turns ratio times the primary current: $(1500/500)(3) = 9$ A. When this current flows through the 12-Ω resistor, it produces a voltage of $9(12) = 108$ V.

15.11 The output stage of an audio system has an output resistance of 2 kΩ. An output transformer provides resistance matching with a 6-Ω speaker. If this transformer has 400 primary turns, how many secondary turns does it have?

The term "resistance matching" means that the output transformer presents a reflected resistance of 2 kΩ to the output audio stage so that there is maximum power transfer to the 6-Ω speaker. Since, in general, the reflected resistance R_r is equal to the turns ratio squared times the resistance R_L of the load connected to the secondary ($R_r = a^2 R_L$), the turns ratio of the output transformer is

$$a = \sqrt{\frac{R_r}{R_L}} = \sqrt{\frac{2000}{6}} = 18.26$$

and the number of secondary turns is

$$N_2 = \frac{N_1}{a} = \frac{400}{18.26} = 22$$

15.12 In the circuit shown in Fig. 15-10, find R for maximum power absorption. Also, find **I** for $R = 3\ \Omega$. Finally, determine if connecting a conductor between terminals d and f would change these results.

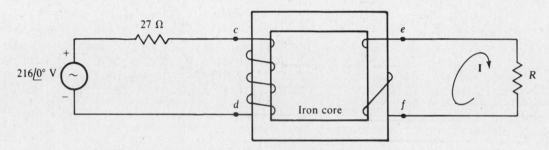

Fig. 15-10

The value of R for maximum power absorption is that value for which the reflected resistance a^2R is equal to the source resistance of 27 Ω. Since the primary winding has 4 turns and the secondary winding has 2 turns, the turns ratio is $a = N_1/N_2 = 4/2 = 2$. And, from $27 = 2^2R$, the value of R for maximum power absorption is $R = 27/4 = 6.75\ \Omega$.

For $R = 3\ \Omega$, the reflected resistance is $2^2(3) = 12\ \Omega$. So the primary current directed into terminal c is $(216\underline{/0°})/(27 + 12) = 5.54\underline{/0°}$ A. If terminal c is dotted, then terminal e should be dotted, as is evident from the right-hand rule. And, since **I** is directed out of terminal e while the calculated current is into terminal c, **I** is just the turns ratio times the current entering terminal c: $\mathbf{I} = 2(5.54\underline{/0°}) = 11.1\underline{/0°}$ A.

A conductor connected between terminals d and f does not affect these results since current cannot flow in a single conductor. For current to flow there would have to be another conductor to provide a return path.

15.13 Find i_1, i_2, and i_3 for the circuit shown in Fig. 15-11. The transformers are ideal.

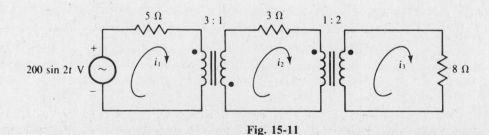

Fig. 15-11

A good procedure is to find i_1 using reflected resistances, then find i_2 from i_1, and last find i_3 from i_2. The 8 Ω reflects into the middle circuit as $8/2^2 = 2$ Ω, making a total resistance of $2 + 3 = 5$ Ω in the middle circuit. This 5 Ω reflects into the source circuit as $3^2(5) = 45$ Ω. Consequently,

$$i_1 = \frac{200 \sin 2t}{5 + 45} = 4 \sin 2t \text{ A}$$

Because i_1 and i_2 both have reference directions into dotted terminals of the first transformer, i_2 is equal to the negative of the turns ratio times i_1: $i_2 = -3(4 \sin 2t) = -12 \sin 2t$ A. Finally, since i_2 has a reference direction into a dotted terminal of the second transformer, and i_3 has a reference direction out of a dotted terminal of this transformer, i_3 is equal to the turns ratio $(1/2 = 0.5)$ times i_2: $i_3 = 0.5(-12 \sin 2t) = -6 \sin 2t$ A.

15.14 Find I_1 and I_2 for the circuit shown in Fig. 15-12.

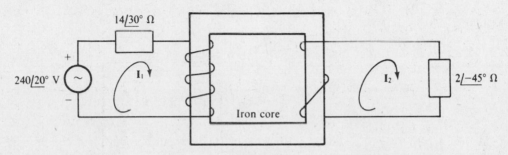

Fig. 15-12

Since the primary has 6 turns and the secondary has 2 turns, the turns ratio is $a = 6/2 = 3$ and the impedance reflected into the primary circuit is $3^2(2\underline{/-45°}) = 18\underline{/-45°}$ Ω. So,

$$I_1 = \frac{240\underline{/20°}}{14\underline{/30°} + 18\underline{/-45°}} = \frac{240\underline{/20°}}{25.5\underline{/-13°}} = 9.41\underline{/33°} \text{ A}$$

If the upper primary terminal is dotted, the bottom secondary terminal should be dotted. Since then both I_1 and I_2 would be into dots, I_2 is equal to the negative of the turns ratio times I_1:

$$I_2 = -3I_1 = -3(9.41\underline{/33°}) = -28.2\underline{/33°} \text{ A}$$

15.15 Find I_1 and I_2 for the circuit shown in Fig. 15-13a.

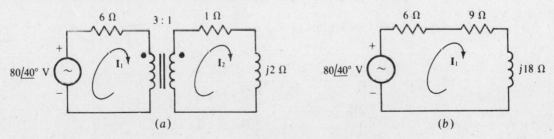

(a) (b)

Fig. 15-13

The 1-Ω resistance and the $j2$-Ω inductive impedance in the secondary circuit reflect into the primary circuit as $3^2(1) = 9$ Ω and $3^2(j2) = j18$ Ω in series with the 6-Ω resistance, as shown in

Fig. 15-13*b*. In effect, these reflected elements replace the primary winding. From the simplified circuit, the primary current is

$$\mathbf{I}_1 = \frac{80\underline{/40^\circ}}{6 + 9 + j18} = \frac{80\underline{/40^\circ}}{23.43\underline{/50.2^\circ}} = 3.41\underline{/-10.2^\circ} \text{ A}$$

Because $\mathbf{I}_1$ is referenced into a dotted terminal and $\mathbf{I}_2$ is referenced out of a dotted terminal, $\mathbf{I}_2$ is equal to the positive turns ratio times $\mathbf{I}_1$:

$$\mathbf{I}_2 = 3\mathbf{I}_1 = 3(3.41\underline{/-10.2^\circ}) = 10.2\underline{/-10.2^\circ} \text{ A}$$

15.16 Find $\mathbf{I}_1$, $\mathbf{I}_2$, and $\mathbf{I}_3$ for the circuit shown in Fig. 15-14.

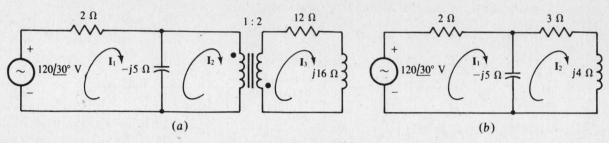

Fig. 15-14

The 12-Ω resistance and the $j16$-Ω inductive impedance reflect into the primary circuit as a $(1/2)^2(12) = 3$-Ω resistance and a series $(1/2)^2(j16) = j4$-Ω inductive impedance in parallel with the $-j5$-Ω capacitive impedance, as shown in Fig. 15-14*b*. The impedance of the parallel combination is

$$\frac{-j5(3 + j4)}{-j5 + 3 + j4} = \frac{20 - j15}{3 - j1} = 7.91\underline{/-18.4^\circ} \ \Omega$$

So,

$$\mathbf{I}_1 = \frac{120\underline{/30^\circ}}{2 + 7.91\underline{/-18.4^\circ}} = 12.2\underline{/44.7^\circ} \text{ A}$$

By current division,

$$\mathbf{I}_2 = \frac{-j5}{3 + j4 - j5} \times 12.2\underline{/44.7^\circ} = 19.3\underline{/-26.8^\circ} \text{ A}$$

Finally, since $\mathbf{I}_2$ and $\mathbf{I}_3$ both have reference directions into dotted terminals, $\mathbf{I}_3$ is equal to the negative of the turns ratio times $\mathbf{I}_2$:

$$\mathbf{I}_3 = -0.5(19.3\underline{/-26.8^\circ}) = -9.66\underline{/-26.8^\circ} \text{ A}$$

15.17 Find $\mathbf{V}$ for the circuit shown in Fig. 15-15*a*.

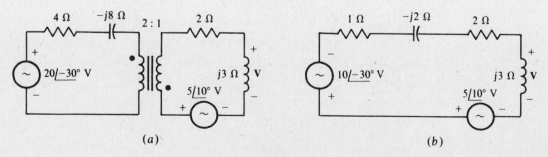

Fig. 15-15

Although reflection can be used, a circuit must be reflected instead of just an impedance because each circuit has a voltage source. And, because a voltage in the secondary circuit is desired, it is slightly preferable to reflect the primary circuit into the secondary. Of course, each reflected impedance is $(1/a)^2$ times the original impedance, and the reflected voltage source is $1/a$ times the original voltage. Also, the polarity of the reflected voltage source is reversed because the dots are located at opposite ends of the windings. The result is shown in Fig. 15-15b. By voltage division,

$$\mathbf{V} = \frac{j3}{1 - j2 + 2 + j3} \times (5\underline{/10°} - 10\underline{/-30°}) = \frac{20.9\underline{/212°}}{3.16\underline{/18°}} = 6.6\underline{/194°} = -6.6\underline{/14°} \text{ V}$$

15.18 An air-core transformer has primary and secondary currents of $i_1 = 0.2$ A and $i_2 = 0.4$ A that produce fluxes of $\phi_{1m} = 100$ μWb, $\phi_{1l} = 250$ μWb, and $\phi_{2l} = 300$ μWb. Find ϕ_{2m}, M, L_1, L_2, and k if $N_1 = 25$ turns and $N_2 = 40$ turns.

By the mutual inductance formulas,

$$M = \frac{N_1\phi_{2m}}{i_2} = \frac{N_2\phi_{1m}}{i_1} \quad \text{from which} \quad \phi_{2m} = \frac{i_2 N_2\phi_{1m}}{N_1 i_1} = \frac{0.4(40)(100)}{25(0.2)} = 320 \text{ } \mu\text{Wb}$$

Also

$$M = \frac{N_1\phi_{2m}}{i_2} = \frac{25(320 \times 10^{-6})}{0.4} \text{ H} = 20 \text{ mH}$$

From the self-inductance formulas,

$$L_1 = \frac{N_1(\phi_{1m} + \phi_{1l})}{i_1} = \frac{25(100 \times 10^{-6} + 250 \times 10^{-6})}{0.2} \text{ H} = 43.8 \text{ mH}$$

and

$$L_2 = \frac{N_2(\phi_{2m} + \phi_{2l})}{i_2} = \frac{40(320 \times 10^{-6} + 300 \times 10^{-6})}{0.4} \text{ H} = 62 \text{ mH}$$

The coefficient of coupling is

$$k = \sqrt{\frac{\phi_{1m}}{\phi_{1l} + \phi_{1m}} \times \frac{\phi_{2m}}{\phi_{2l} + \phi_{2m}}} = \sqrt{\frac{100 \times 10^{-6}}{250 \times 10^{-6} + 100 \times 10^{-6}} \times \frac{320 \times 10^{-6}}{300 \times 10^{-6} + 320 \times 10^{-6}}} = 0.384$$

Alternatively

$$k = \frac{M}{\sqrt{L_1 L_2}} = \frac{20 \times 10^{-3}}{\sqrt{(43.8 \times 10^{-3})(62 \times 10^{-3})}} = 0.384$$

15.19 What is the greatest mutual inductance that an air-core transformer can have if its self-inductances are 0.3 and 0.7 H?

From $k = M/\sqrt{L_1 L_2}$ rearranged to $M = k\sqrt{L_1 L_2}$ and the fact that k has a maximum value of 1, $M_{max} = \sqrt{0.3(0.7)} = 0.458$ H.

15.20 For each of the following, find the missing quantity—either self-inductance, mutual inductance, or coefficient of coupling:

(a) $L_1 = 0.3$ H, $L_2 = 0.4$ H, $M = 0.2$ H

(b) $L_1 = 4$ mH, $M = 5$ mH, $k = 0.4$

(c) $L_1 = 30$ μH, $L_2 = 40$ μH, $k = 0.5$

(d) $L_2 = 0.4$ H, $M = 0.2$ H, $k = 0.2$

(a) $k = \dfrac{M}{\sqrt{L_1 L_2}} = \dfrac{0.2}{\sqrt{0.3(0.4)}} = 0.577$

(b) $k\sqrt{L_1 L_2} = M$ from which $L_2 = \dfrac{M^2}{L_1 k^2} = \dfrac{5^2}{4(0.4)^2} = 39.1$ mH

(c) $M = k\sqrt{L_1 L_2} = 0.5\sqrt{30(40)} = 17.3$ μH

(d) $L_1 = \dfrac{M^2}{L_2 k^2} = \dfrac{0.2^2}{0.4(0.2)^2} = 2.5$ H

15.21 An air-core transformer has an open-circuited secondary winding with 50 V across it when the primary current is 30 mA at 3 kHz. If the primary self-inductance is 0.3 H, find the primary voltage and the mutual inductance.

Since phasors are not specified or mentioned, presumably the electric quantities specified and wanted are rms. Because the secondary is open-circuited, $I_2 = 0$ A, which means that $\omega M I_2 = 0$ and $\omega L_2 I_2 = 0$ in the voltage equations. So, the rms primary voltage is

$$V_1 = \omega L_1 I_1 = 2\pi(3000)(0.3)(30 \times 10^{-3}) = 170 \text{ V}$$

Also, the secondary voltage equation is $V_2 = \omega M I_1$, from which

$$M = \frac{V_2}{\omega I_1} = \frac{50}{2\pi(3000)(30 \times 10^{-3})} \text{ H} = 88.4 \text{ mH}$$

15.22 An air-core transformer has an open-circuited secondary with 80 V across it when the primary carries a current of 0.4 A and has a voltage of 120 V at 60 Hz. What are the primary self-inductance and also the mutual inductance?

Because the secondary is open-circuited, there is no current in this winding and so no mutually induced voltage in the primary winding. As a consequence, the rms voltage and current of the primary are related by the primary winding reactance: $\omega L_1 = V_1/I_1$, from which

$$L_1 = \frac{V_1}{\omega I_1} = \frac{120}{2\pi(60)(0.4)} = 0.796 \text{ H}$$

Because the open-circuited secondary carries zero current, the voltage of this winding is the result solely of the mutually induced voltage: $V_2 = \omega M I_1$, from which

$$M = \frac{V_2}{\omega I_1} = \frac{80}{2\pi(60)(0.4)} = 0.531 \text{ H}$$

15.23 Find the voltage across the open-circuited secondary of an air-core transformer when 35 V at 400 Hz is applied to the primary. The transformer inductances are $L_1 = 0.75$ H, $L_2 = 0.83$ H, and $M = 0.47$ H.

Because the secondary is open-circuited, $I_2 = 0$ A, which means that the rms primary voltage is $V_1 = \omega L_1 I_1$ and the rms secondary voltage is $V_2 = \omega M I_1$. The ratio of these equations is

$$\frac{V_2}{V_1} = \frac{\omega M I_1}{\omega L_1 I_1} \quad \text{from which} \quad V_2 = \frac{M V_1}{L_1} = \frac{0.47(35)}{0.75} = 21.9 \text{ V}$$

15.24 An air-core transformer with an open-circuited secondary has inductances of $L_1 = 20$ mH, $L_2 = 32$ mH, and $M = 13$ mH. Find the primary and secondary voltages when the primary current is increasing at the rate of 0.4 kA/s.

With the assumption of associated references,

$$v_1 = L_1 \frac{di_1}{dt} \pm M \frac{di_2}{dt} \quad \text{and} \quad v_2 = L_2 \frac{di_2}{dt} \pm M \frac{di_1}{dt}$$

In the first equation, di_2/dt is zero because of the open circuit, and di_1/dt is the specified 0.4 kA/s. So, $v_1 = (20 \times 10^{-3})(0.4 \times 10^3) = 8$ V. Similarly, the secondary voltage is $v_2 = \pm M \, di_1/dt = \pm(13 \times 10^{-3})(0.4 \times 10^3) = \pm 5.2$ V. Since the reference for v_2 is not specified, the sign of v_2 cannot be determined.

15.25 A transformer with a short-circuited secondary has inductances of $L_1 = 0.3$ H, $L_2 = 0.4$ H, and $M = 0.2$ H. Find the short-circuit secondary current I_2 when the primary current is $I_1 = 0.5$ A at 60 Hz.

Because of the short circuit,

$$\mathbf{V}_2 = j\omega L_2 \mathbf{I}_2 \pm j\omega M \mathbf{I}_1 = 0 \qquad \text{from which} \qquad j\omega L_2 \mathbf{I}_2 = \pm j\omega M \mathbf{I}_1 \qquad \text{and} \qquad L_2 \mathbf{I}_2 = \pm M \mathbf{I}_1$$

Since only rms quantities are of interest, as must be assumed from the problem specification, the angles of $\mathbf{I}_1$ and $\mathbf{I}_2$ can be neglected and the + sign of $\pm$ used, giving $L_2 I_2 = M I_1$. From this, the short-circuit secondary current I_2 is

$$I_2 = \frac{M I_1}{L_2} = \frac{0.2(0.5)}{0.4} = 0.25 \text{ A}$$

The same result would have been obtained by dividing $\omega M I_1$, the rms induced generator voltage, by ωL_2, the reactance that the short-circuit secondary current I_2 flows through.

15.26 When connected in series, two windings of an air-core transformer have a total inductance of 0.4 H. With the reversal of the connections to one winding, though, the total inductance is 0.8 H. Find the mutual inductance of the transformer.

Because the windings are in series, the same current i flows through them during the inductance measurement, producing a voltage drop of $L_1 \, di/dt \pm M \, di/dt = (L_1 \pm M) \, di/dt$ in one winding and a voltage drop of $L_2 \, di/dt \pm M \, di/dt = (L_2 \pm M) \, di/dt$ in the other. If the windings are arranged such that i flows into the dotted terminal of one winding but out of the dotted terminal of the other, both mutual terms are negative. But if i flows into both dotted terminals or out of them, both mutual terms are positive. Since the $M \, di/dt$ terms have the same sign, either both positive or both negative, the total voltage drop is $(L_1 + L_2 \pm 2M) \, di/dt$. The $L_1 + L_2 \pm 2M$ coefficient of di/dt is the total inductance. Obviously, the larger measured inductance must be for the positive sign, $L_1 + L_2 + 2M = 0.8$ H, and the smaller measured inductance must be for the negative sign, $L_1 + L_2 - 2M = 0.4$ H. If the second equation is subtracted from the first, the result is

$$L_1 + L_2 + 2M - (L_1 + L_2 - 2M) = 0.8 - 0.4 = 0.4$$

from which $4M = 0.4$ and $M = 0.1$ H.

From this result, a method for finding the mutual inductance of an air-core transformer is to connect the two windings in series and measure the total inductance, then reverse one winding connection and measure the total inductance. The mutual inductance is one-fourth of the difference of the larger measurement minus the smaller measurement. Of course, the self-inductance of a winding can be measured directly if the other winding is open-circuited.

15.27 An air-core transformer has a 3-mH mutual inductance and a 5-mH secondary self-inductance. A 5-Ω resistor and a 100-μF capacitor are in series with the secondary winding. Find the impedance coupled into the primary for $\omega = 1$ krad/s.

The coupled impedance is $(\omega M)^2 / \mathbf{Z}_2$ where $\mathbf{Z}_2$ is the total impedance of the secondary circuit. Here, $\omega M = 10^3 (3 \times 10^{-3}) = 3 \, \Omega$ and

$$\mathbf{Z}_2 = R + j\omega L + \frac{-j1}{\omega C} = 5 + j10^3(5 \times 10^{-3}) + \frac{-j1}{10^3(100 \times 10^{-6})} = 5 + j5 - j10 = 5 - j5 = 7.07\underline{/-45°} \, \Omega$$

and so the coupled impedance is

$$\frac{(\omega M)^2}{\mathbf{Z}_2} = \frac{3^2}{7.07\underline{/-45°}} = 1.27\underline{/45°} \, \Omega$$

Notice that the capacitive secondary impedance couples into the primary circuit as an inductive impedance. This change in the nature of the impedance always occurs on coupling because the secondary circuit impedance is in the denominator of the coupling impedance formula. In contrast, there is no such change in reflected impedance with an ideal transformer.

15.28 A 1-kΩ resistor is connected across the secondary of a transformer for which $L_1 = 0.1$ H, $L_2 = 2$ H, and $k = 0.5$. Find the resistor voltage when 250 V at 400 Hz is applied to the primary.

A good approach is to first find $\omega M I_1$, which is the induced mutual secondary voltage, and then use it to find the voltage across the 1-kΩ resistor. Since both M and I_1 in $\omega M I_1$ are unknown, they must be found. The mutual inductance M is

$$M = k\sqrt{L_1 L_2} = 0.5\sqrt{0.1(2)} = 0.224 \text{ H}$$

With M known, the coupled impedance can be used to obtain I_1. This impedance is

$$\frac{\omega^2 M^2}{R_2 + j\omega L_2} = \frac{(2\pi \times 400)^2 (0.224)^2}{1000 + j(2\pi \times 400)(2)} = 61.6\underline{/-78.7°}\ \Omega$$

The current I_1 is equal to the applied primary voltage divided by the magnitude of the sum of the coupled impedance and the primary winding impedance:

$$I_1 = \frac{250}{|j(2\pi \times 400)(0.1) + 61.6\underline{/-78.6°}|} = \frac{250}{191} = 1.31 \text{ A}$$

Now, with M and I_1 known, the induced secondary voltage $\omega M I_1$ can be found:

$$\omega M I_1 = (2\pi \times 400)(0.224)(1.31) = 735 \text{ V}$$

Voltage division can be used to find the desired voltage V_2 from this induced voltage. The voltage V_2 is equal to this induced voltage times the quotient of the load resistance and the magnitude of the total impedance of the secondary circuit:

$$V_2 = 735 \frac{1000}{|1000 + j2\pi(400)(2)|} = \frac{735 \times 10^3}{5.13 \times 10^3} = 143 \text{ V}$$

15.29 Find v for the circuit shown in Fig. 15-16a.

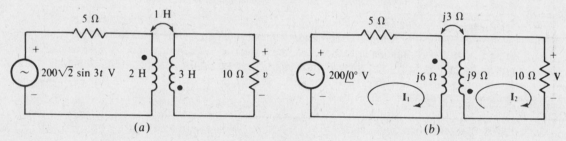

Fig. 15-16

The first step is the construction of the frequency-domain circuit shown in Fig. 15-16b. Next, the mesh equations are written:

$$(5 + j6)I_1 + \qquad j3I_2 = 200$$
$$j3I_1 + (10 + j9)I_2 = 0$$

Notice that the mutual terms are positive because both I_1 and I_2 have reference directions into dotted terminals. By Cramer's rule,

$$I_2 = \frac{\begin{vmatrix} 5+j6 & 200 \\ j3 & 0 \end{vmatrix}}{\begin{vmatrix} 5+j6 & j3 \\ j3 & 10+j9 \end{vmatrix}} = \frac{-j3(200)}{(5+j6)(10+j9) - (j3)^2} = \frac{600\underline{/-90°}}{5 + j105} = \frac{600\underline{/-90°}}{105\underline{/87.3°}} = 5.71\underline{/-177.3°} \text{ A}$$

And $\mathbf{V} = 10I_2 = 57.1\underline{/-177.3°}$ V. The corresponding voltage is

$$v = 57.1\sqrt{2} \sin(3t - 177.3°) = -80.7 \sin(3t + 2.7°) \text{ V}$$

15.30 Find I_2 for the circuit shown in Fig. 15-17.

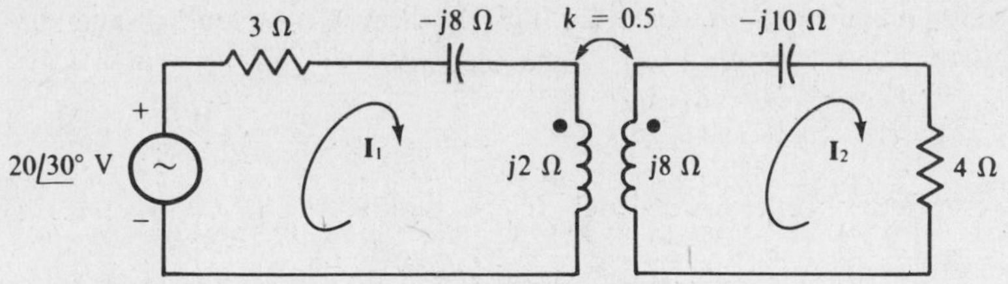

Fig. 15-17

Before mesh equations can be written, the magnitude ωM of $j\omega M$ must be determined. From multiplying both sides of $M = k\sqrt{L_1 L_2}$ by ω,

$$\omega M = k\sqrt{(\omega L_1)(\omega L_2)} = 0.5\sqrt{2(8)} = 2 \ \Omega$$

Now the mesh equations can be written:

$$(3 - j8 + j2)\mathbf{I}_1 - \qquad\qquad j2\mathbf{I}_2 = 20\underline{/30°}$$
$$-j2\mathbf{I}_1 + (4 + j8 - j10)\mathbf{I}_2 = 0$$

Notice that the mutual voltage terms have an opposite sign (negative) from that (positive) of the self-induced voltage terms because one current reference direction is into a dotted terminal and the other one is not. These equations simplify to

$$(3 - j6)\mathbf{I}_1 - \qquad\qquad j2\mathbf{I}_2 = 20\underline{/30°}$$
$$-j2\mathbf{I}_1 + (4 - j2)\mathbf{I}_2 = 0$$

By Cramer's rule,

$$\mathbf{I}_2 = \frac{\begin{vmatrix} 3 - j6 & 20\underline{/30°} \\ -j2 & 0 \end{vmatrix}}{\begin{vmatrix} 3 - j6 & -j2 \\ -j2 & 4 - j2 \end{vmatrix}} = \frac{-(-j2)(20\underline{/30°})}{(3 - j6)(4 - j2) - (-j2)^2}$$

$$= \frac{(2\underline{/90°})(20\underline{/30°})}{12 - 12 - j24 - j6 + 4} = \frac{40\underline{/120°}}{30.3\underline{/-82.4°}} = 1.32\underline{/202.4°} = -1.32\underline{/22.4°} \ \text{A}$$

15.31 What is the total inductance of an air-core transformer with its windings connected in parallel if both dots are at the same end and if the mutual inductance is 0.1 H and the self-inductances are 0.2 and 0.4 H?

Because of the mutual-inductance effects, it is not possible to simply combine inductances. Instead, a source must be applied and the total inductance found from the ratio of the source voltage to source current, which ratio is the input impedance. Of course a frequency-domain circuit will have to be used. For this circuit the most convenient frequency is $\omega = 1$ rad/s, and the most convenient source is $\mathbf{I}_S = 1\underline{/0°}$ A. The circuit is shown in Fig. 15-18. The transformer impedances should be obvious from the specified inductances and the radian frequency of $\omega = 1$ rad/s. As shown, $\mathbf{I}_1$ of the $1\underline{/0°}$-A input current flows through the left-hand winding, leaving a current of $1\underline{/0°}$-$\mathbf{I}_1$ for the right-hand winding.

The voltage drops across the windings are

$$\mathbf{V} = j0.2\mathbf{I}_1 + j0.1(1\underline{/0°} - \mathbf{I}_1)$$
$$\mathbf{V} = j0.1\mathbf{I}_1 + j0.4(1\underline{/0°} - \mathbf{I}_1)$$

The mutual voltage terms have the same signs as the self-induced voltage terms because both current reference directions are into dotted ends. Upon rearrangement and simplification, these equations become

$$-j0.1\mathbf{I}_1 + \mathbf{V} = j0.1$$
$$j0.3\mathbf{I}_1 + \mathbf{V} = j0.4$$

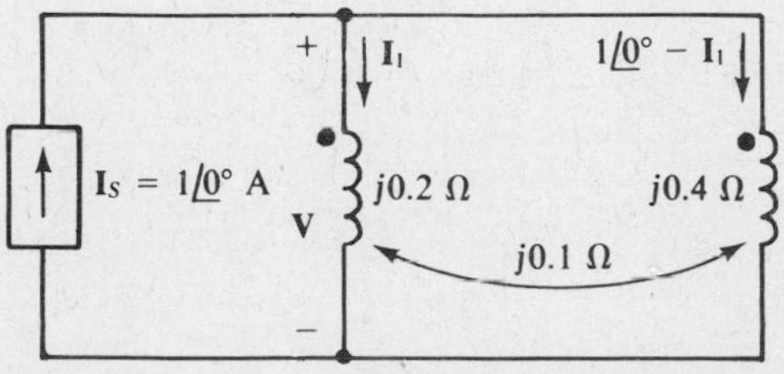

Fig. 15-18

The unknown I_1 can be eliminated by multiplying the first equation by 3 and adding corresponding sides of the equations.   The result is

$$3V + V = j0.3 + j0.4 \qquad \text{from which} \qquad V = \frac{j0.7}{4} = j0.175 \text{ V}$$

But

$$j\omega L_T = \frac{V}{I_S} = \frac{j0.175}{1\underline{/0°}} = j0.175$$

Finally, since $\omega = 1$ rad/s,   the total inductance is $L_T = 0.175$ H.

15.32   Find i_2 for the circuit shown in Fig. 15-19a.

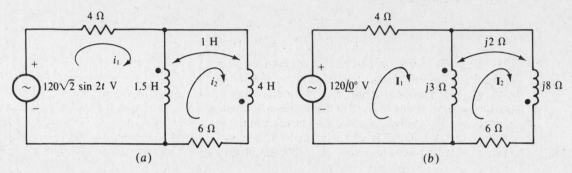

Fig. 15-19

The first step is the construction of the frequency-domain circuit shown in Fig. 15-19b, from which mesh equations can be written.   These are

$$(4 + j3)I_1 - j3I_2 - j2I_2 = 120\underline{/0°}$$
$$-j3I_1 - j2I_1 + [j3 + j8 + 6 + 2(j2)]I_2 = 0$$

In the first equation, the $4 + j3$ coefficient of I_1 is, of course, the self-impedance of mesh 1, and the $-j3$ coefficient of I_2 is the negative of the mutual impedance.   The $-j2I_2$ term is the voltage induced in the left-hand winding by I_2 flowing in the right-hand winding.   This term is negative because I_1 enters a dotted terminal but I_2 does not.   In the second equation, the $-j3I_1$ term is the mutual-impedance voltage, and $-j2I_1$ is the voltage induced in the right-hand winding by I_1 flowing in the left-hand winding.   This term is negative for the same reason that $-j2I_2$ is negative in the first equation, as has been explained.   The $j3 + j8 + 6$ part of the coefficient of I_2 is the self-impedance of mesh 2.   The $2(j2)$ part of this coefficient is from a voltage $j2I_2$ induced in each winding by I_2 flowing in the other winding.   It is positive because I_2 enters undotted terminals of both windings.

These equations simplify to

$$(4 + j3)I_1 - \qquad j5I_2 = 120$$
$$-j5I_1 + (6 + j15)I_2 = 0$$

By Cramer's rule,

$$I_2 = \frac{\begin{vmatrix} 4+j3 & 120 \\ -j5 & 0 \end{vmatrix}}{\begin{vmatrix} 4+j3 & -j5 \\ -j5 & 6+j15 \end{vmatrix}} = \frac{-(-j5)(120)}{(4+j3)(6+j15) - (-j5)^2} = \frac{j600}{4 + j78} = 7.68\underline{/2.94°} \text{ A}$$

The corresponding current is

$$i_2 = 7.68\sqrt{2} \sin(2t + 2.94°) = 10.9 \sin(2t + 2.94°) \text{ A}$$

15.33   Find **V** for the circuit shown in Fig. 15-20.   Then replace the 15-Ω resistor with an open circuit and find **V** again.

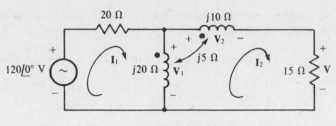

Fig. 15-20

The mesh equations are

$$(20 + j20)\mathbf{I}_1 - j20\mathbf{I}_2 + j5\mathbf{I}_2 = 120\underline{/0°}$$
$$-j20\mathbf{I}_1 + j5\mathbf{I}_1 + [j20 + j10 + 15 - 2(j5)]\mathbf{I}_2 = 0$$

All of the terms should be apparent except, perhaps, those for the mutually induced voltages. The $j5\mathbf{I}_2$ in the first equation is the voltage induced in the vertical winding by $\mathbf{I}_2$ flowing in the horizontal winding. It is positive because both $\mathbf{I}_1$ and $\mathbf{I}_2$ enter dotted terminals. The $j5\mathbf{I}_1$ term in the second equation is the voltage induced in the horizontal winding by $\mathbf{I}_1$ flowing in the vertical winding. It is positive for the same reason that $j5\mathbf{I}_2$ is positive in the first equation. The $-2(j5)\mathbf{I}_2$ term is the result of a voltage of $j5\mathbf{I}_2$ induced in each winding by $\mathbf{I}_2$ flowing in the other winding. It is negative because $\mathbf{I}_2$ enters a dotted terminal of one winding, but not of the other. These equations simplify to

$$(20 + j20)\mathbf{I}_1 \qquad\quad - \; j15\mathbf{I}_2 = 120$$
$$-j15\mathbf{I}_1 + (15 + j20)\mathbf{I}_2 = 0$$

from which

$$\mathbf{I}_2 = \frac{\begin{vmatrix} 20 + j20 & 120 \\ -j15 & 0 \end{vmatrix}}{\begin{vmatrix} 20 + j20 & -j15 \\ -j15 & 15 + j20 \end{vmatrix}} = \frac{-(-j15)(120)}{(20 + j20)(15 + j20) - (-j15)^2} = \frac{j1800}{125 + j700} = \frac{1800\underline{/90°}}{711\underline{/79.9°}} = 2.53\underline{/10.1°} \text{ A}$$

Finally,

$$\mathbf{V} = 15\mathbf{I}_2 = 15(2.53\underline{/10.1°}) = 38\underline{/10.1°} \text{ V}$$

If the 15-Ω resistor is removed, then $\mathbf{I}_2 = 0$ A and $\mathbf{V}$ is equal to the sum of the voltage drops across the two windings. The only current that flows is $\mathbf{I}_1$, which is

$$\mathbf{I}_1 = \frac{120\underline{/0°}}{20 + j20} = \frac{120\underline{/0°}}{28.3\underline{/45°}} = 4.24\underline{/-45°} \text{ A}$$

Across the vertical winding, $\mathbf{I}_1$ produces a self-inductive voltage drop of

$$\mathbf{V}_1 = j20\mathbf{I}_1 = j20(4.24\underline{/-45°}) = 84.8\underline{/45°} \text{ V}$$

referenced positive on the dotted end. Across the horizontal winding, $\mathbf{I}_1$ produces a mutually induced voltage of

$$\mathbf{V}_2 = j5\mathbf{I}_1 = j5(4.24\underline{/-45°}) = 21.2\underline{/45°} \text{ V}$$

Like the other induced voltage, it also has a positive reference on a dotted end since part of the same flux produces it. (Of course, actually a changing flux produces the corresponding voltages v_1 and v_2.) Finally, since the dotted ends of the two windings are adjacent, $\mathbf{V}$ is equal to the difference in the two winding voltages:

$$\mathbf{V} = \mathbf{V}_1 - \mathbf{V}_2 = 84.8\underline{/45°} - 21.2\underline{/45°} = 63.6\underline{/45°} \text{ V}$$

15.34 What is the turns ratio of a two-winding transformer that can be connected as a 500/350-kV autotransformer?

As can be seen from Fig. 15-7, the lower voltage is the voltage across one winding, and the higher voltage is the sum of the winding voltages. So, for this transformer, one winding voltage rating is

350 kV and the other is $500 - 350 = 150$ kV. The turns ratio is, of course, equal to the ratio of these ratings: $a = 350/150 = 2.33$ or $a = 150/350 = 0.429$, depending upon which winding is the primary and which is the secondary.

15.35 Compare the winding currents of a fully loaded 277/120-V, 50-kVA two-winding transformer and an autotransformer with the same rating.

The high-voltage winding of the conventional transformer must carry $50\,000/277 = 181$ A, and the low-voltage winding must carry $50\,000/120 = 417$ A. So, one winding carries the source current and the other winding carries the load current. In contrast, and as shown in the circuit of Fig. 15-21, part of the autotransformer winding must carry only the difference in the source and load currents, which is $417 - 181 = 236$ A, as compared to the 417 A that the low-voltage winding of the conventional transformer must carry. Consequently, smaller wire can be used in the autotransformer, which results in a saving in the cost of copper. Also, the autotransformer can be smaller and lighter.

15.36 A 12 470/277-V, 50-kVA transformer is connected as an autotransformer. What is the kilovoltampere rating if the windings are connected as shown in Fig. 15-7a? And what is this rating if the windings are connected as shown in Fig. 15-7b?

For either connection the maximum input voltage is the sum of the voltage ratings of the windings: $12\,470 + 277 = 12\,747$ V. Since, for the connection shown in Fig. 15-7a, the source current flows through the low-voltage winding, the maximum input current is the current rating of this winding, which is $50\,000/277 = 181$ A. So, the kilovoltampere rating for this connection is $12\,747 \times 181$ VA = 2300 kVA. For the other connection, that illustrated in Fig. 15-7b, the source current flows through the high-voltage winding. Consequently, the maximum input current is the current rating of this winding, which is $50\,000/12\,470 = 4.01$ A, and the kilovoltampere rating is only $12\,747 \times 4.01$ VA = 51.1 kVA.

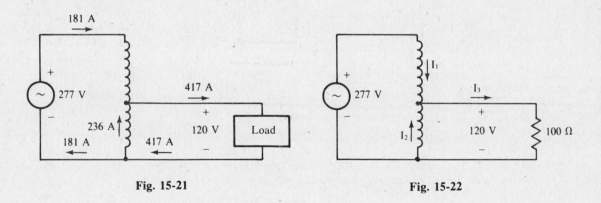

Fig. 15-21   Fig. 15-22

15.37 Find the three currents I_1, I_2, and I_3 for the circuit shown in Fig. 15-22.

The resistor current is obviously $I_3 = 120/100 = 1.2$ A. And the resistor receives $120 \times 1.2 = 144$ VA. Since this is also the voltamperes supplied by the source, $277I_1 = 144$ and $I_1 = 144/277 = 0.52$ A. Last, from KCL applied at the transformer winding tap, $I_2 = I_3 - I_1 = 1.2 - 0.52 = 0.68$ A.

Supplementary Problems

15.38 In the transformer shown in Fig. 15-23, what is the direction of flux produced in the core by current flow into (a) terminal a, (b) terminal b, (c) terminal c, and (d) terminal d?
Ans. (a) Clockwise, (b) counterclockwise, (c) counterclockwise, (d) clockwise

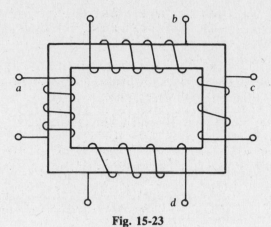

Fig. 15-23

15.39 Supply the missing dots for the transformers shown in Fig. 15-24.
Ans. (*a*) Dot on terminal *d*; (*b*) dot on terminal *b*; (*c*) dots on terminals *b*, *c*, and *g*.

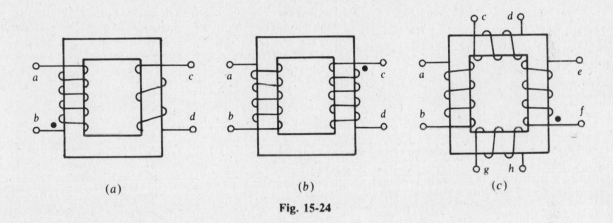

(*a*)   (*b*)   (*c*)

Fig. 15-24

15.40 What is the turns ratio of a power transformer that has a 6.25-A primary current at the same time that it has a 50-A secondary current?   *Ans.* $a = 8$.

15.41 Find the turns ratio of a power transformer that transforms the 12 470 V of a power line to the 480 V used in a factory.   *Ans.* $a = 26$.

15.42 What are the full-load primary and secondary currents of a 7200/120-V, 25-kVA power transformer? Assume that the 7200-V winding is the primary.
Ans. 3.47-A primary current and 208-A secondary current

15.43 A power transformer with a 13 200/480-V rating has a full-load primary current rating of 152 A.   Find the transformer kilovoltampere rating and the full-load secondary current rating if the 480 V is the secondary voltage rating.   *Ans.* 2000 kVA, 4.18 kA

15.44 A 7200/120-V, 60-Hz transformer has 1620 turns on the primary. What is the peak rate of change of magnetic flux? (*Hint*: Remember that the voltage ratings are in rms.)   *Ans.* 6.29 Wb/s

15.45 An iron-core transformer has 3089 primary turns and 62 secondary turns. If the applied primary voltage is 13 800 V rms at 60 Hz, find the secondary rms voltage and the peak magnetic flux.
Ans. 277 V, 16.8 mWb

15.46 If a 27-turn transformer winding has 120 V rms applied, and if the peak coupling flux is 20 mWb, what is the frequency of the applied voltage? *Ans.* 50 Hz

15.47 An iron-core transformer has 1620 primary turns and 54 secondary turns. A 10-Ω resistor is connected across the secondary winding. Find the resistor voltage when the primary current is 0.1 A. *Ans.* 30 V

15.48 What should be the turns ratio of an output transformer that connects a 4-Ω speaker to an audio system that has an output resistance of 1600 Ω? *Ans.* $a = 20$

15.49 In the circuit shown in Fig. 15-25, what should a and X_C be for maximum average power absorption by the load impedance, and what is this power? *Ans.* 3.19, -4.52 Ω, 376 W

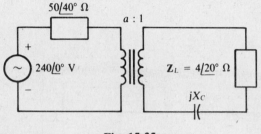

Fig. 15-25

15.50 Find i_1, i_2, and i_3 for the circuit shown in Fig. 15-26.

Ans. $i_1 = 4 \sin (3t - 36.9°)$ A
$i_2 = 8 \sin (3t - 36.9°)$ A
$i_3 = -24 \sin (3t - 36.9°)$ A

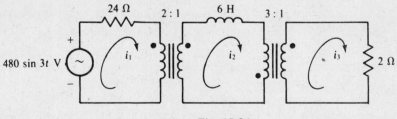

Fig. 15-26

15.51 Find V for the circuit shown in Fig. 15-27. *Ans.* $-312\underline{/60.7°}$ V

15.52 Find I_1, I_2, and I_3 for the circuit shown in Fig. 15-28.
Ans. $I_1 = 1.49\underline{/-23.5°}$ A
$I_2 = 4.46\underline{/-23.5°}$ A
$I_3 = -8.93\underline{/-23.5°}$ A

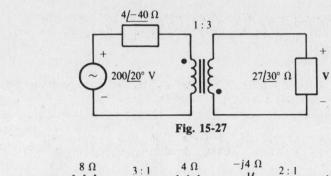

Fig. 15-27

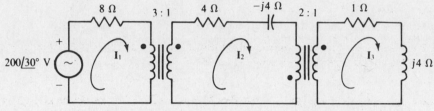

Fig. 15-28

15.53 What is v for the circuit shown in Fig. 15-29? *Ans.* $-23.7 \sin (2t - 6.09°)$ V

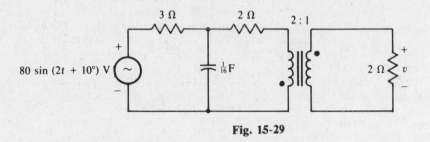

Fig. 15-29

15.54 Find **I** for the circuit shown in Fig. 15-30. *Ans.* $2.28\underline{/-39.7°}$ A

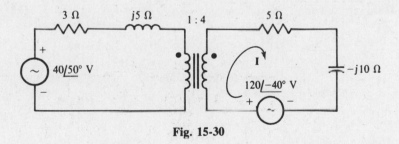

Fig. 15-30

15.55 An air-core transformer has a primary current of 0.2 A and a secondary current of 0.1 A that produce fluxes of $\phi_{1l} = 40 \ \mu$Wb, $\phi_{2m} = 10 \ \mu$Wb, and $\phi_{2l} = 30 \ \mu$Wb. Find ϕ_{1m}, L_1, L_2, M, and k if $N_1 = 30$ turns and $N_2 = 50$ turns.
Ans. $\phi_{1m} = 12 \ \mu$Wb, $L_1 = 7.8$ mH, $L_2 = 20$ mH, $M = 3$ mH, $k = 0.24$

15.56 What is the greatest possible mutual inductance of an air-core transformer that has self-inductances of 120 and 90 mH? *Ans.* 104 mH

15.57 For each of the following, find the missing quantity—either self-inductance, mutual inductance, or coefficient of coupling.

(a) $L_1 = 130$ mH, $L_2 = 200$ mH, $M = 64.5$ mH

(b) $L_1 = 2.6$ μH, $L_2 = 3$ μH, $k = 0.4$

(c) $L_1 = 350$ mH, $M = 100$ mH, $k = 0.3$

Ans.   (a) $k = 0.4$,   (b) $M = 1.12$ μH,   (c) $L_2 = 317$ mH

15.58   An air-core transformer has an open-circuited secondary winding with 70 V induced in it when the primary winding carries a 0.3-A current and has a 120-V, 600-Hz voltage across it.   What is the mutual inductance and the primary self-inductance?   Ans. $M = 61.9$ mH, $L_1 = 106$ mH

15.59   An air-core transformer with an open-circuited secondary has inductances of $L_1 = 200$ mH, $L_2 = 320$ mH,   and $M = 130$ mH.   Find the primary and secondary voltages, referenced positive at the dotted terminals, when the primary current is increasing at the rate of 0.3 kA/s into the dotted terminal of the primary winding.   Ans. $v_1 = 60$ V, $v_2 = 39$ V

15.60   An air-core transformer has inductances of $L_1 = 0.3$ H, $L_2 = 0.7$ H,   and $M = 0.3$ H.   The primary current is increasing into the dotted primary terminal at the rate of 200 A/s, and the secondary current is increasing into the dotted secondary terminal at the rate of 300 A/s.   What are the primary and secondary voltages referenced positive at the dotted terminals?   Ans. $v_1 = 150$ V, $v_2 = 270$ V

15.61   An air-core transformer with a shorted secondary has a 90-mA short-circuit secondary current and a 150-mA primary current when 50 V at 400 Hz is applied to the primary.   If the mutual inductance is 110 mH, find the self-inductances.   Ans. $L_1 = 199$ mH, $L_2 = 183$ mH

15.62   An air-core transformer with a shorted secondary has inductances of $L_1 = 0.6$ H, $L_2 = 0.4$ H,   and $M = 0.2$ H.   Find the winding currents when a primary voltage of 50 V at 60 Hz is applied.   Ans. $I_1 = 265$ mA, $I_2 = 133$ mA

15.63   A transformer has self-inductances of 1 and 0.6 H.   One series connection of the windings results in a total inductance of 1 H.   What is the coefficient of coupling?   Ans. $k = 0.387$

15.64   The transformer windings of a transformer are connected in series with dotted terminals adjacent.   Find the total inductance of the series-connected windings if $L_1 = 0.6$ H, $L_2 = 0.4$ H,   and $k = 0.35$.   Ans. 0.657 H

15.65   An air-core transformer has an 80-mH mutual inductance and a 200-mH secondary self-inductance.   A 2-kΩ resistor and a 100-mH inductor are in series with the secondary winding.   Find the impedance coupled into the primary for $\omega = 10$ krad/s.   Ans. 178$\underline{/-56.3°}$ Ω

15.66   Find **V** for the circuit shown in Fig. 15-31.   Ans. $-80\underline{/-37.4°}$ V

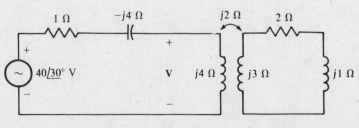

Fig. 15-31

15.67 A 6.8-kΩ resistor is connected across the secondary of a transformer having inductances of $L_1 =$ 150 mH, $L_2 = 300$ mH, and $M = 64$ mH. What is the resistor current when 40 V at 10 krad/s is applied to the primary?   *Ans.* 2.33 mA

15.68 Find i for the circuit shown in Fig. 15-32.   *Ans.* 103 sin (1000t − 73.1°) mA

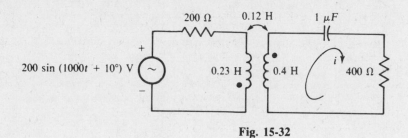

Fig. 15-32

15.69 What is the total inductance of the parallel-connected windings of an air-core transformer if the dots are at opposite ends and if the mutual inductance is 100 mH and the self-inductances are 200 and 400 mH?   *Ans.* 87.5 mH

15.70 Find i for the circuit shown in Fig. 15-33.   *Ans.* 24 sin (2t − 76.5°) A

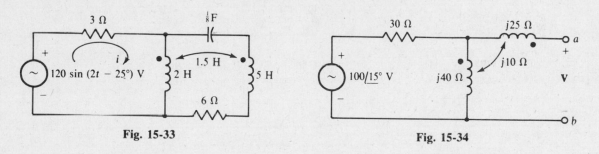

Fig. 15-33   **Fig. 15-34**

15.71 Find **V** for the circuit shown in Fig. 15-34. Then switch the dot on one winding and find **V** again.   *Ans.* 100/51.9° V, 60/51.9° V

15.72 In the circuit shown in Fig. 15-34, place a short circuit across terminals a and b and find the short-circuit current directed from terminal a to terminal b.   *Ans.* 1.85/−4.44° A

15.73 For the circuit shown in Fig. 15-34, what load connected to terminals a and b absorbs maximum power and what is this power?   *Ans.* 54.1/−56.3° Ω, 83.3 W

15.74 Find **I** for the circuit shown in Fig. 15-35.   *Ans.* 7.38/39.4° A

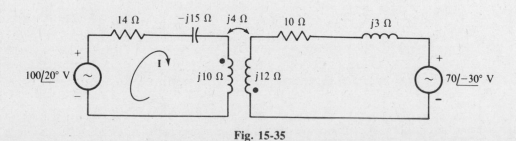

Fig. 15-35

15.75 What is the turns ratio of a two-winding iron-core transformer that can be connected as a 277/120-V autotransformer?   *Ans.* $a = 1.31$ or $a = 0.764$

15.76 A 4800/240-V, 75-kVA power transformer is connected as an autotransformer.   What is the kilovolt-ampere rating of the autotransformer for the connection shown in Fig. 15-7*a*?   What is the kilovolt-ampere rating for the connection shown in Fig. 15-7*b*?   *Ans.* 1575 kVA, 78.75 kVA

15.77 Find the currents I_1, I_2, and I_3 for the circuit shown in Fig. 15-36.
Ans. $I_1 = 800$ A, $I_2 = 343$ A, $I_3 = 1.14$ kA

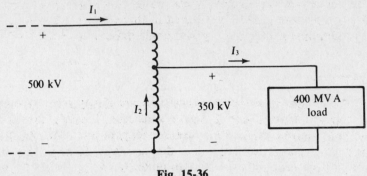

Fig. 15-36

Chapter 16

Three-Phase Circuits

INTRODUCTION

Three-phase circuits are important because almost all electric power is generated and distributed three-phase. A three-phase circuit has an ac voltage generator, also called an *alternator*, that produces three sinusoidal voltages that are identical except for a phase angle difference of 120°. The electric energy is transmitted over either three of four wires, more often called *lines*. Most of the three-phase circuits presented in this chapter are *balanced*. In them, three of the line currents are identical except for a phase angle difference of 120°.

SUBSCRIPT NOTATION

The polarities of voltages in three-phase circuits are designated by double subscripts, as in V_{AB}. As may be recalled from Chap. 2, these subscripts identify the nodes that a voltage is across. Also, the order gives the voltage reference polarity. Specifically, the first subscript specifies the positively referenced node and the second subscript the negatively referenced node. So, V_{AB} is a voltage drop from node A to node B. Also, $V_{AB} = -V_{BA}$.

Double subscripts are also necessary for some current quantity symbols, as in I_{AB}. These subscripts identify the nodes between which I_{AB} flows, and the order of the subscripts specifies the current reference direction. Specifically, the current reference direction is from the node of the first subscript to the node of the second subscript. So, the current I_{AB} has a reference direction from node A to node B. Also, $I_{AB} = -I_{BA}$. Figure 16-1 illustrates the subscript convention for I_{AB} and also for V_{AB}.

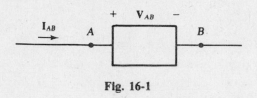

Fig. 16-1

Double subscript notation is also used for some impedances, as in Z_{AB}. The subscripts identify the two nodes that the impedance is connected between. But the order of the subscripts has no significance. Consequently, $Z_{AB} = Z_{BA}$.

THREE-PHASE VOLTAGE GENERATION

Figure 16-2a is a cross-sectional view of a three-phase alternator having a stationary stator and a counterclockwise rotating rotor. Physically displaced by 120° around the inner periphery of the stator are three sets of armature windings with terminals A and A', B and B', and C and C'. It is in these windings that the three-phase sinusoidal voltages are generated. The rotor has a field winding in which the flow of a dc current produces a magnetic field.

As the rotor rotates counterclockwise at 3600 r/min, its magnetic field cuts the armature windings, thereby inducing in them the sinusoidal voltages shown in Fig. 16-2b. These voltages reach peaks one-third of a period apart, or 120° apart, because of the 120° spatial displacement of the armature windings. As a result, the alternator produces three voltages of the same rms value, which may be as great as 30 kV, and of the same frequency (60 Hz), but phase-shifted by 120°. For example, these voltages might be

$$v_{AA'} = 25\,000 \sin 377t \text{ V}, \quad v_{BB'} = 25\,000 \sin (377t - 120°) \text{ V}, \text{ and } v_{CC'} = 25\,000 \sin (377t + 120°) \text{ V}$$

If the voltages shown in Fig. 16-2b are evaluated at any one time, it will be found that they add to zero. This zero sum can also be shown by vector graphical addition of the phasors correspond-

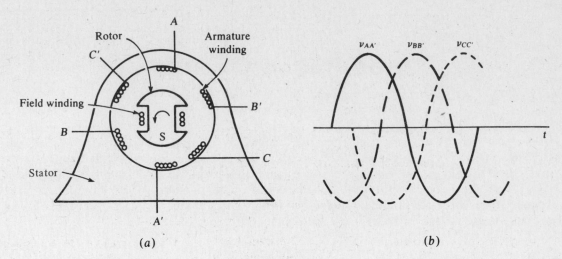

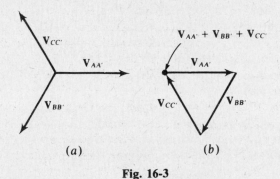

Fig. 16-2

ing to these voltages. Figure 16-3a is a phasor diagram of the three phasors $V_{AA'}$, $V_{BB'}$, and $V_{CC'}$, corresponding to the generated voltages. These three phasors are added in Fig. 16-3b by connecting the tail of $V_{BB'}$ to the tip of $V_{AA'}$, and the tail of $V_{CC'}$ to the tip of $V_{BB'}$. Since the tip of $V_{CC'}$ touches the tail of $V_{AA'}$, the sum is zero. And since the sum of the phasor voltages is zero, the sum of the corresponding instantaneous voltages is zero for all times.

Fig. 16-3

In general, three sinusoids have a sum of zero if they have the same frequency and peak value but are phase-displaced by 120°. This is true regardless of what, if anything, that the sinusoids correspond to. In particular, it is true for currents.

GENERATOR WINDING CONNECTIONS

The ends of generator windings are connected together to decrease the number of lines required for connections to loads. The primed terminals can be connected together to form the Y (wye) shown in Fig. 16-4a, or primed terminals can be connected to unprimed terminals to form the Δ (delta) shown in Fig. 16-4b. The primed letters are included this once to show these connections. But since the terminals at which they are located also have unprimed letters, the primed letters are not necessary. In circuit diagrams, sometimes circular ac generator symbols are used instead of the shown coil symbols. These Y and Δ connections are not limited to generator windings but apply as well to transformer windings and load impedances. There are some practical reasons for preferring the Y connection for alternator windings, but both the Y and Δ connections are used for transformer windings and for load impedances.

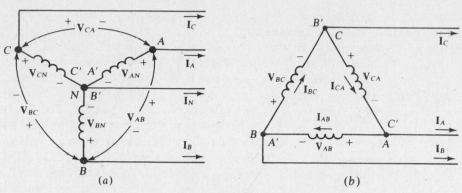

Fig. 16-4

In the Y connection shown in Fig. 16-4*a*, the primed terminals are joined at a common terminal marked *N* for *neutral*. There may be a line connected to this terminal, as shown, in which case there are four wires or lines. If no wire is connected to the neutral, the circuit is a three-wire circuit. The Δ connection illustrated in Fig. 16-4*b* inherently results in a three-wire circuit because there is no neutral terminal.

Note for the Y connection, that the *line currents* are also the winding currents, also called *phase currents*. A line current is a current in one of the lines and by convention is referenced from the source to the load. A phase current is a current in a generator or transformer winding or in a single impedance of a load.

A Y connection of windings or of impedances has two sets of voltages. There are the voltages V_{AN}, V_{BN}, and V_{CN} from terminals *A*, *B*, and *C* to the neutral terminal *N*. These are *phase voltages*. These differ from the *line-to-line* voltages, or just *line voltages*, V_{AB}, V_{BC}, and V_{CA}, across terminals *A*, *B*, and *C*. There are three other line voltages that have a 120° angle difference. These are V_{AC}, V_{BA}, and V_{CB}, which are the negatives of the other line voltages. In each set of line voltages, no two subscripts begin or end with the same letter. Also, no two pairs of subscripts have the same letters.

For the Δ shown in Fig. 16-4*b*, the line voltages are the same as the phase voltages. But the line currents I_A, I_B, and I_C differ from the phase currents I_{AB}, I_{BC}, and I_{CA} that flow through the windings. There is another suitable set of phase currents: I_{AC}, I_{BA}, and I_{CB}, which are the negatives of the currents in the first set. The comments on line voltage subscripts apply as well to Δ phase current subscripts.

PHASE SEQUENCE

The *phase sequence* of a three-phase circuit is the order in which the voltages or currents attain their maxima. As an illustration, Fig. 16-2*b* shows that $v_{AA'}$ peaks first, then $v_{BB'}$, then $v_{CC'}$, then $v_{AA'}$, etc., which is in the order of . . . *ABCABCAB* . . . . Any three adjacent letters can be selected to designate the phase sequence, but usually the three selected are *ABC*. This is sometimes called the *positive phase sequence*. If in Fig. 16-2*a* the labels of two windings are interchanged, or if the rotor is rotated clockwise instead of counterclockwise, the phase sequence is *ACB* (or *CBA* or *BAC*), also called the *negative phase sequence*. Although this explanation of phase sequence has been with respect to voltage peaking, phase sequence applies as well to current peaking in a balanced circuit.

Phase sequence can also be related to the subscripts of voltage and current phasors. If, for example, V_{AN} has an angle 120° greater than that of V_{BN}, then v_{AN} must lead v_{BN} by 120°, and so the phase sequence must be *ABC*. Incidentally, the terms "lead" and "lag" are often applied to the voltage phasors as well as to the corresponding instantaneous voltages. For another example, if V_{CN} leads V_{BN} by 120°, then in the phase sequence the first subscript *C* of V_{CN} must be immediately ahead of the first subscript B of $\mathbf{V}_{BN}$. Consequently, the phase sequence must be CBA, or ACB, the negative phase sequence.

Phase sequence can also be related to either the first or second subscripts of the line voltage phasors. This can be verified with an example. Figure 16-5a shows a phasor diagram of phase voltages $\mathbf{V}_{AN}$, $\mathbf{V}_{BN}$, and $\mathbf{V}_{CN}$ for an ABC phase sequence. Also included are terminals A, B, C, and N positioned such that lines drawn between them give the correct corresponding phasors. Drawn between terminals A, B, and C are a set of line voltage phasors: $\mathbf{V}_{AB}$, $\mathbf{V}_{BC}$, and $\mathbf{V}_{CA}$, which are redrawn in the phasor diagram shown in Fig. 16-5b. Note that $\mathbf{V}_{AB}$ leads $\mathbf{V}_{BC}$ by 120° and that $\mathbf{V}_{BC}$ leads $\mathbf{V}_{CA}$ by 120°. On the basis of this leading, the order of the first set of subscripts is ABC, in agreement with the phase sequence. The order of the second set of subscripts is BCA, which is equivalent to ABC, also in agreement with the phase sequence. This order can also be found by using a reference point R on the phasor diagram, as shown. If the phasors are rotated *counterclockwise* about the origin, the first subscripts pass the reference point in the order of the phase sequence, as do the second subscripts.

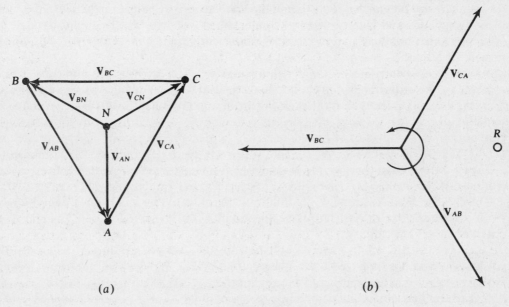

(a)   (b)

Fig. 16-5

In a similar manner it can be shown for a balanced circuit that the line current phasor subscripts correspond to the phase sequence order in the same way as explained for the voltage phasor subscripts. Also, the same is true for either the first or the second subscripts of the phase current phasors for a balanced Δ load. (A balanced Δ load has three equal impedances.)

BALANCED Y CIRCUIT

Figure 16-6 shows a *balanced Y circuit* that has a balanced Y load (a Y load of identical impedances) energized by a generator having Y-connected windings. Instead of generator windings, the windings could as well be the secondary windings of a three-phase transformer. A neutral wire connects the two neutral nodes.

A balanced three-phase circuit is easy to analyze because it is, in effect, three interconnected but separate circuits in which the only difference in responses is an angle difference of 120°. The general analysis procedure is to find the desired voltage or current in one phase, and use it with the phase sequence to get the corresponding voltages or currents in the two other phases. For example, in the circuit shown in Fig. 16-6, the line current $\mathbf{I}_A$ can be found from $\mathbf{I}_A =$

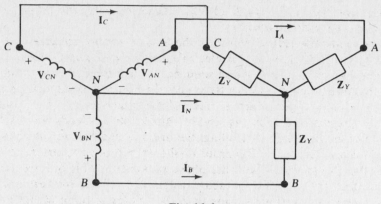

Fig. 16-6

$\mathbf{V}_{AN}/\mathbf{Z}_Y$.   Then, $\mathbf{I}_B$ and $\mathbf{I}_C$ can be found from $\mathbf{I}_A$ and the phase sequence: They have the same magnitude as $\mathbf{I}_A$ but lead and lag $\mathbf{I}_A$ by 120°, as determined from the phase sequence.

Since the three currents $\mathbf{I}_A$, $\mathbf{I}_B$, and $\mathbf{I}_C$ have the same magnitude but a 120° angle difference, their sum is zero: $\mathbf{I}_A + \mathbf{I}_B + \mathbf{I}_C = 0$.   And from KCL, $\mathbf{I}_N = -(\mathbf{I}_A + \mathbf{I}_B + \mathbf{I}_C) = 0$ A.   Because the neutral wire carries no current, it can be eliminated to change the circuit from a four- to a three-wire circuit.   A further consequence of the zero neutral current is that *the two neutral nodes are at the same potential*, even without the neutral wire.   In practice, though, it may be a good idea to have a small neutral wire to ensure balanced phase voltages in case the load impedances are not exactly the same.

The set of phase voltages and either set of line voltages for a balanced Y load have certain angle and magnitude relations that are independent of the load impedance.   These relations can be obtained from one of the triangles shown in Fig. 16-5a.   Consider the triangle formed by $\mathbf{V}_{BN}$, $\mathbf{V}_{CN}$, and $\mathbf{V}_{BC}$.   The largest angle is 120°, leaving $180° - 120° = 60°$ for the two other angles.   Since these two are opposite sides of equal length, they must be equal and so 30° each as shown in Fig. 16-7a.   It can be seen that there is a 30° angle between line voltage $\mathbf{V}_{BC}$ and phase voltage $\mathbf{V}_{BN}$, as is better shown in Fig. 16-7b.   As should be evident from Fig. 16-5a, there is also a 30° angle difference between $\mathbf{V}_{AB}$ and $\mathbf{V}_{AN}$ and between $\mathbf{V}_{CA}$ and $\mathbf{V}_{CN}$.   In general, in the voltage phasor diagram for a balanced Y load, there is a 30° angle between each phase voltage and the nearest line voltage.   This 30° can be either a lead or a lag, depending on the set of line voltages and also the phase sequence.

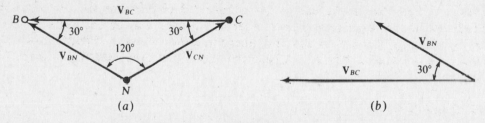

Fig. 16-7

Figure 16-8 has all the possible phasor diagrams that relate the Y phase voltages and the two sets of line voltages for the two phase sequences.   Thus, all angle relations between the line and Y phase voltages can be determined from them.   From the subscripts it should be apparent that Fig. 16-8a and b are for an *ABC* phase sequence and Fig. 16-8c and d for an *ACB* phase sequence.   Only relative angles are shown.   For actual angles, the appropriate diagram would have to be rotated until any one phasor is at its specified angle.

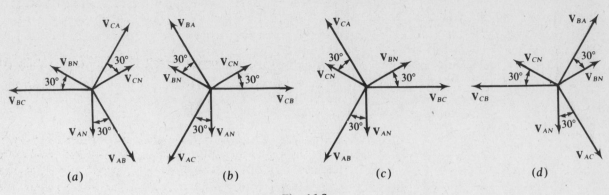

(a)   (b)   (c)   (d)

Fig. 16-8

There is also a relation between the magnitudes of the line and phase voltages. From Fig. 16-7a and the law of sines,

$$\frac{V_{BC}}{V_{BN}} = \frac{\sin 120°}{\sin 30°} = \frac{\sqrt{3}/2}{1/2} = \sqrt{3}$$

or $V_{BC} = \sqrt{3}\,V_{BN}$.   In general, for a balanced Y load the line voltage magnitude V_L is $\sqrt{3}$ times V_p, the phase voltage magnitude: $V_L = \sqrt{3}\,V_p$.

Incidentally, a voltage in the description of a three-phase circuit is the rms line voltage and not the line-to-neutral voltage, unless so specified.

BALANCED Δ LOAD

Figure 16-9 shows a balanced Δ load connected by three wires to a three-phase source. As a practical matter, this source is either a Y-connected alternator or, more probably, a Y- or Δ-connected secondary of a three-phase transformer. There is, of course, no neutral wire because a Δ load has only three terminals.

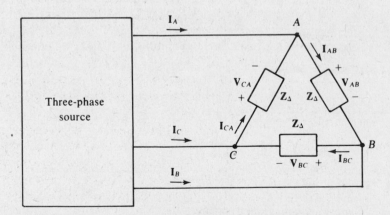

Fig. 16-9

The general procedure for finding the Δ phase currents is to first find one phase current and then use it with the phase sequence to find the other two. For example, the phase current I_{AB} can be found from $I_{AB} = V_{AB}/Z_\Delta$ and then I_{BC} and I_{CA} from I_{AB} and the phase sequence: These have the same magnitude as I_{AB}, but lead and lag I_{AB} by 120° as determined from the phase sequence.

The set of line currents and either set of phase currents for a balanced Δ have certain angle and magnitude relations that are independent of the load impedance. These can be found by applying

KCL at any terminal in the circuit shown in Fig. 16-9. If done at terminal A, the result is $\mathbf{I}_A = \mathbf{I}_{AB} - \mathbf{I}_{CA}$. Figure 16-10$a$ is a graphical representation of this subtraction for an ABC phase sequence. Since this is the same form of triangle as for the phase and line voltages of a balanced Y load, the results are similar: On a phasor diagram there is a 30° angle difference between each phase current and the nearest line current, as shown in Fig. 16-10b. This 30° can be either a lead or a lag, depending on the particular set of phase currents and on the phase sequence. Also, the line current magnitude I_L is $\sqrt{3}$ times I_p, the phase current magnitude: $I_L = \sqrt{3}I_p$.

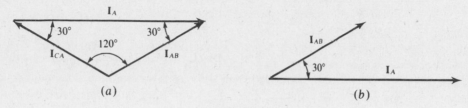

Fig. 16-10

Figure 16-11 has all the possible phasor diagrams that relate the line currents and the two sets of phase currents of balanced Δ loads for the two phase sequences. Thus all angle relations between the line and Δ phase currents can be determined from them. From the subscripts it should be evident that Fig. 16-11a and b are for an ABC phase sequence and that Fig. 16-11c and d are for an ACB phase sequence. Only relative angles are shown. For actual angles, the appropriate diagram would have to be rotated until any one phasor is at its specified angle.

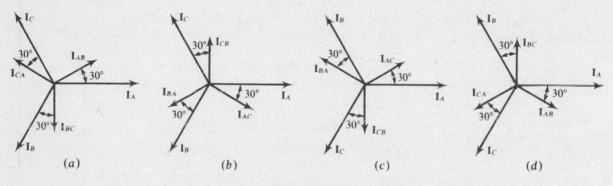

Fig. 16-11

PARALLEL LOADS

If a three-phase circuit has several loads connected in parallel, a good first step in an analysis is to combine the loads into a single Y or Δ load. Then, the analysis methods for a single Y or Δ load can be used. This combining is probably most obvious for two Δ loads, as shown in Fig. 16-12a. Being in parallel, corresponding phase impedances of the two Δ's can be combined to produce a single equivalent Δ.

If there are two Y loads, as shown in Fig. 16-12b, and if there is a neutral wire (not shown) connecting the two neutral nodes of the loads, corresponding phase impedances of the two Y's are in parallel and can be combined to produce a single equivalent Y. Even if there is no neutral wire, the corresponding phase impedances are in parallel provided that both Y loads are balanced because then both neutral nodes are at the same potential. If the loads are unbalanced and there is no neutral wire, corresponding impedances of the two Y's are *not* in parallel. Then, the two Y's can be converted to two Δ's, and these combined into a single equivalent Δ.

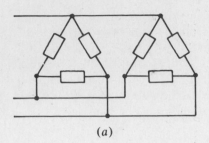 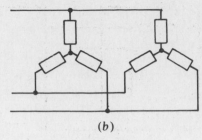 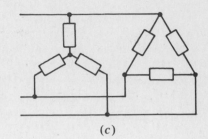

(a)　　　　　　　　　　　(b)　　　　　　　　　　　(c)

Fig. 16-12

Sometimes a three-phase circuit has a Y load and a Δ load, as shown in Fig. 16-12c. If the loads are balanced, the Δ can be converted to a Y and then the two Y's combined. If the loads are unbalanced, the Y can be converted to a Δ and then the two Δ's combined into a single equivalent Δ.

POWER

The average power absorbed by a balanced three-phase Y or Δ load is, of course, just three times the average power absorbed by any one of the phase impedances. For either a balanced Δ or a Y load, this is $P = 3V_pI_p \cos \theta$. The power formula is usually expressed in terms of the rms line voltage V_L and the rms line current I_L. For a Y load, $V_p = V_L/\sqrt{3}$ and $I_p = I_L$. And for a Δ load, $V_p = V_L$ and $I_p = I_L/\sqrt{3}$. With either substitution the result is the same:

$$P = \sqrt{3}V_LI_L \cos \theta$$

which is the formula for the total average power absorbed by either a balanced Y or Δ load. It is important to remember that θ is the load impedance angle and not the angle between a line voltage and line current.

Formulas for complex power S and reactive power Q can be readily found using the relations with average power presented in Chap. 14. For a balanced three-phase load, the result is

$$S = \sqrt{3}V_LI_L\underline{/\theta} \qquad \text{and} \qquad Q = \sqrt{3}V_LI_L \sin \theta$$

THREE-PHASE POWER MEASUREMENTS

If a three-phase load is balanced, the total average power absorbed can be measured by connecting a wattmeter into a single phase and multiplying the wattmeter reading by three. Of course, the wattmeter current coil should be connected in series with a phase impedance and the wattmeter potential coil should be connected across this impedance. If the load is unbalanced, three wattmeters can be used, one in each phase.

Frequently, though, it is impossible to connect a wattmeter into a phase. This is true, for example, for the common three-phase electric motor that has just three wires extending from it. For such an application, the *two-wattmeter method* can be used, provided that there are just three wires to the load.

Figure 16-13 shows the wattmeter connections for the two-wattmeter method. Notice that the current coils are in series in two of the lines and that the respective potential coils are connected between these two lines and the third line. The ± terminals are connected such that each wattmeter is connected as if to give an upscale reading for power absorbed by the load.

It can be shown that the total average power absorbed by the load is equal to the *algebraic* sum of the two wattmeter readings. So, if one reading is negative, it is added, sign and all, to the other wattmeter reading. (Of course, it may be necessary to reverse a coil to get this reading.) This two-wattmeter method is completely general. The load does not have to be balanced. In fact, the circuit does not have to be three-phase or even sinusoidally excited.

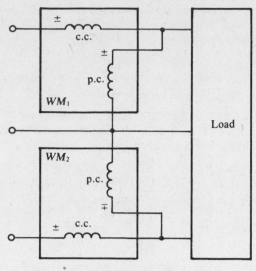

Fig. 16-13

From the line voltage and current phasors, it can be calculated that, for a balanced load with an impedance angle of θ, one wattmeter reading is $V_L I_L \cos(30° + \theta)$ and the other is $V_L I_L \cos(30° - \theta)$. The wattmeter with the $V_L I_L \cos(30° + \theta)$ reading has a current coil in the line corresponding to the phase sequence letter that immediately precedes the letter of the line in which there is no current coil. For example, if there is no current coil in line C, and if the phase sequence is ABC, then, since B precedes C in the phase sequence, the wattmeter with its current coil in line B has the $V_L I_L \cos(30° + \theta)$ reading.

The impedance angle for the phase impedance of a balanced load can be found from the readings of wattmeters connected for the two-wattmeter method. There are six formulas that relate the tangent of the impedance angle to the power readings. The appropriate formula depends on the phase sequence and the lines in which the current coils are connected. If P_A, P_B, and P_C are the readings of wattmeters with current coils in lines A, B, and C, then, for an ABC phase sequence,

$$\tan\theta = \sqrt{3}\,\frac{P_A - P_B}{P_A + P_B} = \sqrt{3}\,\frac{P_B - P_C}{P_B + P_C} = \sqrt{3}\,\frac{P_C - P_A}{P_C + P_A}$$

For an ACB phase sequence, $\tan\theta$ equals the negative of these.

UNBALANCED CIRCUITS

If a three-phase circuit has an unbalanced load, none of the shortcuts for the analysis of balanced three-phase circuits can be used. Conventional mesh or loop analysis is usually preferable. Of course, if the load is an unbalanced Y with a neutral wire, the voltage across each phase impedance is known, which means that each phase current can be readily found. The same is true for an unbalanced Δ load if there are no line impedances. Otherwise, it may be preferable to convert the Δ to a Y so that the line impedances are in series with the Y phase impedances.

Solved Problems

16.1   What is the phase sequence of a balanced three-phase circuit in which $\mathbf{V}_{AN} = 7200\underline{/20°}$ V and $\mathbf{V}_{CN} = 7200\underline{/-100°}$ V? Also, what is $\mathbf{V}_{BN}$?

Since V_{CN} lags V_{AN} by 120°, and the first subscripts are C and A, respectively, C follows A in the phase sequence. So, the phase sequence is ACB, the negative phase sequence. Of course, V_{BN} leads V_{AN} by 120°, but has the same magnitude: $V_{BN} = 7200\underline{/20°} + 120° = 7200\underline{/140°}$ V.

16.2 What is the phase sequence of a balanced three-phase circuit in which $V_{BN} = 277\underline{/-30°}$ V and $V_{CN} = 277\underline{/90°}$ V? And what is V_{AN}?

Since V_{CN} leads V_{BN} by 120°, and the first subscripts are C and B, respectively, C leads B in the phase sequence, which must be CBA, or ACB, the negative phase sequence. Of course, V_{AN} has the same magnitude as V_{CN}, but has an angle that is 120° greater:

$$V_{AN} = 277\underline{/90°} + 120° = 277\underline{/210°} = 277\underline{/-150°} \text{ V}$$

16.3 In a three-phase, three-wire circuit, find the phasor line currents to a balanced Y load in which each phase impedance is $Z_Y = 20\underline{/30°}$ Ω. Also, $V_{AN} = 120\underline{/20°}$ V, and the phase sequence is ABC.

The line current I_A can be found by dividing the phase voltage V_{AN} by the phase impedance Z_Y:

$$I_A = \frac{V_{AN}}{Z_Y} = \frac{120\underline{/20°}}{20\underline{/30°}} = 6\underline{/-10°} \text{ A}$$

The other line currents can be determined from I_A and the phase sequence. These currents have the same magnitude as I_A, and for the specified ABC phase sequence, I_B and I_C, respectively, lag and lead I_A by 120°. So,

$$I_B = 6\underline{/-10°} - 120° = 6\underline{/-130°} \text{ A} \quad \text{and} \quad I_C = 6\underline{/-10°} + 120° = 6\underline{/110°} \text{ A}$$

16.4 What is the phase sequence of a three-phase circuit in which $V_{AB} = 13\,200\underline{/-10°}$ V and $V_{BC} = 13\,200\underline{/110°}$ V? Also, which line voltage has an angle that differs by 120° from the angles of these voltages?

The phase sequence can be found from the voltage angles and first subscripts. Since V_{BC} leads V_{AB} by 120°, and since the first subscripts are B and A, respectively, B is immediately ahead of A in the phase sequence. So the phase sequence must be BAC or equivalently, ACB, the negative phase sequence.

The third line voltage is either V_{CA} or V_{AC} because only A and C of ABC have not been used together in subscripts. The proper third line voltage—the voltage that has an angle differing by 120° from those of V_{AB} and V_{BC}—is V_{CA} since no two line voltages of a set can have subscripts that start with the same letter, as would be the case if V_{AC} were used. Thus, $V_{CA} = 13\,200\underline{/-130°}$ V. This result is also obvious from Fig. 16-8c.

16.5 A balanced three-phase Y load has one phase voltage of $V_{CN} = 277\underline{/45°}$ V. If the phase sequence is ACB, find the line voltages V_{CA}, V_{AB}, and V_{BC}.

From Fig. 16-8c, which is for an ACB phase sequence and the specified line voltages, it can be seen that the line voltage V_{CA} has an angle that is 30° less than that of V_{CN}. Of course, its magnitude is greater by a factor of $\sqrt{3}$. So, $V_{CA} = 277\sqrt{3}\underline{/45° - 30°} = 480\underline{/15°}$ V. Also, from the same figure or from the fact that V_{AB} has an angle that is 120° greater because its first subscript A is just ahead of the first subscript C of V_{CA} in the phase sequence ACB, $V_{AB} = 480\underline{/15°} + 120° = 480\underline{/135°}$ V. Similarly, V_{BC} must lag V_{CA} by 120°: $V_{BC} = 480\underline{/15°} - 120° = 480\underline{/-105°}$ V.

16.6 What are the phase voltages for a balanced three-phase Y load if $V_{BA} = 12\,470\underline{/-35°}$ V? The phase sequence is ABC.

From Fig. 16-8b, which is for an ABC phase sequence and the set of line voltages that includes V_{BA}, it can be seen that V_{BN} leads V_{BA} by 30°. Of course, the magnitude of V_{BN} is less by a factor of $\sqrt{3}$. So,

$$\mathbf{V}_{BN} = \frac{12\,470}{\sqrt{3}} \underline{/-35° + 30°} = 7200\underline{/-5°} \text{ V}$$

Also from this figure, or from the phase sequence and first subscript relation, $\mathbf{V}_{AN}$ leads $\mathbf{V}_{BN}$ by 120°, and $\mathbf{V}_{CN}$ lags it by 120°:

$$\mathbf{V}_{AN} = 7200\underline{/-5° + 120°} = 7200\underline{/115°} \text{ V} \quad \text{and} \quad \mathbf{V}_{CN} = 7200\underline{/-5° - 120°} = 7200\underline{/-125°} \text{ V}$$

16.7 A balanced three-phase, three-wire circuit with an *ABC* phase sequence has one line current of $\mathbf{I}_B = 20\underline{/40°}$ A. Find the other line currents.

Because the circuit is balanced, all three line currents have the same magnitude of 20 A. And because the phase sequence is *ABC*, and *A* precedes *B* in the sequence, $\mathbf{I}_A$ leads $\mathbf{I}_B$ by 120°. For a similar reason, $\mathbf{I}_C$ lags $\mathbf{I}_B$ by 120°. Consequently,

$$\mathbf{I}_A = 20\underline{/40° + 120°} = 20\underline{/160°} \text{ A} \quad \text{and} \quad \mathbf{I}_C = 20\underline{/40° - 120°} = 20\underline{/-80°} \text{ A}$$

16.8 What is the $\mathbf{I}_B$ line current in an unbalanced three-phase, three-wire circuit in which $\mathbf{I}_A = 50\underline{/60°}$ A and $\mathbf{I}_C = 80\underline{/160°}$ A?

By KCL, the sum of the three line currents is zero: $\mathbf{I}_A + \mathbf{I}_B + \mathbf{I}_C = 0$, from which $\mathbf{I}_B = -\mathbf{I}_A - \mathbf{I}_C = -50\underline{/60°} - 80\underline{/160°} = 86.7\underline{/-54.6°}$ A.

16.9 A balanced Y load of 40-Ω resistors is connected to a 480-V, three-phase, three-wire source. Find the rms line current.

Each line current is equal to the load phase voltage of $480/\sqrt{3} = 277$ V divided by the phase impedance of 40 Ω: $I_L = 277/40 = 6.93$ A.

16.10 A balanced Y load of $50\underline{/-30°}$-Ω impedances is energized by a 12 470-V, three-phase, three-wire source. Find the rms line current.

Each line current is equal to the load phase voltage of $12\,470/\sqrt{3} = 7200$ V divided by the phase impedance magnitude of 50 Ω: $I_L = 7200/50 = 144$ A.

16.11 Find the phasor line currents to a balanced Y load of impedances $\mathbf{Z}_Y = 50\underline{/25°}$ Ω energized by a three-phase source. One phase voltage is $\mathbf{V}_{BN} = 120\underline{/30°}$ V, and the phase sequence is *ABC*.

The line current $\mathbf{I}_B$ can be found by dividing the phase voltage $\mathbf{V}_{BN}$ by the phase impedance $\mathbf{Z}_Y$. Then the other line currents can be found from $\mathbf{I}_B$ with the aid of the phase sequence. The line current $\mathbf{I}_B$ is

$$\mathbf{I}_B = \frac{\mathbf{V}_{BN}}{\mathbf{Z}_Y} - \frac{120\underline{/30°}}{50\underline{/25°}} = 2.4\underline{/5°} \text{ A}$$

Since the phase sequence is *ABC*, the angle of $\mathbf{I}_A$ is 120° more than the angle of $\mathbf{I}_B$. Of course, the current magnitudes are the same: $\mathbf{I}_A = 2.4\underline{/5° + 120°} = 2.4\underline{/125°}$ A. Similarly, the angle of $\mathbf{I}_C$ is 120° less: $\mathbf{I}_C = 2.4\underline{/5° - 120°} = 2.4\underline{/-115°}$ A.

16.12 In a three-phase, three-wire circuit, find the phasor line currents to a balanced Y load for which $\mathbf{Z}_Y = 60\underline{/-30°}$ Ω and $\mathbf{V}_{CB} = 480\underline{/65°}$ V. The phase sequence is *ABC*.

From Fig. 16-8b, the phase voltage $\mathbf{V}_{CN}$ has an angle that is 30° greater than that of $\mathbf{V}_{CB}$ and, of course, a magnitude that is less by a factor of $1/\sqrt{3}$:

$$\mathbf{V}_{CN} = \frac{480/65° + 30°}{\sqrt{3}} = 277/95° \text{ V}$$

The line current $\mathbf{I}_C$ is

$$\mathbf{I}_C = \frac{\mathbf{V}_{CN}}{\mathbf{Z}_Y} = \frac{277/95°}{60/-30°} = 4.62/125° \text{ A}$$

Since A follows C in the phase sequence, $\mathbf{I}_A$ lags $\mathbf{I}_C$ by 120°: $\mathbf{I}_A = 4.62/125° - 120° = 4.62/5°$ A. And because B precedes C in the phase sequence, $\mathbf{I}_B$ leads $\mathbf{I}_C$ by 120°:

$$\mathbf{I}_B = 4.62/125° + 120° = 4.62/245° = 4.62/-115° \text{ A}$$

16.13 What is the phase sequence of a balanced three-phase circuit with a Δ load in which two of the phase currents are $\mathbf{I}_{BA} = 6/-30°$ A and $\mathbf{I}_{CB} = 6/90°$ A? And, what is $\mathbf{I}_{AC}$?

Since $\mathbf{I}_{CB}$, with a first subscript of C, has an angle 120° greater than that of $\mathbf{I}_{BA}$, which has a first subscript of B, the letter C precedes the letter B in the phase sequence. Thus the phase sequence must be ACB, the negative phase sequence. From this phase sequence, $\mathbf{I}_{AC}$, with a first subscript of A, has an angle that is 120° less than that of $\mathbf{I}_{BA}$. Of course, the magnitude is the same: $\mathbf{I}_{AC} = 6/-30° - 120° = 6/-150°$ A.

16.14 Find the phase currents $\mathbf{I}_{BC}$, $\mathbf{I}_{AB}$, and $\mathbf{I}_{CA}$ of a balanced three-phase Δ load to which one line current is $\mathbf{I}_B = 50/-40°$ A. The phase sequence is ABC.

From Fig. 16-11a, which is for an ABC phase sequence and the specified set of Δ phase currents, it can be seen that $\mathbf{I}_{BC}$ has an angle that is 30° greater than that of $\mathbf{I}_B$, and, of course, a magnitude that is less by a factor of $1/\sqrt{3}$. Consequently,

$$\mathbf{I}_{BC} = \frac{50/-40° + 30°}{\sqrt{3}} = 28.9/-10° \text{ A}$$

Also, from the same figure or from the fact that $\mathbf{I}_{AB}$ has an angle that is 120° greater because its first subscript A is just ahead of the first subscript B of $\mathbf{I}_{BC}$ in the phase sequence ABC, $\mathbf{I}_{AB} = 28.9/-10° + 120° = 28.9/110°$ A. Then $\mathbf{I}_{CA}$ must have an angle that is 120° less than that of $\mathbf{I}_{BC}$: $\mathbf{I}_{CA} = 28.9/-10° - 120° = 28.9/-130°$ A.

16.15 A balanced three-phase Δ load has one phase current of $\mathbf{I}_{BA} = 10/30°$ A. The phase sequence is ACB. Find the other phasor phase currents and also the phasor line currents.

The two other desired phase currents are those having angles that differ by 120° from the angle of $\mathbf{I}_{BA}$. These are $\mathbf{I}_{AC}$ and $\mathbf{I}_{CB}$, as can be obtained from the relation of subscripts: No two currents can have the same first or second subscript letters, or the same two letters. This is also obvious from Fig. 16-11c. Since the phase sequence is ACB or negative, $\mathbf{I}_{CB}$ must lead $\mathbf{I}_{BA}$ by 120° because in the phase sequence the letter C, the first subscript letter of $\mathbf{I}_{CB}$, precedes the letter B, the first subscript letter of $\mathbf{I}_{BA}$. Also, Fig. 16-11c shows this 120° lead. Therefore, $\mathbf{I}_{CB} = 10/30° + 120° = 10/150°$ A. Then $\mathbf{I}_{AC}$ must lag $\mathbf{I}_{BA}$ by 120°: $\mathbf{I}_{AC} = 10/30° - 120° = 10/-90°$ A.

From Fig. 16-11c, $\mathbf{I}_A$ lags $\mathbf{I}_{AC}$ by 30°, and since it has a magnitude that is greater by a factor of $\sqrt{3}$, $\mathbf{I}_A = 10\sqrt{3}/-90° - 30° = 17.3/-120°$ A. Because the phase sequence is ACB, currents $\mathbf{I}_B$ and $\mathbf{I}_C$, respectively, lead and lag $\mathbf{I}_A$ by 120°:

$$\mathbf{I}_B = 17.3/-120° + 120° = 17.3/0° \text{ A}$$

$$\mathbf{I}_C = 17.3/-120° - 120° = 17.3/-240° = 17.3/120° \text{ A}$$

16.16 What are the phasor line currents to a balanced three-phase Δ load if one phase current is $\mathbf{I}_{CB} = 10/20°$ A and if the phase sequence is ABC?

From Fig. 16-11b, which is for an ABC phase sequence and the set of phase currents that includes $\mathbf{I}_{CB}$, it can be seen that $\mathbf{I}_C$ leads $\mathbf{I}_{CB}$ by 30°. Of course, its magnitude is greater by a factor of $\sqrt{3}$. So $\mathbf{I}_C = 10\sqrt{3}\underline{/20° + 30°} = 17.3\underline{/50°}$ A. From the phase sequence, $\mathbf{I}_B$ leads $\mathbf{I}_C$ by 120° and $\mathbf{I}_A$ lags it by 120°:

$$\mathbf{I}_B = 17.3\underline{/50° + 120°} = 17.3\underline{/170°} \text{ A} \quad \text{and} \quad \mathbf{I}_A = 17.3\underline{/50° - 120°} = 17.3\underline{/-70°} \text{ A}$$

16.17 A 208-V three-phase circuit has a balanced Δ load of 50-Ω resistors. Find the rms line current.

The rms line current I_L can be found from the rms phase current I_p, which is equal to the 208-V line voltage (and also phase voltage) divided by the 50-Ω phase resistance: $I_p = 208/50 = 4.16$ A. The rms line current I_L is greater by a factor of $\sqrt{3}$: $I_L - \sqrt{3}(4.16) = 7.21$ A.

16.18 Find the phasor line currents to a balanced Δ three-phase load of impedances $\mathbf{Z}_\Delta = 40\underline{/10°}$ Ω if one phase voltage is $\mathbf{V}_{CB} = 480\underline{/-15°}$ V and if the phase sequence is ACB.

A good first step is to find the phase current $\mathbf{I}_{CB}$:

$$\mathbf{I}_{CB} = \frac{\mathbf{V}_{CB}}{\mathbf{Z}_\Delta} = \frac{480\underline{/-15°}}{40\underline{/10°}} = 12\underline{/-25°} \text{ A}$$

From Fig. 16-11c, which is for an ACB phase sequence and the set of phase currents that includes $\mathbf{I}_{CB}$, the line current $\mathbf{I}_C$ lags $\mathbf{I}_{CB}$ by 30°. Of course, its magnitude is greater by a factor of $\sqrt{3}$. So

$$\mathbf{I}_C = 12\sqrt{3}\underline{/-25° - 30°} = 20.8\underline{/-55°} \text{ A}$$

Since the phase sequence is ACB, the line currents $\mathbf{I}_A$ and $\mathbf{I}_B$ respectively lead and lag $\mathbf{I}_C$ by 120°:

$$\mathbf{I}_A = 20.8\underline{/-55° + 120°} = 20.8\underline{/65°} \text{ A} \quad \text{and} \quad \mathbf{I}_B = 20.8\underline{/-55° - 120°} = 20.8\underline{/-175°} \text{ A}$$

16.19 A balanced Δ load of impedances $\mathbf{Z}_\Delta = 24\underline{/-40°}$ Ω is connected to the Y-connected secondary of a three-phase transformer. The phase sequence is ACB and $\mathbf{V}_{BN} = 277\underline{/50°}$ V. Find the phasor line currents and load phase currents.

One approach is to find the corresponding $\mathbf{Z}_Y$ and use it to find $\mathbf{I}_B$ from $\mathbf{I}_B = \mathbf{V}_{BN}/\mathbf{Z}_Y$. The next step is to use the phase sequence to obtain $\mathbf{I}_A$ and $\mathbf{I}_C$ from $\mathbf{I}_B$. The last step is to use either Fig. 16-11c or d to obtain the phase currents from $\mathbf{I}_B$. This is the approach that will be used, although there are other approaches just as short.

The corresponding Y impedance is $\mathbf{Z}_Y = \mathbf{Z}_\Delta/3 = (24\underline{/-40°})/3 = 8\underline{/-40°}$ Ω. And

$$\mathbf{I}_B = \frac{\mathbf{V}_{BN}}{\mathbf{Z}_Y} = \frac{277\underline{/50°}}{8\underline{/-40°}} = 34.6\underline{/90°} \text{ A}$$

Since the phase sequence is ACB, the line currents $\mathbf{I}_A$ and $\mathbf{I}_C$ respectively lag and lead $\mathbf{I}_B$ by 120°:

$$\mathbf{I}_A = 34.6\underline{/90° - 120°} - 34.6\underline{/-30°} \text{ A}$$

$$\mathbf{I}_C = 34.6\underline{/90° + 120°} = 34.6\underline{/210°} = 34.6\underline{/-150°} \text{ A}$$

Either set of load phase currents can be found: $\mathbf{I}_{AB}$, $\mathbf{I}_{BC}$, and $\mathbf{I}_{CA}$ or $\mathbf{I}_{BA}$, $\mathbf{I}_{AC}$, and $\mathbf{I}_{CB}$. If the first set is selected, then Fig. 16-11d can be used, which has these currents for an ACB phase sequence. It can be seen that $\mathbf{I}_{AB}$, $\mathbf{I}_{BC}$, and $\mathbf{I}_{CA}$ lag $\mathbf{I}_A$, $\mathbf{I}_B$, and $\mathbf{I}_C$ respectively by 30°. The magnitude of each load phase current is, of course, $34.6/\sqrt{3} = 20$ A. Thus,

$$\mathbf{I}_{AB} = 20\underline{/-60°} \text{ A} \quad \mathbf{I}_{BC} = 20\underline{/60°} \text{ A} \quad \mathbf{I}_{CA} = 20\underline{/-180°} = -20 \text{ A}$$

16.20 Find the rms line voltage V_L at the source of the circuit in Fig. 16-14. As shown, the rms load phase voltage is 100 V and each line impedance is $2 + j3$ Ω.

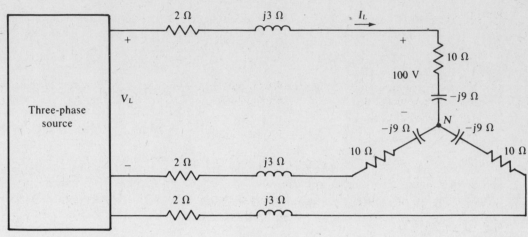

Fig. 16-14

The rms line current I_L can be used to find V_L. Of course, I_L is equal to the 100-V load phase voltage divided by the magnitude of the load phase impedance:

$$I_L = \frac{100}{|10 - j9|} = 7.43 \text{ A}$$

In flowing, this current produces a voltage drop from a source terminal to the load neutral terminal N, which drop is equal to the product of this current and the magnitude of the sum of the impedances that the current flows through. This voltage is

$$I_L|\mathbf{Z}_{\text{line}} + \mathbf{Z}_Y| = 7.43|(2 + j3) + (10 - j9)| = 7.43|12 - j6| = 7.43(13.42) = 99.7 \text{ V}$$

The line voltage at the source is equal to $\sqrt{3}$ times this: $V_L = \sqrt{3}(99.7) = 173$ V.

16.21 Find the rms line voltage V_L at the source of the circuit in Fig. 16-15. As shown, the rms line voltage at the load is 100 V and each line impedance is $2 + j3 \ \Omega$.

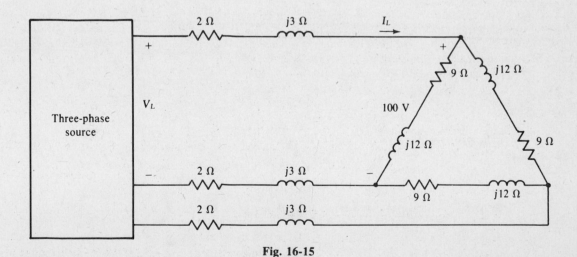

Fig. 16-15

Perhaps the best approach is to convert the Δ to an equivalent Y and then proceed as in the solution to Prob. 16.20. The equivalent Y impedance is $(9 + j12)/3 = 3 + j4 \ \Omega$. Since the line voltage at the load is 100 V, the line-to-neutral voltage for the equivalent Y load is $100/\sqrt{3} = 57.74$ V. The rms line current I_L is equal to this voltage divided by the magnitude of the Y phase impedance:

$$I_L = \frac{57.74}{|3 + j4|} = \frac{57.74}{5} = 11.55 \text{ A}$$

In flowing, this current produces a voltage drop from a source terminal to the Y neutral terminal, which drop is equal to the product of this current and the magnitude of the sum of the impedances that the current flows through. The voltage is

$$I_L|Z_{\text{line}} + Z_Y| = 11.55|(2 + j3) + (3 + j4)| = 11.55|5 + j7| = 11.55(8.6) = 99.3 \text{ V}$$

And the line voltage at the source is equal to $\sqrt{3}$ times this: $V_L = \sqrt{3}(99.3) = 172 \text{ V}$.

16.22 A 480-V, three-phase, three-wire circuit has two balanced Δ loads, one of 5-Ω resistors and the other of 20-Ω resistors. Find the total rms line current.

Because the corresponding resistors of the Δ loads are in parallel, the resistances can be combined to produce an equivalent single Δ of $5\|20 = 4$-Ω resistors. The phase current of this Δ is equal to the line voltage divided by the 4 Ω of resistance: $I_p = 480/4 = 120 \text{ A}$. And of course, the line current is $\sqrt{3}$ times greater: $I_L = \sqrt{3}(120) = 208 \text{ A}$.

16.23 A 208-V, three-phase, three-wire circuit has two balanced Y loads, one of 6-Ω resistors and the other of 12-Ω resistors. Find the total rms line current.

Since the loads are balanced, the load neutral nodes are at the same potential even if there is no connection between them. Consequently, corresponding resistors are in parallel and can be combined. The result is a net resistance of $6\|12 = 4 \text{ }\Omega$. This divided into the phase voltage of $208/\sqrt{3} = 120 \text{ V}$ gives the total rms line current: $I_L = 120/4 = 30 \text{ A}$.

16.24 A 600-V three-phase circuit has two balanced Δ loads, one of $40\underline{/30°}$-Ω impedances and the other of $50\underline{/-60°}$-Ω impedances. Find the total rms line current and also the total average absorbed power.

Being in parallel, corresponding Δ impedances can be combined to

$$Z_\Delta = \frac{(40\underline{/30°})(50\underline{/-60°})}{40\underline{/30°} + 50\underline{/-60°}} = \frac{2000\underline{/-30°}}{64\underline{/-21.3°}} = 31.2\underline{/-8.7°} = 30.9 - j4.7 \text{ }\Omega$$

The rms phase current for the combined Δ is equal to the line voltage divided by the magnitude of this impedance:

$$I_p = \frac{V_L}{Z_\Delta} = \frac{600}{31.2} = 19.2 \text{ A}$$

And the rms line current is $I_L = \sqrt{3}I_p = \sqrt{3}(19.2) = 33.3 \text{ A}$.

The total average power can be found using the phase current and resistance for the combined Δ:

$$P = 3I_p^2 R = 3(19.2)^2(30.9) = 34.2 \times 10^3 \text{ W} = 34.2 \text{ kW}$$

Alternatively, it can be found from the line quantities and the power factor:

$$P = \sqrt{3}V_L I_L \times PF = \sqrt{3}(600)(33.3) \cos(-8.7°) = 34.2 \times 10^3 \text{ W} = 34.2 \text{ kW}$$

16.25 A 208-V three-phase circuit has two balanced loads, one a Δ of $21\underline{/30°}$-Ω impedances and the other a Y of $9\underline{/-60°}$-Ω impedances. Find the total rms line current and also the total average absorbed power.

The two loads can be combined if the Δ is converted to a Y or if the Y is converted to a Δ so that, in effect, the loads are in parallel. If the Δ is converted to a Y, the equivalent Y has a phase impedance of $(21\underline{/30°})/3 = 7\underline{/30°} \text{ }\Omega$. Since the circuit now has two balanced Y loads, corresponding impedances are in parallel and so can be combined:

$$Z_Y = \frac{(7\underline{/30°})(9\underline{/-60°})}{7\underline{/30°} + 9\underline{/-60°}} = \frac{63\underline{/-30°}}{11.4\underline{/-22.13°}} = 5.53\underline{/-7.87°} = 5.47 - j0.76 \ \Omega$$

The rms line current is equal to the phase voltage of $V_p = 208/\sqrt{3} = 120$ V divided by the magnitude of the combined phase impedance:

$$I_L = \frac{V_p}{Z_Y} = \frac{120}{5.53} = 21.7 \ A$$

Since this current effectively flows through the resistance of the combined Y, the total average power absorbed is

$$P = 3I_L^2 R = 3(21.7)^2(5.47) = 7.8 \times 10^3 \ W = 7.8 \ kW$$

Alternatively, the line voltage and current power formula can be used:

$$P = \sqrt{3}V_L I_L \times PF = \sqrt{3}(208)(21.7) \cos(-7.87°) = 7.8 \times 10^3 \ W = 7.8 \ kW$$

16.26 A balanced Y of $20\underline{/20°}$-Ω impedances and a balanced Δ of $42\underline{/30°}$-Ω impedances are connected by three wires to the secondary of a three-phase transformer. If $\mathbf{V}_{BC} = 480\underline{/10°}$ V and if the phase sequence is ABC, find the total phasor line currents.

A good approach is to obtain an equivalent single combined Y impedance, and also a phase voltage, and then find a line current by dividing this phase voltage by this impedance. The other line currents can be obtained from this line current by using the phase sequence. For this approach the first step is find the equivalent Y impedance for the Δ. It is $(42\underline{/30°})/3 = 14\underline{/30°} \ \Omega$. The next step is to find a combined Y impedance $\mathbf{Z}_Y$ by using the parallel combination formula:

$$Z_Y = \frac{(20\underline{/20°})(14\underline{/30°})}{20\underline{/20°} + 14\underline{/30°}} = \frac{280\underline{/50°}}{33.87\underline{/24.1°}} = 8.27\underline{/25.9°} \ \Omega$$

From Fig. 16-8a, which is for an ABC phase sequence, $\mathbf{V}_{BN}$ has an angle that is 30° less than that of $\mathbf{V}_{BC}$ and, of course, it has a magnitude that is less by a factor of $1/\sqrt{3}$:

$$\mathbf{V}_{BN} = \frac{480\underline{/10° - 30°}}{\sqrt{3}} = 277\underline{/-20°} \ V$$

The line current $\mathbf{I}_B$ is equal to this voltage divided by the combined Y phase impedance:

$$\mathbf{I}_B = \frac{\mathbf{V}_{BN}}{\mathbf{Z}_Y} = \frac{277\underline{/-20°}}{8.27\underline{/25.9°}} = 33.5\underline{/-45.9°} \ A$$

Of course, from the phase sequence, $\mathbf{I}_A$ and $\mathbf{I}_C$, respectively, lead and lag $\mathbf{I}_B$ by 120°: $\mathbf{I}_A = 33.5\underline{/74.1°}$ A and $\mathbf{I}_C = 33.5\underline{/-165.9°}$ A.

16.27 A balanced Δ load of $39\underline{/-40°}$-Ω impedances is connected by three wires, with 4 Ω of resistance each, to the secondary of a three-phase transformer. If the line voltage is 480 V at the secondary terminals, find the rms line current.

If the Δ is converted to a Y, the Y impedances can be combined with the line resistances, and the line current found by dividing the magnitude of the total Y phase impedance into the phase voltage. The Y equivalent of the Δ has a phase impedance of

$$\frac{39\underline{/-40°}}{3} = 13\underline{/-40°} = 9.96 - j8.36 \ \Omega$$

Being a Y impedance, this is in series with the line resistance and so can be combined with it. The result is

$$4 + (9.96 - j8.36) = 13.96 - j8.36 = 16.3\underline{/-30.9°} \ \Omega$$

And, the rms line current is equal to the phase voltage of $480/\sqrt{3} = 277$ V divided by the magnitude of this impedance: $I_L = 277/16.3 = 17$ A.

16.28 Find the average power absorbed by a balanced three-phase load in an ABC circuit in which $\mathbf{V}_{CB} = 208\underline{/15°}$ V and $\mathbf{I}_B = 3\underline{/110°}$ A.

The formula $P = \sqrt{3}V_L I_L \times PF$ can be used if the power factor PF can be found. Since it is the cosine of the impedance angle, what is needed is the angle between a load phase voltage and current. With $\mathbf{I}_B$ known, the most convenient phase voltage is $\mathbf{V}_{BN}$ because the desired angle is that between $\mathbf{V}_{BN}$ and $\mathbf{I}_B$. This approach is based on the assumption of a Y load, which is valid since any balanced load can be converted to an equivalent Y. Figure 16-8b, which is for an ABC phase sequence, shows that $\mathbf{V}_{BN}$ leads $\mathbf{V}_{CB}$ by 150°, and so here has an angle of $15° + 150° = 165°$. The power factor angle, the angle between $\mathbf{V}_{BN}$ and $\mathbf{I}_B$, is $165° - 110° = 55°$. So the average power absorbed by the load is

$$P = \sqrt{3}V_L I_L \times PF = \sqrt{3}(208)(3) \cos 55° = 620 \text{ W}$$

16.29 A three-phase induction motor delivers 20 hp while operating at an 85 percent efficiency and at a 0.8 lagging power factor from 480-V lines. Find the rms line current.

The current I_L can be found from the formula $P_{in} = \sqrt{3}V_L I_L \times PF$, in which P_{in} is the input power to the motor:

$$P_{in} = \frac{P_{out}}{\eta} = \frac{20 \times 745.7}{0.85} = 17.55 \times 10^3 \text{ W}$$

and

$$I_L = \frac{P_{in}}{\sqrt{3}V_L \times PF} = \frac{17.55 \times 10^3}{\sqrt{3}(480)(0.8)} = 26.4 \text{ A}$$

16.30 In a 208-V three-phase circuit a balanced Δ load absorbs 2 kW at a 0.8 leading power factor. Find $\mathbf{Z}_\Delta$.

From $P = 3V_p I_p \times PF$, the phase current is

$$I_p = \frac{P}{3V_p \times PF} = \frac{2000}{3(208)(0.8)} = 4.01 \text{ A}$$

Since the line voltage is also the phase voltage, the magnitude of the phase impedance is

$$Z_\Delta = \frac{V_p}{I_p} = \frac{208}{4.01} = 51.9 \text{ } \Omega$$

The impedance angle is the power factor angle: $\theta = -\cos^{-1} 0.8 = -36.9°$. So the phase impedance is $\mathbf{Z}_\Delta = 51.9\underline{/-36.9°}$ Ω.

16.31 Given that $\mathbf{V}_{AB} = 480\underline{/30°}$ V in an ABC three-phase circuit, find the phasor line currents to a balanced load that absorbs 5 kW at a 0.6 lagging power factor.

From $P = \sqrt{3}V_L I_L \times PF$, the line current magnitude is

$$I_L = \frac{P}{\sqrt{3}V_L \times PF} = \frac{5000}{\sqrt{3}(480)(0.6)} = 10 \text{ A}$$

If, for convenience, a Y load is assumed, then from Fig. 16-8a, $\mathbf{V}_{AN}$ lags $\mathbf{V}_{AB}$ by 30° and so has an angle of $30° - 30° = 0°$. Since $\mathbf{I}_A$ lags $\mathbf{V}_{AN}$ by the power factor angle of $\theta = \cos^{-1} 0.6 = 53.1°$, $\mathbf{I}_A$ has an angle of $0° - 53.1° = -53.1°$. Consequently, $\mathbf{I}_A = 10\underline{/-53.1°}$ A and, from the ABC phase sequence,

$$\mathbf{I}_B = 10\underline{/-53.1°} - 120° = 10\underline{/-173.1°} \text{ A}$$

and

$$\mathbf{I}_C = 10\underline{/-53.1°} + 120° = 10\underline{/66.9°} \text{ A}$$

16.32 A 480-V three-phase circuit has two balanced loads. One is a 5-kW resistive heater and the other an induction motor that delivers 15 hp while operating at an 80 percent efficiency and a 0.9 lagging power factor. Find the total rms line current.

A good approach is to find the total complex power S_T and solve for I_L from $|S_T| = S_T = \sqrt{3}V_L I_L$, the apparent power. Since the heater is purely resistive, its complex power is $S_H = 5\underline{/0°}$ kVA. The complex power of the motor has a magnitude (the apparent power) that is equal to the input power divided by the power factor, and it has an angle that is the arccosine of the power factor:

$$S_M = \frac{15 \times 745.7}{0.8(0.9)} \underline{/\cos^{-1} 0.9} = 15.5 \times 10^3 \underline{/25.8°} \text{ VA} = 13.98 + j6.77 \text{ kVA}$$

The total complex power is the sum of these two complex powers:

$$S_T = S_H + S_M = 5 + (13.98 + j6.77) = 20.15\underline{/19.6°} \text{ kVA}$$

Since the apparent power is $|S_T| = S_T = 20.15$ kVA,

$$I_L = \frac{S_T}{\sqrt{3}V_L} = \frac{20.15 \times 10^3}{\sqrt{3}(480)} = 24.2 \text{ A}$$

16.33 If in an ABC, three-phase, three-wire circuit, $I_A = 10\underline{/-30°}$ A, $I_B = 8\underline{/45°}$ A, and $V_{AB} = 208\underline{/60°}$ V, find the reading of a wattmeter connected with its current coil in line C and its potential coil across lines B and C. The $\pm$ terminal of the current coil is toward the source, and the $\pm$ terminal of the potential coil is at line C.

From the specified wattmeter connections, the wattmeter reading is equal to $P = V_L I_C \cos(\text{ang } V_{CB} - \text{ang } I_C)$. Of course, $V_L = 208$ V. Also,

$$I_C = -I_A - I_B = -10\underline{/-30°} - 8\underline{/45°} = 14.3\underline{/-177.4°} \text{ A}$$

From an inspection of Fig. 16-8a and b, it should be fairly apparent that V_{CB} leads V_{AB} by 60° and so here is $V_{CB} = 208\underline{/60° + 60°} = 208\underline{/120°}$ V. Therefore, the wattmeter reading is

$$P = 208(14.3)\cos[120° - (-177.4°)] \text{ W} = 1.37 \text{ kW}$$

16.34 A balanced Y load of 25-Ω resistors is energized from a 480-V, ABC, three-phase, three-wire source. Find the reading of a wattmeter connected with its current coil in line A and its potential coil across lines A and B. The $\pm$ terminal of the current coil is toward the source, and the $\pm$ terminal of the potential coil is at line A.

With the specified connections, the wattmeter has a reading equal to $P = V_L I_L \cos(\text{ang } V_{AB} - \text{ang } I_A)$, for which I_L and the angles of V_{AB} and I_A are needed. Since no phasors are specified in the problem statement, the phasor V_{AB} can be conveniently assigned a 0° angle: $V_{AB} = 480\underline{/0°}$ V. The current I_A can be found from the phase voltage V_{AN} and the phase resistance 25 Ω. Of course, V_{AN} has a magnitude of $480/\sqrt{3} = 277$ V. Also, from Fig. 16-8a, it lags V_{AB} by 30° and so has an angle of $0° - 30° = -30°$. Consequently, $V_{AN} = 277\underline{/-30°}$ V and

$$I_A = \frac{V_{AN}}{R_Y} = \frac{277\underline{/-30°}}{25} = 11.09\underline{/-30°} \text{ A}$$

Since the magnitude of I_A is the rms line current,

$$P = V_L I_L \cos(\text{ang } V_{AB} - \text{ang } I_A) = 480(11.09)\cos[0° - (-30°)] = 4.61 \times 10^3 \text{ W} = 4.61 \text{ kW}$$

Incidentally, this wattmeter reading is just half the total average power absorbed of $\sqrt{3}V_L I_L \times PF = \sqrt{3}(480)(11.09)(1) = 9220$ W. As should be evident from the two-wattmeter formulas $V_L I_L \cos(30° + \theta)$ and $V_L I_L \cos(30° - \theta)$, this result is generally true for a purely resistive balanced load ($\theta = 0°$) and a wattmeter connected as if it is one of the two wattmeters of the two-wattmeter method.

16.35 A balanced Δ load of $j40$-Ω inductors is energized from a 208-V, *ACB* source. Find the reading of a wattmeter connected with its current coil in line *B* and its potential coil across lines *B* and *C*. The $\pm$ terminal of the current coil is toward the source, and the $\pm$ terminal of the potential coil is at line *B*.

With the specified connections, the wattmeter has a reading equal to $P = V_L L_L \cos (\text{ang } \mathbf{V}_{BC} - \text{ang } \mathbf{I}_B)$, for which I_L and the angles of $\mathbf{V}_{BC}$ and $\mathbf{I}_B$ are needed. Since no phasors are specified, the phasor $\mathbf{V}_{BC}$ can be conveniently assigned a $0°$ angle: $\mathbf{V}_{BC} = 208\underline{/0°}$ V. Then $\mathbf{V}_{AB} = 208\underline{/-120°}$ V, as is apparent from the relation between the specified *ACB* phase sequence and the first subscripts. It follows that

$$\mathbf{I}_B = \mathbf{I}_{BC} - \mathbf{I}_{AB} = \frac{\mathbf{V}_{BC}}{\mathbf{Z}_\Delta} - \frac{\mathbf{V}_{AB}}{\mathbf{Z}_\Delta} = \frac{208\underline{/0°}}{j40} - \frac{208\underline{/-120°}}{j40} = 9.01\underline{/-60°} \text{ A}$$

So the wattmeter reading is

$$P = V_L I_L \cos (\text{ang } \mathbf{V}_{BC} - \text{ang } \mathbf{I}_B) = 208(9.01) \cos [0° - (-60°)] = 937 \text{ W}$$

This reading has, of course, no relation to the average power absorbed by the load, which must be 0 W because the load is purely inductive.

16.36 A 240-V, *ABC* circuit has a balanced Y load of $20\underline{/-60°}$-Ω impedances. Two wattmeters are connected for the two-wattmeter method with current coils in lines *A* and *C*. Find the wattmeter readings. Also, find these readings for an *ACB* phase sequence.

Since the line voltage magnitude and the impedance angle are known, only the line current magnitude is needed to determine the wattmeter readings. This current magnitude is

$$I_L = I_p = \frac{V_p}{Z_Y} = \frac{240/\sqrt{3}}{20} = 6.93 \text{ A}$$

For the *ABC* phase sequence, the wattmeter with its current coil in line *A* has a reading of

$$P_A = V_L I_L \cos (30° + \theta) = 240(6.93) \cos (30° - 60°) = 1440 \text{ W}$$

because *A* precedes *B* in the phase sequence and there is no current coil in line *B*. The other wattmeter reading is

$$P_C = V_L I_L \cos (30° - \theta) = 240(6.93) \cos [30° - (-60°)] = 0 \text{ W}$$

Notice that one wattmeter reading is 0 W and the other is the total average power absorbed by the load, as is generally true for the two-wattmeter method for a balanced load with a power factor of 0.5.

For the *ACB* phase sequence, the wattmeter readings switch because *C* is before *B* in the phase sequence and there is no current coil in line *B*. So, $P_C = 1440$ W and $P_A = 0$ W.

16.37 A 208-V circuit has a balanced Δ load of $30\underline{/40°}$-Ω impedances. Two wattmeters are connected for the two-wattmeter method with current coils in lines *A* and *B*. Find the wattmeter readings for an *ABC* phase sequence.

The rms line current is needed for the wattmeter formulas. This current is $\sqrt{3}$ times the rms phase current:

$$I_L = \sqrt{3} I_p = \sqrt{3} \frac{V_p}{Z_\Delta} = \sqrt{3} \frac{208}{30} = 12 \text{ A}$$

Since there is no current coil in line *C*, and *B* precedes *C* in the phase sequence, the reading of the wattmeter with its current coil in line *B* is

$$P_B = V_L I_L \cos (30° + \theta) = 208(12) \cos (30° + 40°) = 854 \text{ W}$$

The other wattmeter reading is

$$P_A = V_L I_L \cos (30° - \theta) = 208(12) \cos (30° - 40°) \text{ W} = 2.46 \text{ kW}$$

16.38 A balanced Y load is connected to a 480-V three-phase source. The two-wattmeter method is used to measure the average power absorbed by the load. If the wattmeter readings are 5 and 3 kW, find the impedance of each arm of the load.

Since the phase sequence and wattmeter connections are not given, only the magnitude of the impedance angle can be found from the wattmeter readings. From the angle-power formulas, this angle magnitude is

$$|\theta| = \tan^{-1}\left[\sqrt{3}\frac{5-3}{5+3}\right] = 23.4°$$

The magnitude of the phase impedance Z_Y can be found from the ratio of the phase voltage and current. The phase voltage is $480/\sqrt{3} = 277$ V. The phase current, which is also the line current, can be found from the total power absorbed, which is $5 + 3 = 8$ kW:

$$I_p = I_L = \frac{P}{\sqrt{3}V_L \times PF} = \frac{8000}{\sqrt{3}(480)(\cos 23.4°)} = 10.5 \text{ A}$$

From the ratio of the phase voltage and current, the magnitude of the phase impedance is $277/10.5 = 26.4 \ \Omega$. So the phase impedance is either $Z_Y = 26.4\underline{/23.4°} \ \Omega$ or $Z_Y = 26.4\underline{/-23.4°} \ \Omega$.

16.39 Two wattmeters both have readings of 3 kW when connected for the two-wattmeter method with current coils in lines A and B of a 600-V, ABC circuit having a balanced Δ load. Find the Δ phase impedance.

For an ABC phase sequence and current coils in lines A and B, the phase impedance angle is given by

$$\theta = \tan^{-1}\left[\sqrt{3}\frac{P_A - P_B}{P_A + P_B}\right] = \tan^{-1}\left[\sqrt{3}\frac{3-3}{3+3}\right] = \tan^{-1} 0 = 0°$$

Because the load impedance angle is 0°, the load is purely resistive. The phase resistance is equal to the phase voltage of 600 V, which is also the line voltage, divided by the phase current. From $P = 3V_pI_p \cos \theta$,

$$I_p = \frac{P}{3V_p \cos \theta} = \frac{3000 + 3000}{3(600)(1)} = 3.33 \text{ A}$$

Finally,

$$R_\Delta = \frac{V_p}{I_p} = \frac{600}{3.33} = 180 \ \Omega$$

16.40 Two wattmeters are connected for the two-wattmeter method with current coils in lines B and C of a 480-V, ACB circuit that has a balanced Δ load. If the wattmeter readings are 4 and 2 kW, respectively, find the Δ phase impedance Z_Δ.

The phase impedance angle is

$$\theta = \tan^{-1}\left[\sqrt{3}\frac{P_C - P_B}{P_C + P_B}\right] = \tan^{-1}\left[\sqrt{3}\frac{2-4}{2+4}\right] = \tan^{-1}\left(-\frac{\sqrt{3}}{3}\right) = -30°$$

The magnitude of the phase impedance can be found by dividing the phase voltage of 480 V, which is also the line voltage, by the phase current. From $P = 3V_pI_p \cos \theta$, the phase current is

$$I_p = \frac{P}{3V_p \cos \theta} = \frac{4000 + 2000}{3(480) \cos (-30°)} = 4.81 \text{ A}$$

This divided into the phase voltage is the magnitude of the phase impedance. Consequently,

$$Z_\Delta = \frac{480}{4.81}\underline{/-30°} = 99.8\underline{/-30°} \ \Omega$$

16.41 Two wattmeters are connected for the two-wattmeter method with current coils in lines A and C of a 240-V, ACB circuit that has a balanced Y load. Find the Y phase impedance if the two wattmeter readings are -1 and 2 kW, respectively.

The impedance angle is

$$\theta = \tan^{-1}\left[\sqrt{3}\,\frac{P_A - P_C}{P_A + P_C}\right] = \tan^{-1}\left[\sqrt{3}\,\frac{-1-2}{-1+2}\right] = \tan^{-1}(-3\sqrt{3}) = -79.1°$$

The magnitude of the phase impedance can be found by dividing the phase voltage of $V_p = 240/\sqrt{3} = 139$ V by the phase current, which is also the line current. From $P = \sqrt{3}V_LI_L \cos\theta$, the line current is

$$I_L = I_p = \frac{P}{\sqrt{3}V_L \cos\theta} = \frac{-1000 + 2000}{\sqrt{3}(240)\cos(-79.1°)} = 12.7 \text{ A}$$

So
$$\mathbf{Z}_Y = \frac{139}{12.7}\underline{/-79.1°} = 10.9\underline{/-79.1°}\ \Omega$$

16.42 A 240-V, ABC circuit has an unbalanced Δ load consisting of resistances $R_{AC} = 45\ \Omega$, $R_{BA} = 30\ \Omega$, and $R_{CB} = 40\ \Omega$. Two wattmeters are connected for the two-wattmeter method with current coils in lines A and B. What are the wattmeter readings and the total average power absorbed?

From the wattmeter connections, the wattmeter readings are equal to

$$P_A = V_{AC}I_A \cos(\text{ang } \mathbf{V}_{AC} - \text{ang } \mathbf{I}_A) \quad \text{and} \quad P_B = V_{BC}I_B \cos(\text{ang } \mathbf{V}_{BC} - \text{ang } \mathbf{I}_B)$$

For the calculations of these powers, the phasors $\mathbf{V}_{AC}$, $\mathbf{V}_{BC}$, $\mathbf{I}_A$, and $\mathbf{I}_B$ are needed. Since no angles are specified, the angle of $\mathbf{V}_{AC}$ can be conveniently selected as $0°$, making $\mathbf{V}_{AC} = 240\underline{/0°}$ V. For an ABC phase sequence, $\mathbf{V}_{CB}$ leads $\mathbf{V}_{AC}$ by $120°$ and so is $\mathbf{V}_{CB} = 240\underline{/120°}$ V. But $\mathbf{V}_{BC}$ is needed:

$$\mathbf{V}_{BC} = -\mathbf{V}_{CB} = -240\underline{/120°} = 240\underline{/120° - 180°} = 240\underline{/-60°} \text{ V}$$

Also, $\mathbf{V}_{BA}$ lags $\mathbf{V}_{AC}$ by $120°$ and is $\mathbf{V}_{BA} = 240\underline{/-120°}$ V. The line currents $\mathbf{I}_A$ and $\mathbf{I}_B$ can be determined from the phase currents:

$$\mathbf{I}_A = \mathbf{I}_{AC} - \mathbf{I}_{BA} = \frac{\mathbf{V}_{AC}}{R_{AC}} - \frac{\mathbf{V}_{BA}}{R_{BA}} = \frac{240\underline{/0°}}{45} - \frac{240\underline{/-120°}}{30} = 11.6\underline{/36.6°} \text{ A}$$

$$\mathbf{I}_B = \mathbf{I}_{BA} - \mathbf{I}_{CB} = \frac{\mathbf{V}_{BA}}{R_{BA}} - \frac{\mathbf{V}_{CB}}{R_{CB}} = \frac{240\underline{/-120°}}{30} - \frac{240\underline{/120°}}{40} = 12.2\underline{/-94.7°} \text{ A}$$

Now P_A and P_B can be determined:

$$P_A = V_{AC}I_A \cos(\text{ang } \mathbf{V}_{AC} - \text{ang } \mathbf{I}_A) = 240(11.6)\cos(0° - 36.6°) \text{ W} = 2.24 \text{ kW}$$

$$P_B = V_{BC}I_B \cos(\text{ang } \mathbf{V}_{BC} - \text{ang } \mathbf{I}_B) = 240(12.2)\cos[-60° - (-94.7°)] \text{ W} = 2.4 \text{ kW}$$

Notice that the two wattmeter readings are not the same, even though the load is purely resistive. The reason they are not the same is that the load is not balanced.

The total power absorbed is $P_A + P_B = 2.24 + 2.4 = 4.64$ kW. This can be checked by summing the V^2/R power absorptions by the individual resistors:

$$P_T = \frac{240^2}{45} + \frac{240^2}{30} + \frac{240^2}{40} \text{ W} = 4.64 \text{ kW}$$

16.43 For a four-wire, ACB circuit in which $\mathbf{V}_{AN} = 277\underline{/-40°}$ V, find the four phasor line currents to a Y load of $\mathbf{Z}_A = 15\underline{/30°}\ \Omega$, $\mathbf{Z}_B = 20\underline{/-25°}\ \Omega$, and $\mathbf{Z}_C = 25\underline{/45°}\ \Omega$.

The three phase currents, which are also three of the line currents, are equal to the phase voltages divided by the phase impedances. One phase voltage is the specified $\mathbf{V}_{AN}$. The others are $\mathbf{V}_{BN}$ and $\mathbf{V}_{CN}$. From the specified ACB phase sequence, $\mathbf{V}_{BN}$ and $\mathbf{V}_{CN}$ respectively lead and lag $\mathbf{V}_{AN}$ by $120°$: $\mathbf{V}_{BN} = 277\underline{/80°}$ V and $\mathbf{V}_{CN} = 277\underline{/-160°}$ V. So the phase currents are

$$\mathbf{I}_A = \frac{\mathbf{V}_{AN}}{\mathbf{Z}_A} = \frac{277\underline{/-40°}}{15\underline{/30°}} = 18.5\underline{/-70°} \text{ A} \qquad \mathbf{I}_B = \frac{\mathbf{V}_{BN}}{\mathbf{Z}_B} = \frac{277\underline{/80°}}{20\underline{/-25°}} = 13.9\underline{/105°} \text{ A}$$

$$\mathbf{I}_C = \frac{\mathbf{V}_{CN}}{\mathbf{Z}_C} = \frac{277\underline{/-160°}}{25\underline{/45°}} = 11.1\underline{/-205°} = -11.1\underline{/-25°} \text{ A}$$

By KCL the neutral line current is

$$\mathbf{I}_N = -(\mathbf{I}_A + \mathbf{I}_B + \mathbf{I}_C) = -(18.5\underline{/-70°} + 13.9\underline{/105°} - 11.1\underline{/-25°}) = 7.3\underline{/-5.53°} \text{ A}$$

16.44 For an ABC circuit in which $\mathbf{V}_{AB} = 480\underline{/40°}$ V, find the phasor line currents to a Δ load of $\mathbf{Z}_{AB} = 40\underline{/30°}$ Ω, $\mathbf{Z}_{BC} = 30\underline{/-70°}$ Ω, and $\mathbf{Z}_{CA} = 50\underline{/60°}$ Ω.

Each line current is the difference of two phase currents, and each phase current is the ratio of a phase voltage and impedance. One phase voltage is the given $\mathbf{V}_{AB} = 480\underline{/40°}$ V. And from the given ABC phase sequence, the other phase voltages, $\mathbf{V}_{BC}$ and $\mathbf{V}_{CA}$, respectively lag and lead $\mathbf{V}_{AB}$ by 120°: $\mathbf{V}_{BC} = 480\underline{/-80°}$ V and $\mathbf{V}_{CA} = 480\underline{/160°}$ V. So the phase currents are

$$\mathbf{I}_{AB} = \frac{\mathbf{V}_{AB}}{\mathbf{Z}_{AB}} = \frac{480\underline{/40°}}{40\underline{/30°}} = 12\underline{/10°} \text{ A} \qquad \mathbf{I}_{BC} = \frac{\mathbf{V}_{BC}}{\mathbf{Z}_{BC}} = \frac{480\underline{/-80°}}{30\underline{/-70°}} = 16\underline{/-10°} \text{ A}$$

$$\mathbf{I}_{CA} = \frac{\mathbf{V}_{CA}}{\mathbf{Z}_{CA}} = \frac{480\underline{/160°}}{50\underline{/60°}} = 9.6\underline{/100°} \text{ A}$$

And, by KCL, the line currents are

$$\mathbf{I}_A = \mathbf{I}_{AB} - \mathbf{I}_{CA} = 12\underline{/10°} - 9.6\underline{/100°} = 15.4\underline{/-28.7°} \text{ A}$$

$$\mathbf{I}_B = \mathbf{I}_{BC} - \mathbf{I}_{AB} = 16\underline{/-10°} - 12\underline{/10°} = 6.26\underline{/-51°} \text{ A}$$

$$\mathbf{I}_C = \mathbf{I}_{CA} - \mathbf{I}_{BC} = 9.6\underline{/100°} - 16\underline{/-10°} = 21.3\underline{/144.9°} = -21.3\underline{/-35.1°} \text{ A}$$

As a check, the three line currents can be added to see if the sum is zero, as it should be by KCL. This sum is zero, but it takes more than three significant digits to show this convincingly.

16.45 In a three-wire, ABC circuit in which $\mathbf{V}_{AB} = 480\underline{/60°}$ V, find the phasor line currents to a Y load of $\mathbf{Z}_A = 16\underline{/-30°}$ Ω, $\mathbf{Z}_B = 14\underline{/50°}$ Ω, and $\mathbf{Z}_C = 12\underline{/-40°}$ Ω.

Since the Y load is unbalanced and there is no neutral wire, the load phase voltages are not known. And this means that the line currents cannot be found readily by dividing the load phase voltages by the load phase impedances, as in the solution to Prob. 16.43. A Y-to-Δ conversion is tempting so that the phase voltages will be known and the approach in the solution to Prob. 16.44 can be used. But usually this is considerably more effort than using loop analysis on the original circuit.

As shown in Fig. 16-16, loop analysis can be used to find two of the three line currents, here $\mathbf{I}_A$ and $\mathbf{I}_C$. Of course, after these are known, the third line current $\mathbf{I}_B$ can be found from them by KCL. Note in Fig. 16-16 that the $\mathbf{V}_{CA}$ generator is not shown. It is not needed because the shown two generators illustrated supply the correct voltage between terminals A and C. Of course, as shown, $\mathbf{V}_{BC}$ lags the given $\mathbf{V}_{AB}$ by 120° because the phase sequence is ABC.

The loop equations are

$$(16\underline{/-30°} + 14\underline{/50°})\mathbf{I}_A + (14\underline{/50°})\mathbf{I}_C = 480\underline{/60°}$$

$$(14\underline{/50°})\mathbf{I}_A + (12\underline{/-40°} + 14\underline{/50°})\mathbf{I}_C = -480\underline{/-60°}$$

which simplify to

$$(23\underline{/6.8°})\mathbf{I}_A + (14\underline{/50°})\mathbf{I}_C = 480\underline{/60°}$$

$$(14\underline{/50°})\mathbf{I}_A + (18.4\underline{/9.4°})\mathbf{I}_C = -480\underline{/-60°}$$

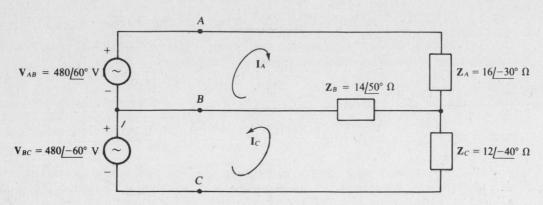

Fig. 16-16

By Cramer's rule,

$$\mathbf{I}_A = \frac{\begin{vmatrix} 480\underline{/60^\circ} & 14\underline{/50^\circ} \\ -480\underline{/-60^\circ} & 18.4\underline{/9.4^\circ} \end{vmatrix}}{\begin{vmatrix} 23\underline{/6.8^\circ} & 14\underline{/50^\circ} \\ 14\underline{/50^\circ} & 18.4\underline{/9.4^\circ} \end{vmatrix}} = \frac{12.1 \times 10^3\underline{/36.2^\circ}}{448\underline{/-9.6^\circ}} = 26.9\underline{/45.8^\circ} \text{ A}$$

$$\mathbf{I}_C = \frac{\begin{vmatrix} 23\underline{/6.8^\circ} & 480\underline{/60^\circ} \\ 14\underline{/50^\circ} & -480\underline{/-60^\circ} \end{vmatrix}}{448\underline{/-9.6^\circ}} = \frac{5.01 \times 10^3\underline{/149.6^\circ}}{448\underline{/-9.6^\circ}} = 11.2\underline{/159.2^\circ} \text{ A}$$

Of course, by KCL,

$$\mathbf{I}_B = -\mathbf{I}_A - \mathbf{I}_C = -26.9\underline{/45.8^\circ} - 11.2\underline{/159.2^\circ} = 24.7\underline{/-110^\circ} \text{ A}$$

16.46 In the circuit shown in Fig. 16-17, in which each line has an impedance of $5 + j8$ Ω, obtain the equations needed for solving for $\mathbf{I}_A$ and $\mathbf{I}_B$.

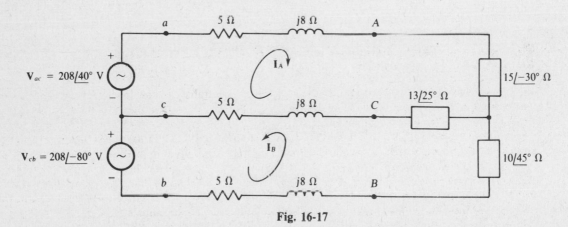

Fig. 16-17

The loop equations are

$$(5 + j8 + 15\underline{/-30^\circ} + 13\underline{/25^\circ} + 5 + j8)\mathbf{I}_A + (5 + j8 + 13\underline{/25^\circ})\mathbf{I}_B = 208\underline{/40^\circ}$$

$$(5 + j8 + 13\underline{/25^\circ})\mathbf{I}_A + (5 + j8 + 10\underline{/45^\circ} + 13\underline{/25^\circ} + 5 + j8)\mathbf{I}_B = -208\underline{/-80^\circ}$$

These simplify to

$$(37.5\underline{/21.9^\circ})\mathbf{I}_A + (21.5\underline{/38.8^\circ})\mathbf{I}_B = 208\underline{/40^\circ}$$

$$(21.5\underline{/38.8^\circ})\mathbf{I}_A + (40.6\underline{/44.7^\circ})\mathbf{I}_B = -208\underline{/-80^\circ}$$

And these solve to $\mathbf{I}_A = 6.41\underline{/-9.14°}$ A and $\mathbf{I}_B = 5.11\underline{/94.1°}$ A. Of course $\mathbf{I}_C = -\mathbf{I}_A - \mathbf{I}_B = 7.22\underline{/-146°}$ A.

Notice in Fig. 16-17 the use of lowercase letters at the source terminals to distinguish them from the load terminals, as is necessary because of the line impedances.

16.47 In a three-wire, *ACB* circuit in which one line voltage at the source is $\mathbf{V}_{ac} = 208\underline{/-30°}$ V, obtain the equations needed for finding the phasor line currents to a Δ load in which $\mathbf{Z}_{AB} = 30\underline{/-40°}$ Ω, $\mathbf{Z}_{BC} = 40\underline{/30°}$ Ω, and $\mathbf{Z}_{CA} = 35\underline{/60°}$ Ω. Each line has an impedance of $4 + j7$ Ω.

From the solution to Prob. 16.46, it should be apparent that a good approach is to convert the Δ to a Y and then use loop analysis. The three Δ-to-Y conversion formulas have the same denominator of

$$\mathbf{Z}_{AB} + \mathbf{Z}_{BC} + \mathbf{Z}_{CA} = 30\underline{/-40°} + 40\underline{/30°} + 35\underline{/60°} = 81.3\underline{/22.4°}$$

With this inserted, the conversion formulas are

$$\mathbf{Z}_A = \frac{\mathbf{Z}_{AB}\mathbf{Z}_{CA}}{81.3\underline{/22.4°}} = \frac{(30\underline{/-40°})(35\underline{/60°})}{81.3\underline{/22.4°}} = \frac{1050\underline{/20°}}{81.3\underline{/22.4°}} = 12.9\underline{/-2.4°} \ \Omega$$

$$\mathbf{Z}_B = \frac{\mathbf{Z}_{AB}\mathbf{Z}_{BC}}{81.3\underline{/22.4°}} = \frac{(30\underline{/-40°})(40\underline{/30°})}{81.3\underline{/22.4°}} = \frac{1200\underline{/-10°}}{81.3\underline{/22.4°}} = 14.8\underline{/-32.4°} \ \Omega$$

$$\mathbf{Z}_C = \frac{\mathbf{Z}_{CA}\mathbf{Z}_{BC}}{81.3\underline{/22.4°}} = \frac{(35\underline{/60°})(40\underline{/30°})}{81.3\underline{/22.4°}} = \frac{1400\underline{/90°}}{81.3\underline{/22.4°}} = 17.2\underline{/67.6°} \ \Omega$$

With the equivalent Y inserted for the Δ, the circuit is as shown in Fig. 16-18. Because of the *ACB* phase sequence, $\mathbf{V}_{ba}$, as shown, leads the specified $\mathbf{V}_{ac}$ by 120°.

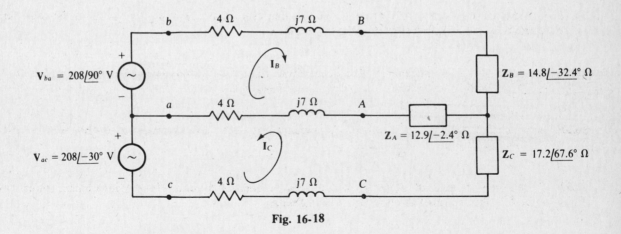

Fig. 16-18

The loop equations are

$$(4 + j7 + 14.8\underline{/-32.4°} + 12.9\underline{/-2.4°} + 4 + j7)\mathbf{I}_B + (4 + j7 + 12.9\underline{/-2.4°})\mathbf{I}_C = 208\underline{/90°}$$

$$(4 + j7 + 12.9\underline{/-2.4°})\mathbf{I}_B + (4 + j7 + 17.2\underline{/67.6°} + 12.9\underline{/-2.4°} + 4 + j7)\mathbf{I}_C = -208\underline{/-30°}$$

These simplify to

$$(33.8\underline{/9.41°})\mathbf{I}_B + (18.1\underline{/20.9°})\mathbf{I}_C = 208\underline{/90°}$$

$$(18.1\underline{/20.9°})\mathbf{I}_B + (40.2\underline{/46.9°})\mathbf{I}_C = -208\underline{/-30°}$$

which solve to $\mathbf{I}_B = 5.4\underline{/54.2°}$ A and $\mathbf{I}_C = 5.11\underline{/130°}$ A. Of course $\mathbf{I}_A = -\mathbf{I}_B - \mathbf{I}_C$, from which $\mathbf{I}_A = 8.27\underline{/-88.9°}$ A.

Supplementary Problems

16.48 What is the phase sequence of a Y-connected three-phase alternator for which $V_{AN} = 7200/-130°$ V and $V_{BN} = 7200/110°$ V? Also, what is V_{CN}? *Ans. ABC*, $V_{CN} = 7200/-10°$ V

16.49 Find the phase sequence of a balanced three-phase circuit in which $V_{AN} = 120/15°$ V and $V_{CN} = 120/135°$ V. Also, find V_{BN}. *Ans. ABC*, $V_{BN} = 120/-105°$ V

16.50 For a three-phase, three-wire circuit, find the phasor line currents to a balanced Y load in which each phase impedance is $30/-40°$ Ω and for which $V_{CN} = 277/-70°$ V. The phase sequence is *ACB*.
Ans. $I_A = 9.23/90°$ A, $I_B = 9.23/-150°$ A, $I_C = 9.23/-30°$ A

16.51 Find the phase sequence of a three-phase circuit in which $V_{BA} = 12\,470/-140°$ V and $V_{AC} = 12\,470/100°$ V. Also, find the third line voltage. *Ans. ACB*, $V_{CB} = 12\,470/-20°$ V

16.52 What is the phase sequence of a three-phase circuit for which $V_{BN} = 7.62/-45°$ kV and $V_{CB} = 13.2/105°$ kV? *Ans. ACB*

16.53 A balanced Y load has one phase voltage of $V_{BN} = 120/130°$ V. If the phase sequence is *ABC*, find the line voltages V_{AC}, V_{CB}, and V_{BA}.
Ans. $V_{AC} = 208/-140°$ V, $V_{CB} = 208/-20°$ V, $V_{BA} = 208/100°$ V

16.54 What are the phase voltages for a balanced three-phase Y load if $V_{CA} = 208/-125°$ V? The phase sequence is *ACB*. *Ans.* $V_{AN} = 120/25°$ V, $V_{BN} = 120/145°$ V, $V_{CN} = 120/-95°$ V

16.55 A balanced three-wire, *ACB* circuit has one line current of $I_C = 6/-10°$ A. Find the other line currents. *Ans.* $I_A = 6/110°$ A, $I_B = 6/-130°$ A

16.56 Find the I_C line current in an unbalanced three-wire, three-phase circuit in which $I_A = 6/-30°$ A and $I_B = -4/50°$ A. *Ans.* $I_C = 6.61/113°$ A

16.57 A balanced Y load of 100-Ω resistors is connected to a 208-V, three-phase, three-wire source. Find the rms line current. *Ans.* 1.2 A

16.58 A balanced Y load of $40/60°$-Ω impedances is connected to a 600-V, three-phase, three-wire source. Find the rms line current. *Ans.* 8.66 A

16.59 Find the phasor line currents to a balanced Y load of $45/-48°$-Ω impedances. One phase voltage is $V_{CN} = 120/-65°$ V, the phase sequence is *ACB*, and there are only three wires.
Ans. $I_A = 2.67/103°$ A, $I_B = 2.67/-137°$ A, $I_C = 2.67/-17°$ A

16.60 For a three-phase, three-wire circuit, find the phasor line currents to a balanced three-phase Y load of $80/25°$-Ω impedances if $V_{AB} = 600/-30°$ V and the phase sequence is *ACB*.
Ans. $I_A = 4.33/-25°$ A, $I_B = 4.33/95°$ A, $I_C = 4.33/-145°$ A

16.61 Find the phase sequence of a three-phase circuit in which two of the phase currents of a balanced Δ load are $I_{AB} = 10/50°$ A and $I_{CA} = 10/170°$ A. Also, find the third phase current.
Ans. ABC, $I_{BC} = 10/-70°$ A

16.62 Find the phase currents I_{AC}, I_{CB}, and I_{BA} of a balanced three-phase Δ load to which one line current is $I_A = 1.4 \underline{/65°}$ A. The phase sequence is *ACB*.
   Ans. $I_{AC} = 0.808 \underline{/95°}$ A, $I_{CB} = 0.808 \underline{/-25°}$ A, $I_{BA} = 0.808 \underline{/-145°}$ A

16.63 A balanced three-phase Δ load has one phase current of $I_{CA} = 4 \underline{/-35°}$ A. If the phase sequence is *ABC*, find the phasor line currents and the other phasor phase currents.
   Ans. $I_A = 6.93 \underline{/175°}$ A $I_{AB} = 4 \underline{/-155°}$ A
           $I_B = 6.93 \underline{/55°}$ A $I_{BC} = 4 \underline{/85°}$ A
           $I_C = 6.93 \underline{/-65°}$ A

16.64 Find the phasor line currents to a balanced three-phase Δ load in which one phase current is $I_{BA} = 4.2 \underline{/-30°}$ A. The phase sequence is *ACB*.
   Ans. $I_A = -7.27$ A, $I_B = 7.27 \underline{/-60°}$ A, $I_C = 7.27 \underline{/60°}$ A

16.65 Find the rms value of the line currents to a balanced Δ load of 100-Ω resistors from a 480-V, three-phase, three-wire source.   *Ans.* 8.31 A

16.66 Find the phasor line currents to a balanced three-phase Δ load of $200 \underline{/-55°}$-Ω impedances if the phase sequence is *ABC* and if one phase voltage is $V_{CA} = 208 \underline{/-60°}$ V.
   Ans. $I_A = 1.8 \underline{/-155°}$ A, $I_B = 1.8 \underline{/85°}$ A, $I_C = 1.8 \underline{/-35°}$ A

16.67 A balanced Δ load of $50 \underline{/35°}$-Ω impedances is energized from the Y-connected secondary of a three-phase transformer for which $V_{AN} = 120 \underline{/-10°}$ V. If the phase sequence is *ABC*, find the phasor line and load currents.
   Ans. $I_A = 7.2 \underline{/-45°}$ A $I_{AC} = 4.16 \underline{/-75°}$ A
           $I_B = 7.2 \underline{/-165°}$ A $I_{BA} = 4.16 \underline{/-195°}$ A
           $I_C = 7.2 \underline{/75°}$ A $I_{CB} = 4.16 \underline{/45°}$ A

16.68 A balanced Y load of $8 + j6$-Ω impedances is connected to a three-phase source by three wires, each of which has $3 + j4$ Ω of impedance. The rms load phase voltage is 50 V. Find the rms line voltage at the source.   *Ans.* 129 V

16.69 A balanced Δ load of $15 - j9$-Ω impedances is connected to a three-phase source by three wires, each of which has $2 + j5$ Ω of impedance. The rms load phase voltage is 120 V. Find the rms line voltage at the source.   *Ans.* 150 V

16.70 A 600-V, three-phase, three-wire circuit has two balanced Δ loads, one of 30-Ω resistors and the other of 60-Ω resistors. Find the total rms line current.   *Ans.* 52 A

16.71 A 480-V, three-phase, three-wire circuit has two balanced Y loads, one of 40-Ω resistors and the other of 120-Ω resistors. Find the total rms line current.   *Ans.* 9.24 A

16.72 A 480-V three-phase circuit has two balanced Δ loads, one of $50 \underline{/-60°}$-Ω impedances and the other of $70 \underline{/50°}$-Ω impedances. Find the total rms line current and the total average power absorbed.
   Ans. 16.8 A, 13.3 kW

16.73 A 600-V three-phase circuit has two balanced loads, one a Δ of $90 \underline{/-40°}$-Ω impedances and the other a Y of $50 \underline{/30°}$-Ω impedances. Find the total rms line current and the total average power absorbed.   *Ans.* 15.4 A, 15.4 kW

16.74   A balanced Y of $30/-30°$-Ω impedances and a balanced Δ of $90/-50°$-Ω impedances are connected by three wires to the secondary of a three-phase transformer. If $\mathbf{V}_{BA} = 208/-30°$ V and the phase sequence is ACB, find the total phasor line currents.
   Ans. $\mathbf{I}_A = 7.88/-140°$ A, $\mathbf{I}_B = 7.88/-20°$ A, $\mathbf{I}_C = 7.88/100°$ A

16.75   A balanced Δ load of $60/50°$-Ω impedances is connected to the secondary of a three-phase transformer by three wires that have $3 + j4$ Ω of impedance each. If the rms line voltage is 480 V at the secondary terminals, find the rms line current.   *Ans.* 11.1 A

16.76   Find the average power absorbed by a balanced three-phase load in an ACB circuit in which one line voltage is $\mathbf{V}_{AC} = 480/30°$ V and one line current to the load is $\mathbf{I}_B = 2.1/80°$ A.   *Ans.* 1.34 kW

16.77   A three-phase induction motor delivers 100 hp while operating at an 80 percent efficiency and a 0.7 lagging power factor from 600-V lines. Find the rms line current.   *Ans.* 128 A

16.78   In a 480-V three-phase circuit, a balanced Δ load absorbs 5 kW at a 0.7 lagging power factor. Find the Δ phase impedance.   *Ans.* $96.8/45.6°$ Ω

16.79   Given that $\mathbf{V}_{AC} = 208/-40°$ V in an ACB three-phase circuit, find the phasor line currents to a balanced load that absorbs 10 kW at a 0.8 lagging power factor.
   Ans. $\mathbf{I}_A = 34.7/-107°$ A, $\mathbf{I}_B = 34.7/13°$ A, $\mathbf{I}_C = 34.7/133°$ A

16.80   A 600-V three-phase circuit has two balanced loads. One is a synchronous motor that delivers 30 hp while operating at an 85 percent efficiency and a 0.7 leading power factor. The other is an induction motor that delivers 50 hp while operating at an 80 percent efficiency and a 0.85 lagging power factor. Find the total rms line current.   *Ans.* 70.2 A

16.81   If $\mathbf{I}_B = 20/40°$ A, $\mathbf{I}_C = 15/-30°$ A, and $\mathbf{V}_{BC} = 480/-40°$ V in a three-wire, ACB circuit, find the reading of a wattmeter connected with its current coil in line A and its potential coil across lines A and B. The $\pm$ terminal of the current coil is toward the source, and the $\pm$ terminal of the potential coil is at line A.   *Ans.* 13.6 kW

16.82   A balanced Y load of 50-Ω resistors is connected to a 208-V, ACB, three-wire, three-phase source. Find the reading of a wattmeter connected with its current coil in line B and its potential coil across lines A and C. The $\pm$ terminal of the current coil is toward the source, and the $\pm$ terminal of the potential coil is at line A.   *Ans.* 0 W

16.83   A balanced Δ load of $9 + j12$-Ω impedances is connected to a 480-V, ABC source. Find the reading of a wattmeter connected with its current coil in line A and its potential coil across lines B and C. The $\pm$ terminal of the current coil is toward the source, and the $\pm$ terminal of the potential coil is at line C.   *Ans.* -21.3 kW

16.84   A 600-V three-phase circuit has a balanced Y load of $40/30°$-Ω impedances. Find the wattmeter readings for the two-wattmeter method.   *Ans.* 5.2 kW, 2.6 kW

16.85   A 480-V, ACB circuit has a balanced Y load of $30/-50°$-Ω impedances. Two wattmeters are connected for the two-wattmeter method with current coils in lines B and C. Find the wattmeter readings.   *Ans.* $P_B = 4.17$ kW, $P_C = 770$ W

16.86   A 600-V, ACB circuit has a balanced Δ load of $60/20°$-Ω impedances. Two wattmeters are connected for the two-wattmeter method with current coils in lines B and C. Find the wattmeter readings.   *Ans.* $P_B = 6.68$ kW, $P_C = 10.2$ kW

16.87   A balanced Y load is connected to a 208-V three-phase source. The two-wattmeter method is used to measure the average power absorbed by the load. If the wattmeter readings are 8 and 4 kW, find the Y phase impedance.   *Ans.* Either $3.12\underline{/30°}$ or $3.12\underline{/-30°}$ Ω

16.88   Two wattmeters both have readings of 5 kW when connected for the two-wattmeter method in a 480-V three-phase circuit that has a balanced Δ load. Find the Δ phase impedance.   *Ans.* $69.1\underline{/0°}$ Ω

16.89   Two wattmeters are connected for the two-wattmeter method with current coils in lines A and B of a 208-V, ABC circuit that has a balanced Δ load. If the wattmeter readings are 6 and -3 kW, respectively, find the Δ phase impedance.   *Ans.* $8.18\underline{/79.1°}$ Ω

16.90   Two wattmeters are connected for the two-wattmeter method with current coils in lines B and C of a 600-V, ABC circuit that has a balanced Y load. Find the Y phase impedance if the two wattmeter readings are 3 and 10 kW, respectively.   *Ans.* $20.3\underline{/-43°}$ Ω

16.91   A 480-V, ACB circuit has an unbalanced Δ load consisting of resistances $R_{AC} = 60$ Ω, $R_{BA} = 85$ Ω,   and $R_{CB} = 70$ Ω. Two wattmeters are connected for the two-wattmeter method with current coils in lines A and C. What are the wattmeter readings?   *Ans.* $P_A = 4.63$ kW, $P_C = 5.21$ kW

16.92   For a four-wire, ABC circuit in which $V_{BN} = 208\underline{/65°}$ V,   find the four phasor line currents to a Y load of $Z_A = 30\underline{/-50°}$ Ω, $Z_B = 25\underline{/38°}$ Ω,   and $Z_C = 35\underline{/-65°}$ Ω.
   Ans. $I_A = 6.93\underline{/-125°}$ A, $I_B = 8.32\underline{/27°}$ A, $I_C = 5.94\underline{/10°}$ A, $I_N = 9.33\underline{/175°}$ A

16.93   For an ACB circuit in which $V_{AC} = 600\underline{/-15°}$ V,   find the phasor line currents to a Δ load of $Z_{AC} = 150\underline{/-35°}$ Ω, $Z_{BA} = 200\underline{/60°}$ Ω,   and $Z_{CB} = 175\underline{/-70°}$ Ω.
   Ans. $I_A = 1.8\underline{/-24.7°}$ A, $I_B = 5.27\underline{/82.7°}$ A, $I_C = 5.04\underline{/-117°}$ A

16.94   In a three-wire, ACB circuit in which $V_{CB} = 208\underline{/-40°}$ V,   find the phasor line currents to a Y load of $Z_A = 10\underline{/30°}$ Ω, $Z_B = 20\underline{/60°}$ Ω,   and $Z_C = 15\underline{/-50°}$ Ω.
   Ans. $I_A = 2.53\underline{/88.8°}$ A, $I_B = 10.7\underline{/133°}$ A, $I_C = 12.6\underline{/-54.8°}$ A

16.95   In a three-wire, ACB circuit in which one source line voltage is $V_{bc} = 480\underline{/-30°}$ V,   find the phasor line currents to a Y load of $Z_A = 12\underline{/60°}$ Ω, $Z_B = 8\underline{/20°}$ Ω,   and $Z_C = 10\underline{/-30°}$ Ω. Each line has an impedance of $3 + j4$ Ω.
   Ans. $I_A = 15.2\underline{/-165°}$ A, $I_B = 27.3\underline{/-33.9°}$ A, $I_C = 20.9\underline{/113°}$ A

16.96   In a three-wire, ABC circuit in which one source line voltage is $V_{ab} = 480\underline{/60°}$ V,   find the phasor line currents to a Δ load of $Z_{AB} = 40\underline{/-50°}$ Ω, $Z_{BC} = 35\underline{/60°}$ Ω,   and $Z_{CA} = 50\underline{/40°}$ Ω. Each line has an impedance of $8 + j9$ Ω.   *Ans.* $I_A = 7.44\underline{/27.8°}$ A, $I_B = 14\underline{/-112°}$ A, $I_C = 9.64\underline{/97.8°}$ A

Index

ac (alternating current), 9, 140
ac circuit, 140
ac generator (alternator), 141, 307
Admittance, 183
  conductance of, 184
  mutual, 207
  self-, 207
  susceptance of, 184
Admittance diagram, 184
Admittance triangle, 184
Air-core transformer, 284
Algebra, complex, 162–165
Alternating current (ac), 9, 140
Alterating current circuit, 140
Alternator (ac generator), 141, 307
Ampere, 9
Analysis:
  loop, 56, 206
  mesh, 55, 205
  nodal, 56, 207
Angle, phase, 143
Angular frequency, 141
Angular velocity, 141
Apparent power, 260
Associated references, 11
Autotransformer, 286
Average power, 144, 257
Average value of periodic wave, 144

Balanced bridge, 83, 230
Balanced three-phase load, 310, 312
Branch, 35
Bridge balance equation, 84, 230
Bridge circuit, 83, 230
  capacitance comparison, 249
  Maxwell, 250
  Wheatstone, 83

Capacitance, 104
  equivalent, 105
  total, 105
Capacitance comparison bridge, 249
Capacitive circuit, 181
Capacitive reactance, 146
Capacitor, 104
  energy stored, 106
  sinusoidal response, 146
Charge, 9
  conservation of, 9
  electron, 9
  proton, 9
Choke, 125
Circuit, 9
  ac, 140
  capacitive, 181

Circuit (continued)
  dc, 35
  frequency-domain, 178
  inductive, 181
  three-phase, 307–334
  time-domain, 178
Circular mil, 7
Coefficient of coupling, 284
Coil, 125
Color code, resistor, 24
Complex algebra, 162–165
Complex number:
  angle, 164
  conjugate, 164
  exponential form, 164
  magnitude, 164
  polar form, 164
  rectangular form, 163
Complex plane, 163
Complex power, 259
Conductance, 21
  of admittance, 184
  equivalent, 37
  mutual, 57
  self-, 57
  total, 37
Conductivity, 22
Conductor, 21
Conjugate, 164
Conservation of charge, 9
Controlled source, 57
Conventional current flow direction, 9
Conversion:
  Δ-Y, 82, 230
  source, 54, 205
Cosine wave, 143
Coulomb, 9
Coupled impedance, 286
Coupling, coefficient of, 284
Cramer's rule, 53
Current, 9
  ac, 9, 140
  dc, 9
  loop, 56
  mesh, 55
  phase, 309
  short-circuit, 80, 229
Current direction, 9
  reference, 9
Current division rule, 38, 184
Current source, 10
  controlled, 57
  dependent, 57
  independent, 57
  Norton, 80, 229
Cycle, 140

dc (direct current), 9
dc circuit, 35
dc source, 10, 11
Delta (Δ) connection, 82, 230, 308
Δ-Y conversions, 82, 230
Dependent source, 57
Derivative, 106
Determinant, 53
Dielectric, 104
Dielectric constant, 105
Digit grouping, 1
Dimensional analysis, 4
Direct current (dc), 9
Direct current circuit, 35
Direction, current, 9
Dot convention, 282
Double-subscript notation, 10, 307
Drop, voltage, 10
Dual, 69

Effective value, 144
Efficiency, 12
Electron, 9
Electron charge, 9
Elimination method, 59
Energy, 10, 12
    stored by a capacitor, 106
    stored by an inductor, 127
Equivalent circuit:
    Norton's, 80, 229
    Thevenin's, 79, 228
Equivalent sources, 54, 205
Euler's identity, 164
Exponential form of complex number, 164

Farad, 104
Faraday's law, 125
Ferromagnetic material, 124
Flux:
    leakage, 283
    magnetic, 124
    mutual, 281
Flux linkage, 125
Force, 10
Frequency, 140
    angular, 141
    radian, 141
Frequency-domain circuit, 178

General transformer equation, 289
Generator:
    ac, 141, 307
    Δ-connected, 308
    Y-connected, 308
Giga-, 3
Ground, 37
Grouping of digits, 1

Henry, 125
Hertz, 140
Horsepower, 11

Ideal transformer, 282
Imaginary number, 162
Impedance, 180
    coupled, 286
    equivalent, 181
    input, 181
    mutual, 206
    output, 236
    reactance of, 181
    reflected, 283, 286
    resistance of, 181
    self-, 206
    Thevenin, 228
    total, 181
Impedance angle, 181
Impedance diagram, 181
Impedance plane, 181
Impedance triangle, 182
Independent source, 57
Induced voltage, 125, 285
Inductance, 125
    equivalent, 126
    mutual, 285
    self-, 285
    total, 126
Inductive circuit, 181
Inductive reactance, 145
Inductor, 125
    energy stored, 127
    sinusoidal response, 145
Inferred zero resistance temperature, 22
Input impedance, 181
Input resistance, 84
Instantaneous current, 106
Instantaneous power, 144, 257
Instantaneous voltage, 106
Insulator, 22
Internal resistance, 25
International System of Units (SI), 3
Ion, 9
Iron-core transformer, 282

Joule, 10

Kilo-, 3
Kilowatthour, 12
Kirchhoff's laws:
    current law (KCL), 36, 207
    voltage law (KVL), 35, 205

Lagging power factor, 258
Lattice circuit, 83
Leading power factor, 258
Leakage flux, 283

Line current, 309
Line voltage, 309
Linear circuit, 79
Linear circuit element, 79
Linear transformer, 284
Load:
    balanced, 310, 312
    Δ-connected, 82, 230, 312
    parallel three-phase, 313
    unbalanced, 315
    Y-connected, 82, 230, 310
"Long time," 116
Loop, 35
Loop analysis, 56, 206
Loop current, 56

Magnetic flux, 124
Matching, resistance, 81, 290
Maximum power transfer theorem, 80, 229
Maxwell bridge, 250
Mega-, 3
Mesh, 35
Mesh analysis, 55, 205
Mesh current, 55
Mho, 21
Micro-, 3
Mil, 6
Milli-, 3
Millman's theorem, 81
Model, 71
    transformer, 282
    transistor, 71
Mutual admittance, 207
Mutual conductance, 57
Mutual flux, 281
Mutual impedance, 206
Mutual inductance, 285
Mutual resistance, 55

Nano-, 3
Negative charge, 9
Negative phase sequence, 309
Network (see Circuit)
Network theorem (see Theorem)
Neutral, 309
Neutron, 9
Newton, 10
Nodal analysis, 56, 207
Node, 35
    reference, 37
Node voltage, 37
Nominal value of resistance, 23
Norton's theorem, 80, 229

Ohm, 21
Ohm's law, 21
Open circuit, 24
Open-circuit voltage, 79, 228

Oscillator, 108
Output impedance, 236
Output resistance, 80, 84

Parallel connection, 25, 37
Period, 109, 140
Periodic quantity, 140
    effective value, 144
Permeability, 124
Permittivity, 105
Phase angle, 143
Phase current, 309
Phase difference, 143
Phase relation, 143
Phase sequence, 309
Phase voltage, 309
Phasor, 165
Phasor diagram, 166
Pico-, 3
Plane, complex, 163
Polar form of complex number, 164
Polarity, reference voltage, 10
Polarity, voltage, 10
Positive charge, 9
Positive phase sequence, 309
Potential drop, 10
Potential rise, 10
Power, 11, 257
    apparent, 260
    average, 144, 257
    complex, 259
    instantaneous, 144, 257
    maximum transfer of, 80, 229
    reactive, 259
    real, 259
    resistor, 23
    three-phase, 314
Power factor, 257
    lagging, 258
    leading, 258
Power factor angle, 257
Power factor correction, 260
Power measurement:
    single-phase, 258
    three-phase, 314
    two-wattmeter method, 267, 314
Power triangle, 260
Powers-of-10 notation, 1–3
Primary winding, 281
Proton, 9

Radian, 141
Radian frequency, 141
Rationalizing, 163, 164
RC time constant, 107
RC timer, 108
Reactance:
    capacitive, 146

Reactance (continued)
   of impedance, 181
   inductive, 145
Reactive factor, 259
Reactive power, 259
Real number, 162
Rectangular form of complex number, 163
Reference current direction, 9
Reference node, 37
Reference voltage polarity, 10
References, associated, 11
Reflected impedance, 283, 286
Relative permeability, 124
Relative permittivity, 105
Resistance, 21
   equivalent, 35
   of impedance, 181
   input, 84
   internal, 25
   mutual, 55
   nominal value, 23
   output, 80, 84
   self-, 55
   Thevenin, 79
   tolerance, 24
   total, 35
Resistance matching, 81, 290
Resistivity, 21
Resistor, 23
   color code, 24
   linear, 23
   nonlinear, 23
   sinusoidal response, 144
Resonant frequency, 186
Right-hand rule, 124, 281
Rise, voltage, 10
RL time constant, 127
rms (root-mean-square) value, 144

Scientific notation, 2
Secondary winding, 281
Self-admittance, 207
Self-conductance, 57
Self-impedance, 206
Self-inductance, 285
Self-resistance, 55
Semiconductor, 22
Series connection, 25, 35
Short circuit, 24
Short-circuit current, 80, 229
SI (International System of Units), 3
Siemens, 21
Sine wave, 140, 141
Sinusoid, 143
   average value, 144
   effective value, 145

Source:
   ac, 140, 307
   controlled, 57
   current, 10
   dc, 10, 11
   dependent, 57
   equivalent, 54, 205
   independent, 57
   Norton, 80, 229
   practical, 25
   Thevenin, 79, 228
   voltage, 11
Source conversion, 54, 205
Step-down transformer, 283
Step-up transformer, 283
Subscript notation:
   current, 307
   voltage, 10, 307
Superposition theorem, 81, 229
Susceptance, 184

Temperature coefficient of resistance, 23
Tera-, 3
Theorem:
   maximum power transfer, 80, 229
   Millman's, 81
   Norton's, 80, 229
   superposition, 81, 229
   Thevenin's, 79, 228
Thevenin's theorem, 79, 228
Three-phase circuits, 307–334
   balanced, 307, 310, 312
   unbalanced, 315
Three-phase power, 314
Three-phase power measurement, 314
Time constant, 107
   RC, 107
   RL, 127
Time-domain circuit, 178
Time-varying voltages and currents, 106
Timer, RC, 108
Tolerance, resistance, 24
Transformation ratio, 283
Transformers, 281–306
   air-core, 284
   ideal, 282
   iron-core, 282
   linear, 284
   step-down, 283
   step-up, 283
Transient, 107
Turns ratio, 283
Two-wattmeter method, 267, 314

Unbalanced three-phase circuit, 315
Unit symbol, 3
Units, SI, 3

var, 259
Volt, 10
Voltage, 10
  induced, 125, 285
  node, 37
  open-circuit, 79, 228
  phase, 309
Voltage difference, 10
Voltage division rule, 36, 182
Voltage drop, 10
Voltage polarity, 10
  reference, 10
Voltage rise, 10
Voltage source, 11
  controlled, 57
  dependent, 57
  independent, 57

Voltage source (continued)
  Thevenin, 79, 228
Voltampere, 259
Voltampere reactive, 259

Watt, 11
Wattmeter, 258
Weber, 124
Wheatstone bridge, 83
Winding:
  primary, 281
  secondary, 281
Work, 10

Y (Wye) connection, 82, 230, 308
Y-Δ conversion, 82, 230

Catalog

If you are interested in a list of SCHAUM'S
OUTLINE SERIES send your name
and address, requesting your free catalog, to:

SCHAUM'S OUTLINE SERIES, Dept. C
McGRAW-HILL BOOK COMPANY
1221 Avenue of Americas
New York, N.Y. 10020